COMMUNITY ECONOMICS

Linking Theory and Practice

Second Edition

COMMUNITY ECONOMICS

Linking Theory and Practice

Second Edition

Ron Shaffer
Steve Deller
Dave Marcouiller

Blackwell
Publishing

©2004 Blackwell Publishing

Cover photo, courtesy of John Duesterhoeft

Blackwell Publishing Professional
2121 State Avenue, Ames, Iowa 50014, USA

Orders:	1-800-862-6657
Office:	1-515-292-0140
Fax:	1-515-292-3348
Web site:	www.blackwellprofessional.com

Blackwell Publishing Ltd
9600 Garsington Road, Oxford OX4 2DQ, UK
Tel.: +44 (0)1865 776868

Blackwell Publishing Asia
550 Swanston Street, Carlton, Victoria 3053, Australia
Tel.: +61 (0)3 8359 1011

First edition, 1989 Iowa State University Press
Second edition, 2004

Library of Congress Cataloging-in-Publication Data

Shaffer, Ron, 1945–
  Community economics : linking theory and practice /
Ron Shaffer, Steven Deller, and Dave Marcouiller.—
2nd ed.
    p. cm.
  ISBN -13: 978-0-8138-1637-1
  ISBN-10: 0-8138-1637-8 (alk. paper)
  1. Regional economics. 2. Economic development.
3. Economic geography. I. Deller, Steven C.
II. Marcouiller, David W. III. Title.
  HT388.S5 2004
330.9—dc22
                                                    2003024733

The last digit is the print number: 9 8 7 6 5 4 3 2

About the Authors

Ron Shaffer is Professor Emeritus with the Department of Agricultural and Applied Economics at the University of Wisconsin-Madison. He is also a community development economist with the University of Wisconsin–Extension. Ron teaches and conducts research in the area of community economics. One of his long-taught courses is Community Economic Analysis, and he was instrumental in organizing a new masters program in community development planning. His extension efforts emphasize working with communities in Wisconsin to create economic development strategies.

Professor Shaffer was a faculty member at the University of Wisconsin-Madison from January 1972 to October 2001, and he served as Director of the University of Wisconsin Center for Community Economic Development. He directed the National Rural Economic Development Institute, which was part of the National Rural Development Partnership. He has worked on rural development issues in various states in the United States and internationally.

Steven Deller is a Professor with the Department of Agricultural and Applied Economics at the University of Wisconsin-Madison. He is also a community development economist with the University of Wisconsin–Extension. He holds joint appointments with the Department of Urban and Regional Planning, UW–Madison, and the Center for Community Economic Development and the Center for Local Government, UW–Extension. His research, teaching, and outreach education focus on community economic development, regional economic modeling, regional growth, and local public finance.

His work has appeared in the *Review of Economics and Statistics, American Journal of Agricultural Economics, Land Economics, Journal of Planning Literature, Rural Sociology, Economic Development Quarterly, Public Productivity and Management Review, The Gerontologist,* and *Journal of Community Development Society.* Along with Dave Marcouiller, he serves as co-editor of *The Journal of Regional Analysis and Policy.*

Dave Marcouiller is an Associate Professor of Urban and Regional Planning at the University of Wisconsin–Madison and also serves as a resource economist with joint appointments in the Department of Forest Ecology and Management, the Institute of Environmental Studies, and the Center for Community Economic Development. His work focuses on the linkages between natural resources and rural economic development, with a particular interest in the mechanisms behind income generation and distribution to rural households.

His recent work has been published in the *Journal of Planning Literature, Society and Natural Resources, Land Economics, Tourism Economics, Economic Development Quarterly, Growth and Change, Forest Science, Wood and Fiber Science, The Canadian Journal of Forest Research, Northern Journal of Applied Forestry, Review of Regional Studies,* and *American Journal of Agricultural Economics.* Along with Steven Deller, he serves as co-editor of *The Journal of Regional Analysis and Policy.*

To the memory of Glen Pulver, whose insights on and constant expressions of concern for the economics of smaller communities were a guiding light for many of us.

Contents

Foreword

The long and fascinating story of the evolution of economics from the early days of political economy to present-day modern economics is marked by several milestones. We have seen the introduction of the theoretical dimension, the temporal dimension, the mathematical dimension, and more recently (mainly postwar) the spatial dimension. The spatial dimension spawned the twin areas of international economics and regional economics, and the latter became one of the cornerstones of the incredibly prolific multidisciplinary field of regional science. In 1989, another important dimension, the community dimension, was added to economics with the publication of Ron Shaffer's book *Community Economics: Economic Structure and Change in Smaller Communities* by Iowa State University Press.

Shaffer's *Community Economics* was important for two reasons. First, it demonstrated that economics, which had been traditionally concerned with the "large" in terms of economies, firms, and development policies, had a legitimate and important function to fulfill in terms of the "small," that is, at the small region or community level. *Community Economics* was the leader in establishing the professional integrity of small economies as important in their own right and worthy of attention as integral units, even though they are small. Second, *Community Economics* provided a substantial professional framework for the area of work known generally as regional or community economic development, an area often observed to lack a sufficient conceptual or theoretical base. Shaffer's book provided this conceptual and theoretical base and established the professional basis for community economics.

Community Economics trawled through the economics literature, both theoretical and applied, to identify the concepts and the practices that are rele-

vant at the small-economy level. The result was an important and valuable book that helped give the idea of "community" the status it deserved. Now, this volume by Shaffer, Deller, and Marcouiller provides a new, updated, and expanded version of the seminal work.

The new *Community Economics* has been carefully written to provide important insights to two substantial audiences. First, it will appeal to the professional economists who wish to gain an appreciation of the contribution economics can make to the understanding of region and community. The familiar economic concepts are applied in a pragmatic context to an extent seldom seen in the economics literature, and one sees these concepts living in a real environment. Second, for those without a substantial economics background, it will provide a useful appreciation of economic concepts and the way in which they can advance the work of regional and community understanding and development.

Readers will be impressed with some important new dimensions in the new *Community Economics,* including perspectives on growth theory, the role of land markets, the amenity values of regions/communities, and the significance of local government. Chapters have been significantly expanded to include important new concepts and principles. These include technology, institutions, and decision-making.

Students of economics, regional science, planning, and public administration will be well served by this book. The informal style of writing, the breadth of topics covered, the obvious professional standing and experience of the authors, and their knowledge of the professional literature combine to ensure that this book will also become a seminal part of the literature.

Shaffer was joined in this new venture by Steven Deller and Dave Marcouiller. Deller brings to the partnership the specialist perspective of public sector economics at the community level. Marcouiller brings the perspective of a planner and resource economist. Both have extensive experience with the more advanced tools of community economics, including regional modeling. The merging of these three talents and their effective teamwork is evident on each page. It has enhanced the discussion from the original book and produced what is, without doubt, the most significant work so far in community economics.

Rod Jensen
Emeritus Professor of Economics
Past President, Regional Science Association
International
University of Queensland, Australia

Preface

Since the first edition was written in 1989, two contrasting phenomena continue to evolve. The first deals with an ever-expanding base of understanding about the functioning and change in small open economies. The second has to do with how much we still do not understand. As we gain more insight into community economics, the more we realize how much we do not understand. The more questions we answer, the more questions arise. Community economics remains a work in progress. As a result, we have completely updated the text and added several new chapters.

We have progressed in our understanding of community economic development from both theoretical and practical perspectives. A significant shift in this edition is that it more fully challenges the student to better understand the linkage between theory and practice. For this to occur, we require the student to better appreciate the theoretical foundations of community economics. Conversely, by better understanding the practical aspects of community economic development, we can reinforce existing theory and develop more grounded theoretical approaches.

This book was written with two audiences in mind. The first audience is upper-level undergraduates and graduate students in the disciplines of economics, sociology, political science, and geography and the applied professional programs in planning, public administration, policy studies, and development studies. The second audience is the practitioner who is faced with day-to-day, real-world problems. The book has restated problems faced in the real world within specific theoretical frameworks. Making clear distinctions among alternative theories helps one think through real-world problems that are seldom neat. Theory gives some organization to the mass of confusion that seems to surround the problem. Theory provides some guidance to policy-makers in thinking comprehensively. There is a need to understand community economic development theories because they help us understand policy options. Theory helps us understand the limits of narrowly conceived policy suggestions and enables us to offer sound alternatives.

Although this edition is more technical, it was written with the mathematically challenged in mind. While some math is included, it is minimal; it should be viewed as a tool to help think through a problem. Graphical and verbal arguments are used to flesh out intuitive appreciation of the theories.

Our primary objective is to improve economic literacy in community economic development. That means helping people make wise decisions. Wise decisions include understanding economic change, learning how to analyze economic conditions, examining what can and cannot be done, and reviewing cause and effect of policy choices. Ultimately this book is written to help people make decisions that are more informed.

This edition takes a broader view of community economic development by introducing theories and concepts that are beyond the control of the community. The first edition was limited to a discussion of factors that can be influenced by the community. Communities, however, need to have a broader understanding and appreciation of the larger forces at play because it is precisely these outside influences that provide context for the real-world choices available to local policy-makers.

This book is unique in regional science because we specifically focus on smaller open economies.

The economics at play and the policy options open to the community apply equally from small rural hamlets to larger urban neighborhoods. This is important because most people live and work in smaller open economies.

We would like to extend our appreciation and thanks to Claire, for her great refereeing and grammatical skills, along with her fabulous coffee and take-out lunches. In addition, we would like to thank Melissa, Margaret, and all the kids and grandkids for their patience in the writing of this edition.

Our future may lie beyond our vision, but is not entirely beyond our control. It is the shaping impulse of America. That is not fate, nor is it chance, nor is it the irreversible tide of history that determines our destinies. Rather it is reason and it is principle and it is the work of our own hands. There may be pride in that, even arrogance, but there is also truth and experience. And in any event, it is the only way we can live.

Robert Kennedy
Seek a Newer World, 1975

Your community should be something you aspire to, not settle for.

—Bob Hartford
Nebraska City, NE, 1991

Doing community economic development is not rocket science—it's harder.

—Jim Hite
Richmond, VA, 2000

Section I
Community Economic
Development Theory

Understanding the economics of smaller communities is best done with an appreciation for the broader underlying aspects of economic theory. The four chapters in this section are written to help readers appreciate these broader contextual elements. It is written in a way that allows those with even the most novitiate background in economics to better understand how the economic world works. For more advanced readers, we have included the necessary rudiments to progress deeper into economic theory. The challenge is to gain practical insights from theories that often are abstract. Written with both the student and the practitioner of community economic development in mind, this section provides a contextual basis for understanding the topics found throughout the rest of the book.

In particular, we focus on those aspects of economics that have unique consequences for smaller communities, ranging from isolated rural communities to urban neighborhoods. We begin Chapter 1 with a basic set of definitions that familiarize the reader with the book's semantics and its organizational framework, which is well represented by the five components found in the Shaffer Star. Also presented in Chapter 1 is a basic set of definitions that introduce the reader to key economic concepts, such as efficiency, equity, and spillovers.

Chapter 2 is a historical account of the theoretical progress in understanding regional economic growth. Moving from earlier theories of linear and structural stages, we formally delineate and critique neoclassical economic growth theory from a regional perspective and conclude with some important community-level shortcomings. These shortcomings move us into a discussion of the current thinking on endogenous growth theory.

Given that, in general, economic theory rests on the assumption that economic space lacks heterogeneity, Chapter 3 focuses on the fact that place in space is particularly important in understanding the economics of smaller communities. Our discussion of location and space builds on several key issues of community economics. Namely, we deal with firm location decisions and regional comparative advantage from the perspective of location theory.

Understanding community economics is increasingly focused on understanding markets. Alternative market conceptualizations provide the basis for our discussion in Chapter 4. We move from an overview of the general circular flow of income in a regional economy to focus on the market for goods and services. Export-base theory and central place theory provide a set of organizing principles in our discussion of market intricacies, specifically the top half of the circular flow model.

Presented as a basic primer to the theory supporting community economic development, this section is written to set the stage, if you will. From such a vantage point, we can better understand the intricacies and pressing concerns associated with economic change as played out within the smaller community.

1
Defining Community Economic Development

Concerns raised every day by community residents include the desire to create jobs for high school graduates who wish to remain in the local community, to improve the access to consumer goods and services, to support new small business start-ups, and to identify the types of training that should be provided young workers. These matters are partially determined by national and regional economic conditions, but local residents can influence the specific circumstances of a community. From the broadest perspective possible, these concerns condense into three basic questions: What is the economic situation now? What could the economic situation be? How can the economic situation be changed?

To address these questions, we need to gain an understanding of a community and its economy. This means reviewing economic theory to provide insight to what is happening in the community's economy and how local and nonlocal decisions influence local economic change.

Community economic analysis is the subset of community development that emphasizes the economic rather than the social-political-environmental dimensions of the community. Since community economic analysis is a direct descendant of community development, it is appropriate to define community development before defining community economic analysis. Next, we present the development paradigm that forms the basis of the book. Finally, we review basic concepts of supply, demand, pareto optimality, and market failure that are needed to fully comprehend community economic development. In doing so, we provide readers the fundamental background to understanding the economics presented in this book.

COMMUNITY

Most definitions of community contain some reference to area, commonality, and social and economic interaction. The definition of *community* used here *is a group of people in a physical setting with geographic, political, social, and economic boundaries, and with discernable communication linkages.* These communication linkages need not always be active, but they must be present. People or groups interact in the defined area to attain shared goals (Christenson and Robinson 1993; Freilich 1963; Hillery 1955; Ryan 1994; Summers 1986).

There are at least five different approaches to studying a community (Long, Anderson, and Blubaugh 1973; Sanders 1966; Wilkinson 1992). They include qualitative, ecological, ethnographic, sociological, and economic approaches. The *qualitative approach* is the perspective of a community as a place to live. This approach looks at housing, schools, neighborhoods, and attitudes of individuals in the community. The *ecological approach* is a study of the community as a spatial unit, specifically, the spatial distribution of groups of people, their activities and interactions within the community and among communities. The *ethnographic approach* is the study of the community as a way of life. The emphasis is on the total cultural dimensions of the community, not just its demography, economics, or geography. The *sociological approach* views the community as a social system and concentrates on the social relationships in a community, which are patterned into groups and larger social systems, both inside and outside the community. The *economic approach* examines the linkages between economic sectors such as agriculture, main street merchants, and households. It also examines the types of jobs and skills present in the community and considers the sources and distribution of income and the changes in income over time. Finally, this approach considers the resources (e.g., natural, financial, human, and managerial) found in the community.

The comprehensive study of a community uses parts of each of these approaches. Broad community goals and decisions involve more than just economic dimensions. Failure to recognize the linkages among the various dimensions reduces the prospects of successful action. These alternative approaches to the study of community also highlight different ways to delineate the boundaries of a community. Regardless of the approach, common interests and concerns define community boundaries that generate a series of functional subcommunities. A *functional subcommunity* is a smaller community contained within the boundaries of a larger, more general community, A single issue or function may define the subcommunity. The boundaries of the trade area or the boundaries of the labor shed define the economic boundaries of a community. The trade area is the geographic area from which the community draws a significant portion of its retail trade customers. The labor shed is the geographic area from which people commute into the community for employment.

Political interests define different community boundaries. The municipal limits of the village are one set of political boundaries. The county in which the village is located is another set. School attendance or school district boundaries also define a community. A river basin or watershed or even an air-quality shed define a community based on common physical interests.[1]

Communications networks define community boundaries. The people in a newspaper circulation or other mass media coverage area are part of a community because they receive news and other information from a single source. A community can also be defined by the source of supportive services, such as rural communities dependent on a larger regional medical center for specialized medical services.

Any definition of a community's boundaries must select those associations or common interests most important for the concern being examined. Regardless of how general or specific the criteria used to identify community boundaries, some people will not conform to the criteria. An example would be residents who do not have children in school and thus are less concerned about the prospect of a neighborhood school being closed.

Years ago, a community may have been easily defined: people lived, worked, shopped, and went to church in a well-defined spatial area. The sense of place and community were easily identified. Today people live in one place, work in another, and shop in yet a third. Our notion of place and community has broadened beyond simple municipal boundaries.

Defining a community generally involves identifying some type of geographic area, communications or social interaction linkages, and commonality or mutual interest. Regardless of the criteria used, another set of criteria will generate a different arrangement of people and redefine the community border. The commonality of interest offered in this book is the current and future economic conditions of the community.

We focus attention on communities or municipalities ranging in size from small rural hamlets to small- and medium-size cities. In many respects, the economics we discuss apply equally well to urban neighborhoods. In practice, the uniqueness of the larger city context is the different power structure of local residents to implement solutions. A common theme of this book centers on the ability of local residents to implement solutions. Within the new urban planning paradigm, neighborhoods are taking on an increasingly important role. Although dealt with in other literature (Keating and Krumholz 2000; Peterman 2000), contemporary neighborhood planning attempts, in an analogous way, to return control of the growth and development process to neighborhood residents. Much of this book will also be directly relevant to urban economic development practitioners.

DEVELOPMENT

Development is a concept, somewhat like community, that almost defies definition (Beauregard 1993; Bothroyd and Davis 1993; Christenson and Robinson 1993; Oberle, Stowers, and Darby 1974; Reese and Fasenfest 1997; Summers 1986). *Development is sustained progressive change to attain individual and group interests through expanded, intensified, and adjusted use of resources.* Development is an ongoing process; although it may progress at different rates over time, it is continuous. At times and in some places, the rate of development approaches zero. While development frequently implies the creation of more, it can also mean less. For example, development occurs if the community replaces a large dirty industry with a smaller clean industry, even though total employment declines.

Often the terms *economic growth* and *development* are incorrectly used interchangeably. *Growth* is generally restricted to more of the same: more jobs, more income, more people, or more real estate transactions. *Development,* in its broadest context, simultaneously involves social, environmental, and economic change to enhance quality of life. For example, growth means that we have changed the factors of production, but we probably have the same type of output produced with income distributed the same way. Structural change, within the context of development, means such things as changes in industry mix, product mix, occupational mix, ownership patterns, and technology. Development means that there is a technical and institutional change in the way we increase production and its distribution. It could be that there has been a change in technology, a change in institutions, or a change in cultural/social framework, specifically, changes in attitudes and values of the population. Development is long term, purposeful, and permanent.

How do we distinguish between the concepts of economic development and economic growth? Maybe an analogy will help: You are 12 years old, and you meet an uncle who has not seen you for 3 years. The uncle says, "My, but how you have grown." Now you are 28 years old, you meet an uncle whom you have not seen for 3 years, and he again says, "My how you've grown." In the second instance, you might take umbrage with this long-lost relative because the implication is that your waistline is increasing. As humans, we tend to stop growing sometime before the age of 20 but continue to develop our understanding, insights, and maturity. Growth and development are very similar; they tend to be related to time and the point from which we started.

Development means enhanced capacity to act, to innovate, and to deal with new circumstances (Liechtenstein and Lyons 2001). Development involves transformation, not just change. Liechtenstein and Lyons said transformation is something that must be sustained through time. Transformations are changes in outlooks, attitudes, and behaviors. Transformations take time and involve long-term prospects.

Development is more focused on equity than on equality. *Equality* means everyone has shoes. *Equity* means everyone's shoes fit. Equity means that everyone has a fair chance, not that everyone has the same chance. Equality suggests the economic pie is divided by the number of people in the population and distributed equally. Equity emphasizes access to opportunity. (This distinction is formalized in the discussion of Pareto optimality later in this chapter and in the notions of a social welfare function in Chapter 12.)

Development can be initiated by external shocks (e.g., changing energy prices or technological changes), which create disequilibrium and disruption in the economy that require some type of institutional or structural adjustment. The goal of development policy is less on preventing such impacts and more on enhancing the economy's capacity to respond, which can be thought of as "reduced vulnerability to sudden shifts in production technology and in the market environment" (Pryde 1981, p. 523). This means the economy becomes resilient, diverse, and innovative. Shapero (1981, p. 26) argued development means the area or people in that area

> achieve a state denoted by resilience—the ability to respond to changes in the environment effectively; creativity and innovativeness—the ability and willingness to experiment and innovate; initiative taking—the ability, desire and power to begin and carry through useful projects. . . . [W]ith diversity there is always part of a local economy relatively unaffected by the changes in a single industry or market place or by legal constraints on a given product. Diversity provides a favorable environment for creativity and innovativeness.

Generally, development, if started by some external shock, becomes self-sustaining and internally driven.

Economic development is sustained, progressive change to attain individual and group interests. Important elements of economic development include the need to create goals, identify individuals and groups and their interrelationships, understand present and future effects of decisions made now, consider new combinations of existing resources or pursue new resources, and identify new markets and deliver to them. Economic development can also be defined as those activities that lead to greater resource productivity, a wider range of real choice for consumers and producers, and broader clientele participation in policy formation. Economic development is goal-oriented change, not change for the sake of change. Development is disruptive and non–status quo.

Human welfare (social well-being and/or quality of life) is the end product of the development process. Human welfare is a value-laden concept that affects the economic efficiency and equity (social justice) dimensions that pervade definitions of economic development. Human well-being is a multifaceted idea that includes health, interpersonal relationships, physical environment, housing, education, arts, and numerous other aspects of life. The significance of the different dimensions of human well-being is that, frequently, achieving desired goals in one dimension limits actions or achievements in other dimensions. This potential conflict and the need to make trade-offs are examined in greater detail in Chapter 12.

While not the major focus in this text, we must remember that economic development is a human/social phenomenon. This means that the behavioral/institutional framework of society conditions economic development. The behavioral framework essentially conditions how development is achieved or even the form of development that is sought. The participants in the development process operate within a behavioral framework that defines acceptable and unacceptable activities. (Throughout the book, *process* denotes change over time rather than social interaction per se.)

The nonstandard definition of economic development leaves a multitude of partial measures of community economic development—changes in the volume and type of economic activity in the community, changes in productivity, and changes in economic stability. Each measure has its own advantages and disadvantages, but none completely encompasses the multifaceted nature of economic development. The measure used depends on the goals of the community. For example, if the community desires to improve the status of individuals currently outside the economic mainstream, development could be measured by the responses to several questions: What has happened to people in poverty? What has happened to the unemployed? What has happened to inequity? The simplicity of the questions belies the difficulties of an accurate response. How will people judge a given response?

Development is a multidimensional concept that incorporates more than just market-determined values; it also includes concern about equity and well-being. It refers to the community's ability to adapt to change and its resiliency to accommodate change.

COMMUNITY ECONOMIC DEVELOPMENT AND ANALYSIS

Community economic analysis occurs when people in a community study the economic conditions of that community, determine its economic needs and unfulfilled opportunities, decide what can and should be done to improve the economic conditions in that community, and then move to achieve agreed-on economic goals and objectives. Stated in a simpler fashion, community economic analysis is examining how a community is put together economically and how the community responds to external and internal stimuli. Community economic analysis is not a rationale for maintaining the status quo; it is a comprehensive concept for changing the economic situation within the community.

There are several common elements in community economic analysis and the community development process (Cary 1970). The community is viewed as the unit for action; the community, rather than some smaller (firm, household) or larger (nation) economic unit, is making and implementing decisions. The community's initiative and leadership, rather than externally imposed mandates, are resources for change. Local citizens assume the leadership position. The identification, encouragement, and training of local leadership are basic objectives. The community can use both internal and external resources to achieve change, drawing on its own strengths and capabilities, and looking beyond its boundaries for supplemental resources. The participation of citizens within the community should be as inclusive as possible. Although not all citizens need to participate, inclusive participation means that all segments in a community are given an opportunity and are actively encouraged to participate. Community development and community economic analysis implicitly assume a democratic political system where people have an opportunity to express their preferences. Community development and community economic analysis are holistic in that they comprehensively examine the different dimensions of the community. Finally, changed attitudes in people are as important as the material achievements of community projects during the initial stages of economic development.

Community economic development is a multifaceted comprehensive approach to community change. It is not limited to just poverty programs, nor is it synonymous with industrial recruitment. Basic strategies for pursuing community economic development are reviewed in Chapter 12. Community economic development is not an attempt to exploit resources to yield the maximum economic return. If resources are nonrenewable and/or unsustainable, the short-run gains obtained contribute little to achieving the human welfare goals of community economic development. A historical example is the extensive harvesting of northern Wisconsin forests in the early 20th century, where short-term gain occurred at the expense of long-term stock assets. A current example of natural resource exploitation with a long-run goal is Norwegian North Sea oil development policy. The Norwegian government chose to limit the rate of exploitation of these hydrocarbon resources to reduce the strains on the economic and social fabric of the country.

Many people see community economic analysis as a data exercise. While understanding the local and larger economies through data analysis is part of the process, community economic analysis is much more. Despite that proviso, this would not be a book on economics without some appreciation for analytical tools (Chapters 14, 15, and 16).

To a large extent, two fundamental issues in community economic development are

- understanding the full range of choices available to alter economic circumstances.
- engaging willing (and even unwilling) collaborators in building long-term strategies.

To understand the full range of choices available to alter economic circumstances means that the community is exploring all of the options. It is more than just attracting a manufacturing plant or adding value to agricultural production. For example, community economic development could include improvement to local health services. It could appear that enhancing the health care sector is not related to economic development. If one thinks about it for a moment, however, both direct and indirect economic development are being promoted. The direct economic development is that the health clinic/hospital brings income into the community in the form of Medicare and Medicaid and insurance payments, so it performs the function of bringing new dollars into the community. The indirect economic development is that as the community improves its health care sector, it improves the quality of its labor force and local quality of life.

Often people are more interested in short-term projects than in long-term strategies. Both are important. The long-term strategies provide the overarching direction for the community. The short-term projects provide mile markers that provide tangible outputs that local people need to remain connected with the long-term strategy. Without a long-term strategy within which short-term projects fit, conflicts may arise such that long-term goals or

objectives are not met. One could think of the long-term strategy as an envelope of short-term efforts.

Economic development takes teamwork. We joke that you need both collaborators and unwilling conspirators. While willing collaborators may be easy to find, many people are unwilling to get involved. To engage unwilling conspirators in building a long-term economic development strategy takes skill on the part of the practitioner (Chapters 11 and 12). It means that often people are not interested in building long-term strategies; they are more interested in worrying about short-term projects.

Community economic development, poverty programs, industrial development, resource use, and data analysis are therefore neither mutually exclusive nor identical. Community economic analysis is using analytical tools and economic theory to build and implement strategies. (The terms *community economic analysis* and *community economic development* are used interchangeably throughout the book. The former implies more technical connotation, and the latter, a stronger policy orientation.)

AN ORGANIZING PARADIGM

An interesting way to view community economic development is to think of a star diagram (Fig. 1.1). Around the nodes of the star and in the center, we have three elements that are typically associated with economics: resources, markets, and space. Three additional elements are associated with our broader definition of community economic development: society, institutions/rules, and decision-making. The components of the nodes are discussed within a spatial context.

Let's begin with *space*. Space is included because communities are generally defined within some spatial connotation as well as some form of communication network. It could be this community or that community, the north side of town, or school attendance boundaries. Furthermore, every community must move product and resources over some physical distance and some form of communication must occur. Generally, this is noted in terms of community boundaries that imply some people are members of a community and some people are not. (More on the role of distance in community economic development can be found in Chapters 3–7.)

Resources are the primary factors used in production. They include land (Chapter 5); labor (Chapter 6); capital, both private (Chapter 7) and public (Chapter 10); and the technology (Chapter 8) that the community uses to produce output. Increasingly

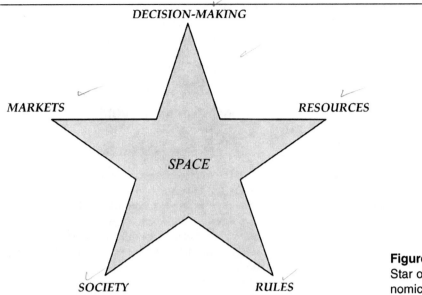

Figure 1.1. The Shaffer Star of community economic development.

important in the functioning of local economies is an amenity base, which we include as the fifth factor of production (Chapter 9).

Land refers to the finite resource on which production activities take place. *Labor* refers to the number of people and skills embodied in people who are actually working and the people who could be working. *Private capital* includes liquid assets, such as financial capital, and nonliquid assets, such as buildings and machinery. *Public capital* refers to roads, schools, parks, and landfill sites, among others. Social capital often is referred to as the glue that holds communities together (Chapter 12). *Technology* is how land, labor, and capital are combined to produce output. Technology can involve the latest innovation or something that has been around for some time, but it generally is regulated by management or community processes. It can be industry sector and/or business specific, or it can be product specific, such as how we produce a good or service by combining inputs, or process specific, such as new ways of dealing with workers or decision-making. *Amenities* can involve cultural, historic, natural, or built environmental resources that increasingly contribute to our notion of quality of life.

The markets node refers to the spatial boundaries of goods and services on which a community relies for production (supply) and consumption (demand). The local market is comprised of businesses buying and selling locally to other businesses and house-

holds. It is important to remember that the local market is composed of two distinct parts: households and businesses. The *nonlocal market* refers to those goods and services that the community produces locally and sells to nonlocal households and businesses. The critical element of the nonlocal market is that it is essentially an external source of sales and income. Thus, it can be the local production of manufactured goods or agricultural products that are sold outside the community or health care that is paid by a third party, such as Medicare. It can also be the local production of goods and services that are sold to nonlocal residents, including tourists and in-commuters. (Goods and service markets are covered in Chapter 4.)

The remaining three nodes—society, rules, and decision-making—are noneconomic. Social norms drive the larger environment in which the economy functions. *Society* is a subtle background force that defines the norms, values, and ethics that determine right from wrong, good from bad. Any economy operates in a social context that is built by tradition, historical precedence, and cultural mores. These are not determined by the economic system, although there can be subtle feedbacks between the economic system and society. For example, the goals of economic policy are often driven by social norms (Chapters 11 and 12).

The rules of the game often are an assumed given element and thus overlooked, but they are critically important to community economic development. *Rules* are important because they govern what can be done with markets, resources, and space. Rules of the game often focus on the rights and responsibilities of ownership and their respective enforcement. In a market-driven economy, the notion of property rights is fundamental to the functioning of the economy.

A rule that prevents a community's business from selling product to Cuba is a rule that limits access to a potential market. The rule that prevents the use of child labor is a rule that governs the types of resources that are available to the community. The recent legislation on telecommunications means that some communities, especially rural communities, now do not have universal access to some elements of telecommunications technology, such as fiber optics and the Internet. These rules are human-made limits or openings that guide the use of community resources and exploitation of markets (Chapter 11).

The decision-making capacity of the community is the ability to distinguish between problems and symptoms and to implement solutions. A symptom is a visible sign that there is an underlying problem, but treating the symptom does not correct the problem. For example, in many communities, local leaders may claim there is a "problem" with the lack of affordable housing. This is really the symptom to what could be at least two possible problems. First, the problem could be that the cost of land and houses is far too high for the types of jobs and wages available locally. Second, the problem could be that the rules and regulations regarding the building of a house are far too restrictive, thus raising the cost of housing. Every community faces symptoms and problems like these. People involved in community decision-making really need to focus on the problem rather than just addressing a symptom. In this example, people have difficulty buying a home.

Implicit in decision-making is that the community needs to establish its values and set priorities. Each community, at any given time, is faced with a range of issues, and effective decision-making requires the community to not only identify issues, but also rank them in terms of priority. How these issues are identified and ranked hinges on the values that the community possesses. These values are driven by the larger society. If you feel the market has the only say then the housing example might not even appear on your radar screen until firms can not hire local labor. Each community faces a plethora of problems, and community priorities go a long way in determining which one will be addressed first,

second, or not at all. (See Chapters 11 and 12 for further discussion.)

For community economic development, keep all of these elements (nodes) in mind, for any one of them or some combination of them can be the focus of a correction strategy.

ECONOMICS: BACKGROUND FUNDAMENTALS

Market Efficiency and Pareto Optimality

To fully understand community economics, it is necessary to understand how free markets allocate resources among the production of various goods and services and allocate those goods and services among consumers. In addition, how will this allocation mechanism affect the economic well-being of both producers and consumers? For the community practitioner, the concepts of economic efficiency and social optimality (optimality from the viewpoint of the community as a whole) are fundamental to the problem at hand. Will the notion of Adam Smith's *Invisible Hand*, which assumes that markets are perfectly competitive and consumers and producers act in their own self-interests, result in a market equilibrium that will also be in society's interests, such that no possible change in that allocation could lead to increased output or individual well-being without harming another economic agent?

To the community practitioner, economic efficiency (also referred to as Pareto optimality) and social optimality (often called a social welfare optimum) form the foundation of our understanding of the economy and how markets function. *Pareto optimality* refers to the situation where resources are allocated efficiently with respect to production and consumption; *social optimality* refers to the situation where resources are not only allocated efficiently but also in a manner that is deemed to be fair by society. One example of a widely accepted social optimal allocation is one in which incomes are evenly distributed. A social welfare optimum must necessarily be Pareto optimal, but the converse need not necessarily be true; a Pareto optimal (i.e., efficient) allocation of resources need not be optimal from a social welfare perspective.

A Pareto optimum is said to exist when resources are allocated in such a way that no individual can be made better off without making at least one individual worse off, and the production of another good or service cannot be increased without reducing the production of another good or service. Any allocation that meets this characteristic is said to be Pareto optimal. In theory, an infinite number of possible allocations can meet this characterization. A social welfare optimum is said to exist when the allocation of resources is not only Pareto optimal but also represents the highest level of social welfare given resources, technology, and tastes and preferences.

The difficulty facing the community practitioner is that a Pareto optimum is a positive statement, whereas a social welfare optimum is a normative statement that is brimming with value judgments. To the economist, positive economics is the study of conditions and cause-and-effect relationships, but normative economics is the study of what should be as opposed to what is. The economist prefers to provide positivist analysis in the decision-making realm. In the society node in Figure 1.1, value judgments concerning social welfare should remain in the realm of politics, not economics. We postulate that community economic development is how we combine the normative value-laden judgments with the strict market efficiency arguments.

Much of this book looks at the characteristics of and interaction between firms and households. Indeed, this is where economic theory is rooted. Economists focus most of their attention on issues that pertain to the allocation of scarce resources among competing uses. In a market economy, this allocation occurs through the interaction of supply and demand. (Although we assume readers have a rudimentary understanding of these concepts, we feel it necessary to reinforce some key aspects of supply and demand.)

Given that we focus mainly on the positive dimensions of local market efficiencies and inefficiencies,

how do we know that Adam Smith's *Invisible Hand* will lead to a Pareto efficient allocation of resources? To answer that question, let's turn to efficiency in consumption, efficiency in production, and global efficiency, which is the interplay of consumption and production over all goods and services.

Demand

Demand represents the theory of consumption by individuals. In specific situations, demand can also reflect the firm's use of factor inputs such as land, labor, and capital. Regardless of the application, *demand* theory rests on the notion of maximizing utility from the perspective of the consumer and profits from the perspective of the firm, subject to a budget or cost constraint.

Let us focus first on the consumer. *Utility* reflects the trade-off in consumption of goods and services. It is typically measured using an indifference curve that reflects points of indifference in consumption of two goods or services. A *budget constraint* reflects the available resources of consumers to purchase goods and services. It is measured in terms of income (m), prices of goods (P_1, P_2), and quantities of goods (q_1, q_2). Steepness in the budget constraint infers price differences. Shifts in the budget constraint refer to income changes. Graphically, consumers maximize utility subject to a budget constraint at the point of tangency between the indifference curve and the budget line (Fig. 1.2).

To see how consumers allocate resources, we make the following assumptions:

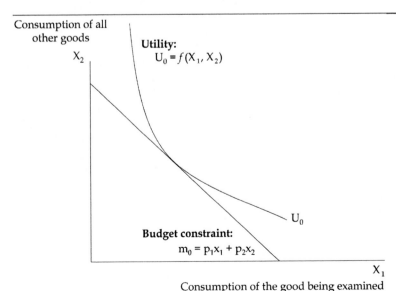

Consumption of all other goods
X_2

Utility:
$U_0 = f(X_1, X_2)$

U_0

Budget constraint:
$m_0 = p_1x_1 + p_2x_2$

X_1
Consumption of the good being examined

Figure 1.2. The concepts of utility and budget constraint.

- Each consumer allocates a given budget among several alternatives in a rational and comprehensive manner that maximizes his or her own utility or welfare.[2]
- Each consumer's tastes and preferences can be described by a utility function that relates his or her satisfaction (utility, welfare) to the levels of consumption of each combination of goods and services. The utility function is characterized by diminishing marginal rates of substitution; that is, as the consumption of one good increases, consumers are willing to forgo less of the other goods and services to maintain the same level of satisfaction (utility, welfare).[3]
- Markets are perfectly competitive, which means that the consumer is a price taker and his or her actions cannot affect market prices.

Under these assumptions, each consumer will select an allocation of goods and services such that the marginal benefit of each good is exactly equal to the marginal cost, or price of the good. If the benefit (utility) of consuming one more unit of a good is greater than the cost of the additional unit of the good, the consumer will purchase and consume the good. If the marginal benefit, or marginal utility, is less than the cost of the good, then the consumer will not purchase the additional unit of the good. The consumer will maximize utility if and only if the marginal rate of substitution (MRS)—that is, the ratio of the marginal utilities between two goods—is exactly equal to the ratio of prices. Stated more formally, utility maximization requires

$$MRS_{1,2} = P_1/P_2 \qquad (1.1)$$

which is the point of tangency between the indifference curve and budget constraint in Figure 1.2. This condition will hold for any consumer and any set of goods and services.

If this condition holds true for each individual, the allocation among consumers must be efficient in the sense that no change is possible that increases the welfare of one individual without decreasing the welfare of another. In our simple world of two goods (X_1 and X_2) and perfect competition in which all consumers face prices P_1 and P_2, all consumers will maximize their individual welfares (utilities) by equating their individual marginal rates of substitution to the ratio of prices

$$MRS^a_{X1,X2} = P_{X1}/P_{X2} = MRS^b_{X1,X2} \qquad (1.2)$$

for consumers *a* and *b*. It is also clear that if the equality outline in equation (1.2) does not hold for any one individual, then consumers can increase their individual welfares (utility) by reallocating their consumption without affecting the welfare of other individuals. But how do we know that if all individuals allocate their consumption choices along the decision rule outlined in equation (1.2), no change is possible that can increase the utility of one individual without reducing the utility of another?

To see this, we must turn to the notion of a budget constraint and realize that consumers are not only constrained by the price they face in the market but also by the income they possess. Given a budget constraint, determined by the individual's income level, consumers will maximize utility by allocating consumption within the budget constraint until the MRS equals the price ratio. In essence, the optimality condition does not change with a budget constraint imposed. The only way to increase the utility of any one individual, given the correct allocation of goods, would be by increasing income or by relaxing the budget constraint by some small amount. But in a world of fixed resources, the only way to increase the income of one individual is to reduce the income of another individual and transfer it to the first individual.

The above situation is Pareto optimal by definition. A *consumption equilibrium* is Pareto optimal if it is only possible to increase the welfare of one individual by reducing the welfare of another. Given perfect competition (consumers are price takers) and a well-behaved utility function (diminishing marginal rate of substitution), individuals that act in a selfish pursuit of utility maximization will lead to efficiency in consumption.

The economic notion of demand can take alternative forms that build from this concept of consumption equilibrium. Two leading alternative forms are *compensated demands* (which imply the rights to constant utility) and *ordinary demands* (which imply constant income levels). Mathematically, both are analogous representations of maximizing utility subject to a budget constraint. Ordinary demands derive their shape as a function of prices and income, while compensated demands are derived as a function of prices and utility. (See intermediate microeconomic textbooks for a more thorough discussion of demand.)

Supply

Supply, which represents the theory of production, most often focuses on the firm's problem of maximizing profits. In specific situations, supply can also be represented as the individual's ownership and supply of primary factor inputs, such as labor, land, and capital. Regardless of the application, we need to distinguish between two aspects to profit: costs and revenues. To maximize profit, firms attempt to minimize costs and maximize revenue.

So what about the production side of the economy? Will firms make production decisions such that it is not possible to increase the production of one good or service without decreasing the production of another? How can society measure the opportunity cost of increasing the production of one good or service relative to another? For insights into these fundamental questions, we make the following assumptions:

- The production of any good or service is determined by a production function that characterizes the use of factor inputs (land, labor and capital) and technology is known and fixed.
- Production is subject to diminishing marginal rates of substitution between factors of production and decreasing marginal rates of production.[4]
- All producers are price takers in both the factors of production and output goods markets, or perfect competition is present.
- Information is readily available and comprehended by firms.

- Profits are maximized by minimizing the costs of producing a given level of output.

Competitive firms will attempt to minimize costs of inputs given some production technology (Figure 1.3.)

Under these assumptions, firms will allocate factors of production such that the marginal rate of technical substitution between factors (MRTS) will be equal to the ratio of input factor prices. This is shown as the point of tangency between the isoquant and the input cost line in Figure 1.3. If the MRTS between firms differ from the ratio of factor prices, firms will be able to lower costs, hence increase profits, by reallocating factor inputs. Note the parallel in the analysis between the consumer and producer sides of the market. The condition for cost maximization is

$$MRTS_{L,K} = w/r \qquad (1.3)$$

where w is the wage rate, or price of labor, and r is the interest rate, or the price of capital. Given that each firm has the same production technology and faces identical factor prices, the MRTS will be equalized throughout all firms producing the product Y:

$$MRTS^a_{L,K} = w/r = MRTS^b_{L,K} \qquad (1.4)$$

with a and b representing our two firms.

As in the case of the consumer, firms will make identical allocation decisions in the factor markets.

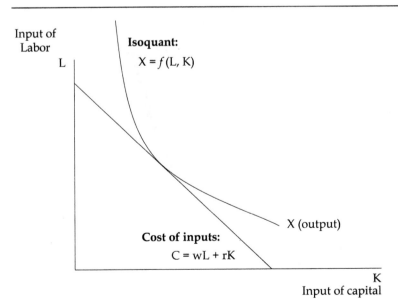

Figure 1.3. The competitive firm attempts to minimize the costs of inputs.

No change in factor allocation can lower costs for a given output level. Conversely, if all firms producing the good or service Y are producing at their cost minimization for a fixed supply of factor inputs, then the only way the output of the good or service could be increased is by diverting factors of production away from the production of another good or service. But this would require that production of the second good would have to decrease. Therefore, the allocation of factor resources that results from self-interested profit-maximizing behavior is efficient in the sense that no more of one factor can be used without reducing that of another and the production of one good or service cannot be increased without reducing the production of another good—the requirements for Pareto optimality in production.

The conclusion outlined in equation (1.4) gives part of the answer to the production side of the market, but we still need to address the question about how much to produce and in what combination. In our discussion of the efficiency of factor markets, we assumed that firms allocate resources in a manner such that, for a given level of output, costs are minimized. Now let's consider the case where firms must decide how much to produce and in what combinations. We make the following additional assumptions on production:

- Marginal costs are rising; that is, as production increases the additional costs incurred per unit of output rise.

- A firm is able to sell at the market-determined price all of the good or service that it is able to produce.

Typically, the competitive firm's supply function is comprised of the upward-sloping portion of the marginal cost curve lying above the average variable cost curve (Fig. 1.4). At prices below this level, competitive firms will decide not to produce.

The problem we face is to show that for profit-maximizing firms the rate of exchange between any two goods or services must equal their price ratio, the same price ratio we saw on the consumer side of the market. First, consider the problem of profit-maximizing on the part of a given firm. To maximize profits, firms will produce at a point where the marginal cost of producing and selling an additional unit of the good or service will be exactly equal to the market price ($MR = P$). If the marginal cost of producing another unit of the good or service is less than the market price, the firm can increase revenues by more than costs will increase; in this situation the firm can increase profits by expanding output. If marginal costs are greater than the market price, the firm could increase its profits by reducing output. Profits will be maximized only at the point where marginal costs (MC) equal the market price (MR) of the good or service. Hence, for any given good or service X, profit maximization requires

$$P_X = MC_X \qquad (1.5)$$

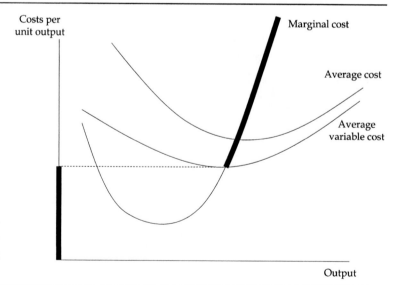

Figure 1.4. The supply function as determined by the competitive firm's cost structure.

Now consider the decision between the production of our two goods or services X_1 and X_2: Which combination of the two should be produced for the economy? To accomplish this, we define the marginal rate of transformation (MRT) as the rate at which the production of one good or service can be exchanged for the production of the other. The MRT can be viewed as opportunity costs because it provides insight into how much of one good or service we must forgo to produce one more of the other good or service.

The total cost of production measures the costs incurred with the cost-minimizing (i.e., efficient) bundle of factor resources for any given level of output. *Marginal cost* is simply the change in the total cost associated with a one-unit change in production. In other words, how much does the total cost change if we increase or decrease production by one unit? The rate of exchange between the two goods or services produced in our simple economy, or the MRT, must also equal the ratio of their marginal costs:

$$MRT_{X1,X2} = MC_{X1}/MC_{X2} \qquad (1.6)$$

But if firms are profit-maximizing, marginal costs must be equal to product price, as stated in equation (1.5). Hence, we combine equations (1.5) and (1.6) to see that

$$MRT_{X1,X2} = MC_{X1}/MC_{X2} = P_{X1}/P_{X2} \qquad (1.7)$$

By definition, the MRT characterizes the rate at which the output of one good or services can be altered in exchange for a change in output of the other good or service. It reflects the boundaries or the production possibilities frontier (PPF) that defines the possible combinations of output that are available to society given a certain level of factor resources. Indeed, the MRT is simply the slope of the PPF, and the profit-maximizing combination of goods and services will be the combination where the slope of the PPF is exactly equal to the ratio of output prices.

What we have now is the condition of Pareto optimality in production. When firms that are price takers in both factor and output markets (i.e., perfect competition) make allocation decisions in a selfish profit-maximizing manner, any change in output, either in levels or combination, will lower the profits of one of the two firms in our simple economy. Profit maximization on the part of individual firms leads to an allocation that maximizes profits for the

whole economy, and any change by any single firm away from the allocation defined in equation (1.7) will lower overall profits in the economy, hence at least one firm will be worse off.

We have now established that individuals acting in selfish utility-maximizing ways will result in an allocation of goods and services that will be Pareto optimal: no one consumer's well-being (utility, welfare) can be increased without causing another consumer's well-being to be lowered. Similarly, if firms act in a manner that maximizes profits through minimizing costs, that also will be Pareto optimal: No one firm can increase its profits without lowering the profit of another firm. But how do we tie the two sides of the market (demand and supply) together to ensure that we have global efficiency (Pareto optimality)?

The key to this question hinges on whether or not consumers and producers face the same prices in the output markets (P_{X1} and P_{X2}). In perfectly competitive markets, prices adjust in a manner that ensures that markets clear, or that supply and demand are exactly equal. As a result, the output markets dictate that, in equilibrium, there is one price to which both consumers and producers respond. In our simple world, keep in mind that consumers and producers are price takers, and that prices are determined through the interaction of market forces, specifically the forces behind the supply and demand relationships. These forces can be captured through selfish utility maximization on the part of the consumer and profit maximization on the part of the producing firm. Given that consumers and producers are reacting to the same output market prices, we can combine equations (1.2) and (1.7) to yield

$$MRS_{X1,X2} = P_{X1}/P_{X2} = MRT_{X1,X2} \qquad (1.8)$$

We have just demonstrated that competitive markets will ensure that the resulting allocation will be Pareto optimal and that markets yield an efficient allocation of resources. This result is often referred to as the *first fundamental theorem of welfare economics.*

For the community practitioner, this theorem explains why so many economists argue that competition is good for the economy, that market forces and self-interested behavior move the economy to an efficient allocation of resources. But as we will see throughout our discussion of community economics, the market forces that drive Adam Smith's *Invisible Hand* (which is embodied in equation 1.8) often break down. The *Invisible Hand* does not discuss social optimality.

Two more fundamental assumptions of the perfectly competitive model must be recognized prior to moving on to social optimality (Chapters 9 and 12). The first is that we must accept the initial distribution of endowments, resources, and income as given. The second is that information is relatively cheap, widespread, and comprehended.

Market Failure

A market brings together consumers and producers of goods, services, and resources and permits them to negotiate a mutually agreeable transaction. If the actors cannot reach an agreement, the market needs to send signals so demanders or suppliers change their behavior. Furthermore, signals need to indicate the type of change required. Not only must the signal be sent, but also the actors within the market must receive, respond, and adjust appropriately to that signal (thus economists' fascination with markets). Markets are the mechanism in which signals are sent to the actors, who then make appropriate adjustments and re-establish equilibrium. Our idealized world of perfect competition and Pareto optimality can fail us for two primary reasons: (1) the unrealisticness of the assumptions in the real world and (2) the presence of what we call externalities.

We turn to the first of these sources of market failure. As we have seen in our discussion of Pareto optimality and market efficiencies, some fairly strong assumptions must be in place. In addition to our explicit assumptions about perfect competition and self-interested behavior, implicit are assumptions about rational behavior, access to perfect and costless information, as well as that there is no risk associated with decisions and actions and, perhaps most important, all goods and services transactions take place within the market. Only if all of these explicit and implicit assumptions hold true will markets produce efficient results. Unfortunately, if any one of these assumptions is invalid, the adjustments associated with a market operating smoothly and perfectly are impeded. The discussion of market failure recognizes that many of the assumptions of the neoclassical perfectly competitive market model are violated consistently. Violation of these assumptions can lead to a divergence of the model's prediction and real world experience. This section examines market failure and its implications for development activity. Market failure can occur under three separate situations: performance failure, structural failure, and the presence of externalities. *Performance market failure* occurs when the market economy, while functioning well in the structural sense of the neoclassical model, fails to yield a socially desirable distribution of income and output. Performance failure arises from a disagreement about the distribution of resource ownership and income. Either form of market failure often justifies government intervention in the market (Chapters 9 and 10).

Structural market failure occurs when one or more of the underlining assumptions do not hold. For example, consumers or producers do not possess full information, or they cannot process and/or understand the information presented to them. Consumers may not be operating under utility maximization, or firms may be attempting to maximize market share as opposed to profit maximization. Structural market failure may also arise from immobility of labor or capital among communities and uses, causing a misallocation of capital or labor among uses and places.

Factor prices, product prices, and profits are not equal over space for several reasons. The most obvious reason is differences in transportation costs. Transportation costs are not uniform in every direction or among commodities. This distorts the factor and product prices and sends an unintended market signal (as seen below this is a pecuniary externality). Other causes of market failure include resource immobility, imperfect information, transactions costs, increasing returns to scale, externalities, market power, second best, and public intervention (Bartik 1990).

Some immobility of natural resources (e.g., land, minerals, and forests) is acceptable, but immobility of other resources (e.g., capital, labor, technology, and management) prevents equalizing factor and product prices over space or among products. There are two forms of resource immobility. The first occurs when external resources fail to perceive and respond to long-run economic signals from the community, represented by the failure of capital and labor to move into a community that offers a higher return. The second form of immobility occurs when community resources are not used in their most productive manner, for example, when labor or capital in a community continues to be used to produce something of lower value to society. This could occur when labor remains unemployed or the price of other outputs increases.

Beyond the general question of type of resource immobility is recognition that the propensity to move in response to market signals varies among resources. For one resource—people—factors that influence the rate of migration (mobility) include distance, information flows, psychic forces, age, occupation, and family status. For capital, another resource, the rate of migration varies with the form of the capital, information flows, historical investment patterns, and uncertainty.

Returning to the suggested causes of structural market failure, we take labor markets as an example. Imperfect information prevents the unemployed worker from being aware of job opportunities or of the types of skills to develop. Transaction costs are the unemployed worker's cost of finding out about job opportunities or acquiring an appropriate skill. Increasing returns to scale become evident in the efficiencies gained by having the federal and state governments provide a job information service rather than each worker individually maintaining job information. Concentrated market power affects the labor market by reducing the incentive for product innovation and development because of altered behavioral response to market signals or because resources are unavailable to smaller economic units. The concept of *second best* means that an imperfection in one market, such as the capital market, causes less-than-optimal conditions in a second market, such as employment for certain individuals.

Public intervention, while designed to improve the functioning of the market, can also generate perverse signals. For example, the regulation of financial institutions to protect depositors discourages investment in high-risk investments even with high-growth potential.

Externalities

The third primary source of market failure is embodied in what economists term *externalities of production*. Four unique types of externalities are significant to community economics as described throughout this text: (1) technical externalities, (2) public goods externalities, (3) ownership externalities, and (4) pecuniary externalities. The first two deal with unique situations of production, while the latter two deal with the manner in which the production by firms is interrelated.

Technical Externalities

Typically, industries operate in a manner that covers both fixed and variable costs, and they do so until the marginal cost just equals the price of the output. In the long run, these points where marginal costs are equated to price help define an industry's supply function. In microeconomics, the industry's supply of a good or service is counterbalanced by the demands for that good or service by consumers. Demands are determined by an individual's preference conditions for consumption of the specific good or service relative to substitutes. Equilibrium is reached when the demand for an industry's output is equal to its supply. This equilibrium determines both the market price of the output and the quantity of output produced. A technical externality occurs when demand for the industry's output crosses a declining portion of the industry's long-run average cost for producing output (Fig. 1.5).

Technical externalities are exemplified by the existence of natural monopolies. Examples include railroads, electric power production and distribution, telephone service, and other public utilities. If unchecked, a natural monopoly once established will exert undue influence on the situation (Fig. 1.5). A natural monopoly will act to set output levels based on equating marginal returns with marginal costs. Notice from the figure, though, that at this level of production, the monopolist can use demand to determine a monopoly price that is significantly above the marginal cost–marginal revenue level to generate monopoly profits.

Technical externalities are clear market failures that have been the focus of a lengthy history of public regulation. Specifically, natural monopolies that exist under this situation are either highly regulated as quasi-public utilities or are run by units of government and exist as wholly owned public utilities. Regulation of quasi-public utilities often acts to control price such that natural monopolists produce higher levels of output at lower prices. This point would be found at an output level where demand crosses the long-run marginal cost curve (Fig. 1.5). Through regulation, quasi-public utilities cover long-run average costs and allow shareholders to generate an equitable return. Another possibility to correct for technical externalities exists in public provision of the good or service. An example is wholly publicly owned utilities, such as community-operated water systems, electric generation systems, and other publicly run endeavors. These operations often cover large fixed costs of setting up the system by using public funds and operate the utility (at a loss) where demand crosses the long-run marginal costs function (Fig. 1.5).

Public Goods Externalities

Many goods and services we expect to be available are rarely provided within a market context. These

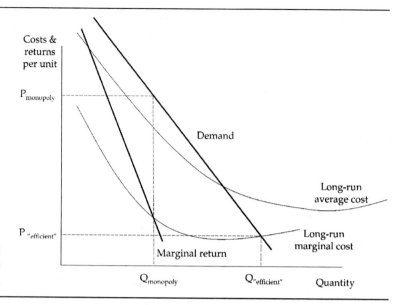

Figure 1.5. A technical externality and alternative levels of corrective action.

goods and services neither have readily defined rights that allow firms to limit access (exclusivity) nor operate in a competitive (rival) environment (Chapter 9). Examples include a healthy, clean environment, many types of outdoor recreation sites, highways, public safety from fire and crime, and national security. These types of goods include both *common-pool* resources and purely *public* goods. The provision of these goods and services exists within the context of public goods externalities.

Public goods externalities come about because of the complete lack of a market structure within which trading can take place, prices can be discovered, and optimal quantities can be determined. Without a market, there is no mechanism for limiting the amount of good that will be consumed. This type of a market failure provides another clear indication for public involvement. Without public provision of these types of goods and services, they would not be provided. Not providing these goods publicly would often result in significant degradation of the resource foundation on which communities operate.

Ownership or Technological Externalities

The last two types of externalities exist because of the interaction of one individual's production or consumption on another's. The first type of interaction externality is known as an ownership or technological externality. In this type of externality, the activities of one individual directly affect the production or consumption of a nearby individual, but

the interaction is *physical*, not economic. The externality results from a general lack of ownership rights to the elements involved in interaction.

The classic examples of this type of externality focus on the ill effects of one individual's pollution on another individual's consumption or production. For instance, if an individual enjoys loud music, pursuing this endeavor could easily impact others in the neighborhood that might enjoy peace and quiet. From a production perspective, an example would include two firms. For instance, if a farmer spreads manure on her fields, she might easily affect the sales of ice cream cones in a nearby open-air restaurant. Again, these are interactions marked by physical linkages, not economic linkages. The key element involved is a lack of assignment to the rights associated with the offensive element.

Our examples are situations where no one clearly owns the rights of silence or smell. The manner in which we address ownership externalities often takes the form of strict regulation of societally determined offenses. In the recent past, we have also developed pseudomarkets within which these rights to offend can be traded.

Pecuniary Externalities

The final type of externality exists when two individuals or firms interact with one another but do so within economic markets. Pecuniary externalities explain shifts resulting from purposive actions of individuals or firms that are not necessarily *market failures* in the strict sense of the term. Indeed, market forces help explain the existence of locational decisions that result because of the interaction of firms.

Pecuniary externalities are important because they provide a good explanation why firms locate in proximity to other firms. The cost advantage of operating a condominium development in close proximity to a ski hill would exemplify a pecuniary externality. The presence of a ski hill attracts people who need to have overnight accommodations. This spillover effect (or interaction) between the operation of a ski hill and the operation of overnight accommodations is a pecuniary externality. A good example from a manufacturing angle is the positive influence that automotive assembly firms exert on parts manufacturers located nearby. Indeed, these pecuniary externalities are a significant portion underlying the theoretical basis of agglomeration in firm location (Chapter 3).

While the importance of these sources of market imperfections must be recognized, they can be overemphasized. Specifically, in a dynamic context, these sources of market failure create opportunities for entrepreneurs. Thus, if the community fails to develop, it may be less a problem of reducing market failures and more a problem of providing incentives or disincentives to entrepreneurs to exploit opportunities in such a way as to attain community objectives. Also, public intervention designed to overcome market failures may be the barrier to development.

In a dynamic economy, some barriers to development include the lack of entrepreneurship; the high cost of adjustment, such as the cost of creating additional highly skilled labor or sophisticated machinery; uncertainty about such things as governmental fiscal policy and monetary policy; institutional rigidity, such as bureaucratic behavior; or a lack of capacity of both institutions and human resources. Other barriers are a lack of key resources or key organizations to support the development process and a lack of integration or coordination between key parts of the economy and political systems, including but not limited to an adversarial relationship among the public and private sectors. These may be more encompassing than the traditional quasi-static market failure concepts listed earlier. When you think in terms of dynamics, you are concerned with creating new products, mobilizing new resources, improving the quality of existing resources, and altering structural and institutional

arrangements that impede the effective utilization of resources.

SUMMARY

Community development differs from community economic analysis in terms of its major emphasis. Community development is about local leadership, citizen participation, collective decision-making and community organization. Community development means that our central thrust is to increase the capacity of the local population to pursue its own interest in a collective fashion.

Boothroyd and Davis (1993) defined *community economic development* as the sum of the three words. *Community* is a group of people who know each other, who plan together over time for their long-term betterment (strengthening local community). They are concerned with nonmonetized as well as monetized improvements. For example, they are concerned about how wealth is distributed. They are concerned that justice and economic institutions are organized to promote cooperation rather than just competition. They are concerned about empowering different segments of the community. The *economy* is a system of human activity directed toward meeting human wants by the deliberate allocation of scarce resources. Generally, this means that you are concerned about monetized economics returns (e.g., real estate transactions, employment growth). *Development* is the deliberate quantitative and qualitative changes in a system. It is concerned about stability, quality, and local control. It is concerned about local desires (especially the environment), diversity of ownership, local control, and stability.

Development and growth are two very separate conceptual issues. *Growth* tends to be more jobs, more buildings, more equipment. Basically, growth is a replication of more of what and how we're doing it now. *Development*, however, is change in the capacity to act and innovate; it involves transformations. Development is longer term, purposeful, and permanent. Development in most cases is more than just economics. Development tends to imply more understanding, more insight, more learning, more nuances. Development tends to imply that we have had structural change. Structural changes mean changes in technology, ownership patterns, occupational mixes, product mixes, industry mixes, and institutions. Development tends to reduce vulnerability to changes outside the community. Development is disruptive; it is not maintenance of the status quo. Development is progressive change to attain individual and group interests.

Community economic analysis is about how economic forces and theory explain community change. It includes how economic structure influences the choices that we can make and how movement or flow across boundaries influences choices. It's about increasing community wealth in both monetary and nonmonetary forms. It's about how dynamics and the resultant disequilibrium or changing circumstances create tensions within the community that require choices. Implementing decisions and strategies means the people are intervening in the economy, and more, with the idea that they can achieve some type of desired outcome.

To understand the fundamental aspects of community economic development, it is critical to have a firm understanding of supply, demand, and Pareto optimality because these are at the core of an understanding of economic efficiency. In addition, in focusing on smaller communities, the ultimate economic dilemmas end up revolving around market failure in its various forms.

The general purpose of community economic analysis is to improve economic opportunity and quality of life through group decisions and actions. Community economic analysis is an action-oriented study of how a community is put together economically and how it responds to internal and external stimuli. This requires identification of specific problems, resources, and alternative actions. Essentially, community economic analysis is the problem-solving steps applied to community economic problems. These steps are as follows:

1. Where are we now?
2. Where do we want to be?
3. Why aren't we there now?
4. What needs to be done to get us there?
5. Who is going to do it?
6. When is it going to be done?
7. How will we know we got there?

Community economic analysis represents a conscious attempt to improve the decision-making associated with community economic development rather than just being subject to fortuitous circumstances. The emphasis is on the technical and structural analysis of community versus individual decisions. This means consideration of the interaction of the economic, political, social, and institutional components of a community. Community

economic development is a dynamic concept concerned with movement and change, with overcoming obstacles and capturing opportunities. As a field of study, community economic development is practice based and practice driven, but that practice is really based on theory, which is the purpose of this book.

STUDY QUESTIONS

1. What does community mean to you?
2. Which is more important in defining community: geography/place or social interaction?
3. Development is often used interchangeably with growth. How are they similar or different concepts?
4. What is community economic analysis? Is a distinction made between it and community economic development?
5. What do community development and community economic development/analysis have in common? How do they differ?
6. How are the scientific method and community economic analysis linked?
7. Why is the concept of equity important in community economic development?
8. In defining community boundaries, why are political boundaries important for community economic development?
9. Why is the premise that a community is an economic decision-making unit important? Why is decision-making capacity so critical? Does the concept of "bounded rationality" play a role?
10. Some suggest community economic development is simply jobs and income-level changes. Is this sufficient given the discussion of development?
11. Why do you believe it is important that community economic development place so much emphasis on group decisions, participation, and indigenous leadership?
12. Describe your paradigm of community economic analysis/development?
13. How can economics contribute to the understanding of community change?
14. How is economic efficiency and Pareto optimality linked?
15. What does the first fundamental theorem of welfare economics tell us about perfectly competitive markets and self-interested behavior?

16. Externalities can take on alternative forms. Identify two types of externality-related market failures.

17. How do certain forms of externalities represent market-based phenomenon and help explain firm location decisions?

18. Is pursuing individual self-interest good for the community?

NOTES

1. While the definition presented earlier implies a strong physical-geographic connotation, there are two extremes to this perspective. The first would be a community that has no physical setting (e.g., a community of scholars). The second would be a community that literally consists of only its physical setting. This would be a community intended for human settlement but temporarily vacant (e.g., a platted subdivision prior to the building of the first home).

2. Implicit in the assumption is that initial distribution of the ownership of resources and income is acceptable to society.

3. Stated more formally, assume that an individual consumes two goods (X_1 and X_2) and their utility function takes the general form $U(X_1, X_2)$. The marginal rate of substitution between the two goods is defined as the ratio of the two marginal utilities: $MRS_{1,2} = (\partial U/\partial X_1)/(\partial U/\partial X_2)$.

4. Assume that a firm uses two inputs, labor (L) and capital (K), and the production function takes the general form $f(L, K)$ to produce good Y. The marginal rate of technical substitution is defined as the ratio of the two marginal products: $MTRS_{L,K} = (\partial Y/\partial L)/(\partial Y/\partial K)$. Note the similarity between the production and consumption sides of the problem.

2
Growth Theory

One of the fundamental questions that economists struggle with deals directly with the causes of economic growth. Why does one economy grow while another struggles and may actually decline? Will economies tend to move together, growing in a way that poor regions will catch up to rich regions, or will poor regions always lag behind? What forces are behind economic growth, and can policy be crafted to influence growth patterns in a way that is more acceptable to the desires of society?

While economic theory can provide insights into these questions, often more questions are raised than are answered. Ideas and concepts that were accepted as "truth" 30 years ago are now questioned. Insights into these basic questions are fundamental to community economic analysis because the level of economic growth sets the tone for nearly all discussions within the community.

It is important to understand the factors affecting growth that are beyond the influence of the community and those that local residents can influence. As we move through the history of economic growth theory, much of the literature that we refer to deals with issues that are beyond the influence and control of smaller communities. We will relate that which is applicable to smaller communities in the developed world. In Chapter 1, "small" in terms of communities ranged from rural hamlets to mid-sized metropolitan areas. Many of the concepts also apply to neighborhoods in larger urban places. Remember that the possibilities at the community level are often constrained by the forces of the larger economy in which the community functions.

While the economic forces at play are similar between developing and developed economies, the fundamental differences for our purposes are the unique characteristics of institutions and society (Chapter 11). Our focus is limited to the developed world context.

Although we present theories of economic growth in this chapter, keep in mind that community economic development is broader than just growth (Chapter 1). *Growth* is associated with more jobs, more income, and more business profit. *Development* captures notions of economic opportunities, equity, and quality of life in the most extensive sense. Some would argue, however, that economic growth is necessary for development to occur. In one sense, development speaks to how economic growth is allocated across economic agents. Without a dynamic growing economy, it is difficult to consider issues related to development. As we address economic growth, the concept of economic development is ever in the background.

The growth literature can be broken into two general approaches: deductive, which focuses heavily on theoretical modeling and attempts to establish paradigms that predict how the economy grows, and inductive, which tends to focus on empirical observation to gain insights to help explain the growth process. Our development of market efficiency embodied in Pareto optimality (Chapter 1) is a deductive approach to economic theory. In our discussion of growth theory, we concentrate on the deductive approach with the goal of providing insight into how economists approach the growth process.

Over the past 50 years, economic growth theory has moved through four periods of thinking.[1] Progression from one period to the next reflects not only our ability to think about the growth process more completely and in more realistic ways, but also the changing economy itself. The historical context during the 1950s begins with the Rostow-Kuznets *stages of growth,* in which capital accumulation plays an important role. Concerns about unique differences and the linkage between rural and urban areas led to the *structural change* models of the 1960s. Extensions of these theories led to the Harrod-Domar model and the more fully developed *neoclassical* theories that dominated much of the thinking on economic growth during the 1960s and 1970s.[2] Today, attention is on lifting the assumption of perfect competition and the ensuing theories of *endogenous growth*. Not withstanding the fundamental differences between inductive and deductive approaches, one could argue that each of these theories is a natural progression from the previous theory.

ROSTOW-KUZNETS STAGES OF ECONOMIC GROWTH

As World War II drew to an end, many economists warned of a return to the dire economic conditions that existed worldwide prior to the war. The return of soldiers to local labor markets and the structural shift from a wartime to a peacetime economy were setting the stage for difficult economic times. What happened, however, was that the postwar rebuilding effort in many countries and the release of pent-up demand for consumer goods resulted in strong economic growth. The theoretical context that much of this discussion took place within was termed the *stages of economic growth* theory, originally brought forward by Rostow, which is best summarized in his classical treatise of 1961 (Rostow 1961, updated 1991).

Rostow suggested that economies progress through five stages of growth: (1) the traditional society, (2) the establishment of the preconditions for takeoff, (3) the takeoff itself, (4) the drive to maturity, and (5) the age of high mass consumption. The *traditional society stage* is one of subsistence economies, where farmers provide for their own households and there is little if any trade. The forces of comparative advantage drive farmers into specialization. Some farmers find that they are more productive at raising livestock, while other farmers are more productive at raising crops. Because of comparative advantage, farmers will move away from subsistence farming and specialize in the product with which they have a comparative advantage. Now that farmers have specialized, the need for trade becomes paramount. Given specialized farmers, economies of scale start to take effect.[3] Here two farmers working together can produce more than the farmers working independently. Combined, specialization and economies of scale in production increase production levels significantly above production levels in a subsistence economy. We have started the process of economic growth and have moved into Rostow's second stage.

The second stage focuses attention on the *preconditions necessary for takeoff*. This has to do with the formation of a commerce class who broker trading

in the young market economy, the differentiation of production and consumption, the development of transportation and communication networks and financial institutions. Farmers are no longer producing products for their own consumption; they are producing products for sale in the market. Products must be transported to markets, and information about what is happening in the markets must be available for farmers to make rational production decisions.

Takeoff, Rostow's third stage, is based on the accumulation of productive capital. One could say that the movement from the second to the third stage hinges on the maturity of the preconditions just described. The key to capital accumulation is the maturity of financial institutions, where money is valued and traded. Firms seeking to specialize require access to funds to purchase capital used in production. The logic of the Marshall Plan to rebuild post–WWII Europe and Japan was embedded in Rostow's concept of the third stage. Post–WWII Europe and Japan needed vast injections of money to rebuild their infrastructures and manufacturing bases.

The *drive to maturity*, the fourth phase, builds on the notions of specialization and access to investment funds to purchase new technologies. Economies of scale come into play in that productivity increases as production increases. In the drive to maximize profits, firms have incentives to try new production processes and to introduce new products. Unfortunately, the source of these new technologies is not clear, and it is a key limitation to growth theories in general. An attempt to formalize this key stage to sustained takeoff in a stylized model is embodied in the Harrod-Domar theories of capital accumulation (Hamberg 1971; Rostow 1965).

The final stage is *high mass consumption* embodied in the shift from agricultural and manufactured production to more of a service-based economy. Appealing to the logic of Engle's law, which states that as income increases, the share of that income spent on food declines, Rostow was able to explain how new markets for consumer goods begin and expand.[4] As a society becomes wealthier through higher income, people have more income at their disposal to spend on what might be deemed luxury items. In essence, the economy has matured.

Income Distribution and the Stages of Development

In his presidential address to the American Economic Association, Kuznets (1955) raised a fundamental question that Rostow did not consider: What is the impact of growth, or movement through the stages, on income distribution? Kuznets's expansion of Rostow's stages of economic growth to include income distribution issues drew much attention by economists because of the venue used by Kuznets. Kuznets began his discussion by presenting empirical evidence that seemed to apply to all developing economies. In subsistence economies, income distributions tended to be equal; income levels were low and equally distributed. As the economy begins to grow, however, the distribution of income becomes less equal, or skewed, with a small number of people benefiting disproportionately. In the terminology of the literature, economic growth is divergent. As the economy continues to grow, the benefits of that growth filter down to people in the lower end of the income distribution. Income distribution starts to become more equal or incomes begin to converge. But what is the economic process that explains or, better yet, predicts this pattern of divergence and convergence?

Kuznets offered his variation on Rostow's stages of growth theory as a potential explanation and proposed that an economy moves from a *primary* stage, where the economy is described as subsistence agriculture, to a *secondary* stage when the economy grows into a manufacturing-based economy, and then to a *tertiary* stage that is a service-based economy. Building on the ideas of comparative advantage, economies of scale, and Engle's law, Kuznets presented a coherent theory explaining the pattern of divergence/convergence observed in the data.

Not all farmers may possess the skills or insights necessary to take advantage of this specialization process. Because not all farmers are equally endowed with levels of human capital, the benefits of this initial growth process is not equally distributed across farmers. Even if all farmers were initially endowed with equal levels of land and water, differences in human capital lead to divergence. As farms specialize and grow in size, there is an increase in the demand for specialized tools and/or inputs used in the agricultural production process. Demand for new products opens markets for manufacturing and mining. Specialization and economies of scale also impact these new industries, increasing not only output but also driving down costs of production. The result is increased profits and incomes. Simultaneously, as agriculture, mining, and manufacturing become more productive and profitable, labor in general is more valuable and can demand higher wages.

If wages are set equal to the contribution of labor to the value of production, wages will increase as the economy grows. These market forces, Kuznets argued, will lead to income convergence.

Like Rostow, Kuznets also observed that consumers' pattern of spending changes as income increases. Again using the logic of Engle's law, Kuznets was able to explain how new markets for consumer goods begin and expand. By tracing through the logic of Engle on agricultural, manufactured, and service goods, Kuznets maintained that the economy would transition from the second stage of development to the tertiary stage (a service-based economy). Again, the economy has matured.

When one considers the history of economic growth for many countries, Rostow-Kuznets stages theory makes logical sense. Take for example the economic history of the United States. Settled primarily by subsistence farmers, many of the settlers were able to specialize and grow their enterprises in the manner described by Rostow-Kuznets. Because of the large scale of many farm enterprises, there was an explicit market for specialized tools and equipment such as the cotton gin, which separated seed from the cotton; the reaper, which mechanized grain harvesting; and the steel plow. These innovations were driven by the demands of the agricultural economy. The inventors of these new technologies, and particularly the entrepreneurs that shepherded these inventions to the market place, spearheaded the U.S. economy into Kuznets's second stage. The United States witnessed the shifting of surplus agricultural labor into manufacturing and, in time, all workers benefited from higher profits and wages.

Today, the U.S. economy is commonly called a service-based economy. This is exactly the transition one would expect given Rostow-Kuznets stage of development theory of economic growth. As those who remained in agriculture became more profitable and manufacturing workers earned higher incomes, the market for services, such as recreational services, expanded. The typical household needs only so many dishwashers and refrigerators. Once those demands are met, additional income is spent on services such as lawn care and landscaping services, restaurants, and recreational activities to name but a few. According to Rostow and Kuznets, one could argue that the U.S. economy has matured.

In the short term, there are winners and losers as the economy makes these transitions. Farmers who are unable to take advantages of specialization,

The Lorenz Curve

The impact of economic growth theory income distribution has been a central question throughout this discussion. Indeed, the founders of classical economics—Adam Smith and David Ricardo—were concerned with the distribution of income among what were then the three great social classes: workers, capitalists, and landowners. Each of these classes owned the key factors of production: labor, capital, and land, respectively. How was income to be distributed across these factors of production?

In neoclassical economics, income is distributed according to its marginal productive value (Chapter 1). In other words, the more productive the factor of production is, the greater its share of income. Karl Marx argued that capitalists (the owners of capital) would become relatively better off at the expense of labor until the economy collapsed on itself.

But how do economists think about and measure income distribution? A common means is with the aid of the Lorenz curve, which graphs the cumulative percentage of income against the cumulative percentage of individuals. If income were distributed equally to every individual, the cumulative percentage of income received by the cumulative percentage of individuals would fall along the straight line labeled *line of equality*. Here 10 percent of individuals would receive 10 percent of all income, 50 percent of all individuals would receive 50 percent of all income, and so on.

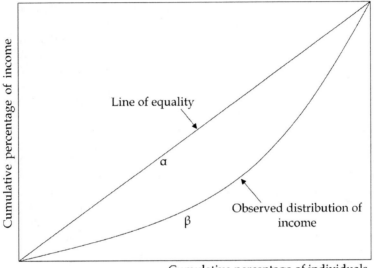

The actual distribution of income for a given economy falls along the line labeled *observed distribution of income*. The farther the line plotting the observed distribution bends away from the diagonal line of equality, the more unequal the distribution of income.

If we call the area between the line of equality and the observed distribution of income α and the area below the line of observed distribution of income β, we can get a numerical measure of inequality by dividing α by α + β. This ratio, termed the *gini coefficient*, ranges between zero (complete equality) and one (complete inequality). Much of the empirical work on economic growth tracks whether the gini coefficient is growing (more inequality or divergence) or shrinking (less inequality or convergence).

economies of scale, and new technologies lag behind. As new technologies make old technologies obsolete, the firms involved with the old technologies lag behind and in many cases go bankrupt. This process is known as Schumpeter's *creative destruction*, where new ideas and technologies push out and destroy old ideas and technologies. While the community economic developer is often worried about the winners and losers of these continuous shifts in the economy, they must keep the larger picture in mind. In the long term, the whole economy is lifted to a higher level of wealth and social well-being.

The challenge to the community economic developer is to position the community to minimize the hurt associated with the losers and to maximize the gain of the winners. According to Kuznets, all economic agents will win in the long term.

Critique of the Stages of Development Theory

The Rostow-Kuznets stages of development theory of economic growth has many critics. First, the stages theory is inductive, trying to draw inferences from historical data. Economists prefer to advance deductive theories, where the behavior of the economy is predicted from a set of premises about human and firm behavior. The primary rationale for the rejection of inductive reasoning is that it draws on conjectures and does not lend itself to rigorous hypothesis testing. As argued by Robbins (1935), drawing inferences from history alone cannot be taken to imply a definitive causal relationship. In essence, drawing on history helps tell a story about the economy and economic processes, but the story seldom tells us where the economy will be in the future.

A second criticism of the stages of development theory is that many economies "jump" stages. For example, take Saudi Arabia and its vast oil wealth. Today Saudi Arabia is one of the richest countries in the world, but it never started as an agrarian economy, nor did it ever develop a manufacturing base. Another example might be Las Vegas, which exists almost solely on a recreational-based economy. A general theory of economic growth should not be subject to special cases where the theory does not apply.

A third and perhaps more fundamental problem with the linear stages of development approach to growth theory relates to its prediction on income convergence. In a modern economy (Rostow's age of high mass consumption or Kuznets's tertiary stage), economic growth should foster convergence in income distribution. Since the late 1970s and early 1980s, however, income growth patterns have been one of divergence. The linear stages approach not only poorly handles this phenomenon, but also cannot predict it.

STRUCTURAL CHANGE THEORIES OF ECONOMIC GROWTH

A major shortcoming of the stages theory of economic growth is that it is essentially an aspatial view of the world. Specifically, the stages way of thinking about growth does not consider the unique economic characteristics of specific regions. Structural change theories attempted to build on the Rostow-Kuznets stages of growth theory by explicitly introducing the unique economic character of regions into the analysis. Like the stages of growth approach, the structural change theory focuses on mechanisms by which lagging regions transform their structures from agrarian to more modern, urbanized, industrially diverse manufacturing and service economies. Space is introduced in a narrow context, specifically the interrelationship between rural and urban economies.

The theory builds on the notion that structural differences exist between rural and urban areas (Chenery 1979). One example of these differences is in labor markets.[5] Early structural change theory characterized rural labor markets as being relatively more informal and engaged in a barter-type exchange economy (Chapter 6). Rural agrarian labor markets tend to have too many laborers and suffer from excess labor usage, which leads to rural labor markets operating closer to a point where the marginal product of labor is zero.

Structural change theory rests on relative differences in how returns to labor are distributed in an urban-rural context. Specifically, rural areas are more likely to be characterized by average returns to labor use, whereas labor markets in urban areas, being relatively more competitive, base returns to labor on marginal conditions. As we saw in our discussion of Pareto optimality (Chapter 1), profit maximization focuses on marginal versus average conditions. Because there tends to be underutilized labor in rural areas, excess labor migrates to urban areas to fill a limited demand for more competitive, higher wage jobs in more modern, industrial sectors. Expansion occurs in modern industrial sectors until all surplus rural labor supplies are exhausted.

The Lewis Model of Structural Change

In the early 1950s, Lewis (1954, 1955) proposed a theory of structural change where rural production was dominated by agriculture with production characterized by fixed technology and capital levels, leaving labor as the only variable input. Furthermore, Lewis assumed that rural workers were less individually focused and more apt to look at labor decisions in the form of family units. In essence, Lewis was envisioning subsistence family farming. Again, rural wages were apt to be determined by average returns, not marginal returns to labor inputs. Finally, Lewis assumed that traditional agriculture suffered from excess labor use (surplus labor). Technically, because of excessive labor existing in the market, adding one more rural laborer added very little if anything to agricultural production. The primary reason for this excess of rural labor was that large families dominated rural areas and tended to outpace the demand for labor. This may be due to high relative birth rates in rural areas compared to urban areas. The logic was that if the family is the sole source of labor for operating the farm, there is a strong economic incentive to have large families.

On the other hand, Lewis assumed that modern urban labor markets reacted to this excess rural labor situation in a manner that created sustained growth. In essence, there were still productivity gains to labor in urban markets; unlike rural markets where the marginal productivity of labor is zero, marginal productivity of labor in urban markets is positive. Given wages set equal to the value of marginal product, wages in urban markets must be greater than in rural markets (Chapter 6).

Finally, Lewis assumed that labor supply in urban areas was linked to unlimited supplies of rural labor through migration. Urban firms saw a large untapped supply of labor in the rural labor markets. Economic growth occurred in a sustainable fashion due, in large part, to a steady stream of rural migrants moving to urban areas in search of jobs.

From a macro perspective, the movement of labor from an area where its marginal productivity is zero (rural areas) to areas where its marginal product is positive results in a net gain to the economy's production level. In other words, increasing urbanization was seen as a growth-inducing process. The higher wages paid in urban markets due to positive marginal productivity in essence pulled cheap surplus rural labor into urban markets.

Lewis saw sustained economic growth as being fundamentally driven by reinvestment of profits in the urban industrialized sector into the expansion of physical plants (capital) that still operated with the same technology. The combined effects of new and expanding plant capacity through investment of profits were complemented with an inflow of cheap labor from rural markets.

Patterns of Development

As economists dissected the base assumptions of Lewis's two-sector theory, some began to argue that Lewis's view of structural change theory did not sufficiently lay the groundwork for the "takeoff" stage in a Rostow-Kuznets world. A slight modifi-

cation of structural change theory extended Lewis's rural-urban relationships to capture interrelated changes in economic structure. These additional changes were argued to be a more comprehensive set of general requirements for a transition from traditional economies to modern economies to take place. This line of argument became known as the *patterns of development* extension to structural change theory. Perhaps the most notable of these extended structural change theorists is Hollis Chenery, who identified both the theory and empirical evidence associated with broader components of structural change (Chenery 1979; Chenery, Robinson, and Syrquin 1986).

Indeed, as regional economic structures move from rural-based agrarian economies to modern manufacturing-based urban economies, both physical capital and human capital accumulation are required. Pattern of development theorists argue that this accumulation alone is necessary but not sufficient for growth to occur. For economies to modernize, they must have capital, but the availability of capital by itself does not ensure that growth will occur.

While structural change theory accounted for the sequential process of economic, industrial, and institutional transition, it neglected to incorporate demand aspects of people, both within and outside of the region, that are required for a takeoff to modernity to occur. Other items that must be present include changes in consumer demand and international demand (trade) of manufactured goods, and transformations of production to these changing consumer demands.

Critiques of Structural Change Theories

While structural change theories provided an interesting historical discussion of the transformation from a rural to urban economy, they are a gross oversimplification of unique characteristics that distinguish rural and urban areas. Specifically, the Lewis model assumed that reinvestment in production took place under fixed technology. The Lewis model breaks down when we relax this assumption and realize that, in the real world, reinvestment can take the form of labor-saving technology. In other words, technology is not fixed. New capital is invested in new technologies that tend to be labor saving, but labor-saving capital reinvestment would stop sustained growth in the Lewis model. If new technologies are available, the need for rural labor to move into urban markets breaks down.

Were capital to be reinvested in technologies that make labor more productive, the marginal product of labor would change. Reinvestment of capital uses the same amount of labor to produce more output. If this is the case, there is no longer a need for rural labor to migrate to urban areas. The theory also assumes that surplus labor exists in rural areas and that full employment is the norm in urban areas. Actually, empirical evidence has shown that the opposite is often true; urban areas suffer from substantial unemployment. The theory also assumes a highly competitive modern-sector labor market; this is what guarantees a continued positive gap of urban wages above rural wages. Again empirical realism suggests that wage rates (both absolute and real) have risen despite the existence of substantial open unemployment. As John Maynard Keynes would argue, *wages are sticky downward.*

Another critique of structural change theory is the location where profits are reinvested. The theory assumes that urban profits are reinvested locally and not reallocated to another urban market. For example, the movement of profits to corporate headquarters and their redistribution as dividends to stockholders violates the local reinvestment assumption. Thus economic growth in the original location is unsustainable.

The final critique is the inductive nature of the theory. The theory describes the historical patterns that have been observed fairly well, but it does not adequately lay a foundation on which to predict future growth. Although conceptually valuable, the Lewis two-sector model needs considerable modification to fit the reality of contemporary economic growth.

NEOCLASSICAL SOLOW-SWAN GROWTH THEORY

A key to economic growth in both the stages of development and structural change theories centers on the relationship between income and investment in capital. The first family of models that tried to formalize the income-investment relationship are known as the Harrod-Domar models (Hamberg 1971). In Harrod-Domar models, capital accumulation clearly becomes the explicit force of economic growth. As income and profit grow, some portion of them is set aside in the form of savings. These savings are not removed from the economy; rather they serve as a pool of funds used to finance investments. These investments fuel the accumulation of new capital.

While the models do not speak directly to financial institutions, they perform the critical function of bringing about the equality of savings and investment. When a household saves a part of its income, it has many options, including a savings account at a bank or stock markets. Both financial institutions recirculate these savings in the form of investment. The Harrod-Domar models clearly attempted to bridge the gap between the inductive models of Rostow-Kuznets and Lewis and the more deductive theories of today.

As an alternative to the Harrod-Domar-type models, Solow (1956) and Swan (1956) broke from previous discussions of growth theory and offered a completely new approach. They hypothesized that the economy can be represented by a traditional production function where output (Y) is a function of technology (A), labor (L), and capital (K): $Y = f(A, L, K)$. Output is defined as income in this theory of economic growth. Given perfectly competitive markets, constant returns to scale, and a closed economy—specifically, no international trade—a unified and useful theory of economic growth can be laid out. What drives growth in this simple theory is growth in the available supply of labor and investment in capital and technology.

Because we are concerned with growth of the economy over time, we need to think about what the economy looks like at a given date, t. So the supply of labor at time t is expressed as $L(t)$, real production or income is $Y(t)$, the index of technology is $A(t)$, and $K(t)$ is the stock of capital at time t. We assume that there is an exogenous and given rate of growth of labor and that it can be expressed as $L(t)e^{nt}$. Labor grows at an exponential rate, n. We also assume that technology growth is exogenous and follows a similar growth pattern as labor, or $A(t)e^{gt}$ with growth rate g. Real production, $Y(t)$, is given by

$$Y(t) = f[A(t)e^{gt}, L(t)e^{nt}, K(t)] \qquad (2.1)$$

Assuming no labor-leisure trade-offs and full employment, population and labor force are equivalent and both grow at rate n.

To simplify the model, we can speak in terms of technology-augmented labor, specifically, effective labor, A_tL_t. Here, technology affects labor productivity but not capital.

We can simplify the model again if we speak in terms of per capita, or effective labor augmented form: y is defined as Y/AL, and k is defined as K/AL. Because the supply of labor and the level of technology are exogenous, or determined outside the model, we want to focus our attention on capital. The production function, assuming a Cobb-Douglas form, can be expressed as

$$Y(t) = K(t)^\alpha [A(t)L(t)]^{1-\alpha} \qquad (2.2a)$$

or

$$y(t) = k(t)^\alpha \qquad (2.2b)$$

in per capita or *intensive* form, and α is a production parameter ranging between zero and one. In a simple accounting framework, α is the share of total income that accrues to capital. Conversely, $1 - \alpha$ is the share of income that accrues to labor.

Given the Cobb-Douglas specification, the production function has the normal curvature properties traditionally known as the Inada conditions, which become important to the stability of the theory below. In simple terms, the Inada conditions are that the production function is upward sloping but at a decreasing rate. Economic growth in this stylized model reduces to growth in capital, which can be expressed as

or
$$\begin{aligned} \Delta K(t) &= sY(t) - \delta K(t) \\ \Delta k(t) &= sy(t) - \delta k(t) \end{aligned} \qquad (2.3)$$

where $\Delta K(t)$ is the change in the stock of capital over time and s and δ are savings and depreciation rates, respectively. If aggregate savings are greater than depreciation, the stock of capital will grow and the economy will grow. If savings fall below depreciation, the stock of capital declines and the economy will shrink.

Substituting equation (2.2b) into the intensive form of (2.3), given our exogenous growth of labor and technology, yields

$$\Delta k(t) = sk(t)^\alpha - (n + g + \delta)k(t) \qquad (2.4)$$

This is what economists call a differential equation with five parameters (s, α, n, g, δ); it is the fundamental equation of the neoclassical model of growth. Capital accumulation is fundamental to the growth process because technology and labor force growth are exogenous.

The question now is, will the economy reach a point where growth is stable or in a *steady state* in which some level of $k(t)$ is achieved such that $\Delta k(t) = 0$? Essentially the economy has reached a point where it is no longer adding to its stock of capital

above depreciation replacement. The economy has "matured" and is no longer growing. Whether or not the model converges to a steady state is important in that if it does not converge, the model is "explosive" and the economy spins out of control. A model that does not converge to a steady state is sometimes referred to as a *knife-edge* model because small deviations away from the equilibrium (the edge of the knife) cause the economy to fall off the edge of the knife. Using mathematics called first-order differential equations one can show that a steady state can be achieved at the level:

$$k(t)^* = [s/(n + g + \delta)]^{1/(1-\alpha)} \qquad (2.5)$$

where $k(t)^*$ denotes the steady state level itself.

This process of convergence to a steady state can best be seen graphically (Fig. 2.1). The relationship of interest is the difference between investment (sY) and capital depreciation (δK), or net capital investment. The straight line in Figure 2.1 is a constant rate of depreciation, and the curved line is savings or investment. The curvature of the investment curve is due to the characteristics of the production function: diminishing rates of return, which follows from our assumption of constant returns to scale.

Note that the initial endowment of capital influences growth rates. Suppose that the economy starts with an initial endowment of capital, K^0. At this point, the level of savings (sY) is greater than depreciation (δK) and net capital investment (ΔK) is positive, and the economy grows, moving the stock of capital toward the steady state, K^*. Conversely, if the stock of capital is greater than the steady state level K^*, then investment is not sufficient to offset depreciation and the economy declines. In other words, regardless of where an economy starts, it will move toward the steady state solution. Economies that have initial endowment close to K^* will have slower growth rates than those with smaller endowments, say K^0. The neoclassical model predicts convergence of economies to a common level where poorer economies will grow more rapidly than richer economies.

The number of empirical studies attempting to model the convergence process predicted by the neoclassical model is vast, and a review of it is beyond the scope of this discussion. The preponderance of these studies, most of which examine developed economies, have indeed found strong evidence of convergence (Barro and Sala-i-Martin 1992,

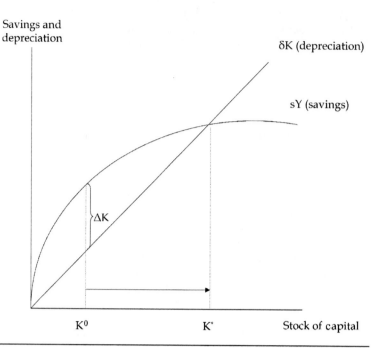

Figure 2.1. Neoclassical dynamic growth adjust-ment processes.

1995). It is important to note that the majority of these studies have found *conditional convergence,* that is, the empirical results depend on the specification of the empirical model that is estimated. From a theoretical perspective, convergence is conditional on the growth rates of labor and, in particular, technology; this is supported in the empirical evidence. Studies found that levels of human capital, trade policies, and a host of other factors can and do influence the convergence conclusion.

The policy ramifications of the Solow-Swan neoclassical growth model are significant. Notice that the economy moves to a stable growth level independent of public policy. Much like Adam Smith's *Invisible Hand,* self-interested behavior of firms and individuals in a market economy drives the economy to not only an efficient allocation of resources (i.e., Pareto optimality) but also to a shared equilibrium. Because the economy on its own dictates convergence, one could argue that the *Invisible Hand* also results in an equal distribution of income. The policy implication is clear: Because the markets work in not only an efficient manner but also in an equitable manner, there is no need for policy intervention.

One potential policy prescription that might follow from the neoclassical model is related to savings. If through some type of policy the government can promote savings which can affect levels of

investment, it can spur growth. For years, policymakers looked at the Japanese economy with its high saving rates and its rapid growth as evidence supporting this argument.

The difference between U.S. and Japanese savings rates was often pointed to as the reason for sluggish domestic growth. Unfortunately, a shift in the savings rate will not alter the long-term growth dynamics of the economy. Although a shift up in the savings rate function in Figure 2.1 may cause a short-term jump in investment, it does not affect the long-term patterns. More currently, the high savings rate of the Japanese people has not seemed to help the stagnation of the Japanese economy over the past several years. The community economic developer must keep in mind that there are no "magic bullets" such as the savings rate.

Spatial Interpretation of the Neoclassical Theory

Although the neoclassical model as presented by Solow and Swan implicitly discusses space in terms of different economies converging on a common growth path, the model is structured in a purely aspatial world. Smith (1975) presented an alternative specification of the theoretical model in which space—or regions or communities—is explicitly recognized and space becomes instrumental in the convergence growth process. In Smith's spatial neoclassical model, factor resources—labor and capital—are free to migrate between regions or communities seeking the highest rate of return (wages for labor and interest for capital). Through resource migration across regions, regional incomes will converge.

To see this, we need to only revisit our Cobb-Douglas aggregate production function and the dynamic means in which factor resources are changed. For simplicity, let us remove technology from the discussion. Our production function for the economy is

$$Y(t) = K(t)^\alpha L(t)^{1-\alpha} \qquad (2.6)$$

and all terms are the same as in our aspatial model above. In the Solow-Swan model, the supply of labor was assumed to be determined outside the model, and we assumed a growth rate of $L(t)e^{nt}$, and investment in capital outlined in equations (2.3) and (2.4) determined growth. In Smith's spatial model, the labor supply, or more correctly the change in labor, is now predicted by the model or becomes endogenous.

Focusing first on capital, the change in the supply of capital is again the net difference between investment, $sY(t)$, and depreciation, $\delta K(t)$, as outlined in equation (2.3). But in this spatial world, capital is free to move between regions, so we must augment equation (2.3) with the net of the flow of capital into and out of the region:

$$\Delta K(t) = sY(t) - \delta K(t) + NKM(t) \qquad (2.7)$$

where $NKM(t)$ is net capital movement. This latter component is of particular interest to regional economic growth because this component not only explicitly recognizes regions but also plays a fundamental role in the phenomena of growth convergence or divergence. Specifically, capital is perfectly mobile (i.e., zero transportation costs) between regions, and the driving factor is the rate of return earned within the region:

$$NKM(t) = \rho(r - r_A)K(t) \qquad (2.8)$$

Here r, the rate of return to capital within the region, is a function of capital's share of income, $r = \gamma(Y/K)$; r_A is an average rate of return within the nation, or $r_A = \gamma(Y_A/K_A)$, where Y_A and K_A are national income and capital levels, respectively, at time t.

The key to capital movement across regions is again the rate of return to capital. If the regional rate of return, r, is greater than some national average,

r_A $(r > r_A)$, then profit-maximizing owners of capital will shift capital to the region with the highest rate of return, and capital will flow into the region:

$$r > r_A \rightarrow NKM(t) > 0 \qquad (2.9)$$

If local investment, $sY(t)$, and net capital movement, $NKM(t)$, are large enough to offset natural depreciation, $\delta K(t)$, then the growth in regional capital, $\Delta K(t)$, is positive, placing upward pressure on growth of the regional economy.

On the other hand if the regional rate of return is below the national average, profit-maximizing owners of capital will move from the region and relocate capital to the region with the greatest rate of return:

$$r < r_A \rightarrow NKM(t) > 0 \qquad (2.10)$$

If local investment is not sufficient to offset negative net capital movement and natural depreciation, the stock of capital will decline, placing downward pressure on regional growth.

Labor in this model is also assumed to be perfectly mobile, and people are again assumed to be maximizing utility. Within this model, maximizing utility means that labor will seek out the region that pays the highest wages. The regional change in labor can be expressed as

$$L(t) = L(t)e^{nt} + M(t) \qquad (2.11)$$

where $M(t)$ is net migration and $L(t)e^{nt}$ is again the natural change in labor. Following the logic of the capital flow equation (2.8), net migration can be expressed as

$$M(t) = \theta(w - w_A)L \qquad (2.12)$$

As in the net capital movement equation, w is the local wage rate and w_A is a national average. If the regional wage rate is below the national average, utility-maximizing individuals will seek out a region that pays higher wages and net migration will be negative:

$$w < w_A \rightarrow M(t) < 0 \qquad (2.13)$$

Unless the natural change in the labor supply, $L(t)e^{nt}$, is not sufficiently large to offset the outflow of labor to migration, downward pressure will be placed on growth rates. Conversely, if regional wages are higher than the national average, the region will be attractive to workers and will experience a net inflow of population, or positive net migration, placing upward pressure on regional growth rates.

We can see that by simply introducing the concept of space, migration of capital and labor become clear. But what does this simple expansion of the model have to say about economic convergence? As the national economy grows, should we see regions that make up the national economy grow together or apart? Will some regions continuously lag behind others? To answer these questions within our regional neoclassical growth model, we need to examine the effects of migration on regional wages and rates of return.

In Figure 2.2, we lay out the regional labor market. Assume that the regional economy begins by paying a wage, w_0, which is greater than the national average, w_A, and that we have a level of employment equal to l_0. Given our net migration rule stated by equation (2.12), we have a situation where there is an incentive for labor in other regions to move into the region paying wage w_0, or as described in equation (2.13), net migration will be positive. As people move into the region to benefit from the region's higher wages, the supply of labor will start

to expand, pushing the local labor supply curve outward. Suppose that the inflow of people is so great that the regional supply curve moves from its initial position, S_0, to a new position, S_1. The amount of labor employed moves from l_0 to l_1, output expands, and the regional economy grows.

The outward shifting of the supply curve not only affects the level of employment, but also the wages that are paid in the regional economy. With more labor bidding for jobs, downward pressure is put on wages, in this case pushing wages from the initial level of w_0 to a new level, w_1. Before the immigration of labor, the regional wage level was above the national average, but now the region is paying below the national average. In essence, too much labor moved into the region to take advantage of the region's higher wages. Now local labor has an incentive to move out of the region in search for higher wages. Net migration now becomes negative, and we see a shift in the regional supply of labor to the left.

This type of dynamic adjustment process continues to the point where the local wage rate is identical to the national average, or where w_2 is equal to w_A. At this point, there is no incentive for people to

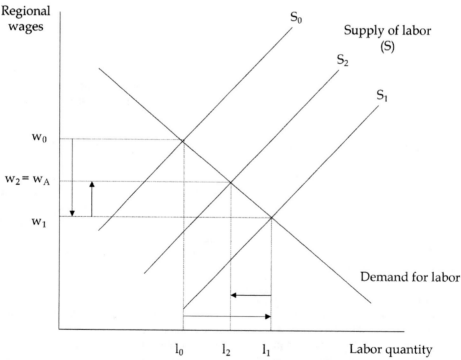

Figure 2.2.   Spatial neoclassical dynamic adjustment processes for labor.

move into or to leave the region. The economy has reached equilibrium; all wages are equal across the nation.

An identical adjustment process occurs with capital; all one needs to do is substitute capital's rate of return, r, for wages, w, and capital for labor. As with the aspatial neoclassical model, the spatial neoclassical model predicts that regional income will converge over time.

The policy implications of the spatial neoclassic model are just as strong as with the aspatial model: Government and policy have very minimal roles. In the end, the economy not only functions efficiently in a Pareto sense but also equitably in that overtime incomes and economic well-being will converge. The only real role for government and policy is to enforce the rules of perfect competition and, in a spatial world, ensure that transportation costs are as low as possible. Any other types of intervention will interfere with the efficient, and now equitable, functioning of the economy.

Critiques of the Neoclassical Theory

The neoclassical model of economic growth dominated much of the debate for several decades and is still the foundation for many empirical studies. The framework was easily understandable because the production function approach outlined in equation (2.1) is familiar and comfortable ground for economists. In addition, it provided a more rigorous foundation for the pattern of convergence in developed countries identified by Kuznets. Perhaps most importantly, nearly all of the available empirical evidence supported the convergence predication of the model. Because of the "cleanness" of the model, the strong empirical evidence, and the lack of any role for policy, growth theory was placed on the back burner for economists for a number of years.[6] It seemed as though we had the answers to the questions, and that was that.

The one thing that dogged the neoclassical model was its treatment of changes in technology. Economic historians are keen to point out that changes in technology drive economic growth. The introduction of the cotton gin and the mechanized tractor, for example, revolutionized the farming economy. The invention of the automobile at the turn of the past century has had huge impacts on the costs of transportation and has radically changed the notions of local and regional economies. The recent advancements in computing technologies and the Internet are still working their way through the economy.

But what causes these jumps in technology that have significant impacts on the growth prospects of the economy?

The neoclassical model does not help us come to an answer to this question or provide us with any understanding of the processes at work. In our simple presentation of the neoclassical model, technology is deemed to be determined outside of the model, or exogenously. Change in technology just seems to happen. Arrow (1962) suggested that changes in technology are an unintended consequence of the experience of producing capital, a phenomenon labeled *learning by doing*. Learning by doing was purely external to the firm producing the capital or the firm using the capital. Workers may see new ways of using old technology and, hence, foster a new level of technology. But why would workers do this? How does this explain radical jumps that go far beyond reworking old technologies? In short, the neoclassical model does not handle technology very well, and this has been troublesome for many economists.

A second problem with the neoclassical model has emerged over the past few decades. The empirical evidence for income convergence has weakened significantly, and there is strong evidence that incomes are diverging: The rich are getting richer and the poor are getting poorer (Bernat 2001). This is not only occurring across income groups in an aspatial sense, but also across regions in a spatial sense. Numerous studies across the world have documented that since the late 1970s and early 1980s incomes are diverging across and within countries. This is very troublesome for the neoclassical model because it predicts strict convergence. For the past several years, the trend has been away from rather than toward k^* in Figure 2.1. Although short-term bouts of divergence are not sufficient to cast aside the neoclassical model, extended periods of divergence cast serious doubts on the model.

A third critique of the model centers on the strong assumptions on which the model is based: perfect competition, constant returns to scale, and no externalities. While simplifying assumptions in our stylized models are common, attempts to make the neoclassical model more general by lifting these assumptions have proven difficult. As noted in Chapter 1 and argued throughout this book, externalities are predominating all through the economy. Market failure is almost a fact of life that the neoclassical model does not address in any reasonable manner.

These problems with the neoclassical model have spurred economists to look for alternative theories and ways of thinking about economic growth. By directly tackling the third critique of the neoclassical model head-on, a new theory of economic growth has been advanced that has revolutionized our way of thinking about the growth process.

ENDOGENOUS GROWTH THEORY

Kaldor (1957, reprinted in 1979; 1970) was the first to challenge the earlier work of Solow and Swan along several fronts. First, Kaldor questioned the convergence conclusion of the neoclassical model. If convergence is the only possible outcome of economic growth, why is there evidence of divergence in Kuznets's data? Kaldor also challenged the *laissez-faire* policy implications of the neoclassical model. Kaldor was a British economist writing during the post-WWII period of reconstruction. The British economy was struggling and not experiencing the growth of Europe or Japan, and there was significant political pressure for the government to "do something." If the neoclassical model was indeed correct, there is no real role for government. Not a politically acceptable answer. But Kaldor's main thrust of attack came in the role of constant returns to scale and diminishing marginal returns.

Kaldor noted that Kuznets had correctly put economies of scale at the forefront of his stage of development theory and pointed to the Hicksian notion of an *accelerator effect* in the macroeconomy. In essence, once a firm starts growing, economies of scale result in the firm having the potential to grow exponentially. Kaldor argued that identical to a firm, once a region gains a growth advantage, it will tend to sustain that advantage through the increasing returns that growth itself induces. Mydral (1957) described this growth-inducing growth as *circular and cumulative causation effect*.

Kaldor argued that there are two types of economies of scale: static scale economies that are internal to the firm, and dynamic scale economies composed of three parts. First, there is a learning-by-doing element that enhances the productivity of labor. While similar in spirit to Arrow's (1962) notion, Kaldor was speaking only of labor productivity and not induced technological change by doing things differently. Second, there are increasing returns that are brought about by *induced* technological progress. Here, changes in technology are endogenous to the growth process, not exogenous,

as is the case with the neoclassical model. Unfortunately, Kaldor was not very explicit in describing this process but hinted at Kuznets's logic about demand-induced technological progress. Third, there are *external economies* or positive spillovers between firms and industries. Firms of a similar type benefit from locating in close proximity because they can benefit from common business services infrastructure and common pool resources, predominately labor. This is also sometimes referred to as localization economies.

This latter element of dynamic scale economies is very important to the community practitioner because it has explicit implications on the structure of regional economic growth. In a Kaldor-type world, firms that share resources, whether it be business services or labor pools, have a comparative advantage over firms that do not share resources. Financial services on Wall Street in New York or The City in London or LaSalle Street in Chicago are a clustering of businesses of a similar type in one geographic location in which individual firms benefit by being part of the cluster. Workers who socialize together may share ideas or ways of doing things that can be brought back to individual firms. Firms that cluster together build synergies that directly impact growth (Chapter 3). The forces of economies of scale that Kaldor advanced are what regional economists call *agglomeration economies*.

Unfortunately, attempts to formalize the verbal arguments made by Kaldor in a stylized model of economic growth have fallen short. The forces of agglomeration economies can be readily described in detail, but an underlying model that predicted agglomeration economies had been lacking. In discussing the frustration of regional economists and the followers of Kaldor, Krugman (1995) quoted a sarcastic physicist who made the comment "So what you are saying is firms agglomerate because of agglomeration effects." The implications of dynamic scale economies or agglomeration economies on the economy are clear, but the forces at work that drive or cause agglomeration economies remained a bit of a *black box*.

The work of Romer (1986, 1987) and Lucas (1988), however, radically changed how economists thought about the growth process and the underlying factors causing growth. Following a trend in economic theory, they lifted the assumption of perfect competition, which had already occurred in the industrial organization and international trade literatures. Romer asked, what would happen if we abandoned the neoclassical model and started with a clean slate in which the economy is not held to perfect competition? Romer suggested an economy that had the following characteristics:

1. There are many firms in a market economy.
2. Discoveries of new ideas differ from other inputs in the sense that many people can use them at the same time. In other words, ideas are public goods (see Chapters 9 and 10 for a detailed discussion of public goods).
3. It is possible to replicate physical activities. If one firm can produce a good or service in a certain way, there is nothing preventing another firm from replicating the first firm.
4. Technology advances from things people do. Technological advances do not fall from heaven, as in the neoclassical model.
5. Many individuals and firms have market powers and earn monopoly rents on discoveries.

The last characteristic is the linchpin of the Romer view of the world. Earning monopoly rents on discoveries cannot occur in a perfectly competitive economy, but it is the striving to capture these rents that spurs economic growth. How does this process play out?

Grossman and Helpman (1994) point to the key idea that profit-seeking investments in knowledge play a critical role in long-term growth. Investment in knowledge has two components. The first is investment in human capital through education. The second is investment in research and development of new products and technologies in an attempt to capture monopoly rents on those discoveries. The latter leads us to *endogenous technological progress* or *endogenous growth*. Unlike the neoclassical model of growth in which technological progress just seems to happen with no rhyme or reason, in endogenous growth theory, firms have a profit-maximizing incentive to invest in research and development. (For a more complete discussion of technological progress and the process of innovation within a community setting, see Chapter 8.)

Put another way, if research and development of new products and technologies are fundamental to the growth process, why would firms or people invest in research and development? The key is short-term monopoly rents that can be gained on the new technology. Before this can take place, however, certain institutional rules need to be in place and enforced, specifically patents. A *patent* is a legal

mechanism to ensure restricted use to the holder of the patent for a certain period of time. A firm or individual that develops a new product or technology has the ability to patent that product or technology and become sole supplier of that good. Because of the exclusive supply of the good by the holder of the patent, the firm or individual can charge monopoly prices and earn monopoly profits. This is a strong incentive for firms and individuals to invest in research and development.

There is an important distinction here between ideas and products or technologies.[7] An idea or way of thinking about a problem generally cannot be patented. If a business reorganizes itself to gain internal synergies, that firm cannot prevent a competitor from following its lead. Ideas and knowledge are embodied in people who are mobile and can move from one firm to another, taking their knowledge with them. Trade secrets, or the way a particular firm goes about its business, cannot be patented and at times are fiercely protected by firms. Innovations, or new ways to do old things, are almost immediately part of the public domain. Products and technologies, on the other hand, are tangible "things" that can be patented and protected from competitors.

Two classic examples that document the difference between ideas and things, and the use of the patent power on things are videotapes (beta versus VHS) and personal computers (IBM versus Apple). Sony Corporation developed the idea of placing movies on videotapes for use in home viewing. While the technology of videotapes had been in use in the television industry for years, Sony developed the technology for home use. Sony patented its videotape format and called it beta. The demand for Sony video machines was enormous, and profit levels were high. The "idea" was video taped movies for home use; the "thing" was the beta-formatted machines. Because Sony has many competitors, other firms attempted to enter this market created by Sony. But Sony's patent on the beta format precluded competitors from duplicating Sony's product. An alternative video format, called VHS, was soon developed and was distinctly different enough that Sony could not enforce its patent on these competing firms. The floodgates for competition were opened and soon Sony's preferred beta format machines were taken to the landfill.

The Apple-IBM wars in the 1980s over the personal computer market are another example of the interplay between innovations, research and development, monopoly rents, and competition. The idea or innovation of computers had been around for years, and IBM had been producing mainframe computers for businesses and universities for years. Apple, a small start-up firm in the late 1970s, had an idea for what they called a personal computer (PC) that would take the power of large bulky mainframe computers and put it in a format that could be used by individuals. While the technology was old, the innovation was to put it into a more user-friendly format. What Apple was able to bring to market, however, was not just the personal computer itself, but more importantly, the software that allowed it to work. Apple held its patents for its hardware and software very closely, and profits were strong.

IBM, fearing that it might be losing out to a new market it had previously ignored, brought out its own personal computer that was sufficiently different from Apple that Apple's patents were not infringed upon. The key difference between Apple and IBM is that IBM licensed its patented technology to other companies. Here, for a fee paid to IBM, firms were able to replicate IBM's product under their own name. Because of the number of firms producing "IBM clones," competition drove prices downward. For most users, the difference between an Apple and IBM clone was so small that consumers went with the cheaper product. It was only after IBM clones dominated the market that Apple attempted to license its products to regain market share. The decision was too late and Apple Computers, the company identified as the founder of the personal computer, was almost forced into bankruptcy.

What Chenery, Kaldor, and Kuznets neglected to fully develop was the idea of demand-induced technological progress as originally put forward in Structural Change Theories of Economic Growth, Patterns of Development above. With the presence of patents and short-term monopoly rents, firms and individuals have a profit-maximizing incentive to invest in new technologies and products through research and development. But profit-driven incentives for businesses to invest in researching new ideas and the development of those ideas into marketable products are not sufficient to ensure endogenous growth. An integral part of the process is Schumpeter's idea of entrepreneurial activity. Endogenous growth takes more than a scientist working with an engineer to develop a product; it takes an entrepreneur to bring it to market and earn those monopoly rents. (These ideas and how they relate to community economic development are more fully clarified in Chapter 8.)

Spatial Implications of Endogenous Growth Theory

Krugman (1991a, 1991b, 1995, 1999) explicitly incorporated space into the new endogenous growth theory by asking a basic question: What are the economic forces at play that result in the creation of megalopolises such as New York City and Tokyo? He suggested that many of the notions common to regional and urban economics could be reconsidered in the new light of endogenous growth theory. Specifically, the notion of agglomeration economies that is widely used in regional economics takes on a new meaning in the new growth theory. Within a spatial world, agglomeration economies have a significant impact on the location of economic activity. Krugman maintained that without the new endogenous growth theory there is no mechanism that moves the economy beyond a series of smaller rural hamlets. The agglomeration forces that drive rural hamlets to become cities are assumed in prior theories.

Revisiting the older ideas of *centrifugal* and *centripetal* forces in a spatial world allows for a better understanding of a system of places. Centripetal forces create urban centers and describe the economic forces that pull economic activity together. Krugman offered three broad types of centripetal forces: market-size external economies, natural site advantages, and pure external economies. Examples of market-size externalities include forward and backward linkages between firms and labor markets with many different occupations. Natural site advantages include such things as natural harbors or access to navigable rivers, or central locations. Pure external economies are knowledge spillovers. Each type of force has the tendency to pull economic activity together into one location.

Centrifugal forces drive or spread economic activity away from the urban center. Krugman again offered three broad types of centrifugal forces: dispersed natural resources, market-mediated forces, and nonmarket forces. Market-mediated forces include transportation costs and urban land rent. Nonmarket forces include negative externalities, such as pollution and congestion. The resulting system of places reflects the balance between centrifugal and centripetal forces.

Drawing on the ideas of the new endogenous growth theory, Krugman was keen to make a sharp distinction between natural advantages and acquired advantages that are self-reinforcing through the market processes. He also made a distinction between technological (nonmarket) and pecuniary (market) externalities. The behavioral forces are that monopoly profits are allowed and that they are increased via agglomeration economies. Firms locating within the same general area create the agglomeration economies. In the same logic as Romer, these agglomeration economies appear as the scale of firms increases and they become self-reinforcing as more firms tend to locate together.

The growth of the City of Chicago offers an example of the processes outlined by Krugman. Chicago was originally settled because of its location at the southernmost point of the Great Lakes and the Illinois River, which connected the Great Lakes to the Mississippi River. Chicago's central location between the agriculture of what is now called the Corn Belt of the United States and the urban markets in the eastern United States gave it a transportation advantage. In addition, Chicago was centrally located between the coalfields of Kentucky and southern Illinois and the iron ore mines of Michigan and Minnesota, so the location was perfectly suited for a new and growing steel industry. The economies of scale of meat processing, transshipment of grains, and production of steel create pecuniary externalities. The multiple locations of firms in the same general industrial sector lead to technological innovations much greater than if the firms had located in isolation.

The inevitable results of Krugman's stylized model are that economic activities tend to cluster within the same location. If the economy begins with a random scattering of hamlets, centripetal forces will result in economic growth being clustered in a small number of cities. These cities, once they gain a growth advantage, will maintain that growth advantage until centrifugal forces begin to come into play. The implication of the Krugman model of this system of cities on smaller communities is not pleasant. In essence, the forces of economies of scale play to the favor of large places, many times at the expense of smaller places.

Critiques of Endogenous Growth Theory

Endogenous growth theory is a relatively new development within the economics literature. As such, no real empirically based criticism has yet been offered. There are, however, two emergent theoretical critiques that give pause to community development practice. The first is simply that the theory leaves the growth prospects for smaller communities and rural areas in doubt. We see in the real world, however, that there are numerous rural areas that are experiencing growth. The role of natural amenities is playing a greater role in economic growth and the new endogenous growth theory is silent on natural amenities (Chapter 9). Second, the theory has become so abstract in its development that advanced mathematics is the language required to follow the current permutation. For example, because of the importance of patents to the new growth theory, much of the current literature is emphasizing strategic behavior within a game theoretic approach. Unfortunately, much of this literature is beyond the grasp of anyone who is not pursuing a Ph.D. in economics.

POLICY IMPLICATIONS AND SUMMARY

We have traced the historical development of growth theory. It is important to remember that most of this theory has a limited direct application to community economic development because it is macro in scope and is conceptual. Much of the theoretical discussion has tended to focus on developing economies and the process of how an economy progresses from a simplistic to a mature, advanced economy. The discussion of growth theory sets the context for much of what we try to do in communities.

These theories provide insights into the growth process when viewed from the local perspective. What drives overall growth sets the tone of discussion at the local level. It was not that long ago that many of the now developed economies of the world were near subsistence level. Many local areas in developed economies think of themselves as being trapped at near subsistence level or are, at a minimum, less than mature. The key is whether these theories of economic growth can provide some insights into meaningful ways to break out or to recognize more reasonable expectations.

What we have learned from these theories and their implications is significant. First and foremost are the common themes that run through all of these theories of economic growth. Clearly defined institutional rules, perhaps the most important of which are property rights, are necessary for economic growth (Chapter 11). Comparative advantage is a key concept, and a focus of community economic development hinges on identifying and acting on a community's comparative advantage. The visions of a community must be realistically in line with the community's comparative advantage (Chapter 3).

Technological progress, including human capital, is an important engine of economic growth. The community has many roles it can play in fostering technological progress at the local level (Chapter 8). In Chapter 6 we discuss ways in which investment in human capital can spur community economic development. The accumulation of capital and the importance of financial markets to community economic development are discussed in detail in Chapter 7.

Within the community setting, one must be sensitive to the *false paradigm* offered by neocolonial dependency theorists (Todaro 2000). This false paradigm caveat represents the real-world situation of blind application of theories and policies without a local context. Classic examples include the rush toward radical transformation of the post–Soviet Russian economy to a capitalist economy or the application of urban theories to rural places. Often, there are unique characteristics of communities that make the application of ungrounded theories problematic. The importance of the culture of individual communities cannot be overstated (Chapter 11).

In the aggregate, the role of policy is not all that clear. Generally, the theories seem to suggest a *laissez faire* approach by government. This type of a doctrine advocates a minimal role for government in interfering with economic affairs beyond the minimum necessary for peace and property rights. It advocates freedom to let people do as they choose. Such things as freeing factor markets to reduce rigidities; deregulation of industries such as airlines, trucking, and telecommunication; and opening of trade markets by reducing barriers to trade (NAFTA, GATT, WTO) all position the economy to build on itself.

But at the same time, the theories emphasize the importance of getting the institutional rules right and seeing that the *rules of the game* are clearly defined and enforced. In addition, the theories clearly state a proactive role in investment in human capital and basic research that firms may find too risky. Investments in transportation infrastructure also play an important role in a spatial economy. Indeed, a lot of our discussion has talked about space in a very general sense; in the next chapter, we more fully develop the comparative advantage of place.

STUDY QUESTIONS

1. What are some critiques of the Rostow-Kuznets stages of development theory of growth from a community's perspective?
2. Do all economies progress through the same growth process?
3. In the Lewis theory of growth, how does the economy progress?
4. As an economy grows, will incomes tend to move together (converge) or move apart (diverge)?
5. What role does technological progress have in economic growth?
6. Which of these theories of growth can be described as deductive? Which as inductive? What is the difference?
7. Why do economists find such comfort in the Solow-Swan theory?
8. Given the current thinking on endogenous growth theory, what does the future of small, remote rural areas hold?
9. What motivates technological change in the Solow-Swan versus the endogenous growth theories?
10. What role does uncertainty and risk play in the Solow-Swan and endogenous growth theories?
11. What role does perfect competition play in economic growth?
12. What is the role of factor resource mobility in each of the major theoretical approaches to economic growth?

NOTES

1. In this discussion we will not cover several theories of economic growth that have been offered by economists. For example, we draw on the implications of neocolonial theory of growth on community economic development in the Summary to this chapter but we do not fully develop the theory itself.
2. The neoclassical theory of the 1960s and 1970s became popularized in the 1980s through the ascendancy of conservative political regimes such as the Reagan, Thatcher, Mulrony, and Kohl governments.
3. The notion of economies of scale is very important throughout our discussion of community economics. We encourage students interested in the microeconomics of

economies of scale to review this concept from any intermediate microeconomic text.

4. Engle's curve is defined as the locus of tangent points where the indifference curve meets the budget constraint. The curve itself is mapping the tangent points as income increases or the budget constraint moves up and to the right.

5. While our discussion here focuses on labor, structural change theory also examines differences in rural and urban factor endowments, such as capital and land.

6. The neoclassical model does not support active intervention for the promotion of growth, but it does support active intervention for the promotion of competitive markets.

7. Within the community setting, the means by which the community approaches problems is called *process technologies* (Chapter 8). This is similar to the ideas discussed here.

3
Space and Community Economics

In our paradigm of community economic development, spatial and geographic dimensions of the community and the larger economic area in which the community is located are fundamental to understanding how the economy of the community functions. Central to this understanding is the economics of firm and market locations. In this chapter, we provide the basic elements of location theory. Location theory focuses on the attributes of space, such as the location of resources, the location of production, the location of markets, and the transportation system. These attributes explain where economic activity occurs. We are now set to discuss the center of the Shaffer Star identified as *space* in Figure 1.1.

Traditional courses in economics tend to assume the spatial elements of the economy away because living in a spaceless world is much easier to understand. But in community economics, space is a fundamental element of our thinking. Communities function in economic space, complementing and, at times, competing with neighboring communities. This is also an increasingly important component associated with the new economic geography (Audretsch 2003; Fujita, Krugman, and Venables 1999) and in the analysis of spatial issues (Anselin 2003; Fingleton 2003).

From a historical perspective, most communities are physically located where they are because of the economics of space and transportation factors. Most major cities were born around some type of water transportation way. As we will see, many villages began life because of their location relative to farmers spread across the landscape. During the expansion of the western United States, communities fought over the location of railroad lines. Communities that won those political wars often thrived, while those that lost withered.

The study of space within economics has historically focused on firm and market location analysis within a neoclassical framework. As we will see, the profit-maximizing behavior of firms and their underlying cost structures dictate much of our thinking. Under the general premise of profit maximization, we examine two special cases: the least cost approach and the demand maximization approach. Each will be reviewed for their contribution to community economic analysis. We then present some alternative approaches to the neoclassical view of location. The final section of the chapter deals with regional comparative advantage and Porter's approach to agglomeration, which focuses on firm location and regional competitive advantage.

SIGNIFICANCE OF LOCATION THEORY IN COMMUNITY ECONOMICS

Location theory explains how spatially separated economic units interact among themselves and their input and output markets. While the traditional focus has been firms, we need to remember that communities represent sites of output markets and input markets. The transferability of location theory to the concepts of community economics builds on this spatial interaction. Households and laborers face spatial decisions, such as where to shop, where to work, and where to live. But by focusing on the firm, we gain a fundamental understanding of the economic forces at play.

The contribution of specific resources to community economic development depends on where that use occurs in space. Likewise, the shifts in demand for products affect specific geographic locations of production, including communities. A community represents the operating environment for economic units interacting in space: businesses and households buying and selling output, labor, raw materials, and capital. Location theory provides insight into how location decisions are made and why economic activities occur where they do. With this information and insights about what forces influence location decisions, communities can consciously try to influence those decisions.

The narrow perception of location theory is it explains decisions to initially locate or relocate a business. But businesses face numerous other location decisions. Since every economic transaction has a spatial dimension, each represents a location decision. Relevant business location issues include where to start a business, where to expand as growth occurs, where to relocate, where to subcontract surges in production, where to merge to acquire capital or achieve market penetration or acquire sources of supply, where to buy inputs, where to market production. Each has a spatial connotation, and different location factors will influence the decision. Thus, the phrase *location decision* refers to any economic transaction with a spatial dimension, not just the traditional relocation decision.

PROFIT MAXIMIZATION

The most general problem facing the firm is a situation where their consumers and suppliers are scattered across a homogeneous economic plane. Here, an economic plane is a featureless surface that is not complicated by natural barriers such as mountains, rivers, or valleys that create transportation bottlenecks or by institutional barriers such as political boundaries. The firm is faced with the locational choice that places the firm somewhere on the economic plane in a manner that maximizes profits. The firm does this by minimizing the transportation costs of shipping input supplies to the firm and maximizing the potential market demand for their good or service. In other words, the profit maximization approach to location decisions declares that businesses select the site from which the number of buyers whose purchases are required for maximum sales can be served at the least possible total cost (Greenhut 1956; Gabszewicz and Thisse 1986; McCann 2002). This site need not be the lowest total cost site possible; it can be a site from which monopolistic control over buyers makes it more profitable than a lower-cost site. In other words, an individual business can offer a delivered price to buyers at lower than competitors' prices. This approach recognizes the interaction between demand (locational interdependence) and the cost of production in site selection.

The profit maximization approach examines both the total revenues and the total costs portion of the profit equation:

$$\text{Profits} = \text{Total revenues} - \text{Total costs} \quad (3.1)$$

The firm is faced with balancing two factors: the location of customers, which drives the revenue side of the profit-maximizing equation, and the location of suppliers, which drives the costs side of the equation. Typically, the firm believes one of these factors is more important than the others, and it focuses on either maximized revenue or minimized costs first. Other factors enter the decision only after that initial choice has been made.

Revenue factors either increase revenues (demand) or ensure that revenues remain at the previous level. The *demand factors of location* include the socioeconomic characteristics of the market, such as income, family composition, and population growth. (The analytical tools used to understand these demand and market characteristics are discussed in Chapter 14.)

The *revenue-increasing factors* arise from the gains the business experiences because of an increase in demand from either agglomeration or deglomeration forces or personal consideration. Agglomeration sales gains appear from the location of the business close to similar businesses or supporting businesses. An example would be the spatial proximity of shoe stores within a shopping mall. The convenience and ease of comparative shopping that shopping malls offer are powerful locational factors. The deglomeration sales gains come from avoiding a site too close to competitors or being the first business in a new geographic market area, and selling at a lower delivered price than more-distant competitors. Purely personal considerations become demand factors when they indicate new and expanding markets (e.g., amenity resources and a way of life).

The *cost factors of location* include transportation and processing costs. While transportation costs are significant in a business location decision, their importance varies with the nature of the business, their contribution to total costs, and the ability to change them. If transportation costs are a major part of total costs or if the business can affect its transportation costs significantly among different sites, then they will be a prime location factor. If there are only minimal differences in transportation costs among sites, then transportation costs will have little influence on the location decision. Processing costs include labor, capital, taxation, and insurance related to production. Processing costs become important in the location decision when transportation costs and demand factors vary little among sites. It is important to remember that the firm seeks to minimize total costs (transportation plus processing).

Cost-reducing factors accrue to the business from agglomerating or deglomerating locations. A pool of skilled labor and other infrastructure needed by a business provide an agglomerating location. A deglomerating location avoids congestion and competition for scarce labor skills and other inputs.

Personal cost-reducing factors are the gains to the business, such as personal contact with individuals in other organizations (e.g., clients or supporting businesses), reduced commuting time for workers in smaller communities, or willingness to substitute the psychic elements of the environment for monetary rewards.

To help see how the profit maximization problem facing the firm plays out, let's formalize the problem and examine two special cases: costs minimization and demand maximization. Assume that a single firm produces one good by using a number of inputs shipped from different locations and the output is shipped to a number of markets. This firm produces a good that is also offered for sale by a large number of competing firms; hence, the firm is in a market that can be described as competitive. In our spatial world, however, firms have some flexibility in setting their own prices. In a spatial world, firms compete through effective prices, where effective prices reflect not only the costs of production but also transportation costs. For example, the effective price of a gallon of milk is composed of two parts: the price at the store plus the cost of traveling to the store to make the purchase. Stores offering milk for sale compete directly by paying attention to the price at the store, but imbedded in the price to the consumer is the cost of traveling to the store. Which store will the consumer select? The store with the lowest travel cost, which is likely the store closest to the customer.

To formalize the firm's problem and define demand, production, and transportation costs, we use the following terms:

$\prod$ = profit

P_i = price charged at market, $i = 1 \ldots m$

$D_i(P_i)$ = demand for the firm's product at market, $i = 1 \ldots m$

s^i = spatial location of market, $i = 1 \ldots m$

$t(s,s^i)$ = cost of transporting one unit of the good from firm location s to market location s^i

f = fixed costs facing the firm to produce the good

v = constant marginal cost of producing one unit of the good

x_i = production inputs from market, $i = 1 \ldots n$

$d(s,s^i)$ = cost of transporting one unit of input x_i from market location s^i to firm location s

$q(x_i)$ = output level of the firm

The firm uses n separate inputs (x_i) shipped from different markets to produce one good (q) that it sells in m separate markets (s^i). In a spaceless or aspatial world, the firm's maximized profits are expressed as

$$\prod = \sum_{i=1}^{m} P_i D_i(P_i) - f - vq(x_i) \quad (3.2)$$

The firm has one decision: what price (P_i) to charge at each separate market. Once a price has been established, say P^*, the amount of the good sold at each market is determined by its respective demand function $D_i(P^*)$. Total revenue is simply the sum of all sales across the m separate markets, or $\sum_{i=1 \ldots m} P_i D_i(P_i)$. The total cost of production is the fixed cost of production (f) plus marginal cost (v) times the quantity produced (q), or $f + vq(x_i)$. What we have is simply price times quantity minus costs of production.

Now let's place our firm in a spatial world, where it must balance not only prices (P_i) at each of the output markets but also transportation costs of shipping both inputs to the firm and output to markets. The firm does this by selecting a location (s) somewhere on our economic plane that minimizes transportation costs. We can express transportation costs as

$$\sum_{i=1}^{m} t(s,s^i) D_i(P_i) + \sum_{i=1}^{n} d(s,s^i) x_i \quad (3.3)$$

which is the sum of total transportation costs of shipping the firm's product to m separate markets plus the total transportation costs of shipping n inputs to the firm from n separate markets. This is traditionally known as Webber's problem.

To better understand the transportation problem, assume that the firm has three output markets ($m = 3$) as well as three input markets ($n = 3$) and that those markets overlap. Graphically, the firm is looking at what is known as the Webber triangle on our economic plane (Fig. 3.1) (McCann 2002). The firm is selecting a location (s^*), somewhere between the three markets (s^1, s^2, s^3), that will maximize profits. In this simple example, the firm is shipping inputs from three markets [$d(s,s^i)$] to a centrally located physicality, then shipping its product in this case to the same three markets [$t(s,s^i)$].

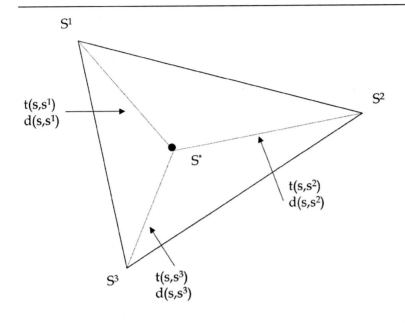

Figure 3.1. Transportation on an economic plane —Webber's problem.

Profit Maximization

We now have all the pieces to formally state our problem by combining equations (3.3) and (3.2) to restate (3.1) as

Webber's Problem/Triangle

$$\prod = \sum_{i=1}^{m} P_i D_i(P_i) - f - vq(x_i)$$

$$- \sum_{i=1}^{m} t(s,s^i) D_i(P_i) \qquad (3.4)$$

$$- \sum_{i=1}^{n} d(s,s^i) x_i$$

where the firm selects a set of prices (P_i) that maximizes demand at each market and a location (s) that minimizes transportation costs. Clearly, the number of output markets (m) need not be equal to the number of input markets (n), and the cost of transporting output [$t(s,s^i)$] need not be the same as the cost of shipping inputs [$d(s,s^i)$]. We could make our problem even more general by allowing for multiple outputs (q_i) and multiple firm locations (e.g., multiple plants) (s^j), and we encourage the interested reader to modify equations (3.2) and (3.3) to allow for those cases.

When thinking about the location problem, the community practitioner must keep in mind that the profit maximization problem as expressed in equation (3.4) is a general theory of firm location. The profit maximization approach provides a general framework to think about the problem of location. By looking at cost minimization and demand maximization, two special cases of the general profit maximization problem, we can gain a better understanding of location theory.

Least Cost or Cost Minimization

The *least cost approach* of location theory parallels the neoclassical model of economic development (Greenhut 1956; McCann 2002; Moses 1958; Smith 1971). The theory contends that firms seek to minimize their total cost of production and transportation. The decision sequence in the least cost approach is first to minimize transportation costs, then include other costs (labor, land, and capital costs) to determine whether adding the other costs alters this minimum transportation cost site. Since location theory focuses on the spatial distribution of economic activity, the initial emphasis on transportation costs has great appeal.

The least cost approach makes several basic assumptions. First, the firm exists in a purely competitive world with no monopolistic gains achievable from a specific location choice. Second, the firm can sell all it produces because its demand is perfectly elastic and unaffected by location. Third, the firm's market consists of separate points, each with a given location and size. Fourth, the geographic distribution of materials is given, and some raw materials are found only in specific locations; the supply of raw materials at these specific locations is perfectly elastic. Fifth, there are several fixed locations where a labor supply exists; at these points the supply of labor is unlimited at any given wage. Sixth, there are no site-specific institutional factors such as taxes, political systems, insurance, or culture to affect the location choice. This means location incentives (e.g., tax rebates) do not enter into the decision process because they do not exist.

From our general profit maximization problem presented in equation (3.4), we assume that the firm has no influence over price. As a result, price (P_i) is removed from the optimization problem. This reduces the profit maximization problem to the special case of cost minimization, and we focus our attention on equation (3.3).

Consider a simple case where the firm produces one output and ships it to one market and uses only one input that is purchased from one supply source. We can see this simple problem using Figure 3.2; our Webber triangle collapses to a single-dimension line. Here, the lower left-hand and right-hand corners of the graph represent the location of the input and output markets, respectively. Reading left to right captures the cost of shipping the input, and reading right to left captures the cost of shipping the output. The firm must select a location between the input supplier and output markets in such a way that transportation costs are minimized.

The costs of shipping the input and/or output are generally a nonlinear relationship because of high fixed costs (a positive intercept) but increasing scale economies. For example, rail transportation has very high fixed costs associated with terminal and rail infrastructure and the trains themselves. Yet, once these high fixed costs are incurred, the cost of shipping a unit of materials an additional mile is relatively low. This suggests an upward-sloping transportation cost curve [$d(s,s^i)$ and $t(s,s^i)$], but increasing at a decreasing rate. The firm is interested in total transportation costs, or $d(s,s^i)$ plus $t(s,s^i)$. In the graphic, the total transportation cost curve is the vertical summation of individual cost curves. At location d_o, the cost of shipping the input to the firm is a and the cost of shipping the output is b, and total cost is a plus b. In this simple example, the least cost location would be for the firm to locate at the location of the input supplier.

Generally, the firm looks at the marginal cost of shipping inputs and outputs. Firms will tend to gravitate to markets where the marginal cost of shipping is the highest. In the simplest form, firms that use weight-losing processes will tend to locate closest to the input markets, while weight-gaining processes will locate closest to output markets. Bottling plants, for example, tend to have weight-gaining processes. The base syrup is concentrated and relatively inexpensive to ship. The final product is a combination of the syrup and water. Because of the

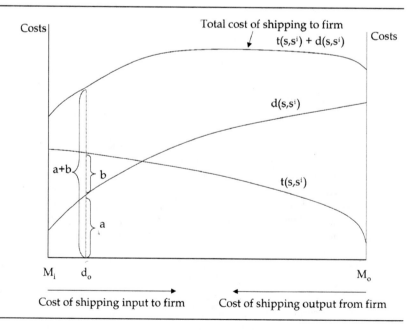

Figure 3.2. Transportation cost minimization.

relative high costs of shipping water from a central location to several markets, it is cheaper for the firm to have multiple plants located at different markets where water can be added to the syrup. An example of a weight-losing process is cheese production, where vast quantities of milk are required. Because it takes several gallons of milk to make a pound of cheese, it is cheaper for the firm to locate near the source of milk production and ship the final product to consumers.

A firm's location decision attempts to minimize total cost, not just transportation costs. Once the firm selects its minimum transportation cost site, there may be another site with labor available at a sufficiently lower cost to overcome the higher transportation costs associated with that alternate site (Smith 1971). The factors of production, including land, labor, and capital, which are of utmost importance to the economic development path of a community, also come into play in firm location theory. Consider a firm that identifies s^* as an optimal location from a transportation cost minimum perspective, but there are several possible locations within a small distance from s^* that may be appealing to the firm. How does the firm select the final location?

In the cost minimization approach, as well as with the demand maximization approach discussed below, the firm has not necessarily made its final decision when it identifies the transportation cost minimization location s^*. If we lift the strict assump-

tion of a homogeneous economic plane and allow for some economic variations across locations, the cost-minimizing firm is said to go through a two-step process. The first step centers on the general location based on transportation costs. The important thing to remember is that once transportation costs are minimized, the firm attempts to minimize the factor of production costs, which may change the location from the minimum transportation costs site. Suppose that the firm makes the decision that it wants to be closer to its input markets than to its output markets. Once this decision is made, it must pick a specific location. Take our cheese producer as an example. Once it makes the decision that it wants to be located in, say, the upper Midwest, it must find a specific location for its plant. This is the second step of the two-step process.

One way to think about the second step of the location process is by using *space cost curves,* or what are widely referred to as *isocosts.* Space cost curves are a spatial representation of total cost, including transportation, labor, land, and capital costs (Smith 1971). Space cost curves represent variations in total costs and profitability over space, given fixed demand. Any location where total costs exceed total revenues does not represent a viable long-term location.

Consider Figure 3.3, where we map out hypothetical isocosts on our Webber triangle. Each isocost is akin to a production *isoquant* or indifference curve

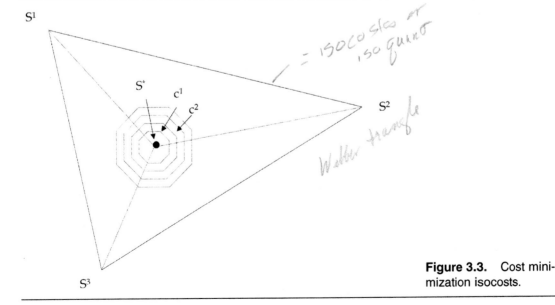

Figure 3.3. Cost minimization isocosts.

from the individual utility maximization problem. The cost to the firm is identical at each site along a particular isocost. For space cost curve (isocost) c^1, all locations along c^1 incur the same cost to the firm. But we do not know *a priori* the relationship between the separate isocosts; specifically, c^2 may be greater than or less than c^1. Because of variations in labor, land, and capital markets, a specific location on c^2 may have lower overall costs to the firm than the transportation cost minimization location s^* or a closer site on c^1.

In this second stage of the location decision-making process, communities often have some influence on firm location decisions. Communities can offer high-quality infrastructure, skilled labor, building locations, and characteristics of a high quality of life generally as well as low cost alternatives for the firm, which may give a community a viable comparative advantage over other locations. (Discussion of policy options for communities that follow from this theoretical view of firm location decisions is provided later in this chapter and throughout the remainder of the book.)

The least cost approach to location decisions assumes the business faces a demand not affected by the business's location choice. The business sells to a point market. Costs can vary among production sites. The location decision substitutes nontransportation and transportation costs among different sites until the firm minimizes total costs. Since demand is constant, the least cost site yields maximum profits. The weakness of this approach is that if demand varies with the site chosen, the least cost site may not be a maximum profit site; it may only be a minimum cost site.

This approach to location decisions indicates communities must be sensitive to the total cost of production in their community compared with other communities. Thus, the community seeks to keep transportation rates low or to offset higher transportation costs by reducing the nontransportation costs, such as low wages, inexpensive land, and tax concessions. The community may attempt to create some agglomeration economies through a fully serviced industrial park.

Demand Maximization

The least cost approach to location selects the site by assuming that the firm sells its total output to a given point market (i.e., effectively eliminating demand from the location decision). The *demand maximization approach*, however, reaches the location decision by explicitly incorporating demand into the decision (Greenhut 1956). The demand maximization approach to location decisions contends that each seller will select a site to control as large a market area as possible. The seller exercises some monopoly control over that portion of the market area she can supply at a lower price than her rivals can (see Fig. 3.4). Consumer behavior and the

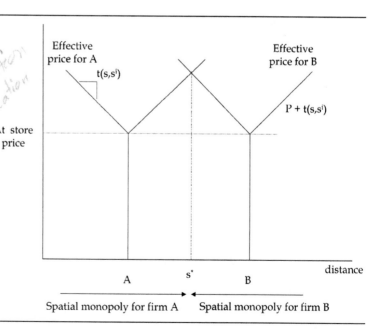

Figure 3.4. Demand maximization, or locational interdependence.

location decisions of competitors determine the size of the market the firm controls. Concern about the location of competitors gives rise to the alternate description of this approach as the *locational interdependence approach.*

In our spatial world, we have a unique situation where the market has characteristics of both competitive markets and spatial monopolies. Over the whole economic plane, competition determines the *at-store* price of the good or service. Thus, all firms in essence charge the same at-store price. Consumers, however, are looking at what is called the *effective price* of the good or service. The effective price is composed of two parts: the at-store price plus the cost of transportation to the firm, or $P + t(s,s')$. For consumers located next to the firm, the effective price and the at-store price are essentially the same. As the consumer moves away from the location of the business, the effective price becomes larger than the at-store price. As the effective price increases, the demand for the good or service declines. In other words, more income is devoted to transportation costs and less is left over for the purchase of the good or service itself. This defines the range of a good or service (Chapter 4).

Basic assumptions of this approach (Greenhut 1956) include the following:

1. The firm sells to a spatially distributed market, not to a point, as was the case with the least cost approach.
2. Customers and resources are uniformly distributed over a homogeneous plane.
3. Customers make their purchase decisions on the basis of minimizing the delivered/effective price.
4. There are uniform transportation rates in all directions from any site; these rates can vary among sites, but they are the same in every direction from any given site.
5. Abnormal profits can exist and will attract competitors.
6. There are no barriers to entry.

The classic example of the demand maximization problem is known as the case of Hotelling's beach vendors. Imagine a beach that is bordered on both ends by cliffs and has a number of beach goers evenly distributed across the beach. On this beach are two refreshment vendors selling identical products at identical prices. From the perspective of the patrons, they will select the vendor that is physical-

ly closest to them. The vendors are mobile and can locate their shop at any location on the beach each morning. Given this structure, one can view the Hotelling problem in a game theoretic framework: How will the two vendors react to each other's location decision with the objective of the game to maximize the number of buying beach patrons?

The game begins with the beach divided into four equal parts: *A, B, C* and *D*. The vendors locate at the point between *A-B* and *C-D*, and all patrons pick the vendor closest to them. In this initial case, all people in sections *A* and *B* of the beach go to the first vendor, and all people in sections *C* and *D* go to the second vendor. The person on the exact center of the beach, the break point between sections *B* and *C*, is indifferent between the two vendors located an equal distance from this person.

On the second day, vendor 1 notes that vendor 2 located at the center of the beach, the point between *B* and *C*, will capture more of the market. Presume this is the second vendor, who was located at the breakpoint between sections *C* and *D*. Now the first vendor returns to the starting point between sections *A* and *B*, but the second vendor locates at the center of the beach. Now the second vendor captures everyone in section *C* and *D* of the beach and splits the patrons in section *B* with the first vendor. The first vendor still captures all of market *A*. When the game began, the beach market was equally divided between the two vendors. In the second round of the game, the second vendor is capturing more than half the market by encroaching into the first vendor's market, specifically part of market *B*.

In the third round of the game, the first vendor realizes that she has lost market share and that the only way to recapture market share is to move closer to the second vendor. The locational game ends when both vendors are located side by side in the center of the beach. Here, all of the patrons in sections *A* and *B* of the beach go to one vendor and all patrons in sections *C* and *D* go to the second vendor. It is of interest to note that while this locational demand maximization game provides for a stable market solution, the solution is not considered a social optimal. If we consider total costs (the price of the goods purchased plus transportation costs) as a proxy for social welfare, then we can see that the final location at the center of the beach does not minimize total costs; it maximizes them. In this example, the initial location of the vendors at the beginning of the game minimized total costs and hence maximized total welfare.

Let us formalize our simple beach vendor example and consider two firms: *A* and *B*. Our simple linear world can be represented in Figure 3.4; the point where the effective price is equal determines the market boundary of the individual firm. Customers that are located where the effective price for firms *A* and *B* are identical (point *s**) are indifferent between firms. Within the distance between the location of the business and the market boundary (*s**), the business has a *spatial monopoly*. The firm, if it so desires, can increase the at-store price, exerting spatial monopoly power, but it does so at the cost of increasing effective price and shrinking the spatial size of its market. This type of location behavior can help provide insights into why big-box stores, such as Wal-Mart and Target, elect to locate on the edges of communities where they attempt to draw not only the customers of the community itself but also customers from surrounding communities. This is particularly true when previously existing big-box stores leap frog to the edge of town as growth moves beyond the store's initial location. The initial store may elect to relocate farther on the edge of the community to retain its locational preference to customers in surrounding communities.

Two unique spatial characteristics of the demand maximization approach warrant comment. First,

because the demand maximization approach hinges on the market power of a particular location, the idea of agglomeration economies discussed below come into play. If we lift the homogeneous economic plane assumption and realize that some communities are larger than others, it becomes clear that firms will be drawn to the larger communities. Return to the Webber triangle in Figure 3.1 and assume that market S^1 is significantly larger than markets S^2 and S^3. In this case, the demand-maximizing firm will tend to migrate to market S^1 and move away from markets S^2 and S^3. In this instance, market S^1 is called a *dominant market* and it tends to pull businesses toward it. In its simplest sense, we start to see agglomeration effects in the demand maximization model.

Second, if we move from a linear world to an economic plane, we begin to understand the spatial size of markets and, indirectly, of communities. Moving the single-dimensional world of Figure 3.4 into a two-dimensional plane helps us understand the spatial layout of markets. In Figure 3.5, suppose that a firm locates at location *s** and that effective price is measured vertically from the location of the business. For a customer located next to the business, say, *s* = *s**, then effective price is equal to the store price. As the customer moves away from the busi-

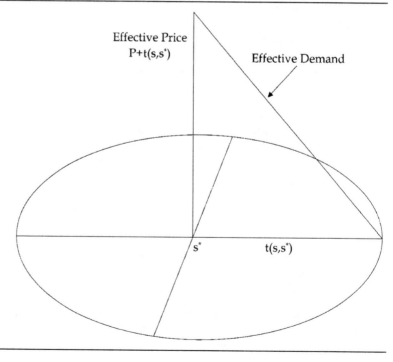

Figure 3.5. Spatial markets: The Lösch demand cone.

ness, effective price rises and the amount of income that can be used to purchase the good or service and effective demand declines. The point where effective demand reaches zero is the boundary of the market for the business. If we now rotate the effective demand curve 360 degrees around the location of the business, we have what is called the Lösch demand cone (Fig. 3.5). The volume of the demand cone is the market potential for the business in a spatial world.

Looking at the demand side of the demand maximization approach provides only part of the picture. To complete it, we need to introduce the cost of the firm. In our simple profit maximization problem stated in equations (3.2) and (3.4), we have assumed that the marginal cost of producing one more unit of the good or service is fixed. In other words, there are no scale economies. Consider a demand-maximizing firm that does not have constant marginal cost but rather has the traditional U-shaped average and marginal cost curves (Fig. 3.6). In the demand maximization problem, we assume that all transportation costs are borne by the customer, so the cost structure does not include transportation costs.

Now suppose that the firm is facing three potential effective demand curves (AR_1, AR_2, and AR_3). From the point of view of the firm, the effective demand curve can be interpreted as an average revenue curve. It is important to realize that demand is community demand and is directly influenced by the population and income of the community, along with other socioeconomic characteristics. Suppose that the firm is facing AR_1. In this case, demand is not sufficient to cover the costs of the firm. The population size of the community or the income level may not be sufficient to support this particular business, or other characteristics of the community may not line up with this particular business.

Suppose now the firm faces demand structure AR_3, which is larger than AR_1. At this level, average revenue is greater than average cost, so not only is the level of demand large enough to support the firm, it also is large enough to earn the firm an economic profit.[1] In a perfectly competitive market, those economic profits will attract firms into the market and chip away at the market of the initial firm. The level of demand where average revenue is just sufficient to cover average costs is also known as the *demand threshold*. In Figure 3.6, demand structure labeled AR_2 is large enough to cover costs but not so large as to generate excessive profits and attract firms into the markets. The spatial market is said to be in a *spatial equilibrium*.

This more complete picture of the demand maximization approach provides the community practitioner with powerful insights into the microeconomics that drives local retail and service markets. The theory behind the demand threshold of a good or service becomes readily clear. Because different types of firms have different cost structures, different spatial markets will be required to support them. Indeed, we can take these simple

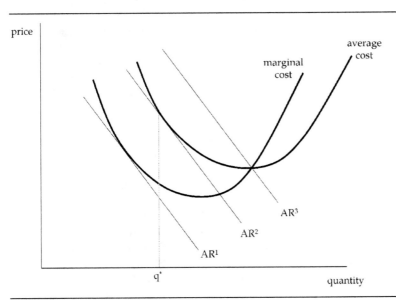

Figure 3.6. Microeconomic foundation of thresholds.

microeconomic concepts and move to a system of communities or central places. (See Chapter 4 for a detailed discussion of central place theory and its implications for community economic development.)

Summary and Critique of Classical Location Theory

Classical location theory conceives the location decision being made by a profit-maximizing firm located on a homogeneous economic plane. Firms and customers have perfect information and can fully process that information. In addition, firms are perfectly mobile and able to enter and exit spatial markets with no costs. The approach, while abstract, provides a rigorous framework to begin to think about space in a community economics setting. As a deductive theory of location, it provides an excellent foundation for more in-depth analysis.

Our current economy, however, is characterized by large-scale, mass production operations that are part of complex multiproduct and multi-establishment businesses. With a multiproduct, multilocation, multidivisional firm, the questions of optimality are complicated by optimality for the total firm or a particular division, total product line or a particular product, internal flows of inputs or markets being served, or the interest of management or the owners. The firm's multiple objective function includes objectives other than profit maximization. These other objectives arise partially from the interaction between the firm and the communities (society).

BEHAVIORAL APPROACH

The classical profit maximization approach yields models with relatively low predictive powers, thus preventing its ready adoption. There are several reasons for this low predictive ability. First, the model does not handle personal preferences and psychic incomes/costs related to location decisions. These personal considerations cause the decision-maker to maximize total (money and psychic) satisfaction rather than just monetary profit. Second, it assumes that the individual making a location decision has perfect knowledge about the future, which is, of course, not possible. Differences in opinions about risk and profit potential associated with various locations lead to different location decisions. A risk-averse owner or firm with limited financial resources may choose a site with less potential prof-

it but less risk of loss. Third, location decisions are typically made infrequently during the career of a business owner/manager. This infrequency, coupled with imperfect knowledge, often yields site selection criteria, such as long-run sales growth, with reasonable profits or space for expansion. The cost of acquiring additional information about alternative sites deters the business from further inquiry. The result is the selection of a satisficing rather than a profit-maximizing site.

The profit maximization approach represents the culmination of the rational economic person as a location decision-maker, with the associated assumptions that he desires to maximize profits, has information available on current and future events and conditions, and reviews and comprehends all necessary information.

The behavioral approach proposes a significant relaxation of these assumptions (Massey 1975; Pellenbarg, van Wissen, and van Dijk 2002; Shepard 1980; Tornquist 1977). It allows for personal goals other than profit maximization, inadequate and inappropriately used information, and uncertainty about current and future conditions of markets, rivals, and inputs. The approach uses game theory and the concept of bounded rationality to analyze location decisions. The behavioral approach attempts to explain why the selection of certain sites appears to be irrational.

Objective Functions

The behavioral approach explicitly permits the decision-maker to seek some objective other than profit maximization in making location decisions. These alternative objectives could be market penetration/share, some minimal return, or expansion within some geographically bounded area to maintain management control. The different objective function takes the form of seeking a satisfactory, rather than maximum, level of some monetary or nonmonetary goal. Thus, the firm no longer necessarily seeks the location yielding maximum profits but a location meeting some minimal profit standard. This reduces the burden on the firm. The inclusion of nonmonetary goals means the site need not be the highest profit site. But, remember, the site still meets or exceeds some minimally acceptable profit standard and still allows the decision-maker to increase his welfare through other channels.

Including nonmonetary elements in the firm's location choice makes the analysis both more realistic and more complex. The more realistic dimension occurs because it reduces the range of possible locations. Most sites fail to provide the noneconomic characteristics desired, such as view of the mountains or the owner's hometown. The firm calculates costs and rival's responses for a limited number of sites, not all possible sites. The more complex dimension occurs with optimizing objective functions that may or may not include noneconomic criteria. Yet, the behavioral approach to location decision-making, by allowing a broader interpretation of the objective function, offers great potential in explaining some "irrational" location choices.

Uncertainty

Relaxing the assumption of complete geographical and temporal knowledge by the firm permits uncertainty in the location decision. Locational uncertainty occurs in many forms. National uncertainty appears as uncertainty about future general price changes, economic growth, interest rates, and federal monetary and fiscal policies. Regional uncertainty includes population shifts, and regional responses to national economic changes. Local uncertainty might be labor-management relations, water supply, or local governmental actions. Firm uncertainty includes the type of production processes, location of competition, and continued availability of inputs. Despite the allusion, the world facing the firm is not completely chaotic. The firm's intelligence and experience filters much of the apparent chaos and gives patterns to events. There even may be relatively precise estimates of the probability of occurrence of some events (e.g., uncertainty becomes a risk). Uncertainty causes the firm to produce a continuum of conditional decisions rather than a single location decision.

Information

The behavioral approach argues that the firm does not have full and complete information even without uncertainty (Lloyd and Dicken 1977; Pellenbarg, van Wissen, and van Dijk 2002). Bounded rationality says the human limitations of the decision-maker within the firm prevent absorbing all the information available.[2] Furthermore, decision-makers are more sensitive to some sources of information than to others. These sources are business associates, friends, family, and some media. The decision-maker's own receptiveness, experiences, and perspectives further filter the flow of information.

Because the firm does not make such a decision frequently, the amount of information actually incorporated into the decision may be far less than commonly presumed. Take the example of the small manufacturer seeking space for expansion or relocation. The firm reduces the information burden of the decision substantially by limiting, consciously or unconsciously, the choices of possible sites capable of meeting the needs of the firm to the neighborhood or nearby communities. In the process, however, the decision-maker forecloses other critical questions, such as, Should the business move to another region to improve its access to markets or inputs and profits?

Information and uncertainty vary greatly by size of the firm and the distance of the potential location options. Larger firms generally have greater access to information or have the resources to gather the necessary information. On the contrary, small firms may find the cost to gather all the necessary information about all possible locational sites to be prohibitive. In addition, firms generally have more information about sites that are in close proximity to their existing locale. Limited information, limited ability to process information, perceptions, and uncertainty lead to a large spatial bias in location decision-making. In essence, a more-distant location is less well known than closer locations. As a result, firms may not be as freely mobile as presumed in the classical profit maximization problem.

Pred's (1967) behavioral matrix displays the implications of the availability of information and the ability to use information in the location decision (Fig. 3.7). Movement from left to right indicates improved quality and/or quantity of information on the location decision. Movement from top to bottom measures improved ability to process and use information. The upper left-hand cell indicates poor information used in a poor fashion. The lower right-hand cell indicates good information used wisely. This matrix can be projected onto the existing/proposed locations of firms.

Some locations are profitable and many are not. Firms may be in an unprofitable location because they have poor information or used information poorly (Townroe 1974). This can result from a bad initial location choice or changes through time that make initially good choices no longer profitable. In either case, the firm failed to use information properly or did not have good information to use. It is possible, however, for a firm to "get lucky" and

	Little/Low Quality Information	Much/High Quality Information
Information Processed and Used Poorly	Likelihood of locating in a profit maximizing site is *low*	Likelihood of locating in a profit maximizing site is *indeterminate*
Information Processed and Used Well	Likelihood of locating in a profit maximizing site is *indeterminate*	Likelihood of locating in a profit maximizing site is *high*

Figure 3.7. Firm profitability and information use and quality.

poorly process bad information and still end up in a profitable location.

Summary and Critique of the Behavioral Approach

Location theory postulates that competitive firms seek locations that maximize profits. But firms are more limited in their ability to determine the optimal location. At least four factors can be identified as limiting this ability: the volume of information required, the quality of information available, uncertainty surrounding future economic conditions, and uncertainty with respect to the actions taken by rivals. These factors, coupled with differences in personal motivations, make it impossible to state with certainty what location decision a particular firm will make given a set of location information. The behavioral approach to location decisions considers the goals of the decision-maker and the information sources, as well as how information is used and how new information is sought.

The behavioral approach focuses community economic development on the information and objective functions. The community can do little about uncertainty beyond providing a stable public-private environment. The community performs an information role by helping the firm acquire and use information through management counseling and assistance. The community affects the objective function by recognizing the importance of noneconomic factors in the location decision.

The behavioral approach, however, is a deductive theory of the location problem. As such, while intuitively appealing, it does not lend itself to rigorous testing. As Scott (2000) argues, a major drawback to the behavioral approach is that it focuses too much on sociological, psychological, and other "soft" variables and often ignores the economic foundation of the classical profit maximization approach (Pellenbarg, van Wissen, and van Dijk 2002).

INSTITUTIONAL APPROACH

The classical profit maximization and behavioral approaches have one common theme: The firm is at the center of the location decision-making process (Pellenbarg, van Wissen and van Dijk 2002). The firm has to take into consideration both economic and noneconomic factors in decision-making and takes the form of *Homo economicus* or *satisficer person* (Hayter 1997). What is overlooked is that the economic processes involved in space are greatly influenced by society's cultural institutions and value systems. In essence, the community practitioner must not only look at the economics of the firm but also at the social, cultural and political context in which the firm makes its decisions. All dimensions of the Shaffer Star (Fig. 1.1) come into play in a spatial world.

Much of the location decision-making process involves negotiations between businesses, owners of the land, labor unions, and local governments for access to local infrastructure. The institutional approach focuses on the rules that set the parameters for negotiations and contract law, as well as the negotiating power of the firm. Larger firms have greater negotiation leverage than smaller firms. State-level economic development policies are often targeted toward larger firms, and firms have become accustomed to seeking incentives from state and local governments. Indeed, the new "war between the states" has state economic development agencies effectively bidding against each other in attempts to influence firm location and expansion decisions.[3]

While the cost effectiveness of these state-level policies have been widely challenged, larger firms can and do exert market power in negotiations.

Small- and medium-size firms focus mostly on two types of institutions: local governments and real estate markets. Local governments can have significant impacts on the location decisions of firms. This institution determines land use laws, such as zoning regulations and building codes, and in many cases has the power to deny the necessary permits for the firm to move forward. From a conceptual point of view, local governments can and do affect the structure of the isocosts discussed in the cost minimization approach (Fig. 3.3). In the name of economic development, local governments often will try to provide flexibility in zoning, access to public infrastructure, and fiscal incentives. Local governments can have a significant impact on the nature of local economic development by how they structure and enforce these local rules.

The status of the local real estate market has a significant impact on firm location decisions. Small- and medium-size firms more often than not are looking for existing facilities that they can buy or rent on the open market. Again, only the largest firms are in a position to build at every location. Therefore, the spatial supply of the real estate market, such as office space, industrial sites, and commercial floor space, is of prime importance for understanding the locational choice of smaller firms.

AGGLOMERATION IN LOCATION THEORY

Understanding firm-level location decisions provides an important building block for explaining why some regions prosper while others languish. In this section, we link firm location to regional uniqueness. This linkage is critical in understanding community economics. Many problems associated with smaller and more-remote communities are because they are often far from large metropolitan areas that provide substantial profit-maximizing characteristics (either because of least cost in production or of demand maximization in selling output). We begin with an overview of primary factor resources, resulting specialization in production, and resource scarcity that leads into a regional economic approach to explaining the competitiveness of regions.

Regional Comparative Advantage

An aspect of firm location that relates to regional uniqueness is based on the productivity and avail-ability of primary factors of production. The importance of primary factor resources provides a basis for the concept of *regional comparative advantage.*[4] Traditionally, primary factors have been identified to include land, labor, and capital.[5] Regional advantages begin to surface when we continue to relax the spatial assumption of a featureless plane to account for differences in the endowments of primary factors of production. Factor endowments, factor productivities, and factor markets differ from one region to the next.

In this regional advantage framework, different factor resource endowments and productivities lead to relative factor scarcities. These factor productivity advantages reflect unique characteristics of a region's land, labor, and capital endowments. For example, advantages in land productivity can reflect climate, growing season, and/or soil characteristics. Advantages in labor productivity could reflect high skill levels and/or large workforce numbers. Advantages in financial capital productivity could reflect lower regional investment risk because of safer, more secure or supportive political systems. Where factor resources are more productive, the natural tendency is for increased specialization based on production that relies on those factors that are relatively more productive.

The framework of comparative advantage in factor resources provides another reason why regions tend to exhibit production cost differences. In regions with highly productive and large labor forces, this relative production advantage leads to a cost advantage as outlined in the earlier section on cost minimization. Likewise, regions endowed with rich land and low-risk capital enjoy relative advantages. With increased specialization, trade can act to aggregate regional advantages to higher overall levels of productivity across space. The benefits of trade are the ability to increase consumption of the good in which you are relatively disadvantaged in producing. This is the standard argument for free trade where prices for all goods and services equilibrate to the point that the region most advantaged in production determines the price of the goods and services sold (Krugman 1995; Todaro 2000).

The scope of comparative advantage involves several key issues. First is a region's natural endowment of factor resources. These include the initial endowment and availability of land, labor, and capital. They reflect the existing climatic conditions (temperature, precipitation) and topography (mountains, rivers, etc.). Next are favorable production (firm) conditions that have been outlined in earlier sections of this chapter that focus on production inputs (backward linkages or cost-minimizing components) and markets for outputs (forward linkages or demand-maximizing components). Also as discussed earlier, transportation considerations of regions (such as infrastructure) lead to transportation and marketing cost advantages. Proximity to centers of research leads to technological advantages. Good examples of this can be found where research parks founded by institutions of higher education provide centers of innovation and lead the development of high technology sectors.

In a more latent way, two additional aspects delineate the scope of regional comparative advantage. First, as outlined above, institutional advantages can exist that speak to issues of production risk (Chapter 11). These include the underlying stability of political institutions, building on our often-overlooked assumption of law and order. Regions without sound and equitably enforced legal structures often suffer from economic disadvantage. The second latent aspect is quality of life. Amenity factors of regions affect how individuals make locational decisions (as discussed under the behavioral approach above and more completely dealt with in Chapters 9 and 10). Also, firms can use a region's amenity base as an issue in location and thereby affect where labor and capital become employed.

Regional advantages play a part in the previously described process of a firm's decision to maximize profits through minimizing costs and maximizing revenues. The initial endowment and underlying productivity of factor resources, combined with the availability of these resources and the knowledge of their use, lead to competitive opportunities for firms and increased specialization of regional output.

Industry Clusters

A relatively recent insight to location and community economics is the concept of industry clusters. Industrial clusters are geographic concentrations of interconnected companies, specialized suppliers, service providers, firms in related industries, and associated institutions in a particular field that compete but also cooperate (Doeringer 1995; Enright 2000; Fesher and Sweeney 2002; Held 1996; McCann 2002; Porter 1996, 1997; Steiner 2002). A cluster is a geographically bounded concentration of independent businesses that have active channels for business transactions, dialog, and communications and that collectively share common opportunities

and threats. The geography that we are talking about can be a single city, a region, or a state or even be national or multinational in nature. The geography relates to the distance over which informational, transactional, and other efficiencies occur. The boundaries of clusters are flexible and are more of an art than an exact science. Clusters provide a constructive and efficient form for dialog among private businesses, their suppliers, their customers, governments, and other institutions.

Clusters typically encompass an array of linked industries and other entities important for competition.[6] Some of these are suppliers of specialized inputs, such as components, machinery, and services; providers of specialized infrastructure, such as rail and power; and companies and industries related by skills, technologies, or common inputs. They can be trade associations and institutions such as governments and universities, including standard-setting agencies, think tanks, and vocational training providers. They can extend downstream and laterally to manufacturers of complimentary products.

The role of government in cluster development is really removing obstacles, relaxing constraints, or eliminating inefficiencies in productivity and productivity growth. The emphasis is on dynamic improvement more than on market share.

Clusters are critical to competition because modern competition depends on productivity, not on access to inputs or scale of individual enterprises. Productivity depends on how companies compete, not on the particular fields in which they compete. Competition is employing sophisticated methods or using advanced technology, or offering unique products and services. Clusters can affect competition in three general ways. One, they can increase the productivity of companies based in the area. Two, they can increase competition by driving the direction and pace of innovation, which underpins future productivity growth. Finally, clusters affect competition by stimulating the formation of new businesses that expand and strengthen the cluster itself.

There are similarities between agglomeration economies and clusters. The difference is that agglomerations are essentially a static concept that considers cost. A cluster or competitive advantage is a dynamic concept because it involves how firms continually innovate and transfer knowledge among themselves.

Rosenfeld (1997), in referring to industrial clusters, listed several aspects of industrial clusters that need to be considered:

- *Workforce skills:* Do the skills of the labor force fit the needs of the industry? Do these include not only technical skills and competencies, but also general knowledge of the industry and entrepreneurial skills?
- *Human resource development:* Are there opportunities for specialized education and training for the cluster's major occupations, and does the industry itself invest in training?
- *Proximity of suppliers:* Are primary and secondary suppliers and sources of raw material located nearby?
- *Capital availability:* How well do area banks understand the industry, and do they meet the cluster's needs for working and start-up capital and access to seed and venture capital?
- *Access to specialized services:* Are there specialized public and private services such as technology extension, export assistance, small business center, designers, engineering consultants, accountants, and lawyers?
- *Machine and tool builders:* Are companies that design and build machines and tools used by the industry nearby, and are there working relationships that foster innovation?
- *Intensity of networking:* Do firms in the industry cooperate? How often and to what degree? Do they share information or resources? Do they participate in joint production, marketing, or problem solving? How often and to what degree?
- *Intensity of competition:* Are there multiple firms with overlapping capabilities and competencies? Does competition push firms to seek new products or new markets?
- *Social infrastructure:* How strong and active are local businesses and civic associations or chapters of associations in the region? How active are their memberships? How often do they interact with each other?
- *Entrepreneurial energy:* What's the rate for new business start-ups by workers and managers from within the cluster? How successful is the cluster in attracting new firms or suppliers from outside?
- *Innovation:* How quickly are new and enhanced technologies conceived, developed, and adopted? How quickly do products, processes, and services that use these technologies and firms that produce them appear?
- *Shared vision and leadership:* Do firms have a collective identity, plan for and share goals, or have a vision for the future? Do they have

leaders who maintain their collective competitiveness and keep them together?

Alternative Agglomeration Semantics

The discussion above comprises a set of fundamental components that, when combined, begins to explain the tendency for firms and individuals to agglomerate in locations. These agglomerations will naturally tend to merge and concentrate economic activity in larger and larger centers. Another perspective of the basis for agglomeration relates to our discussion of externalities in Chapter 1. Remember that externalities can be either negative (e.g., charcoal manufacturing and computer assembly) or positive (e.g., real estate development and golf courses). Firms that have positive externalities benefit from locating in close proximity to one another because it represents cost minimization and/or demand maximization. Another term for this positive interaction is *pecuniary externality.*

We can restate the earlier discussion within an agglomeration context: Location decisions of firms will tend to regionally agglomerate to maximize pecuniary externalities that affect the inputs to production. These sometimes are referred to in the agglomeration literature as *localization effects.* This form of agglomeration addresses input costs related to the availability of and competition for factor resources (land, labor, capital, public goods, amenities, technology) or intermediate purchased inputs (backward linkages in production). These localization-related pecuniary externalities are often infrastructure dependent and reflect regional advantages in transportation costs.

Another dimension to localization effect relates to the internal economies of scale to firms. If we realize that many industries exhibit economies of scale, these firms can influence the spatial location of linked firms, such as input suppliers. Consider Figure 3.1, where we have a three-pointed Webber locational problem. Assume that the three points are the location of three firms within the same industry, say, auto manufacturing. If one of the three firms benefits from significant economies of scale, input suppliers will tend to be drawn spatially toward the larger firm. In the demand maximization approach, we noted that one place, because of its size, can be termed a dominant market. We can have a parallel situation with firms that are characterized by scale economies.

The other type of agglomeration-related pecuniary externality relates to firm location based on an advantage in the ability to sell units of output. Earlier, we labeled this as demand maximization. In the agglomeration literature, these often are referred to as *urbanization effects; they* relate to the mass (or size) of markets as key determinants of overall levels of consumption. Competition among buyers provides strong demand for output that generally leads to price advantages. Again, infrastructure and size of community are key determinants that provide regional advantage in the cost of marketing and sale of products.

Porter's Approach to Regional Competitive Advantage

The tendency of firms to seek and sustain unique technological and organizational advantages has been well summarized by Porter (1990) in his treatise on industries involved in global markets. His now infamous schematic known as the *Porter diamond*, reproduced in Figure 3.8, focuses on regional competitiveness with respect to firm location. According to Porter, firms engaged in producing for global markets will locate production based on four primary issues: factor conditions, related and supporting industries, demand conditions, and firm strategy. Each of these has elements that relate to many of the issues already dealt with in this chapter. Indeed, Porter's real contribution is the manner in which he pulled these elements of regional location decisions together into a coherent framework. Furthermore, much of what Porter raised as factor con-

ditions, strategy, and institutional/government policy is dealt with in later chapters.

Factor Conditions

A region's position in terms of its endowment of factors of production and the relative productivity of these regional resources is captured in what Porter refers to as factor conditions. These include the extent and productivity of important factors of production including land, labor, and financial capital (each of these is more fully discussed in Chapters 5, 6, and 7 respectively). In addition to these traditionally defined factors of production, it is important to also add more latent factor inputs such as technology/management, amenities, and locally provided public goods and services (more fully discussed in Chapters 8, 9, and 10 respectively). The factor condition node of Porter's diamond emphasizes the importance of factor endowments, their relative productivities, and market intricacies in regional location decisions within industries involved in global markets.

Related and Supporting Industries

The regional presence or absence of supplier and related industries that are competitive in the broader geographical realm reflects important components previously described under the cost minimization problem. The presence or absence of supporting firms is important for assisting firms in maintaining and improving their own production

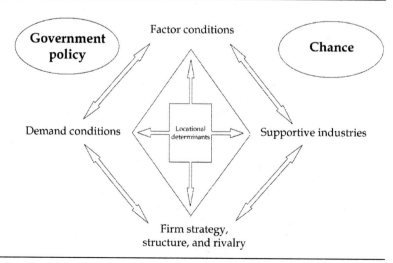

Figure 3.8. The Porter diamond.

processes. The presence of large firms that are characterized by economies of scale can also affect other regional firms by providing cost advantages for inputs. These internal agglomeration economies reflect what Porter calls related and supporting industries.

Demand Conditions

The nature of the market for an industry's product or service is important in determining the location of firms and the competitiveness of regions. These reflect issues raised earlier in the profit maximization section, specifically those dealing with demand maximization. The close proximity of people willing to buy goods and services produced in the region provides a competitive advantage to firms operating in these sectors. These demand conditions parallel our prior discussion of urbanization economies.

Firm Strategy, Structure, and Rivalry

The conditions in the region governing how firms are created, organized, and managed, and the nature of competition help explain why some regions are more competitive than others. This competitiveness (or rivalry) acts on local industry structure to strongly motivate managers to strengthen their firm's competencies. This level of strong competition forces successful firms to be flexible and adaptable by constantly retooling and reworking themselves to remain efficient and thus competitive with their rivals. A sports analogy clearly demonstrates this aspect of the Porter approach. Leagues that are more competitive tend to attract the best athletes and produce sporting events that are of higher quality than smaller, less competitive leagues. This ability of regions to foster competitive industrial structures and the resulting strategies that firms use in producing goods and services relates to an important component that explains why some regions tend to be more competitively advantaged than others. What Porter discussed are the notions of the behavioral approach to location theory outlined above and, to a large extent, the factors influencing firm clustering.

In addition to the four components of regional competition, Porter added government policy, as in the institutional approach, and chance. These play subsidiary roles in location decisions of firms and the competitiveness of regions in producing goods and services that compete within global markets. Although we do not specifically address chance, the role of government in producing and enforcing rules,

as well as serving as a key provider of public goods and services, is dealt with in Chapters 10 and 11.

PRODUCT LIFE CYCLE

It is also important to point out that the spatial location of alternative labor demands or industrial mixes can be addressed in the conceptual realm of *product life cycle theory* (Mack and Schaeffer 1993). According to this theory, products move through three phases: new products, maturing products, and finally standardized products. As products "cycle" through these phases, the spatial location and types of labor demands move from technologically advanced regions to locations where labor costs are minimized.

New products require the maximum amount of flexibility by producers to adapt in getting the right production mix. Figuring out this mix requires extensive use of highly skilled technical labor. Furthermore, as new products and their production technologies evolve, there is a need for extremely close ties to management and advanced business services. Thus, the location of new products takes place in technologically advanced regions, often with close ties to the types of research and management fostered by institutions of higher education. More generally, new products need ready access to the types of inputs found in urban centers.

The second phase of product cycles is found in *maturing products*. This intermediate phase is marked by the tendency for firms to move production toward economies of scale. As the product's demand grows, standardized production methods are put into place and the location of production now depends on minimizing traditional shipment and factor input costs, including labor and land. This phase starts the movement of production, and hence the demand for labor, away from technically advanced regions and loosens the firm's ties to urban centers.

The final phase of production in product life cycle theory is *standardized products*. This marks the end of innovation in design and production, with industrial organization and production processes characterized as stable. The product now exists in a situation of considerable price competition for both inputs and outputs and requires a significant amount of unskilled labor. The tendency for spatial location of mature production is a general movement of production to low-wage labor supplies. Traditionally, these have been located in rural, nonmetropolitan regions where land also tends to be cheaper than in urban areas. With rapid globalization during the past 30 years, this spatial shift in labor demand is increasingly being absorbed by foreign workers in developing countries. The North American Free Trade Agreement, for example has stimulated significant mature product manufacturing in Northern Mexico, where labor and land costs are minimized.

LIMITATIONS OF LOCATION THEORY

Much of the discussion of location theory has focused on the behavior of firms, and little attention has been given to the location of households and labor. From the household perspective, the spatial decisions revolve around a completely different set of factors than for businesses. For example, young families with children may focus on the quality of the local school or area crime rates or perhaps neighborhoods with other young children. Retirees, on the other hand, look for a completely different bundle of characteristics in making their location decisions.

Firm location theory also tends to view the customer in a very simplifying light. Little attention is paid to the idea of multiple-purpose shopping, where the consumer makes one trip and buys several products. Retail and service business clustering builds on multiple-purpose shopping, but the ideas are loosely defined and often assumed away (Chapter 4).

A second problem with traditional location theory is that it assumes that firms are perfectly mobile and can move or relocate with little or no cost. Also implicitly assumed is the firm's ability to make marginal changes in the size of the firm's facility. Most often once a firm has made an initial location decision, it becomes somewhat tied to that location because of its investments in the physical property and the local labor market. Firms will find that it is easier to expand at an existing location than it is to pick up and move to a new facility.[7]

A third problem centers on the location process of new start-up businesses. Rather than full consideration of possible profit-maximizing sites, the choice for many new businesses really comes down to one factor: where the owner of the business lives at the time of the start-up. Many "irrational" business location decisions become perfectly understandable when the community practitioner discovers that the owner of the business lived in a certain community when the business was started.

The study of community economies is of necessity a study of spatially separated economic factors. Location theory, however, has tended to emphasize

the relocation decision of firms. We have only begun to study several other spatial decisions. Clearly, our discussion of classical location theory along with the behavioral and institutional approaches is abstract by design. The simple framework provided here provides a theoretical foundation to think about the more complex real world. Indeed, attempts to rigorously introduce the ideas of agglomeration economies by authors such as Krugman and Porter have only recently been woven into the location literature.

SUMMARY

Location theory as reviewed in this chapter includes the driving motivations of firms to maximize profit (through least cost and demand maximization). The least cost approach assumes demand is fixed and thus cost minimization leads to maximum profits. The demand maximization approach allows costs to vary among areas, but not *within* an area. The firm's demand is variable, and the market captured by the firm is influenced by the location of the firm's competitors. The profit maximization approach allows both total revenues and total costs to vary as the firm changes locations and output.

We also provided important behavioral and institutional approaches to firm location. The behavioral approach permits uncertainty, objective functions other than profit maximization, and nonrational choices. Uncertainty is a major force preventing profit-maximizing location. Decision-makers try to reduce the uncertainties and simplify irregularities in the economy so the location decision process can be comprehended. Generally, demand becomes a prime factor in the initial decision to move. At this point, costs at alternative sites are typically ignored. As the search process narrows, specific cost estimates are made for the remaining sites. This need not yield a profit-maximizing, least cost, or demand-maximizing site, but it invariably leads to a site that yields an adequate profit. This decision process reflects a mixture of objective and subjective assignments of weights to the various factors in the decision.

Firms tend to go through a two-step process in which they identify a general region and then a specific site. Often, the community has little if any influence in the first stage of the decision-making process, but it can influence the second stage. Several critical factors that come into play in the second stage vary by specific community, the type of business, and the type of location change contemplated (new start, branch, or relocation). Firms can be categorized as input oriented or market oriented or somewhat indifferent. Orientation means that certain characteristics become extremely important in the location decision and are prerequisites before the plant can exist profitably. Failure to meet prerequisites means the firm will not locate at a particular site. As the location decision of the firm moves from the general to the more-specific community level, the subjectivity of decision-making increases.

Some recent trends in location factors significantly affect communities. First is the declining importance of the linkage between an industry and its source of raw materials or markets and the cost of transportation of heavy and bulky goods. Of increased importance is high-speed flexible transportation of higher-valued goods and the transmission of information and intangible services. Second is the access to markets, which for most industries has increased in importance relative to their access to raw materials. Third is the access to energy, which is increasingly important and, in some cases, the most important criteria (e.g., the availability of natural gas). Finally, there is an increasing dependence on services supplied by other industries, by divisions of the same firm, or by institutions and public bodies.

The new conception of industrial clusters and how the interrelationship among businesses and industries is spatially linked is changing how we think about location theory. It is important for communities to recognize these spatial linkages in their economic development strategies. While location theory fails to provide the comprehensive explanation of how economic units interact in space, it does provide sufficient insight to assist communities in their economic development efforts.

It also is important to note that much of the microeconomics of firm location has tended to focus on manufacturing firms. As economic structure makes a transition from goods- to a service-producing orientation, community practitioners need to broaden their thinking beyond the traditional Webber problem of cost minimization to the more general approach of profit and utility maximization. While the technology important in the old goods-producing era focused on manufacturing and transportation technology, the technology that becomes dominant in a service-oriented economy tends to focus on technologies of communication and the flow of ideas. In the modern service-oriented era, the remaining goods-producing sectors tend to be much more reliant on communication and just-in-

time inventory policies. Firms will tend to locate within a day's shipping time to backward-linked input suppliers, and there will be strong communication links between the input suppliers and the output producers.

Economic space is a dynamic phenomenon, particularly with respect to its implications for location. Consider the dramatic change that has been experienced during the past 50 years in transportation and telecommunication technology and its impact on firm location. As we continue to shift economic structure from goods-producing activity to service-sector activity, it is important to realize there will be an important shift in economic location. The service sector, particularly that which is professionally oriented, is less tied to traditional place-based inputs, and tends to be more footloose in where it locates. This has the potential to truly transform 21st century community economic activity; it will represent a fundamental reshaping of location theory as economic space replaces geographic space as a driver of individual and firm location decisions.

Amenities such as climate, housing, community facilities, and cultural and recreational opportunities (Chapter 9) are influencing decisions to a much greater degree than they once did. As economic space replaces geographic space, the critical decision point for footloose individuals and firms now focuses on which location maximizes quality of life, leisure, and the presence of amenities. The transforming implications this has for regions endowed with high-quality amenities need to be emphasized.

Virtually every economic decision has a spatial component. Sometimes this is as simple as here or there. But, more importantly, most location theory has focused on the relocation of firms. Yet, for location theory to be relevant in the 21st century and in community economics, it must include other forms of economic decisions.

STUDY QUESTIONS

1. Several different ways of thinking about location theory are discussed in the chapter. What are they and generally what do they emphasize in location decisions? Are the approaches linked, and if so, how?
2. The least cost approach seeks to minimize total costs. Total costs consist of three types of costs. What are they? How do they relate to each other and how are they incorporated in the location decision?

3. Uncertainty, satisficing, and information are important elements of the behavioral approach to location decisions. What is their significance? How are they handled by the profit maximization, institutional, and behavioral approaches?
4. What is involved in determining the comparative advantage of a region with respect to the types of industrial structure found active in the region?
5. How do localization and urbanization economies differ? What specific aspects of profit maximization relate to localization and urbanization economies?
6. How does a pecuniary externality help explain firm location?
7. Describe the various elements of Porter's diamond of regional competitiveness.
8. Do different types of location decisions (start-ups, expansions, etc.) rely on different location factors? If so, how do they differ?
9. A community can influence some location decisions. Which ones, and how might the community exert that influence?
10. Describe clusters and how they affect community economic development.

NOTES

1. By construction, the average costs curves used here include returns to land, labor, and capital. In the business accounting world, returns to land and capital are sometimes called *normal profit*. *Economic profit* exists when long-run average revenue exceeds long-run average costs. Economic and normal profits are distinct and separate ideas. In a competitive world, normal profits are positive, but economic profits are zero. Economic profits are sometimes referred to as *excessive profit*.
2. Decision-makers have a tendency to repeat prior successful decisions. For communities, this means that businesses continue to use the same production technique or location rather than trying new ones (Lloyd and Dicken 1977; Rees 1974). Townroe (1974) described three management conditions associated with location decisions: (1) lack of experience with the type of decision and no precedents to follow; (2) ignorance of all the relevant location possibilities; and (3) uncertainty about what decision criteria to use. Location decisions are made in a dynamic environment

in which the firm and community affect each other. This learning and adaptation process is continuous. The adaptation may arise as altered production scheduling and processes, reduced sales and profits, or it may be relocation or closure. Townroe (1974) argued that the adjustment/adaptation process occurs because of unforeseen circumstances at the time of the initial decision, inadequate information was sought and used, critical factors for operation were not fully accounted for, and, finally, poor judgment was used initially.

3. This situation points out the asymmetry of information in the location decision process. The firm knows whether or not a subsidy is truly necessary to make the site either profitable or the site of first choice. The community does not know this information, thus there is a constant danger that it will oversubsidize the already profitable decision.
4. Although the term *regional* is used, it is important to remember that *community* could also be used. *Regional* is used because we are generally referring to area-wide comparative advantage.
5. These primary factor resources are dealt with more completely in the second section of this text. As the astute reader will note, land, labor, and capital are overly limiting in explaining regional comparative advantage. Later in the book we add to this list more latent primary factors of production that are increasingly important in explaining the locational decisions of firms and households. These latent factors include technology (Chapter 8), amenities (Chapter 9), and publicly provided goods and services (Chapter 10).
6. Clusters defy being classified according to some NAICS or SIC code. They are broader than a single NAICS category.
7. Typically, a firm seeks to change its location because it is experiencing some form of stress. This stress can be that the market has geographically moved, the source of input factors has geographically moved, or the firm faces a surge in demand for its output. A firm will generally seek many alternatives before changing location. These alternatives include expanding production by adding a second and third shift, expanding facilities on the present site, or partnering with another firm.

4
Concepts of Community Markets

Community economies are dynamic entities that do not operate in isolation. They represent inextricably linked components that relate internal markets to the outside world. The vibrancy of a local community's economy can be thought of in terms of how effective its internal and external linkages are. This set of linkages provides the focus for this chapter and reflects the important aspects of the *markets* node of the Shaffer Star (Fig. 1.1).

We can think about community markets in two fashions. First, we can think of them in terms of *goods and services* markets and *factor of production* markets. Second, we can also think about internal and external markets. These latter markets can be goods and services or factors of production. In this chapter, we focus on the *goods and services* markets and defer the discussion of the *factors of production* markets to later chapters.

Market-oriented community economic development theories study the forces that affect the demand for the goods and services the community produces and how that is translated into community income or employment. If a community has a comparative cost advantage in the production and distribution (Chapter 3) of a good or service demanded in the external and/or internal market, then the community will attract the capital and labor necessary to produce the good or service. Over the long run, competitive market forces create an optimal spatial distribution of economic activity by selecting those production sites most profitable for the market served.

Local unemployment, low income, and slow growth represent short-run symptoms of a decline or shift in demand for the community's output. The market responds to this decline in demand by shifting capital and labor to more-productive uses within the community or in other communities. The presence of underutilized factors of production, such as unemployed labor, attracts other lines of production

to the community. Persistent symptoms of local economic distress may indicate that the national economy may be improved if underutilized local factors of production relocate out of the community.

External markets are markets outside of the spatial boundaries of the community, while internal markets are markets inside the spatial boundaries of the community, which is that geographic area from which it draws the majority of its retail and service trade. Each of these represents a completely different theoretical perspective on how the economics of the community work. The external market is modeled by the export base theory. The internal market is modeled by central place theory and local market analysis.

We begin the chapter with the circular flow model of the local economy and then move on to export base theory, where we focus on how export or external markets are related to internal markets. Next, we narrow our discussion to internal markets and build on the ideas of the demand-maximizing firm outlined in Chapter 3. We do this because community economic development policy often overlooks the one market on which the community can often have the greatest impact: internal markets.

CIRCULAR FLOW

A useful organizational framework within which to better understand these economic linkages is the circular flow concept. This conceptual approach, which outlines how a community economy operates, is presented in its regionalized form in Figure 4.1. The general circular flow is made up of the basic actors involved in economic transactions and the markets within which supply and demand interact to discover prices and quantities. The circular flow diagram is linked to the outside world by imports/exports of goods and inflows/outflows of factor resources. Because we respect inflows and outflows of goods

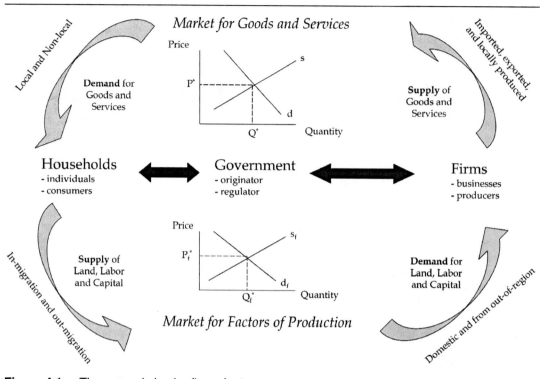

Figure 4.1.   The general circular flow of an open economy.

and factors, this is a representation of an open economy, not a closed economy. This typically means that we are talking in terms of a regional economy rather than just an isolated community.

The basic factors involved in community economics revolve around consumers (households) and business interests (firms). *Households* are individuals and/or family units that own factor resources employed to generate income, which in turn is spent in consumption of goods and services. *Firms* are local entities engaged in producing goods and services by using primary factors of production. *Governments* regulate, create legal processes, and produce public goods that affect how households and firms interact (see Chapters 10 and 11).

Households and firms interact in two distinct markets that track how we produce (factor markets) and consume (goods and services markets). At the top of Figure 4.1, we present the market for goods and services in which households are consumers (demanders) and firms are producers (suppliers). Examples of this market are easy to find in communities: local merchants, manufacturers, or commercial interests selling goods and services to a customer base. At the bottom of Figure 4.1 is the market for factors of production.

These primary factors of production are traditionally viewed as land, labor, and capital (Chapters 5, 6, and 7.) Firms demand these factors of production that are, in their most basic form, owned by (or supplied by) households.

Regional issues that link the community to the outside world are captured in both markets and include both households and firms. In the market for goods and services, households can make conscious decisions about consuming goods and services from local sources or purchasing them from the outside. This demand provides the impetus for imports of consumable goods into the region. Firms, on the other hand, can decide to sell their products locally or export them to the outside world.

In the factor market, households and firms are likewise linked to the outside world. Households can decide to sell their labor resources locally or commute to other regions to sell their time. Certainly, households often choose to employ their capital resources in stock markets that are often far distant from their local communities. Likewise, firms can decide to employ factors that are locally available or import them from the outside. Thus, in the factor market, inflows and outflows of primary factor resources are important determinants of regional linkage.

In our oversimplification of the world, the physical circular flow is represented by the arrows in Figure 4.1. Consider the linkage between households and the factor markets. Households own land, labor, and capital and offer it for sale (supply) in the factor markets. Firms demand these factors of production and pay for them in the forms of rent, wages, and interest. These payments represent income to households that, in turn, is used for consumption (and savings) in the goods and services market.

EXTERNAL MARKETS: EXPORT BASE THEORY

As in the circular flow model, there are internal as well as external markets in which the community functions. Community economic development policy has historically focused on only one small part of the circular flow representation, specifically the external market for goods and services. This focus has followed from a simple model of economic growth known as *export base theory* or *economic base theory*. In our broader view of the local economy in the circular flow model, narrowly focusing on exports limits the options available to the community. Export base theory, however, does provide a useful means of thinking about the community's economy and how it changes.

Export base theory argues that the community's economy can be divided into two sectors (Andrews 1970a, 1970b, 1970c; Blumenfeld 1955; North 1955, 1956; Tiebout 1956a, 1956b). The first sector is the *export, or basic, sector.* The export sector consists of that portion of the community's goods and services market that trades with other areas. The export sector brings dollars into the community because someone outside the community purchases goods and services produced in the community. The second sector, termed the *nonexport, nonbasic, or residentiary sector,* sells its product within the boundaries of the community (internal markets) and exists to support the export sector. Equation (4.1) displays the division of the total local economy into its basic and nonbasic sectors. It is a fairly simple theory claiming that the local economy can be divided into two parts: basic (or export) and nonbasic (or nonexport).

$$\text{Total} = \text{Basic} + \text{Nonbasic}$$

or
$$E_T = E_B + E_{NB} \tag{4.1}$$

The nonexport base (E_{NB}) component of the local economy (E_T) is generally larger than the export sector (E_B) but is dependent on the export sector. Thus, the export base model argues that any change in the export base leads to some multiple of the change in the total local economy. Specifically, the change in the export sector has a feedback or multiplier impact on the nonbasic sector.

This multiplier effect is represented graphically in Figure 4.2 (see below and Chapter 15 for a fuller discussion of multipliers). The basic sector (E_B) shown on the horizontal axis drives the nonbasic

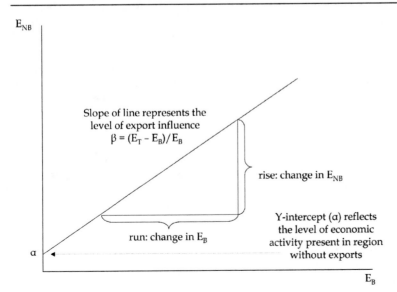

Figure 4.2. A graphical approach to understanding export base theory.

E_{NB}

Slope of line represents the level of export influence
$\beta = (E_T - E_B)/E_B$

rise: change in E_{NB}

run: change in E_B

Y-intercept (α) reflects the level of economic activity present in region without exports

α

E_B

sector (E_{NB}) on the vertical axis. Any change in the basic sector has a two-part impact on the total economy. First is the initial change in the basic sector itself; the second is the impact on the supporting nonbasic sector. This second impact is some multiple of the change in the basic sector. The slope of the curve in Figure 4.2 maps out that multiplier relationship. A relatively flat multiplier relationship (small slope) suggests a weak relationship between the basic and nonbasic sectors of the local economy. A steep line (large slope) implies a strong relationship between the basic and nonbasic sectors.

Export base theory contends the development of a community depends on the vigor of its export or basic industries. The critical force in the community's economic development is external demand, not the community's ability to supply capital and labor or use technology. The timing and pace of the community's economic development depend on the success of its export sector, the characteristics of the export sector, and the disposition of income received from export sales (North 1961).

The export sector carries external economic forces into the community. The characteristics of the export sector and the disposition of its income translate the external forces into community economic development. As we have seen, how changes in the export sector affect the rest of the community depends on the number and strength of linkages between the export and nonexport sectors. Furthermore, the distribution of income from the export sector and the ownership of resources used in the export sector assist in translating export sector changes into community economic development. For example, if the ownership of export base resources is external to the community, then changes in the export sector have a minimal impact on the community because the income is not recirculated or reinvested in the community. Finally, the availability of skills that permit the local labor force to work in the export sector also contribute to the success of translating external demand into local economic change.

The real criterion of the export function is whether the activity brings income into the community. Exports occur when an economic transaction occurs across the community's economic boundaries. An export transaction occurs either as the movement of a good or service to the consumer or purchaser or as the movement of the consumer or purchaser to the good or service (Andrews 1970b, 1970c). The movement of goods to the nonlocal consumer or purchaser is the most commonly perceived type of export transaction. Examples would include the shipment of agricultural, mining, and manufacturing goods to consumers located outside the community. The movement of the consumer to the good or service characterizes the second major type of export transaction. The movement of the consumer to services occurs when the nonresident consumer enters the community and consumes the service within the community. Some examples are recreation/tourism, regional medical services, and university education. In other words, the locally produced good or service can be shipped to the consumer or the consumer can travel into the community to make the purchase.

Historically, export base theory has focused on the top half of our circular flow model, the goods and services market. But the factor resources market can also be a source of exports. The owners of the factors of production are not limited to local markets. In the labor market, households could elect to commute to neighboring markets for employment and transport their wages back into the community. Capital can be invested in any number of markets outside the community; the return on that investment represents an inflow of income into the community. Likewise, the ownership of nonlocal land yields returns to local residents. For example, a retired farmer who moves to a warmer climate but continues to own and rent the farmland represents a source of new monies being injected into her new residence.

The volume of exports from a community can increase or decrease over time. An increase in export demand is caused when the external demand curve shifts right, income levels of nearby areas increase, comparative advantage in the community improves, or a community's factor endowment increases. An example of the shift in external demand for some communities is the increased demand for microprocessors in the United States during the 1990s. A service with a high-income elasticity is the hospitality and recreation industry. The income levels of nearby areas are important because most communities sell more to nearby communities than to distant ones. The improvement of a community's comparative advantage increases its exports because it permits lowering prices more than other communities can. Changes in technology or prices affect the profitability of using the community's factor endowment for exports, for example, the profitable exploitation of tar sands in northern Alberta, Canada, as crude oil prices increase and new technology becomes available. A change in tastes and preferences that increases the demand for specific products favorably impacts some communities because they have the resources, or factor endowment, permitting production of that good or service (see Chapter 9 on amenities for an example of this).

A community could also experience a decline in the volume of its exports over time. This could happen through depletion of its natural resource base or through a relative decline in its comparative advantage because the costs of land, labor, and capital increase, or because technological changes alter input combinations for which this community previously had an advantage. Or a leftward shift in external demand could be caused by a change in consumers' tastes and preferences. If a community does not adjust to the forces decreasing its volume of exports, the community will find itself stranded outside the economic mainstream with a relatively or even absolutely worsening economic position.

Just as communities experience increases or declines in their existing export base, they also experience the creation of new export sectors. The same factors causing an increase in exports can cause new export activity. Two additional forces of particular significance are governmental investments and new technology. Governmental investments in social overhead capital, such as water and sewer systems, industrial parks, and transportation systems, can eventually lead to new export businesses. New technology has a differential effect among communities. Some communities gain a competitive advantage through the early adoption of new technology. An example is the growth of the microprocessor industry and communities capable of supporting those businesses.

While widely used as a foundation for community economic development policy, export base theory requires several assumptions that frequently are not explicitly recognized (Pfister 1976; Richardson 1973; Tiebout 1956a). The simple linear relationship between basic and nonbasic sectors outlined in Figure 4.2 is built on a very strong set of assumptions:

1. Income and employment changes in a community depend totally upon changes in the level of exports, with no other stimulus for local change.
2. The marginal propensity to consume locally, specifically the amount of local income spent for local products, is stable over time and over a relatively wide range of income change.
3. The amount of local income generated by each dollar of local spending does not change, thus the local labor content does not vary over time.
4. There are no changes in the relative prices of capital or labor as their use increases or decreases (i.e., no shift from labor to capital or vice versa in response to changes in export demand).
5. The additional capital and labor required to expand production is available immediately and without any increase in wages or profits since the community has a perfectly elastic supply of capital and labor to meet increases in demand.
6. The economic structure of a community at one time will predict its future economic structure.
7. The homogeneous export sector implies that earnings from jobs and backward linkages, among other factors in separate subsectors of the export sector, are roughly equivalent.
8. None of the local consumption of the goods and services sold for export comes from importing those goods and services (i.e., no *cross-hauling*).[1]

From a policy perspective, these restrictive assumptions limit how useful export base theory is, but even more important are its internal inconsistencies from a purely theoretical perspective. First, all communities are exporting and none are importing. Who is purchasing the exports? Second, in the extreme, according to export base theory, it is impossible for the world economy to grow because there are no export markets. As in most of economics, extremely simple theories or ways of thinking about a problem often are partial and, when taken to a logical conclusion, lead to contradictory results.

Limitations

From the perspective of the community development economist, there are several major weaknesses in export base theory (Blumenfeld 1955; North 1955; Richardson 1969b; Tiebout 1956a, 1956b). First and foremost, export base theory is not a general theory of community economic development; it is more appropriate for smaller communities, simpler economies, and the short run than for larger communities, complex economies, and the long run.

A theory that argues a simple exogenous shift in export demand as the source of economic development borders on naivete; several other forces affect community economic development. The theory does not explain changes in the marginal propensity to consume locally as income changes. The export sector is not homogeneous, and changes in different parts of the export sector have dissimilar multiplier effects within the local economy. For example, think of the different linkages agriculture has with the local economy compared with manufacturing or the recreation industry. The changes in the export sector depend on nonexport or supporting economic activities that help create the comparative cost advantage for the export sector. For example, a change in export activities partially depends on the efficiency of the transportation system. In addition, export base theory is aspatial and the role of market size and location is ignored. Nor does export base theory handle very well how changes in markets can alter the comparative advantages of the community. For example, a community endowed with forest resources may initially export lumber but changing markets suggest that a better use of the resource would be tourism. Export base theory lends no insights into how or what markets might change.

Another shortcoming of the export base approach is its failure to explain community economic development occurring despite a decline in a community's exports. A community experiencing a decline in its exports could still grow because nonexport businesses grow enough to offset the decline. Improving the community's terms of trade, such as the community getting a better deal for the goods it sells, means the community sells fewer goods yet still maintains or improves its total income.

Export base theory is only a partial theory. It fails to provide any insight into why new technologies are adopted or why institutions change over time. It does not explain how communities change because of changing comparative advantages, nor does it provide any insight into how communities are linked through both exports and imports. So for a system of communities, export base theory fails to explain the initial export changes, which are conveniently and justifiably assumed exogenous for the individual community.

The export base approach to community economic development does not account for structural changes in a community. It argues that a community's economy will forever depend on its export sector. As communities grow, they become less

dependent on exports and some nonbasic sectors become self-sufficient and maintain themselves without heavy reliance on export activities, which the theory does not explain.

While export base theory argues that nonbasic activities are a passive component of the community's economy, they need not be. In many cases, such activities such as downtown revitalization programs promote change without a change in the basic sector, which is in direct violation of the fundamental premise of export base theory. The source of growth can flow from the nonexport to the export sector, also in direct contradiction to export base theory. Some examples of this type of change follow:

1. The nonbasic sector becomes self-sustaining and is no longer dependent on the export sector because the local market has become large enough to support the business and there is sufficient trade among nonexport businesses and/or households (i.e., import substitution).

2. Local government investments change the amount and the character of the social overhead capital base of the community and stimulate new export activities.

3. Non-market-stimulated in-migration (e.g., stimulated by amenity resources and retirement) causes residentiary activities to increase without an increase in exports.

4. Improved efficiency in the nonbasic sector induces growth in the basic sector by making the export sector even more competitive relative to other areas.

5. The nonbasic sector experiences a higher rate of technological change than the basic sector. Here a shift of resources into the nonbasic sector increases community growth more than if the resources continue to be allocated to the export sector. Unlike example 4, the basic sector need not grow in this situation.

In summary, export base theory has intuitive appeal and simplicity. It also has a relatively sound theoretical foundation based on the concept that some local economic sectors transmit external economic forces into the community to stimulate further change. Changes in community income depend on changes in export demand. Exports increase because of a rightward shift in demand or an improved competitive position of the community, while exports decline because of a leftward shift in demand or a loss of competitive position. New export sectors appear in the community because of changes in tastes and preferences and in technology. Export base theory is more appropriate in smaller economic areas that are relatively more dependent on external trade. Likewise, it is more appropriate in simpler, less diverse economies. The importance of the export sector to the community declines with increases in the diversity and completeness of the local economy (i.e., self-sufficiency).

Export Base Multipliers

Historically, there are two approaches to economic accounting, or the means of organizing and reporting economic data: the Keynesian and Leontief approaches. It can be shown that the approaches represent two sides of the same coin. Both can be used to develop models of the economy and estimates of economic linkage and interrelationships. (We use the traditional Keynesian approach to income determination in this chapter and defer discussion of the Leontief approach to Chapter 15.[2])

In community economics, these linkages often are thought of through the multiplier effects. The multiplier measures the spending and respending of an exogenous injection of income (export income) that results in a total change in community income exceeding the original change. Rather than trace this spending and respending process for each change in exogenous income, multipliers provide a short cut.

Income (Y) is the sum of consumption, exports, imports, investment, and government expenditures:[3]

$$Y = C + X - M + I + G \qquad (4.2)$$

Total consumption (C) is driven by total income:

$$C = a + cY \qquad (4.3)$$

where a (sometimes called a subsistence level of consumption) is some fixed level of consumption and c is the marginal propensity to consume.

Local consumption (C_L) is the difference between total consumption and imports:

$$C_L = C - M \qquad (4.4)$$

Savings (S) and *investments* (I) are usually defined as being equal,

$$I = S \qquad (4.5)$$

as are *governmental expenditures* (G) and *taxes* (Tx):

$$G = Tx \qquad (4.6)$$

Exports (X = gross exports) are assumed to be exogenous:

$$X = \overline{X} \qquad (4.7)$$

Imports are assumed to be a function of local income:

$$M = b + mY \qquad (4.8)$$

where M = gross imports, b (sometimes called subsistence level of imports) is some fixed level of imports, and m is the marginal propensity to import.

In equation (4.2), exports, investments (which equal savings), and government expenditures (which equal taxes) are exogenously determined in the short run. The specification of equations (4.5) and (4.6) explicitly assumes that the income level of the community is a function of local consumption and exports. How changes in exports and local consumption affect the local economy can be seen by substituting equations (4.3) and (4.8) into equation (4.2), performing the algebra, and solving for income (Y):

$$Y = \frac{a - b + X + I + G}{1 - (c - m)} \qquad (4.9)$$

Differentiating equation (4.9) with respect to exports (X) shows how a change in exports influences income. Specifically,

$$\frac{dY}{dX} = \frac{1}{1 - (c - m)} \qquad (4.10)$$

where $c - m$ can be interpreted as the marginal propensity to consume locally. Thus, the change in local income, in response to a change in exports, depends on local consumption. Alternatively, a similar computation could be made for exogenous changes in government spending and/or investments.

Equation (4.10) and the export base multiplier outlined in Figure 4.2 (β) are identical since the ratio of nonbasic income to total income is a proxy for the propensity to consume locally (Pleeter 1980). Given from equation (4.1), we have

$$\frac{dY}{dX} = \frac{1}{1 - \left(\dfrac{Y_{nonbasic}}{Y_{total}}\right)}$$

$$= \frac{1}{\left(\dfrac{Y_{basic}}{Y_{total}}\right)} \qquad (4.11)$$

$$= \frac{Y_{total}}{Y_{basic}}$$

$$= k$$

and k is the export base multiplier. Equation (4.11) indicates that the multiplier for the economy is the ratio of total to basic income. Equation (4.11) clearly shows why accurate identification of the basic sector is so important. If the basic sector is estimated as smaller than it really is, then the multiplier is biased upward. Since many community economic development practitioners examine the export base only to get an estimate of the multiplier for planning purposes, it is very important that an accurate estimate for the basic sector is created. Several methods have been offered for bifurcating total economic activity into its basic and nonbasic portions; we review them in Chapter 14.

Equation (4.10) suggests that local consumption is the crucial determinant in income change, but every dollar of local consumption does not yield a dollar of local income (Tiebout 1962). Some local consumption dollars are siphoned out of the community to pay for such things as imported inputs, nonlocal taxes, or returns on local investments by nonlocal investors. In our open economy model, there are leakages.

Consumption, investment, or government expenditures generate local income in two steps. First, consumption, investment, or government expenditures occur locally in the community. Second, these expenditures are converted into local income. A straightforward statement of income change is achieved by substituting into equation (4.2):

$$Y_T = X' + C_L' + I' + G' \qquad (4.12)$$

where ' indicates income from the type of expenditure.[4] Specifically, X' is the level of community income generated by export activity, C_L' is the level of income generated by local consumption, and so on. Equation (4.12) can be simplified into

$$Y_T = Y_B + Y_{NB} \qquad (4.13)$$

where Y_B is the change in basic sector income ($Y_B = X' + I' + G'$) and Y_{NB} is the change in nonbasic sector income ($Y_{NB} = C_L'$).

The change in nonbasic income comes from the change in local consumption and its conversion into income:

$$Y_{NB} = C_L' = Y_T(MPC_L)(PSY) \qquad (4.14)$$

where MPC_L is the proportion of income change spent locally (marginal propensity to consume locally) and PSY is the proportion of local consumption expenditures that becomes local income. Recognition of the income from expenditures process leads to refining the earlier multiplier formulation, equation (4.10), into

$$k = \frac{1}{1 - MPC_L * PSY} \qquad (4.15)$$

The power of the Keynesian approach to thinking about economic linkages is its flexibility in defining linkages and capturing them via the multiplier. The export base multiplier is a function of local spending and its conversion into local income. More importantly, this discussion demonstrates the importance of local consumption, local government spending, and local investment to the conceptualization of economic linkages and the simple economic base multiplier. Furthermore, the importance of converting local spending into local income is recognized. Finally, the simple ratio of total to basic activity as an estimate of the multiplier highlights the importance of accurate estimates of the basic sector. The Keynesian approach is but one way to think of economic linkages and multipliers; alternatives are discussed in detail in Chapters 15 and 16.

INTERNAL MARKETS: CENTRAL PLACE THEORY

Export base theory argues that the nonbasic sector is totally dependent on the export base, but in reality, such is not the case. The provision of local goods and services by local firms greatly affects the well-being of community residents. Two dimensions of supplying these goods and services are of particular interest for community economic analysis. First is the interdependence among merchants across different communities. As we saw in the demand-maximizing approach to firm location in Chapter 3, firms are dependent on where their competitors locate. Second is the firm's interpretation of market demand and the socioeconomic characteristics that dictate market demand and how that interpretation influences decisions to provide particular goods and services.

Central place theory (CPT) is one conceptual framework that addresses these two facets of economic activity. CPT does not provide all the detail needed to determine the feasibility of a particular investment in a particular market, but the theory does help to collect the necessary details needed for community analysis. Central place theory offers insight into why specific goods and services are or are not present in a particular community. This theory specifically recognizes that no community's trade, or nonbasic, sector can be viewed in isolation. It is important to remember that central place theory essentially argues there is a hierarchy of communities based on the functions (retail and service) that are provided in the community. In a sense, central place theory attempts to place the local goods and services markets of our circular flow model onto an economic plane.

Central place theory takes our demand maximization problem of Chapter 3 to the next step in defining a spatial system of markets or communities (Mulligan 1984; Parr 2002; Potter 1982). In Chapter 3 we outlined how an individual firm can increase sales through a lower effective price that generates more demand per person or through an increase in the geographic area included in the firm's market. A circular market surrounds each firm (Fig. 3.5). Transportation costs, consumers' willingness to pay the effective price, and competition from other locales determine the limits of each market. This system of circular markets leaves some areas unserved. Perfect competition allows competitors to enter the market, serve presently unserved areas, and compete away the excess profits of current firms. Thus, entry of competitors creates a series of regular, hexagonal market areas (Fig. 4.3).

Two critical behavioral assumptions of central place theory are that businesses will attempt to maximize the area served, and that consumers will attempt to minimize the distance traveled. What we begin to see in Figure 4.3 is how firms will allocate themselves on a homogeneous economic plane in a way that maximizes profits by maximizing demand. We also begin to see the economic foundation for a system of places on the economic plan. As we introduce varying cost structures of different types of firms below, we start to develop a hierarchy of urban places. Two concepts of particular importance to

this hierarchy that play a fundamental role in community economics are the range and the demand threshold of a good or service.

Range of a Good or Service

The *range* of a good or service is the maximum distance people will travel to purchase that good or service at a particular location (Berry and Garrison 1958b; Olsson 1966; Parr and Denike 1970). The range is the outer limit of the geographic market for a good or service from a particular location.[5] In the Lösch demand cone (Fig. 3.5), the radius of the demand cone is the range of the good or service. The circumference of the demand cone defines a trade area, which is the geographically delineated area containing potential customers to purchase goods and services offered for sale by a particular firm (Davies 1977; Huff 1964).

The distance determining the geographic limit of a market is measured in terms of physical separation and travel costs, including time. Other determinants of the outer geographic limit are the ease of access to competitive markets, transportation facilities, and technology (Shepard and Thomas 1980). Ease of access to competitive markets reduces the geographic limits of the market area for a community. Physical features such as mountains and rivers influence access. Transportation facilities and technology are important because better facilities and faster movement permit greater spatial movement by consumers.[6]

Individual characteristics influence the distance people are willing to travel (Shepard and Thomas 1980). Younger, more educated, higher-income people are likely to travel farther and more frequently than those with contrasting characteristics. The distribution of income affects the number of people able to pay transportation costs of greater movement. Movement imagery involves the consumer's perceived options about movement from one place to another in the quest of a desired good or service. The mode of travel, travel time, the cost of overcoming distance, and communication flows affect movement imagery.

Behavior space is that part of the total central place system which the individual perceives as a potential source for satisfying his or her demand for goods or services. This behavior space is influenced by previous shopping experiences in the various central places and by sources of information, such as advertising. For any given consumer, the behavior space can include several communities.

Stage 1: Single firm's circular market area

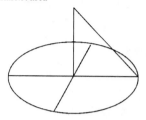

Stage 2: Multiple single firm's circular market areas and some unserved areas

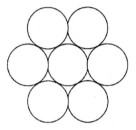

Stage 3: Entry of competing firms forcing smaller market areas

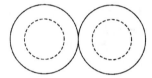

Stage 4: Multiple firms forming hexagonal market areas covering entire market surface

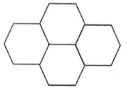

Figure 4.3.   Formation of
hexagonal market areas.

Demand Threshold

Demand threshold is the minimum market required to support a particular good or service and still yield a normal profit for the merchant (Berry and Garrison 1958a, 1958b; Deller and Harris 1993; Henderson, Kelly, and Taylor 2000; Olsson 1966; Parr and Denike 1970; Shonkwiler and Harris 1996; Wensley and Stabler 1998). The concept of demand threshold, based on the internal economies of the firm and the characteristics of consumer demand, is defined where average cost is just equal to or tangent to average revenue (Fig. 3.6). Because of this, demand thresholds are not absolute; they vary with the type of good or service. Demand thresholds usually are measured in terms of population, rather than quantity sold, by assuming consumers are homogeneous in their buying power (income) and tastes.

The internal cost economies of the firm determine the thresholds that will yield a competitive equilibrium (Parr and Denike 1970; Olsson 1966). Suppose we have two firms that offer different goods and services, hence they have two separate cost structures (AC_1 and AC_2 in Fig. 4.4). Clearly, in a competitive market these two firms will require different demand structures to yield a competitive equilibrium. The demand thresholds, or the population required to support the two separate businesses, will be different, with Firm 2 requiring the larger market. If consumers are evenly distributed across our economic plane, the larger market required for Firm 2 will be a larger geographic area, or larger range.

Consider, for example, two establishments: a tavern and a high-end furniture store. The tavern has only a handful of costs associated with its operation: the rent of the building or space, the cost of goods sold, labor, and miscellaneous costs, such as utility and insurance costs. A high-end furniture store, on the other hand, has a much higher cost of goods sold and requires a larger building, higher level of labor-related services, such as delivery and design services, and advertising. In Figure 4.4, the tavern may have a cost structure similar to AC_1, while the furniture store may have a cost structure similar to AC_2. Clearly, the furniture store will require a much larger market area or demand structure (AR_2) to support its operations than will a tavern (AR_1). In other words, the range and threshold of the furniture store will be significantly greater than the tavern. Given different ranges and thresholds for different types of firms, we begin to see a system of overlapping markets. These overlapping markets form the basis of a hierarchy of central places.

Central Place Hierarchy

Recognizing that central places and their tributary areas cover the entire market surface of our economic plane, the elements of a system of central places emerge (Berry and Garrison 1958a, 1958b). The theory provides a framework that helps explain and predict where heterogeneous firms will locate and cluster together.[7] The classic presentation of Christaller's (1933) system of central places shows places (cities) arranged in tiers with each member of the tier, except the highest tier, subordinate to at least one other central place on a higher tier (e.g., hierarchy) (Fig. 4.5). The lower levels of the hierarchy are made up of more central places serving smaller tributary areas. Moving up the hierarchy, there are fewer central places, but each serves an increasingly larger geographic area and population than the communities (central places) at lower levels. In the real world there are numerous small hamlets or villages (lower-level places) but only a small handful of large cities (higher-level places).

A mutual dependence exists between the central place and its tributary or complementary area. The central place provides higher-order goods and services and the tributary area provides a market. The tributary area contains lower-order central places for all but the lowest-order center, which contains just farms or scattered site residences. For example, the market area of a hamlet or village is relatively small,

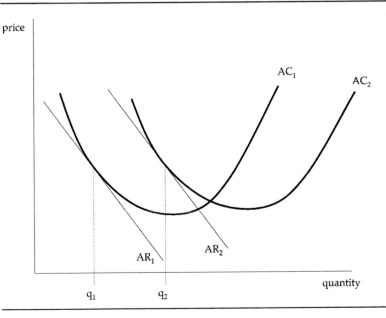

price

AC_1

AC_2

AR_1

AR_2

quantity

q_1 q_2

Figure 4.4. Cost structure determinants of demand thresholds.

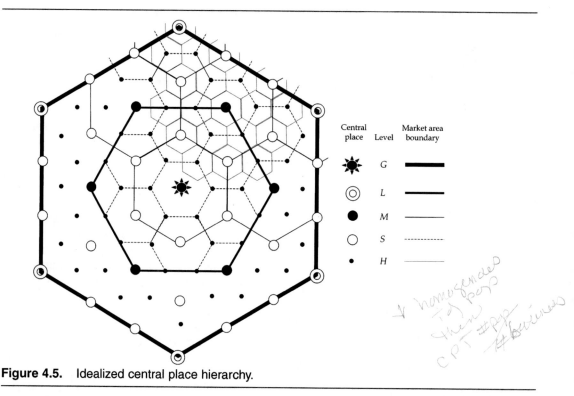

Figure 4.5.   Idealized central place hierarchy.

but the market area for a large city is much larger and includes the small hamlets.

The tributary areas for different central places in a central place system become a collage of overlapping boundaries (Fig. 4.5). The larger cities provide more specialized activities to a larger tributary area. At higher levels of the hierarchy, the number of central functions increases and they become increasingly specialized (Fig. 4.6).[8] These cities also provide the same general services found in the smaller cities. The smaller places offer more generalized goods and services to geographically restricted trade areas. Each level of the central place hierarchy has its own system of hexagonal markets. A result of the demand-maximizing problem from Chapter 3 (in which consumers react to an effective price that includes travel costs, and firms maximize profits) is that the center of many market areas will occur at the same point regardless of the hierarchical level. This offering of several goods and services from one location prevents a totally random settlement pattern and increases the efficiency of providing each good or service.

Suppose that in Figure 4.5 we have five different types of firms that vary in terms of cost structure, with H-type firms having the lowest cost structure, such as a tavern, and G-type firms having the highest cost structure, such as a luxury car dealership. As we have seen, the population required to support H-type businesses will be much smaller than that for G-type businesses. Accordingly, on our homogeneous economic plane, we would expect to find a large number of places that have H-type businesses but only a few that have G-type businesses. Firms will enter the market and locate on the economic plane in such a way that all consumers are captured and profits are maximized. In this example, the spatial economy may be large enough to support only one luxury car dealership, which will locate in the center of the economic plane and draw its consumers from the entire plane. At the same time, the spatial economy can support numerous taverns evenly distributed across the economic plane. If we want to, we can relabel business types with place types: We can call G-type places cities and H-type places small hamlets or villages.

The number of different economic (central) functions available differentiates places. The number of central functions performed depends directly on the cost structure of businesses and the population of the spatial economy. Since costs vary among types of goods or services (i.e., central functions), some goods or services are available only in the higher-level places, while others are available in even the smallest places. There is a pattern of similarity

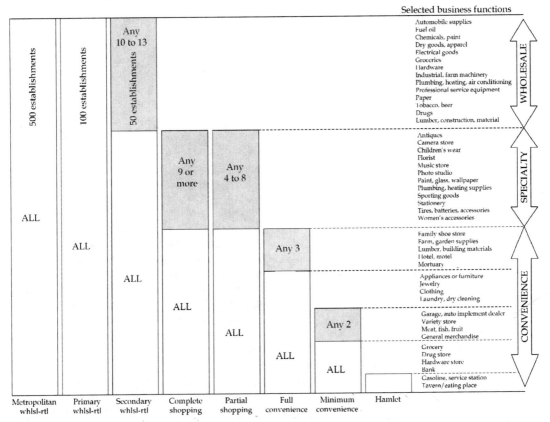

Figure 4.6. Central functions and central place hierarchy.

between market size within a given hierarchical level and the differences between hierarchical levels (Foust and de Souza 1978).

The regularity and flexibility of how places come to be described as central is outlined in Figure 4.6. Every hamlet has a gasoline service station and a tavern/eating place. Every minimum convenience place has all the central functions found in a hamlet plus a grocery, drug store, hardware store, and bank. In addition, a minimum convenience place should have any two or more of these central functions: garage/auto and implement dealer, a variety store, a meat/fish/fruit market, and a general merchandise store. Different authors use different labels for the various levels of the central place hierarchy, but the labels, per se, are less important than the recognition that different levels of the hierarchy provide a different mix of goods and services.

From the point of view of business, one reason for the hierarchy is conditions of entry. With free entry and the minimum market required (threshold) to sup-

port the business available, the business will exist.[9] It cannot exist without the minimum required market. Both minimum market and cost structure of the firm are linked to the number of customers or volume of sales in the community and available to the business (i.e., not already served by competitors).

A key assumption of central place theory (as summarized in Fig. 4.5) is an even distribution of homogeneous people across the economic plane. Clearly, this assumption does not hold in the real world; lifting the assumption greatly complicates our notion of central places, but it does not destroy it. People cluster together into cities and villages, fundamentally altering the shape of our Löschian demand cones. Likewise, external economies of size vary with each community's social and economic characteristics. This community influence is often expressed as external economies or economies of agglomeration that shift the firm's average cost curve. By allowing for something other than a homogeneous economic plane the rather sterile view of central places becomes much more reflective of the real world. The ideas of multipurpose shopping trips, business clustering such as shopping malls, and market segmentation such as entertainment districts as opposed to office buildings begin to make sense. The wealth of these ideas is discussed in a policy context below.

One immediate outcome of lifting the homogeneous economic plane assumption is the use of population as a proxy for differentiating cities within the hierarchy. The logic for substituting population for an actual tabulation of central functions is that the market required to achieve economies of size and make certain goods and services available is linked to population. Returning to Figure 4.5, we now can think of places in terms of population rather than number of businesses.

Consumers play a part in creating the hierarchy of central places as well. First, consumers wish to minimize the distance traveled to purchase any good or service. Therefore, frequently purchased items should be available nearby; less frequently purchased items need not be located nearby. Consumers' desires to minimize total travel means that they go to higher-order centers only for the goods and services not available in lower-order centers. Even people who live in higher-order centers will purchase lower-order goods and services in their neighborhood stores. Consumers make only single-purpose shopping trips to the higher-order centers.

Second, the type of shopping also influences the hierarchy of goods and services. Goods and services which are everyday convenience items (e.g., milk, bread) will be available in smaller central places, but if an item is subject to comparative shopping (furniture, automobiles), it is likely to be available only in larger centers.

Third, the type of transportation system available affects the hierarchy through the frequency of shopping and spatial dimension of the market area. Important elements of the transportation system include whether individual (car) or mass transit is the source of transportation. Consumers dependent on public transportation may find geographical as well as time-of-day limits placed on shopping. Both factors limit the market available to businesses in a central place.

Shifts in the Hierarchy of Central Places

One of the powers of central place theory is its ability to help the community practitioner think through how shocks, shifts, or changes in the economy will filter through the retail and service markets. As we have seen repeatedly, communities do not function in isolation from their surrounding communities. Any change in the economics of one community will affect not only its own place in the hierarchy, but all communities in the hierarchy.

The community practitioner needs to understand how shifts to one community can affect all communities in the system. For example, what happens to the system of central places if the population of one place significantly increases? Consider a case where a small hamlet has a large manufacturing firm locate which in turn draws a significant increase in population. Clearly, the hierarchy of central places will shift, but how? In the simplest sense, the hamlet's place in the hierarchical chain will move upward. The hamlet will attract more businesses because the higher population will satisfy the demand thresholds of more types of businesses. But equally important is the impact that change will have on other communities in the hierarchy. In essence, the spatial competition of firms and markets has changed. Neighboring hamlets will find a sharp increase in competition.

Another change or shift might be a decrease in transportation costs. Clearly, as transportation costs decline the effective price facing the consumer will decline and the consumer will be willing to travel greater distances to make purchases. Not only will a decrease in transportation costs expand the spatial market of one place, but of all places. Investments in transportation infrastructure is a two-way street: Not only do local businesses see an increase in the spatial size of their market, but also local customers can now more easily go to other markets.

We can also use central place theory to help think through the impact of changing household demographics on retail and service markets. With time becoming an increasingly valuable commodity, consumers will look to maximize the effectiveness of their time spent shopping. Consumers will look for places or communities where multiple purchases can be made. As we have seen in our discussion of central place theory, higher-order places will offer a greater variety of goods and services. With the increased pressure on the value of time, these higher-order places will be more attractive. The advent of shopping malls and big-box stores in higher-order places feeds into external economies of size or agglomeration, giving higher-order places an even greater comparative advantage over lower-order places. The prediction of the theory would be that the system of central places becomes more concentrated.

Using the concepts of the range and the threshold of a good or service, we can think through changes to the local community and the impact of those changes on the surrounding communities. Perhaps most important is the improved understanding of how local communities are interconnected in a predictable manner.

Limitations of Central Place Theory

While central place theory offers considerable insight into the spatial allocation of consumer functions, its limitations must be recognized (Parr 1973; Parr and Denike 1970; Shepard and Thomas 1980; Turner and Cole 1980).

In its earliest form, central place theory assumed the physical dimension of space (geographic space) was the dominant element. The consumer's perception of economic space was not considered. In this form, central place theory was limited because, in the real world, physical distance is irrelevant; consumers and producers use travel time and its associated cost. In other words, the real world is driven by economic space, *not* by geographic space.

Originally, Christaller assumed that the hexagonal markets of central places were on a homogeneous plane with a uniformly settled population. Later, Löschian adaptations featured economic distortions that affected the size and shape of these hexagonal markets. The transportation network (roads, railroads) was shown to have dramatic influence on the size and shape of markets. Also, an unequal distribution of natural resources and physical barriers, such as rivers and mountains, affects the shape of the market area. Economies of size and different population densities alter market area size but not shape. This means that new transportation technology which reduces travel time or costs or both and which alters consumers' perceptions of access to a given place will alter the market size because physical distance is unchanged.

Another limitation of central place theory is that it does not consider the qualitative variety among alternative locations (central places). Qualitative dimensions include product selection, parking availability, store hours, clerks' attitudes, community quality of life, and amenities in general. In addition, the idea of a tourist visiting a community's downtown is a completely alien concept. As we will see below, the market for tourism can be a significant part of a community's strategy for development of its internal markets.

The theory implicitly assumes that an equilibrium exists between demand and supply and that this equilibrium will be maintained. Thus, it provides very little insight into the adjustment process that takes place after some economic disturbance. Since the spatial distribution of consumers is given, the theory doesn't explain why changes in population would occur. Fluctuating residential settlement patterns make it difficult to determine if a market threshold exists in a given market area. Central place theory is limited in explaining the causal element behind agglomeration; the theoretical construct is silent on the underlying reasons why change might occur. Central place theory fails to explain the evolution of places among different levels of the hierarchy (Parr 1981). The theory is not dynamic; it uses comparative statics to explain conditions, not changes.

A final limitation of central place theory is the assumption of single-purpose shopping trips. The exclusion of multipurpose shopping trips distorts results from the analysis (Parr 1973). If the consumer makes multipurpose trips, the total cost (purchase price plus transportation) for any item is lowered by sharing the transportation costs between more than one good or service and thus increases the real range of all goods and services.[10] This phenomenon is particularly significant for stores offering lower-order goods and services in higher-order central places; such stores can acquire excess profits because their market is larger than it would be if that good or service had to bear the full cost of consumer travel. This, of course, assumes the bidding for labor or land in the central place does not shift the average cost curve. The spreading of travel costs over several goods and services purchased on a multipurpose shopping trip reduces the price advantage of firms locating near the consumer (i.e., reduces the effective market where the local firm has some monopoly price control). The assumption that consumers have complete knowledge of their spatial options to purchase a given good or service eliminates the consumer risk reduction strategy of bypassing a nearer intermediate center, which may or may not have the item, and making the purchases in a higher center (Shepard and Thomas 1980).

CONCEPTS OF INTERNAL MARKET ANALYSIS

Most communities have at least two geographic areas from which they offer goods and services to the public: downtown and the periphery or edge of the community.[11] Downtowns, historically, have been the center for community and economic activity.[12] They have suffered the loss of retail and other business activities to sites in shopping centers and commercial strips. Many downtowns face high vacancy rates, a poor mix of retail tenants, and multiple property owners. Periphery shopping areas are characterized by newer investment in private and public infrastructure and major land use changes.[13] Communities and individual merchants typically lack the market research support available to the big retailers and shopping center developers. The citizens and business interests of the community need to come together and study and reflect both on market conditions and preferences of the community in determining the type of retail/service they desire.

Understanding market conditions is the first part of any analysis (Chapter 14). This includes analyzing current building uses, business mix, trade area size, economic (including competition) and consumer data, consumer attitudes, and business operator needs. When analyzing internal markets, always remember to analyze the competition from larger and/or nearby areas. This will provide the necessary foundation for more in-depth analysis of different business sectors. Market opportunities are identified by specific business sectors, including retail, service businesses, restaurants, entertainment opportunities, residential units, office space, and lodging facilities.

How Trade Areas Differ

As central place theory suggests, different business types will have different trade areas. That is, people will travel from greater distances to purchase certain goods and services than others. While each individual store may have its own unique trade area, these areas can often be generalized into two different types: convenience-shopping trade areas/goods/services and comparison-shopping trade areas/goods/services. Local *convenience trade areas/goods/services* are based on the ease of access to these types of products. That is, people will obtain these products (e.g., gasoline, groceries) based on travel distances or travel time. Conversely, *comparison-shopping trade areas/goods/services* are based on price, selection, quality, and style. People are more likely to compare these types of goods (e.g., appliances, furniture) and to travel longer distances for their purchases.

In addition to different types of shopping goods, there are also different types, or market segments, of customers frequenting a downtown.[14] Three common market segments are local residents, daytime employees, and tourists. Local residents live within the trade area. As they reside year-round, they provide the majority of spending potential for most downtowns. Daytime employees may live in the

trade area but may also commute from other outside areas. While these employees are in the downtown, however, they provide the potential to stay and make purchases. Furthermore, depending on the community, tourists can provide a large amount of spending potential. While they are not permanent customers, tourists make purchases while they visit the area. Indeed, for many high-amenity areas, the tourist can dominate the internal market.

Identifying High Potential Internal Market Sectors

Communities can influence the form and timing of retail and service sector investment decisions by drawing attention to local market characteristics that might otherwise be overlooked. Economic development professionals can assist by identifying high potential sectors, compiling information of interest to prospects, and marketing to those prospects. The key is to demonstrate that the community is a profitable place to do business (Shaffer and Ryan 1997). The key to any market analysis is that the location and business type must find a competitively advantageous place (i.e. a profitable place to do business). Porter's (1995) fundamental argument is that economic activity "will take root and grow when it enjoys a competitive advantage and occupies a niche that is hard to replicate elsewhere." Porter's analysis reinforces the importance of focusing on unique community characteristics.

To identify appropriate retail and service sectors in a community, it is often useful to first analyze retail and service deficits or opportunities. Analysis of the local context is critical to sustainable retail and service sector development (Ryan and Campbell 1996). We need to ask questions like, Why did this concept work in community X? Are the same factors present in our community? Often we look to other places for ideas on how we can revitalize retail and service sector activity in our community. As a result, we may overlook the unique characteristics of our community that can lead to successful and appropriate retail and service sector development.

If there appears to be demand in certain retail and service sectors, competition in and around the trade area must be carefully evaluated so that an oversupply of a certain type of business is avoided. Do not try to attract businesses to your community if demand for their products is too low or competition is too fierce. Local or regional businesses, particularly those that have branch locations, are often excellent prospects for expansion. They typically have a good knowledge of the market area, and if they already have multiple locations, have demonstrated an interest in expansion. They are often interested in expansion as a way to improve their penetration of the market. Retail and service sector trade shows and conferences or industry newsletters offer a direct means of contacting potential businesses.[15]

Information of Interest to Retail and Service Sector Prospects

Retailers are very interested in knowing the size of the market for their products or services, and how effectively they can penetrate that market. They want to minimize their risk by selecting sites that offer the greatest sales potential. Communities can provide data to prospective retail/service businesses. Data can help potential retailers:

- Demographic data, such as population, age, income, and ethnicity.
- Lifestyle data profiling buying behavior of local residents.
- Local construction trends in housing, commercial and industrial space.
- Local employment trends.
- Transportation data, including traffic volume and parking.
- Mix of existing retail/service, entertainment, and services in the area.
- Mix of residential, lodging, office and industrial space in the area.
- Local and regional competition, including location and size.

When targeting retail and service sector prospects, remember that not all businesses have the same requirements. As we saw in central place theory, a grocery store typically requires different market characteristics than a hotel. Communities should customize information to fit the needs of the prospect.

To attract internal market firms, a community must first make its business district visibly active, attractive, convenient, and safe. This is often more difficult for non–shopping center locations, including downtowns, because they do not operate under central management. It is important to get local merchants organized early on to address issues like hours of operation or sidewalk cleanliness or safety. It is also important to highlight what the community is doing to increase activity in its business district. This might include government incentives, including TIF districts, façade improvements, tax abatements, and lending programs. If available, a master plan for the community should be provided that explains how new development, parking, traffic, security, beautification and cleanliness will be handled.

Retail and Service Clustering

Clustering builds on the agglomeration effects within and across retail and service sector markets.[16] In our context here, clustering is a more narrowly defined concept than that introduced in Chapter 3. Much of the current research on central place theory is focused on clustering of retail and service sector business and spatial competition (Henderson, Kelly, and Taylor 2000; Shonkwiler and Harris 1996; Wensley and Stabler 1998). Retail and service sector clustering is an important but often overlooked feature of business recruitment strategies. *Clustering* is the grouping together of a mix of businesses that enable individual businesses to benefit from each other's sales and customers. Clustering is a technique long used by shopping centers and retail district developers.

Clustering provides consumers with a critical mass of businesses in one location and creates retail and service sector synergy. Clustering can

- provide consumers with a broad selection and variety at a single convenient location.
- enable consumers to make purchases at more than one business and satisfy a number of shopping needs in one trip.
- allow a business district to function as a single economic unit instead of a series of unrelated destination businesses.
- increase spending since the appropriate mix of businesses will offer more goods and services that appeal to targeted shoppers.
- increase impulse buying among clustered stores that offer complementary goods.

For business clustering to be successful, an appropriate business mix is essential. Individual businesses must be able to effectively serve the same or overlapping segments of the market. Clusters also must be physically located so that they are compact and are not interrupted by incompatible space uses. The cluster must encourage the customer to shop the entire cluster.

When developing a cluster policy, it is important to understand that there are three basic types of clusters:

- *Compatible clusters* are groups of businesses that share a particular market segment but offer unrelated goods and services. Outlet malls are an example since their tenants share a market segment that enjoys looking for bargains. Most business districts are classified as compatible clusters.

• *Complementary clusters* are groups of businesses that share customers and market segments but offer complementary goods and services. An office supply store, copy center, and office furniture store together could form a complementary cluster (business services). Retailers must offer goods and services of a similar style, quality and price range. It is interesting to note that department stores are typically organized this way.

• *Comparison clusters* are groups of businesses that carry the same or similar goods and often appeal to the same markets. In some larger regional malls, a clustering of shoe stores can be found. Consumers are able to shop the various lines and compare goods before purchasing them. This also is observed within many department stores.

Developing a Clustering Strategy

Clustering in malls and shopping centers is relatively easy because such facilities have site and/or merchandizing plans in place from day one and, more importantly, have one owner. They have the flexibility to move or resize their tenant's space and replace tenants that no longer fit into the overall mix. Traditional commercial centers, such as downtowns, have multiple property owners, some of which do not live in the community. Business leaders need to overcome this obstacle and show property owners the benefits of clustering, namely the maximizing of real estate values, which occurs in successful clusters.

Hyett-Palma (1989) recommends a four-step clustering strategy for business districts that do not have the centralized control of a shopping center. First one needs to analyze the market served by the business district to determine the targeted markets and appropriate mix of businesses for the district. This should address the trade area, target market purchasing characteristics, competition, character of existing businesses, image of the center, projection of realistic sales capture potential, and appropriate mix of businesses. Customer surveys can be used to reveal underserved retail and service segments within the area.

Second, one needs to prepare business-clustering maps for the business district. This includes maps that display (a) existing businesses and available commercial space, (b) what types of clusters and their locations might be appropriate for the business district, and (c) the specific types of businesses as well as the optimal placement within the center given available space.

Some combinations of retail and service sectors do not work well together (Ray 1996). For example, some apparel retailers are not good together with grocery stores because shopping for clothing and food is seldom done on the same trip. When filling space, it is important to know what types of stores are complementary.

Third, the community needs to gain control of the building space within the business district if possible. This could be done by centralized retail and service sector management by a group of property owners and businesses that provides a coordinated set of activities, including implementation of a leasing plan; having the business district organization obtain the right of first refusal to approve or disapprove new tenants; or obtaining voluntary cooperation by showing the owners that they can benefit from a viable mix of businesses.

Fourth, the community needs to institute an aggressive management mechanism for the business district. The lead organization must have the support of businesses, property owners, and local government officials.

SUMMARY AND POLICY IMPLICATIONS

Community economic analysis rests on a conception of markets. Markets can take many alternative forms. In particular, two market conceptualizations dominate the thinking of how community economic activity is linked: the market for goods and services and the market for factors of production. Households *demand* goods and services in the former and *supply* factor inputs in the latter. Firms, on the other hand, *supply* goods and services and *demand* factor inputs.

The focus in this chapter is on the market for goods and services. Two specific theories relate to this market. Export base theory emphasizes the importance of producing goods for external consumers and the feedback on internal markets. Central place theory, on the other hand, focuses on internal markets and on the interrelationship among communities that make up a hierarchical system of places.

Historically, most communities have pursued a policy of maintaining and expanding businesses that export. In essence, these activities act to attract new dollars into the community from outside. The assumption of these economic development approaches is that the businesses that do not export (i.e., are nonbasic) will grow at a rate determined by the amount of exporting that takes place. In other words, internal markets are completely driven by the external demand for exported goods and services. The spending and linkages that occur between the export and nonexport sectors do not need to appear locally. In essence, from central place theory, the amount of local consumption will be driven by a community's place within the economic hierarchy.

The significance of central place theory to community economic analysis is its recognition that the community is part of a larger urban hierarchy. No community, especially a smaller community, can provide all the goods and services necessary and desired. Residents in smaller communities and their surrounding tributary areas necessarily relate to larger communities for many goods and services. The community economic analyst must recognize the relationship between the range of a good or service and the demand threshold for a good or service. The range of a good or service indicates the geographic limits of the market area for that central function. The threshold of a good or service indicates if there is sufficient demand within that market area to justify offering that particular central function from a certain central place.

Those who perceive the market for goods and services as being dominant in community economic development maintain that the lack of long-run economic development is not the result of inadequate productive capacity. They point out that communities are often faced with significant unemployment, unused capital, and population migration or commuting out of the community. Rather, those that emphasize the market for goods and services contend that the lack of development in a community results from inadequate consumption demand. What communities need is to increase either the internal or external demand for goods and services produced by the community, or change the types of goods and services produced in the community to better match existing demand. This is essentially a structural question.

As communities make decisions about their economic future, it is important that policy-makers understand the practical importance of the two theories discussed in this chapter. Public and private decisions that affect business activity will inevitably create change in markets. The most effective economic development decisions are crafted within a broader theoretical understanding of how markets react. A market approach argues that policies and decisions that solely treat factor markets tend to be incomplete. The real problem is that the community needs to produce goods and services for which there is sufficient demand. These demands can exist as both internal and external to the community. Communities wishing to grow are best served by focusing production on goods and service where demand is growing.

STUDY QUESTIONS

1. The circular flow of a community's economy represents a flow of what economic characteristic?
2. How do households and firms play different roles in the supply and demand structure of alternative markets?
3. What is the significance of the export sector and how does it affect community economic development?
4. How is an export transaction defined for community economic analysis? What forms does it take?
5. What are some of the assumptions made when using export base theory?
6. In export base theory, what role does the assumption of perfectly elastic supply of factors of production play?
7. Why is it necessary to assume a constant structure between export and nonexport sectors?
8. Why is the determination of the size of the export sector so important? What are the implications of having a homogeneous export sector?
9. What are some of the forces that increase the export sector or even allow the creation of new export sectors?
10. Some central place terminology includes *central places, range of a good/service,* and *demand threshold.* What do these terms mean and how are they linked to community economic analysis?
11. What are the implications of central place hierarchies to community economic analysis, especially the provision of goods and services? What behavioral assumptions for firms and households are required?
12. Why might demand thresholds vary among places, over time and among different goods/services?
13. What is a trade area?

14. What affects the range of a good or service? How is it determined?
15. Why is the shape and level of the average cost curve of a firm so critical in determining thresholds?
16. Why is single-purpose travel so important to central place theory's identification of range and urban hierarchy?
17. Distinguish physical distance from economic distance.
18. What are the three types of business clusters discussed in the chapter? Why do you need to consider them in community economic analysis?
19. Why is it important to consider retail/service clusters in any attempt to increase retail/service investment in the community?

NOTES

1. *Cross-hauling* is the phenomenon of a community simultaneously exporting and importing the same good or service.
2. The following presentation uses income. Employment could be the unit of measure, but the discussion would be excruciating.
3. The economy we have specified here is an *open* economy because we allow for imports and exports. A *closed* economy is self-contained and does not allow for imports or exports.
4. An alternative mathematical representation is $Y_T = y(X) + y(C_L) + y(I) + y(G)$, where y is local income.
5. There are two measures of the range of a good or service. The *ideal range* is represented by the definition given. The *real range* is less than the ideal range because competition from other suppliers will enable the consumer to buy at a lower price from a closer source (Parr and Denike 1970). This means the presence of a competitor in some directions from the firm/community causes a shorter range in that direction(s) while the range will be greater in other directions. The end result is noncircular and nonhexagonal market areas.
6. The idea that mail-order catalogues and Internet shopping reduce to nil the difference between the at-store price and the effective price, or travel costs, has greatly altered conclusions of the theory.
7. Keep in mind that businesses, not population (consumers), relocate on the economic plane.

In the purest sense, central place theory is a theory of business location within the demand-maximizing framework of Chapter 3.

8. Variations in the central functions performed by members of a tier will be less than the variations in the functions performed by members of different tiers.
9. Free entry would be characterized by a minimal number of restrictions such as licenses, permits, large capital requirements, or exclusive franchises.
10. Bacon (1984) demonstrated that minimizing aggregate transportation costs requires a shopping strategy mixing single-purpose and multipurpose trips. Consumers must vary their supply points for the same function, making the shopping location choice conditional on other functions purchased concurrently. By sharing transportation costs, a place's effective price differential for a function may be negated completely or reduced to the point where spatial preference overcomes it.
11. Bill Ryan and Matt Kures were involved in the writing of this section. See also http://www.uwex.edu/ces/cced/dma/
12. Here, we really are talking about community-wide market analysis. We focus on the downtown area because of the historic investment in private and public infrastructure and the historic economic declines of downtowns that have tended to cause poor land use allocation in many communities.
13. Some economic reasons why development in the periphery has superceded development in downtowns are land cost, assembly costs of land, infrastructure, proximity to nonlocal consumers, proximity to new residential developments; also, the technology of selling has moved from multistory to single-story structures.
14. Although the term *downtown* is used frequently in this section, it refers to the retail/service sector without any downtown/edge-of-town connotations.
15. Associations such as the International Council of Shopping Centers, the National Retail Federation, and Value Retail News provide opportunities to meet developers and brokers.
16. This section is drawn from Ryan and Muench (1997).

Section II
Community Factor Markets

The building blocks of community economic development exist as resource endowments. This section focuses on community resource markets that are important in understanding how regional and community output is produced. Often referred to as primary factor inputs, this section includes separate discussions related to the standard tangible factors of land, labor, and capital, but also develops an appreciation for the more latent factors that are important to community economic development. These latent factors include technology and management, amenities, and publicly provided goods and services. Another way of looking at these factors is to think of the bottom half of the circular flow model.

Factor input markets are critical aspects of community economics for a variety of reasons. In addition to providing the basis for producing goods and services, their employment represents a critical aspect associated with how households generate income, how society inserts itself within the market framework, and how local decisions can control the extent of community quality-of-life attributes.

The section begins with three chapters that outline the traditional primary factors of production: land, labor and capital. Indeed, the employment of these three inputs provides households with income to spend on consumption goods and services. Our basis of land value focuses on alternative conceptions of land rent. The spatial components of land rent play dual roles of explaining income generation opportunities for landowners while simultaneously explaining the spatial array of alternative land uses with respect to markets. This latter element is critical for developing a basis to explain contemporary suburbanization and exurbanization pressures within which land is a central issue.

Community labor markets and individual job-related decisions made by households are probably the most obvious elements behind income generation. Our discussion of labor markets frames both supply and demand components with a particular focus on key development issues of community labor markets. These include labor mobility, unemployment, exploitation, and discrimination.

Our initial discussion of capital markets focuses on private financial capital. In distinguishing stock assets from flows, we lay the framework for discussing capital from both a debt and an equity perspective. We conclude with a discussion of several key capital market failures that focus attention on development issues such as mobility, risk and uncertainty, and regulatory influences.

Although important in explaining community economic activity, the traditional primary factors of production are insufficient when understanding the unique aspects of development. This is particularly true in smaller communities. The more latent and somewhat less tangible factors associated with technology, amenities, and publicly provided goods and services are critical components of the modern vibrant and increasingly affluent economic plateau that we've achieved at the onset of the 21st century. In order to capture this set of critical latent factors, we've added the last three chapters to the section.

Chapter 8 focuses on how technology and management innovation have transformed the way in which entrepreneurism affects the economic activity of smaller communities. The next chapter outlines key elements of amenity resources that generate both important market-based activities and provide the backdrop to community quality-of-life. We conclude this section with a chapter representing an extension of the amenities discussion. In particular, community decision-makers act within a fiscal structure that allows for the provision of publicly owned goods and services. Roads, crime prevention, fire service, parks, and other critical needs not provided by private markets are discussed within the context of balancing the burden of taxation with the provision of these key community needs.

5
Land Markets

Rapid urbanization during the latter part of the 20th century created significant social, economic, and environmental demands on land and land-based resources. Although these demands and the changes they bring about exist throughout North America, they are particularly acute in high-growth metropolitan areas and amenity-rich rural regions. Residential, commercial, and industrial land needs have pushed outward from urban core areas as economies have grown. These rapid changes in land use often create a confusing array of growth management issues for communities in their quest for further development.

The basis for this rapidly changing land use can be explained in several ways. A useful explanatory approach lies within the theory and application of market economics. Indeed, economic phenomena are critical determinants of land use change. The values we hold for land and land-based resources are complex and are not often directly captured within operating markets. Thus, the combined aspects of productive use value, speculative development value, and amenity value associated with land and land-based resources become important in capturing the essence of rapid land use change. The intent of this chapter is to provide an economic explanation for understanding rapid land use change within both urban and rural settings.

Cities and towns provide markets and trading centers. These centers of dense human population serve as concentrated masses of buyers and sellers, all interacting through the exchange of goods and services. Given competitive human tendencies for efficiency to maximize economic returns and minimize costs, metropolitan regions across North America have grown dramatically during the past century. Cities and towns often enjoy the relative advantages of efficient transportation networks, close proximity between willing buyers and sellers, and the natural

tendencies of regional economies to agglomerate. This is further fueled by public policies that reward entrepreneurism and competition.

Land, as a primary factor of production, has provided a key basis for expansion of human civilizations across the globe. It has served as a critical production input and has been regularly used as a resource for both household subsistence and income generation while acting as an important tradable asset. Only through conscious public and private planning does land use take on attributes of being something other than a private commodity for use as a production input. This is important in understanding land use change and the implications for managing growth and maintaining land-based environmental quality in these regions.

The unintended consequences of market-driven economic forces have left their mark on land and land-based resources. This is particularly true in regions surrounding rapidly growing communities and in areas that contain high levels of amenities.[1] On a widespread scale, open spaces have been converted to residential neighborhoods and commercial uses. Hills have been leveled to build highways. Highways have used up open space for construction of efficient transportation corridors. Efficient transportation corridors provide avenues for continued growth and development and so on. Once-sleepy rural communities have been transformed by newcomers seeking relatively inexpensive land on which to build their homes, relocate their businesses, and conduct exchange. This transformation has been a hallmark of late 20th-century North American economic prosperity. The nature of change with respect to land use has been both rapid and complete.

The productive uses of land for growing commodities have been superceded by other values associated with land and land-based resources. As will be shown, this is due to several economic aspects associated with land productivity, land location, the value of agricultural commodities relative to other goods and services, transportation costs, and a host of other economic variables. The economics associated with growing crops have become dwarfed by the economics associated with open space ready for development of new residential and commercial/ industrial buildings.

This rapid growth within once-rural communities has implications for a broad array of changes in economic, social, and environmental structures. Furthermore, there are distributional effects of change that depend on both initial endowments and current ownership patterns of land. For instance, local residents who own land within a rapidly growing region often enjoy significant windfall benefits associated with land exchange and/or development. On the other hand, local residents on limited incomes who own property or who are tenants in rental properties face steadily rising land values, property tax bills, and rental costs. Furthermore, rapid regional growth often attracts investment from outside the region. Nonresident control can often lead to diminished local perceptions of regional autonomy. Although the substance of this chapter deals primarily with the microeconomics associated with land and land-based resources, it is important to remain aware of the broad scope of change and the distributional implications of who benefits from growth.

This chapter is organized into four additional sections. First, we outline the basis for land-based resource values and the traditional valuation of productive land use with focus on the concept known as land "rent." The next section deals with the basis for land value, encompassing the important determinants associated with distance, productivity, resource endowments, and amenities. We then critique contemporary neoclassical models of economic growth from the standpoint of how well they capture the value of land and land-based resources. Finally, we tie land back into the circular flow of regional economies as a primary factor of production. We are at the *resources* node of Figure 1.1.

USE CAPACITY OF LAND AND ITS OWNERSHIP

The value we place in land and land-based resources takes many forms. Using land for growing crops represents one aspect of land value. This can be classified as *direct use capacity* of land.[2] Increasingly, these productive land uses are being supplanted by two alternative land values—*speculative development values* and *amenity values*—associated with

land and land-based resources. We begin with direct use values associated with productive land use.

Productive Land Use

The economic story of land and land-based resources for production begins with the thinking of two rather famous academics: David Ricardo and Johann Heinrich von Thünen.[3] Both lived during the latter part of the 18th century and the middle part of the 19th century. Given historical precedence and its reliance on classical microeconomics, we begin with the contributions of Ricardo toward an understanding of land rent.

A Fertility-based Concept of Land Rent

David Ricardo based his understanding of returns to land resources on differing levels of agricultural site productivity (fertility levels). An understanding of Ricardian land rent necessarily begins with the classical analysis of production (or value-product analysis) and firm costs (or cost curve analysis). Production can be characterized in terms of input-output relationships. Given differing levels of inputs, a rather standard form for characterizing output is by using a cubic function, which begins increasing at an increasing rate to a point of inflection. At the inflection point, the function still exhibits increasing returns but it does so at a decreasing rate to a point of maximum output (maximum total output), beyond which output diminishes with increasing inputs. A standard value-product analysis is presented in Figure 5.1. This type of growth function often is useful in representing biological growth and the response of land and land-based resources to primary factor inputs. (This is a general extension of material on location theory in Chapter 3.)

Note from Figure 5.1 that, given this perspective, there is a distinct relationship between average production and marginal production. The point of tangency between average and marginal product is found where average production is maximized. This begins the zone where rational producers will logically choose to produce given standard assumptions about costs of production. Rational production will take place from this point to the point of maximum spread between total product and total cost (the *Zone of Rational Action* in Fig. 5.1). Although we could dissect these relationships further using value-product analysis, Ricardian rent is more typically based on an analogous yet slightly different perspective of inputs and outputs known as cost curve

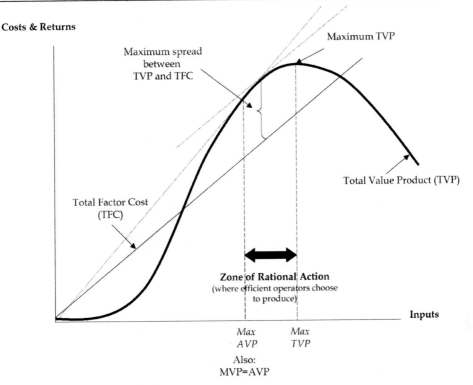

Costs & Returns

Maximum TVP

Maximum spread
between
TVP and TFC

Total Value Product (TVP)

Total Factor Cost
(TFC)

Zone of Rational Action
(where efficient operators choose
to produce)

Inputs

*Max
AVP*

*Max
TVP*

Also:
MVP=AVP

Figure 5.1. Value-product analysis.

analysis. This is important because of the standard classical notion of supply as being representative of the cost structure of firms.

Cost curve analysis is based on a mirror image of the functional production relationship between inputs and outputs. The primary difference is that the focus of attention is now on how costs vary with production of additional units of output.

Costs can be viewed in two ways. First, there are costs required in all productive activities that are fixed across a range of production. An example of this is a water treatment facility that is required for developing residential, commercial, and industrial lands. Investments in public services can be used to produce development up through a range of output, until which a higher level of physical plant is required. The second type of cost is variable cost, which is determined by the level of output itself. A good example of this with respect to development-based land use is the extension of sewer lines and the use of sewerage systems. In sanitation, there are significant fixed costs associated with sewage treatment facilities. The use of these facilities, however, depends on the level of use. Thus, costs associated with treating sewage are variable but rely on the presence of facilities that represent fixed costs.

Graphically, the cost function is another cubic function analogous to a mirror image of the production function represented in Figure 5.1. In the cost curve analysis summarized in Figure 5.2A, costs begin at a level that covers the fixed costs required for production. Variable costs then increase at a decreasing rate until a point of inflection beyond which total costs begin to increase at an increasing rate up to a point where production becomes congested and the need for additional fixed inputs pushes costs upward dramatically.

Ricardian land rent (Fig. 5.2B) consists of the area below the marginal revenue but above average cost at a point identified where the price of the output is equal to the marginal cost of production. This simplification rests on competitive price taking behaviors, thus we can equate marginal revenue with market price. Ricardian land rent represents the total "profit" that can be obtained for producing output under these conditions.

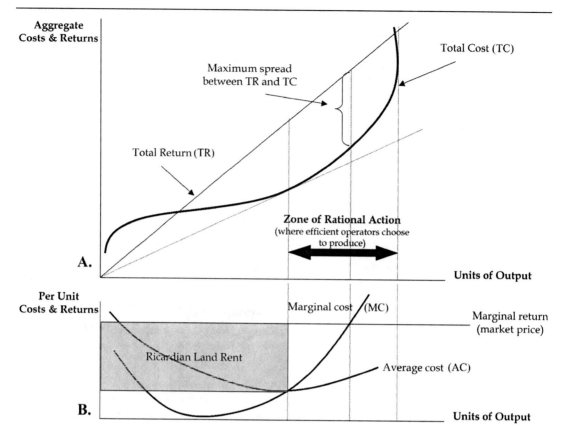

Figure 5.2. Cost curve analysis.

Note from this discussion that production will occur under several differing levels of site productivity. Indeed, this was one of the key features of David Ricardo's arguments for the limits of land's economic productivity. Production will take place on lands until the point of extensive margin is attained. In other words, the extensive margin is the limit of economic productivity; it is defined as the lowest quality of land that can produce output where the price of the output is just equal to the marginal cost of producing output. Ricardian land rent for alternative qualities of land is summarized in Figure 5.3.

Whereas Ricardo provided a useful explanation for land use from a fertility perspective, a serious shortcoming was also painfully evident. What is required to understand land economics from a community perspective is an explanation that incorporates the spatial array of land uses and the locational aspects of land relative to market centers. Although

Ricardian land rent does capture fertility differences and fully utilizes the cost structure of production, it remains essentially aspatial. That is, it does not fundamentally deal with distance and the sprawling nature of cities. These spatial gradients are critical to explaining development of communities and land use alternatives on the urban fringe. Spatial explanations are necessary to develop the basic implications for alternative land uses and the distance decay of land rent as we move away from urban core areas.[4]

A Spatial Perspective of Land Rent

von Thünen was the first to develop a basic analytical model of the relationships between markets, production, land use, and distance to trading centers. Land use, in von Thünen's world, was thought to be determined by the relative costs associated with transporting different commodities to the central market place.[5] Given competition and price taking behavior, land uses that are more productive compete for the closest

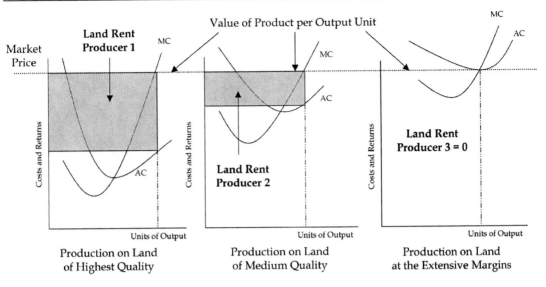

Figure 5.3. Ricardian land rent for alternative land quality.

land, while less productive activities locate farther and farther away from the central place. It is important to note that productivity, in this sense, includes both fertility aspects of land and its scarcity (as reflected in the market price) of the commodity being produced. Thus, distance to central place, the needs of human sustenance, and land productivity (both soil fertility and scarcity/price of commodity) provide key elements associated with explaining alternative land uses.

During his lifetime, von Thünen perceived a rather obvious and regularized pattern of settlement throughout the countryside of rural Germany. He used this perception to develop a generalized conceptual model of alternative land uses in his rural part of Germany. This perception provided the basis for explaining differing land use types as a function of distance from a central market center. Our modern-day understanding of land use around marketplaces follows a strikingly similar pattern to von Thünen's view of the countryside. The regular progression of land use types and change in land use patterns can be explained by their relative productivity. Figure 5.4 outlines a stylized 18th-century German village and the distance from central market in a land use variation gradient.

Humans, like all organisms on the planet, have requirements that are necessary to prosper. These basic staples for life include food, water, and shelter.

The provision of these basic staples lies at the heart of how we combine our labor and meager capital resources with land through production of commodities.[6] Settlement patterns and central business districts originate at crossroads of transportation infrastructure. In industrializing Germany during the late 17th century and early 18th century, this involved the intersection of navigable streams and cart/foot paths. Today, we see this agglomeration of market activities where railroads, highways, and ports exist. This exchange of goods and services in a central market place required permanent residence for the people involved.

At transportation crossroads in 18th-century Germany, one could find retail and service sector businesses engaged in market activities. Just outside of this central place were the homes of merchants, service workers, and local residents. Because refrigeration had yet to be invented, von Thünen realized that the transport of goods sold in the central place required production in close proximity to the market. Thus, just outside of central market places and just beyond the houses for residents, he noticed a pattern of agricultural production. In particular, he noticed production of those products that were highly perishable, like tomatoes, potatoes, carrots, and cabbage. Another key need in sustaining the food needs and industrial development activities of

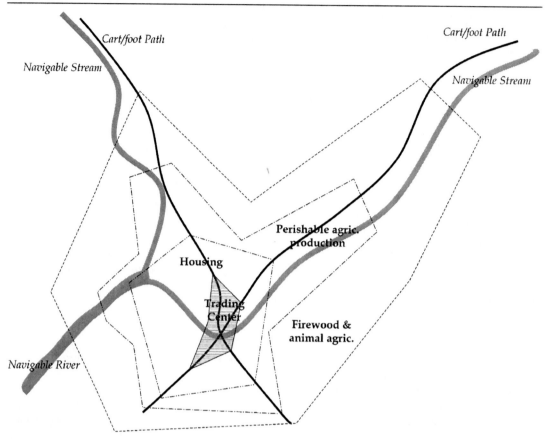

Figure 5.4.   Land use gradients from a typical 18th-century German village center.

human settlements surrounds fuel production for cooking and industry as well as the perishable milk and cheese from dairy cows. This relatively lower productivity level of sites was found just outside of the region where perishable food was produced. This change in land use as distance to market center increases progresses to include the production of grains (swiddle agriculture) and forestry activities associated with timber production. The latter was required for construction materials to build homes, shops, and factories.

The economics behind settlement patterns and alternative uses of land focuses on the production activities of people, both individually and in teams (today's small-scale entrepreneurs and business interests). To simplify the explanation, it is again useful to characterize production from the standpoint of costs. These costs involve the costs of growing or making salable goods, marketing goods to potential buyers, and transporting these goods to

market for final sale. For simplification, we can assume that market price is competitively fixed (determined outside the region) and that the costs of production and marketing of goods remains relatively the same regardless of the distance to market. What changes with distance are the costs of transporting both raw commodities and finished products.

Figure 5.5 outlines the concepts behind production cost. In particular, we're interested in the varying costs of production as this production takes place farther and farther away from the market place. Given a market price determined regionally, distance affects transportation costs and identifies the relative profit associated with producing commodities at different locations. If production and marketing costs remain relatively stable with distance and the price of the product is competitively set, the difference between market price and costs of delivering a good to final market (production, marketing, and transportation costs) identifies what is

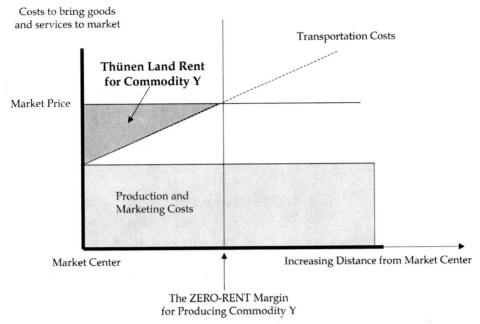

Figure 5.5. von Thünen land rent with specification of alternative production costs.

commonly referred to as *von Thünen land rent*. The amount of land rent for production of a commodity decreases as distance to the market center increases.

Note from Figure 5.5 that it becomes unprofitable for producers to engage in the production of commodity *Y* beyond the distance where total cost of production exceeds the market price. This zone of production is delineated by a region around a central business district with a limit known as the *zero-rent margin*. Beyond this margin, the land rent associated with production of the good or service is either zero or negative (e.g., costs now begin to exceed the returns from its sale).

When we focus on land rent from the market center to the zero-rent margin, we begin to see decreasing profitability (or land rent) in producing goods the farther we move away from the market center. A decrease in land rent by commodity with increasing distance is outlined in Figure 5.6; note that this is simply a transposition of the land rent found in Figure 5.5 up to the zero-rent margin.

Different commodities will experience different costs of delivery to final market. For example, producing tomatoes (a highly perishable commodity) will be much more sensitive to distance from market than producing wheat or timber (which, as commodities, aren't nearly as perishable). Also, when we con-

sider that use of land for housing or industrial/retail uses necessarily requires proximity to market centers and supercedes the productive use of land for agricultural purposes, a gradient of land uses becomes evident (Fig. 5.7). Residences, offices, and factories require a highly intensive use of land (typically measured not in acres but in square feet) that is inextricably tied to the market center. An excellent example is the phenomenon of the skyscraper, which requires very little land per square foot of office space.

Thus, land uses will be arrayed spatially from the market center. In Figure 5.7, the region that surrounds the market center outward to point *a* will be dominated by land use for production of commodity *X* (e.g., housing and space for commercial/industrial activity). Land use will begin to change at point *a* as people realize that higher land rents can be obtained by switching from production of commodity *X* to commodity *Y*. The concentric ring delineated by the distance from point *a* to point *b* will be dominated by production of *Y*. This is simply due to the relatively higher profits or land rents that can be obtained by producing commodity *Y*. Likewise, land uses will switch at point *b* to the production of commodity *Z* because now land rents are higher with production of the alternative commodity.

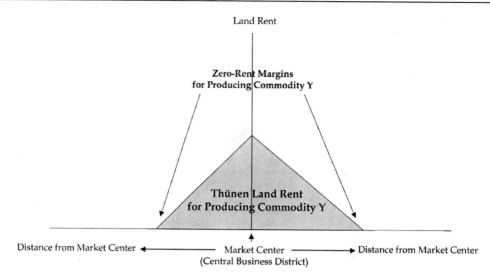

Figure 5.6.   von Thünen land rent for a single land use relative to the market center.

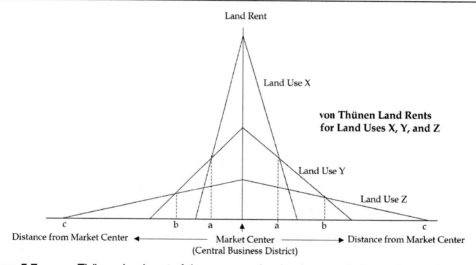

Figure 5.7.   von Thünen land rent of three competing land uses relative to the market center.

In this discussion, we need to bridge the historical presentation of von Thünen land rent with contemporary land use patterns. Although von Thünen developed his theory of land rent with reference to 18th-century German villages, there are analogous patterns to 21st-century North American land patterns in both urban and rural settings. A primary difference, however, is that transportation cost gradients now need to be conceptualized in a broader fashion. Instead of perishability of commodities as a primary explanatory force in distance gradients, we must consider commuting costs, agglomeration, and amenity-based value premiums. In today's regional economies, distance to both production inputs (e.g., labor, technology, intermediate inputs from other industries) and demand sources (e.g., forward-linked industries, final consumers, export markets) are critical in locating the array of economic sectors present in the modern-day economy. Furthermore, amenities provide alternative distance premiums analogous to the von Thünen presentation (amenities are discussed in Chapter 9).

These contemporary forces provide continuing salience to the basic spatial approach of land value as presented for perishable agricultural commodities during von Thünen's time.

In concluding this section, it is important to recognize the continuous nature of distant gradients. If we consider the multitude of alternative uses produced using the primary factor input of land and stylize this presentation, we arrive at the urban *bid-rent cone* (also called a *distance decay function*), which arrays alternative land uses by their respective land rents as a function of distance to the market center (Fig. 5.8). The simple conical land rent gradient as a function of distance provides the basis for analysis of urban land uses and is a fundamental component of the contemporary urban land use model.

Although North American cities have progressed through several variants of this conical representation of land rent,[7] the simple representation helps us understand relative boundaries associated with the spatial array of land use activities surrounding cities. For instance, it helps us understand and spatially define the limits of urban sprawl at any given moment. This is particularly true given the notion that commuting distances, like perishable commodities, have an outward limit. For instance, few people are willing to commute for more than an hour to get to work. Given modern interstate highways, this limit is roughly identified at 50 to 60 miles from the urban center. Of course, this varies with highway congestion, itself a function of time of day that commuting takes place, size and efficiency of transportation networks, and type, or mode, of travel.

The Economic Basis of Urban Sprawl

Given this economic conception of urban influence on alternative land uses as we progress away from the city center, we now have a rather convenient and useful tool with which to examine the issues associated with urban sprawl. To understand land use change in peri-urban regions, we need to realize that market forces sort out alternative land uses by their respective "highest and best" uses (e.g., rational economic competitors will increasingly choose to "produce" the land use that returns the highest land rent). Agriculture and forestry, as land uses, are lower in hierarchical value than residential, commercial, and industrial uses. Furthermore, highly competitive and technology-intensive agricultural producers have driven agricultural commodity prices steadily downward. This, combined with well-developed rural infrastructure, has resulted in bid-rent gradients associated with agriculture and forestry to be nearly flat relative to other more-urban land uses. For simplicity, we can superimpose a flat bid-rent gradient for agriculture and forestry within the urban bid-rent cone (Fig. 5.9). Given static conditions, land use change will occur from rural uses

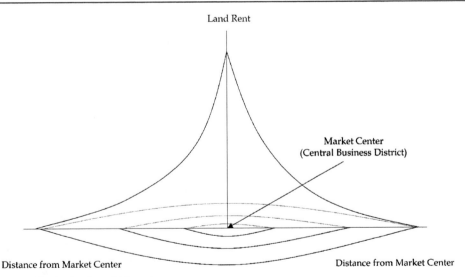

Land Rent

Market Center
(Central Business District)

Distance from Market Center

Distance from Market Center

Figure 5.8. Typical urban bid-rent cone.

(agriculture, forestry, open space) to urban uses (residential, commercial, and industrial), where the bid-rent function yields the highest and best use (denoted as edge of the city in Fig. 5.9).

Now, to illuminate urban sprawl, assume that economic growth occurs in this urban area. Economic growth often stimulates labor markets, resulting in lower unemployment and higher levels of income. With growth, better jobs become plentiful, new residents are attracted to the region, more residents create further demands for residential, commercial, and industrial lands, and so on. That is, increasing demands for land are matched with land supplies that require changing land uses. Thus the supply of land is constrained by conversion, leading to a general upward pressure on market prices for land. This increase in the market price of land has the effect of shifting the bid-rent surface of land upward. As residual land uses, agriculture and forestry experience higher demands but are dramatically affected by land use conversion. The productivity of these lands is outstripped by land uses characterized by more elastic distance decay functions. The edge of the city now moves outward (new edge of the city in Fig. 5.9). The distance between the old city edge and the new city edge is referred to as *sprawl*.

Among traditional urbanists within the planning realm and others, strong public sentiments against unchecked urban sprawl exist. While often taking a normative tone, arguments raised by sprawl opponents have been effective in stimulating several public policy approaches to limit outward expansion of urban areas. From an objective (or positivist) perspective, economists raise several issues of urban sprawl that focus on the existence and influence of important market failures that foster urban sprawl. Economic critiques of sprawl recognize three fundamental market failures that provide a basis for policy intervention (Brueckner 2000). These are (1) an inability of the market to discover prices for open space, (2) an inability of the market to account for the social costs of congestion, and (3) a failure of the market to make new development pay for the infrastructure demands it creates.

While providing a rudimentary explanation of urban sprawl, this discussion overlooks a host of additional locational issues associated with an economic basis for land use decisions. Other issues associated with land rent gradients involve publicly owned goods such as infrastructure, education, and other amenities. Recently, economists have been working to incorporate the presence (or absence) of public goods and amenities into estimates of land rent (Lancaster 1966; Kohlhase 1991; Earnhart 2001). Similar to market centers, amenities can also provide an important initiator of increased land rents. To more fully explore this notion, we must

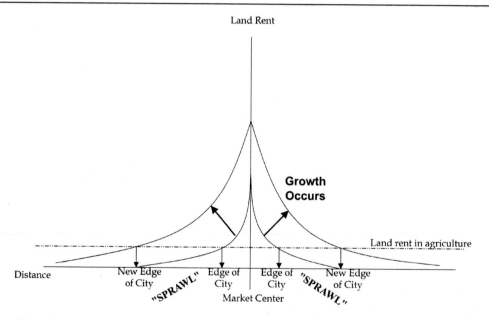

Figure 5.9. An economic perspective of urban sprawl based on growth-induced bid-rent functions.

first briefly recognize some important aspects of how contemporary economic thinking characterizes the amenity basis for land, land-based resources, and the use of land surrounding cities.

Much of the explanation of change in land applications along the fringe of urban areas can be characterized as indirect use and nonuse in nature (or being non–production oriented). That is, the driver of suburban land use change is land use for nonagricultural purposes.

Productive uses of land are just one of a wide variety of uses we put to land. Economists addressing the variety of uses and their respective values have provided a useful categorization scheme within which to view various land uses and the respective land-based resources (Hodge and Dunn 1992; Hodge 1995). (These alternative values represent an important issue in community economics that is covered in Chapter 9.)

The Exurbanization Process

The impacts of land development and land use change in rural regions are relevant to economic development practitioners for several reasons. First, certain types of rural regions have witnessed dramatic growth, while other types have stagnated or declined. This differential rural growth is most notable in regions endowed with high levels of natural and cultural amenities. Indeed, the growth of amenity-rich rural regions has exploded in the past 30 years, catching most communities and local governments unprepared. Second, the comprehensive impacts of development that follow this growth are scarcely understood by these communities. Local planners are often overwhelmed and unable to deal with the totality of demands placed on them in the face of rapid growth. Third, there is an increasing realization that improper application of urban land use planning techniques into rural areas without an understanding of the underlying social and economic implications actually serves to exacerbate the progressive exurbanization process.

This last point is particularly important. Although providing some pause to planners, an understanding of the exurbanization process allows creative approaches unique to rural land use planning to surface. Some argue that exurban land use planning actually tends to exacerbate these problems by treating rural areas as extensions of central cities (Esparza and Carruthers 2000). Regulations placed on land use as a reaction to rural residential demands force exurbanization farther into the hin-

terland. Indeed, exurbanization is an increasingly dominant mode of land development in rural parts of North America.

Recent research into the exurbanization process focuses on the need to treat rural regions as unique entities requiring creative land use planning approaches (Esparza and Carruthers 2000). The exurbanization process occurs in four distinct phases (Fig. 5.10). Stage 1 begins with growing demands for amenity-rich rural lands to develop for residential purposes. Throughout the 20th century, these demands have been readily met across rural North America despite their particular acuteness in amenity-rich regions. The latter part of the 20th century, however, witnessed an increased saturation of high-amenity locations. Supply of high-amenity residential parcels in many rural areas has been constrained by availability. The scarcity of high-amenity sites is an issue that is playing itself out through rapidly escalating rural land prices.

Rural regions containing the highest-amenity rural sites have already experienced what Esparza and Carruthers (2000) referred to as *commodification of place*. Namely, these regions have moved through the consumption of large tracts of rural land and converted these lands into low-density and amenity-driven, privately owned residential parcels (Stage 2). On the part of frontier-seekers, this exurbanized low-density residential development creates a fear of losing the amenity base that provided the original motivation for development. Exurbanized regions possessing significant amenity values experience continually strong demand matched with fear of complete urbanization. This leads to widespread application of urban and suburban land use tools. In many regions, these act to urbanize the rural landscape.

The exurbanization process renews itself as a result of new and in-place demands continually seeking the frontier of residential development. Stages 3 and 4 of the exurbanization process represent the realization that urbanization has occurred, which leads to a continual pressure to develop the frontier. The expanding urbanization phenomenon continues to play itself out as demands expand across relatively lower-amenity, increasingly remote regions. Esparza and Carruthers (2000) empirically assessed exurbanization as it had taken place in Arizona and criticized the application of urban planning tools to rural settings. Ironically, they concluded that land use planners and planning are partially responsible for this exurbanization process. Thus, the next

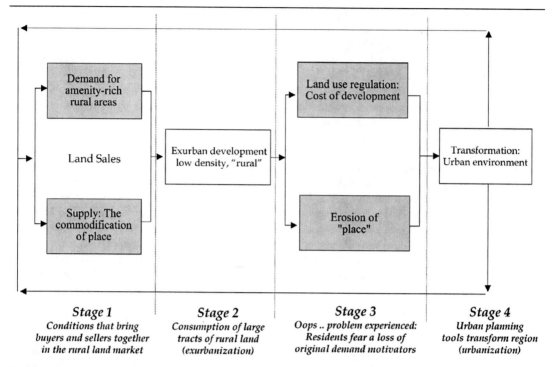

Stage 1
*Conditions that bring
buyers and sellers together
in the rural land market*

Stage 2
*Consumption of large
tracts of rural land
(exurbanization)*

Stage 3
*Oops .. problem experienced:
Residents fear a loss of
original demand motivators*

Stage 4
*Urban planning
tools transform region
(urbanization)*

Figure 5.10. A conceptual model of exurbanization. (Adapted from Esparza and Carruthers 2000.)

section focuses on the role of land use planning as a key mechanism within which society affects the spatial patterns of development.

LAND USE PLANNING: THE ANALYTICAL TOOLBOX

The primary goal of land use planners is to affect the course of land use and the spatial patterns of development. Land use planners typically are public servants of local units of government (cities, towns, counties, and regional planning commissions) or are private consultants working for themselves or for larger engineering firms. Where resources are available, these local units of government will employ a cadre of planners to perform the necessary analytical and process-oriented tasks. Consultants often perform a significant number of the specific tasks associated with land use planning. This is particularly true in smaller governmental units that do not have the resources with which to support a planning staff and in situations where specialized expertise is required.

Land use planning involves a myriad of actors and requires a significant amount of stakeholder involvement, technical analysis, and political decision-making. An important aspect of land use planning involves both fiscal and project-level assessments for costs and benefits. Economists and their analyses provide but one input into a very complex and politically motivated landscape. In the end, it is often the case that politics remains the principal driver of local decision-making regarding land and land use. The role of planners, applied economists, and policy analysts is to provide locally elected officials with the best information possible so decisions can be made objectively and rationally. The ability of land use planning analysts to identify winners and losers among an array of alternative land use decisions is meaningful only to the extent these affected people and groups are involved in the political process. That is to say that estimates of costs and benefits in evaluating specific projects, in and of themselves, provide only one input into how decisions are made.

Community practitioners use several categories of evaluative tools to assess the economics of land use decisions. These tools (broadly defined) include project-level cost-benefit analysis, fiscal impact assessment, economic impact analysis, and an array of accounting tools. Although viewed through various academic lenses as distinct lines of research and academic inquiry, each has a very specific role in assisting planners with analysis of alternatives with respect to land use decisions.

Cost-Benefit Analysis

Cost-benefit analysis (CBA) encompasses a myriad of specific analytical approaches that are applied to evaluate costs and returns at the project level. Ultimately, these approaches are dynamic, in the sense that they account for changes in both resources and the valuation of money over time. Key elements of CBA necessarily employ specific measures to capture how we value goods and services across various time frames. This temporal valuation approach raises important issues involving appropriate discount rates and the compound nature of time preferences with respect to economic value.[8] Given the complex nature of benefits derived from land and land-based resources, we need to incorporate the value of non-market goods and services into benefit estimation. This is particularly true when land use alternatives have differential impacts on the availability of non-market goods and services. The techniques of non-market valuation are slowly beginning to be applied, but they are dependent on the skills available and the specific problems being addressed (Chapter 9). As these techniques become more widely known and as the values they represent become more important, routine cost-benefit analysis to evaluate alternative land uses will inevitably represent a more comprehensive array of benefits and costs with which to base decisions.

Fiscal Impact Assessment

Fiscal impact assessment deals with decisions about land use alternatives from the point of view of local government revenues and costs. Common questions analyzed in fiscal impact assessment include how one development option differs from another with respect to the local tax base. How does the potential for increased property taxes offset the need for increased services provided by the local unit of government?

The public policy tools local governments use to encourage certain forms of development often are determined through fiscal impact assessment. Examples include the determination and imposition of impact fees, levels of tax increment financing, and programs designed to offer property tax relief.[9]

Economic Impact Analysis

Economic impact analysis provides an assessment of the gross benefits associated with land use alter-

natives with particular interest in how changes in land use affect the overall economic activity of the region. Economic impact analysis incorporates specific tools that can be both descriptive and inferential. Descriptive techniques of economic impact analysis contribute useful measures of economic activity that provide context for decision-making. These include descriptive industry measures, such as location quotients, shift-share analysis, and export-base multipliers. Inferential techniques of economic impact analysis attempt to use available data on past and present economic activity to estimate current impacts of change and to forecast future activity resulting from some policy. Examples of these techniques include input-output analysis, social accounting matrix analysis, computable general equilibrium analysis, and an array of econometric forecasting methods.[10]

Several dilemmas associated with economic assessments in land use planning reinforce earlier statements made about the complexity of land use values. Comprehensive assessments of land use value are complex in the sense that nonmarket goods and services of land-based resources are critically important to capture and are elusive to empirically estimate. Since we don't fully understand several aspects of nonmarket goods and services, it is nearly impossible to make meaningful and comprehensive assessments at this point. To be sure, nonmarket valuation techniques are part of cost-benefit analysis and measures of consumer and producer surplus help us understand societal welfare differences, but our ability to compare and integrate these quantitative values within a decision-making framework remains somewhat limited. Furthermore, much of the inter-industry analysis that makes up economic impact analysis remains at the market-based goods and services level. For instance, when we assess employment and income impacts associated with a land use alternative, these remain at the level of earnings derived from labor and productive land/capital usage. Developing usable empirical measures to incorporate societal welfare change at the household level is a continual challenge to both academics and practitioners of economic development.

LAND USE PLANNING: THE DIRECTIVE TOOLBOX

Often an interest in maintaining certain land uses within communities would not be represented if market forces were to predominate. This is the dilemma of land use planners. During the recent past, increasing interest in providing an array of tools has lead to some very interesting and innovative approaches to land use in both urban and rural settings. To a large extent, the progression of tools has lead to strategies to maintain land uses that are deemed socially desirable but do so in a way that mimics market forces. These economic tools provide incentives for people to manage land in certain ways. Again, given totally unfettered market forces, these land uses would be superceded by less socially desirable forms of land use.

Economic land use tools are policies or programs that regulate land use or create incentives to encourage or assist individuals in exchanging rights in land, in a way that is consistent with a set of broader land use policy objectives. Economic incentive policies provide financial rewards (or penalties) for undertaking specified actions that support (or undermine) societal goals for land use.

The toolbox of land use planning tools provides a useful perspective of current land use planning initiatives. Categories in the toolbox are (1) public programs and regulatory policies that attempt to alter land use in accordance with publicly defined goals and objectives, and (2) quasi-public/private interaction tools that also affect land use but do so in accordance with more market-oriented goals and objectives.

Public Ownership

An obvious mechanism that society uses to control land use is through direct control of land by public ownership. Governmental units own and control significant amounts of land and manage these lands based on constituency needs and desires. Federally owned land is significant in both the United States and Canada. For example, across the western United States (and to a lesser extent throughout the Midwest, South, and East) the U.S. Federal government (through the U.S.D.A. Forest Service and the U.S.D.I. Bureau of Land Management, Park Service, and Fish and Wildlife Service) controls vast tracts of land. In Canada, federal land (referred to as "crown lands") dominates the vast expanse of sparsely populated northern wilderness.[11] States also own land, typically managed as state parks and/or forests. Across the Lake States, for example, county governments own and manage significant parcels of land as county forests. Finally, local and regional units of government control lands for a multitude of uses.

Lands come into public ownership via four basic mechanisms. Historically, much of the public land base originated as *residual parcels after settlement* (leftover lands following the initial wave of settlement) during the 1700s, 1800s, and 1900s. This is particularly true for federal lands in the western United States and crown lands across Canada. Second, default of private owners on property tax payments led to *tax reversion* of land to government authorities. These payment-defaulted lands acted to create publicly owned land bases. This has been a significant factor in regions characterized by marginal agriculture (based on fertility or the need for irrigation) where past public policy acted to provide incentives for agricultural conversion but owners eventually couldn't profitably produce agricultural commodities. Third, land comes into public ownership through the exercise of *eminent domain*. This is a typical mechanism for acquiring land for infrastructure, such as roads, utility rights-of-way. Fourth, units of government increasingly are involved in land markets using *outright purchase* as a mechanism to gain control of land. This is significant where stewardship programs have been set up to acquire lands with sensitive ecological resources.

Public Regulations

Zoning is a typical tool used by urban planners to define appropriate land uses. Simply stated, zoning delineates areas, or zones, where certain activities are allowed to take place. In addition, zoning regulations and their related ordinances often specify details about the physical design of residential, commercial, and industrial parcels that must be adhered to by the owner for development to take place.

Although important in directing urban form, zoning has limited value in addressing the sprawling aspects of urban growth. In regions surrounding cities, zoning is often an aftereffect of decisions about land use change (an exception is the urban growth boundary). However, zoning has not been widely applied throughout rural North America. This is probably due, in large part, to political resistance to overt regulations that dictate how land can be used in rural areas. Also, although zoning can be implemented in concert with economic factors, it rarely is considered an "economic" tool for land use planning. Rather, zoning provides a regulatory mechanism for development within already urbanized areas.

Zoning of land assumes that planners delineate fixed boundaries around zones and that land use and development within these zones take place according to some prescribed zoning ordinance. Necessarily, zoning and zoning ordinances affect the value of land within which these tools are enforced. Those who own land within a zoned area can expect values of using that land to follow the regulated uses to which land is allowed to develop. In other words, the regulations associated with zoning dictate land value within the demands for land of that particular type. This is regardless of the market forces that act to allocate land to its highest and best use. The value of land is determined by the zoning in place. Change in value after zoning will result from scarcity of land zoned for that particular use.

In response to dramatic urban growth and an overriding interest in maintaining open space and rurality outside of city limits, a very tight form of urban zoning control known as an *urban growth boundary* (UGB) has seen limited implementation. This relatively unique form of land use control is most widely known from the Portland, Oregon, example. Instituted during the early 1970s, this tool literally entailed drawing a line around where Portland was expected to grow and regulating development both within the delineated region and outside the region according to publicly determined land use objectives. As part of Oregon's Statewide Planning Program, a UGB was defined for all 241 cities in Oregon. Since the 1970s other areas, including the San Francisco, California, metro region; Charleston, South Carolina; and Knoxville, Tennessee, have looked into applications of this regulatory tool.

Urban growth boundaries were developed with multiple objectives in mind. Most advocates of UGBs point to the success of UGBs in providing limits to the extension of costly public services and facilities, the preservation of land for agricultural purposes, greater certainty for people who own, use, and invest in land at the edge of cities, and better coordination between city and county land use planning.

Critics often recognize these benefits but also point to the negative impacts that UGBs have by placing artificial upward pressures on land values within the UGB and artificial depressions on land values outside of the boundary. Many argue that the tight controls of UGBs work against several socially desirable aspects of urban growth, such as the provision of affordable housing, both within the urban core and within the wildland-urban interface surrounding cities. To illustrate the economic critique of urban growth boundaries on affected land markets, it is useful to understand that the distribution of benefits and costs depends on where land is located. The alternative economic effects of a UGB on land markets are illustrated in Figure 5.11, where

the supply and demand relationships of land are represented by a standard upward-sloping supply curve and a corresponding downward-sloping demand curve. The ultimate effect of an UGB is that it constrains the supply of land for development. In effect, the boundary around a city can be represented as a discontinuity of land supply. Given political pressures, it is relatively easy to set a boundary well beyond the current urban fringe. In our conceptualization, this occurs when current demand (D) crosses the supply function where equilibrium is found at P^* and Q^*. This UGB is identified well beyond the equilibrium (where the supply curve is kinked upward, identified as quantity Q^{**}).

Those critical of the UGB concept argue that as time progresses and growth occurs, the market-based equilibrium between supply and demand is eventually affected by the policy. With economic growth, demand for developable land shifts outward. Eventually, demand shifts reach a point where the UGB places artificially upward pressures on land price (beyond the supply "kink" point, where $Q^{*'} = Q^{**}$). At points beyond Q^*, there will be an artificial benefit accruing to those who own land within the boundary and an artificial loss to those owning land outside the boundary. Some refer to this as a *structural market inefficiency* and denote the artificial benefit as a "windfall" and the artificial loss as a "wipeout" (Van Kooten 1993).

The use of UGBs to control sprawl continues to be a topic of contemporary policy discussion. The efficacy of UGBs to attain policy objectives remains a matter of contemporary political debate and provides an important applied research topic. Mixed results of some empirical works (Phillips and Goodstein 2000; Lang and Hornburg 1997) suggest that it's too early to draw conclusions about the efficacy of the UGB as a viable land use tool.

Publicly Provided Incentives

The public, or societal, will in directing the course of land use is often administered through government incentive programs. These can take the form of positive incentives and negative incentives. A good analogy is that the course of land use change can be affected by carrots (positive incentives) or sticks (negative incentives).

Positive Incentives

A wide variety of programs has been established to address land management and land use practice across the United States. In addition to the federal

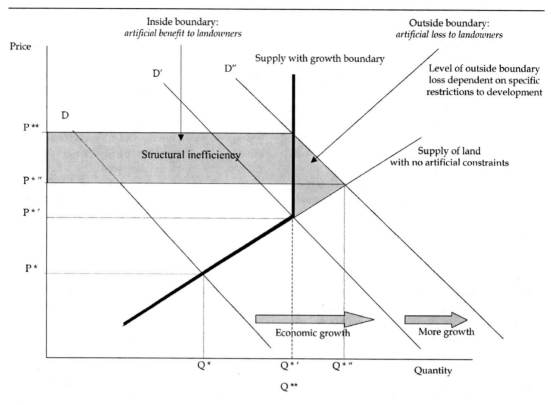

Figure 5.11.   Economic effects of urban growth boundaries on land markets.

government, state and local governments have a variety of incentive programs that target land use, broadly defined. Categories of publicly provided incentives programs include rental payment programs (e.g., the Conservation Reserve Program), subsidies (e.g., Stewardship Incentives Program and state-administered cost-share programs), and land purchase programs (e.g., federally administered Land and Water Conservation programs).

Positive incentives programs, whether they provide up-front rental payments or cost-share payments perform two essential tasks. First, they provide an incentive to landowners by lowering the landowners' relative costs of production. For instance, if, in return for meeting program objectives, government provides financial assistance to landowners in the form of tree seedlings, technical expertise, or annual rental payments, landowners will be more willing to maintain land as a managed forest. This lowered cost of production may now allow land use for producing timber to better compete with alternative land uses from the landowner's

financial perspective. Second, the essential task of any incentive program is to translate societally determined wants and needs into land management action. For example, prior to landowners being given governmental financial assistance, program involvement often entails a landowner agreeing to legally binding contracts or other enforcement mechanisms that specify how land will be used and managed. In this way, societally determined goals can be injected into how land is used.

Negative Incentives

The institution of development impact fees is a good example of negative incentives intent on affecting the spatial pattern of land use. Listed here as an incentive to more rational development, impact fees intend to remove inefficiencies and inequities associated with private decisions (sometimes referred to as market failures, or failures to efficiently and equitably translate true market costs and returns to market clearing levels) by more closely linking benefits with actual costs to local governments. Impact fees are fees assessed to developers in order to more fully capture the true costs of development. As background, it is important to note that new residential developments at the urban fringe (and elsewhere) often create significant additional costs for service provision and facility development that are publicly offered by towns, municipalities, and other smaller units of government. Examples include sewer and water provision, roads, and sidewalks.

Development impact fees, which are fairly standard in most urban centers now, attempt to assess developers the average costs of providing these services and facilities. This is particularly important with sewer and water because any use of excess capacity within a system represents an incremental step toward very high fixed-cost facility upgrades. That is, if developers use up all the excess capacity, the city will be forced to build a new sewage treatment plant or water treatment facility. Development impact fees result in a developer internalizing the negative fiscal externalities of urban expansion. Fees and regulations imposed on developers at the urban fringe can have important effects on the rate of development (Mayer and Somerville 2000; Skidmore and Peddle 1998). They can dampen development pressure along the urban fringe and create more-equitable relationships between cities and developers.

Quasi-public Market-oriented Land Use Tools

Several new initiatives have been developed that attempt to split the various rights and responsibilities of land ownership. These are often initiated in a collaborative way between landowners, private special interest groups, and local units of government, thus they are included here as quasi-public partnerships. There are two categories of land use tools that assign various land use rights for the purpose of attaining open space or conservation demands: transfer of development rights (or TDRs) and purchase of development rights (or PDRs). With both TDRs and PDRs, note that the tools rely on development rights that are independent of land ownership. The two approaches involve severing the right to develop land from the rights to exclusive ownership.

Transfer of Development Rights

TDR programs are useful land use tools for attaining previously identified goals to both maintain open space and foster more highly competitive and dense urban development. TDR agreements provide a market within which development rights can be traded.

For instance, if land use planning has targeted one area as logical for development and another for maintenance of more natural landscapes and open space, a TDR structure can allow landowners in the more restricted zone to sell development rights to landowners in a development zone. The development zone landowner might be required to buy some extra development rights in order to develop the property or to increase the density of development. Thus, TDRs allow development to take place in one area while providing incentives for landowners to make decisions that are more in concert with societally determined wants and desires in another.

A key element of the TDR approach is that it relies on stable and well-recognized long-term plans of a community. The identification of sending areas (restricted) and receiving areas (areas for development) need to be clearly and unequivocally identified through on overall planning initiative (e.g. comprehensive planning) that is accepted, implemented, and under close control (Fig. 5.12).

TDRs rely on the ability to legally sever development rights from the rights to own land. Development rights are transferred by clearly specified and legally binding deed restrictions that permanently separate the rights of owning land from the rights associated with development of that land. Often these development restrictions are specified in the deed as *conservation easements*. Also, TDRs rest on an institutional framework within which the rights can be traded between landowners.

Operationally, TDRs are set up by a regional governing body by establishing a TDR "bank" that is responsible for holding the assets of land ownership and the rights to develop land. Landowners in the sending area "sell" rights to develop their land to the bank, which then turns around and uses these development assets to allow developers within the receiving area to develop. The incentive for both governments and developers to enter into these artificial markets is the ability to develop in the growth zone at higher densities than previously allowed.

The three markets presented in Figure 5.13A, B, and C are the market for land in the receiving zone (urban land), the market for land in the sending zone (open space—rural), and the market for the rights to develop. In this conceptualization, the region is considered as a whole and developable land within the region is artificially constrained (no development can occur in the *conservation zone,* also called the *sending area*). The constraint occurs because some of the land will now be placed in permanent open

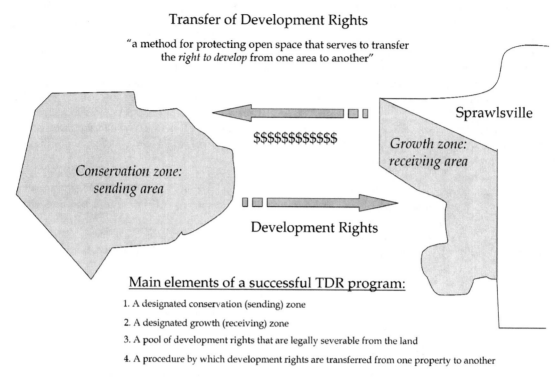

Transfer of Development Rights

"a method for protecting open space that serves to transfer the *right to develop* from one area to another"

Sprawlsville

Growth zone: receiving area

$$$$$$$$$$$$

Development Rights

Conservation zone: sending area

Main elements of a successful TDR program:

1. A designated conservation (sending) zone

2. A designated growth (receiving) zone

3. A pool of development rights that are legally severable from the land

4. A procedure by which development rights are transferred from one property to another

Figure 5.12.   Areas of land affected by a TDR program.

space. Thus, in the region as a whole, the supply of developable land decreases and the supply of open space within the region increases (shown in Fig. 5.13A and B as $S_u \rightarrow S_u{}'$ and $S_o \rightarrow S_o{}'$, respectively). The windfall benefit and the wipeout loss resulting from this policy-generated artificial supply shift provide the basis for trade (Fig. 5.13C). Note that the land bank as an institution allows for the demand for development rights to exhibit itself as a downward-sloping function (developers will internalize the cost of development). The amount of transfer in a TDR program is bounded by the supply of development rights and the demand for development rights.

The ability of TDRs to be successful depends on three critical components. First, the TDR program must be simple and easy for landowners, the public, and developers to understand. Education of affected stakeholders is key to success. Secondly, the TDR program must be a clearly identified growth management component of an overall comprehensive planning program for the region as a whole. Without

a broader regional planning effort, the security required for a market for development rights vanishes. The regional decision to support a TDR clearly must provide predictability to those landowners permanently restricting their land's development potential and to those who purchase the rights to develop within receiving zones. This predictability is critical to the success of TDRs.

Although TDRs have been around for about 30 years, some debate still brews over their efficacy. Clearly though, developing markets for transferring development options presents a more market-driven approach to land use planning. It has had the effect of making development restrictions more palatable to developers while permanently maintaining land as open space in peri-urban regions (Nickerson and Lynch 2001; Plantinga and Miller 2001; Thorsnes and Simons 1999).

Purchase of Development Rights

Another approach to affecting development is to provide financial incentives to landowners for their

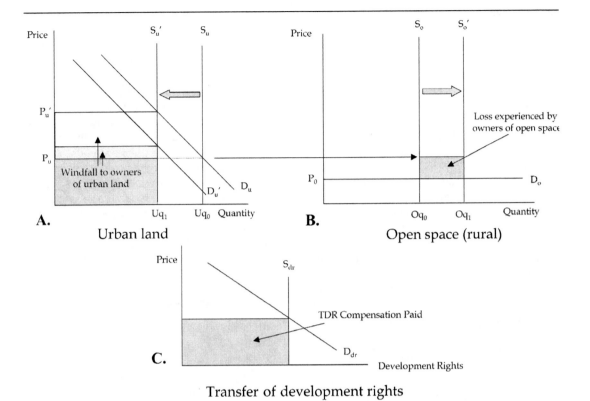

Figure 5.13. The alternative markets for land where the rights to develop land are legally severed from the rights to own land.

rights to develop land. Again, the essence of this approach relies on separating the rights to own land from the rights to develop that land. Now though, instead of trading the rights to develop for other parcels located within a development zone, these rights are purchased, hence, the term *purchase of development rights,* or PDRs. These payments for development rights can be in the form of outright payments to landowners or in the form of long-term tax breaks in return for restrictive deed language. Landowners agree, for a fee, to irrevocably restrict the deed to their land to prohibit certain uses.

The PDR approach has been successfully applied in several communities throughout the United States during the past 20 or so years. Experience, however, has shown that widespread application of PDRs is an extremely costly endeavor for local, regional, or state units of government. With increasingly tight fiscal conditions and the devolution of an array of social programs, very few units of government can

afford to make payments or forgo the tax revenue to implement PDR programs.

SUMMARY

Economic growth has led to an array of land use changes within and around cities, towns, and villages. This is due to increased demands for residential, commercial and industrial uses within agglomerated economies surrounding market centers. The strongly positive correlation between increased demands for land and distance to the urban core helps identify the notion that distance, not soil productivity, plays a primary role in determining land use within rapidly growing regions. Hence, land use that is driven by soil productivity (such as agriculture) will logically be superceded by more profitable uses, particularly in the peri-urban region surrounding rapidly growing cities.

Indeed the essence of growth management is to balance the need for economic growth with other less market-oriented objectives of society. The maintenance of open space in these rapidly growing regions is often the basis for farmland or forestry conservation initiatives. Given rapid economic growth, the market alone will not support these uses. Indeed, many of the open space attributes demanded by society are of a nonmarket nature. The values of these nonmarket goods and services are difficult to quantify but economists have begun to develop models and approaches to assist in decision-making. To make the issue of land use more complex, land is also unique from the standpoint that the rights and responsibilities associated with managing or developing land are often distinct from the rights and responsibilities to own land.

Land use planning is key to rational decisions about land use in and around community centers. Clearly, objective empirical assessments are critical for making good decisions. From an economic perspective, this analysis takes the form of fiscal impact assessment, cost-benefit analysis, and economic impact modeling. Several tools available to the land use planner help in implementing land use decisions. These include programs that offer both carrots (incentives) and sticks (regulations) to accomplish goals set out by policy-makers. The arguments associated with regulation-based market inefficiencies have lead to a couple of specific tools that work to create markets within which public land use decisions can be made in concert with private decisions about development.

Even the best analysis-based toolbox, when implemented, suffers from some inherent spillover effects between land and the socioeconomic activities of urban areas. These unintended consequences of land use planning efforts often create a spiraling effect on land values. Increased prices for land lead to inevitable difficulties for lower-income individuals and families and often work against public policies to foster affordable housing. Thus, land use planning is not simply a problem with market efficiency. There is a fundamental need to simultaneously address both efficiency *and* equity concerns of land use since the distribution of benefits from growth often accrues in a disproportionate manner to a minority of the stakeholders involved.

In closing the chapter, we want to point out that the economic growth witnessed during the 1990s exceeded most predictions and that downturns in growth are inevitable as markets adjust to outside influences. Forestry and agriculture are, in many respects, residual land uses that will be strongly impacted by changing rates of growth. This is most keenly witnessed in those spatial locations near

where growth is most rapid. This growth has been most obviously witnessed throughout urban North America, thus, the peri-urban region surrounding cities is where land use change is most rapid. Rural North America, however, is also beginning to see dramatic economic growth take place, particularly on sites directly adjacent to natural amenities such as waterfronts, vistas, and key natural landscapes. Many of the same economic forces, analytical techniques, and land use tools discussed in this chapter also have direct application to rapidly growing amenity-based rural regions that are distant from urban centers. Our ability to balance economic growth with other societally determined wants and needs provides the challenge associated with managing growth. Sound economic, social, and environmental planning is necessary to maximize societal benefits while ameliorating the societal costs.

STUDY QUESTIONS

1. Compare and contrast land fertility and distance to market center as explanatory aspects of how land is valued.
2. How is agglomeration in economic activity spatially analogous to perishability of agricultural commodities?
3. How does economic growth affect land value in regions just outside rapidly growing communities?
4. What distributional issues arise with rapidly escalating land values?
5. What aspects of land rent are analogous between urban and rural regions?
6. What aspects of land are not well addressed in a purely market-driven land use allocation process characteristic of many North American cities during the latter part of the 20th century?
7. How do societally determined values play a role in how land is developed?
8. What are some unintended social consequences of strict land use regulations?
9. How is the exurbanization process the inevitable result of misapplication of urban land use planning tools?

NOTES

1. Amenities can include natural, cultural, and human developments that affect a region's quality of life. Several complexities related to amenities involve definitional aspects, unique

characteristics, and usefulness as regional factors of production (Chapter 9).

2. The spectrum of economic values includes both use and nonuse elements. The complexities of nonmarketed goods and services are addressed in Chapter 9.

3. These historical figures had several landmark publications that provided a basis for our contemporary understanding of land economics. Interested readers are referred to von Thünen's work entitled *Der Isolierte Staat,* originally published in 1826 (see *Isolated State,* a translation by Wartenburg in 1966), and Ricardo's work *On the Principles of Political Economy and Taxation,* originally published in 1817.

4. In addition to distance decay as we move away from central urban core areas, an analogous spatial concept is required to understand the economic effects of amenities. This amenity-based land rent gradient is further explored in Chapter 9.

5. This and other topics are fully described in the classic text on land economics written by Barlowe (1986).

6. As this story progresses, we'll interject a fourth staple of life that represents improved standards of living, which was beyond von Thünen's grasp but is critical to bring this explanation up to modern times. This fourth staple involves quality of life, or amenities, and is increasingly important in location decisions of firms and households. Land use for amenity values and speculative land values will inevitably dictate land use in an expanding urban-exurban interface.

7. For example, urban decay in some cities during the 1960s and 1970s created a depression in the center of the cone with beltway highways surrounding cities determining maximum land rents.

8. Specific details of these techniques are beyond the scope of this chapter. Readers are referred to classic texts by Mishan (1976) and Pearce (1983) for full descriptions, caveats, and applications of cost-benefit analysis.

9. Details and limitations of specific fiscal impact analysis techniques are beyond the scope of this chapter. Interested readers are referred to a classic text in fiscal impact modeling by Burchell and Listokin (1978). We also elaborate on this general topic in Chapter 10.

10. Modeling economic growth is theoretically presented in Chapter 2, while tools for economic impact assessment are discussed in Chapters 14 and 15. Interested readers also can learn more from classic texts written by Pleeter (1980) and/or Miller and Blair (1985).

11. Crown lands predominate across Canada; they are formally considered federal in ownership, but the provincial governments retain management control. In effect, provincial governments exert dominant interest in using federally owned crown lands in Canada.

6
Labor Markets

Society places great emphasis on jobs and the community's supply of labor is one of its greatest resources (Fig. 1.1). The possession of a job in the American economy provides an income that determines, to a large extent, the capacity to pursue a particular lifestyle. Because jobs are central to society and personal perception of worth, preparing people for work, placing and keeping them in jobs, and providing opportunities for advancement are critical.[1] Thus, a job represents a very valuable element of modern life, with many economic, social, and psychological benefits attached to it (Cherry and Rodgers 2000; McConnell, Brue, and MacPherson 2003). As we have progressed through state and federal welfare reform policies of the 1990s, the safety net for those in poverty was replaced, in large part, with contemporary social policy that emphasizes work first (Brown 1997). In the United States, there is an increasing emphasis on work as a replacement for welfare (Weber and Theodore 2002).

Labor economics traditionally examines the interaction between the worker or group of workers and the firm or group of firms. We follow this approach initially but then we examine the implications of such an analysis in the context of community economic development. If land and capital are fixed within a community economic output, growth and to a larger extent development are a function of the size and quality of the labor force and how it is utilized. We present some of the major theories concerning the role of labor in the economy and its utilization, with specific emphasis on employment and unemployment, corrective actions for labor market barriers, and the emerging concern about job quality. We also review components of labor market theories explaining the demand for and supply of labor. Informal labor, poverty reduction policies, and welfare reform also are discussed.

LABOR MARKETS

The exchange of labor services between worker and employer occurs in the context of demand for and supply of labor, within an overarching institutional framework that affects the interchange. As we saw in the circular flow model of the economy (Fig. 4.1) households own all factors of production and sell them to firms in the factor markets. The case of households owning the factors of production is perhaps clearest with respect to labor. The market for labor displays many of the same characteristics as the market for capital or any other factor input (Kalleberg and Sorensen 1979; Kreps et al. 1980; Clark 1983; Hotchkiss and Kaufman 2002). At the same time, we recognize that labor is not a commodity, but it is embodied in people and therefore requires particular attention not necessary in other markets (McConnell, Brue, and MacPherson 2003).[2] In theory, a labor market is an institution where labor services are bought and sold and therefore allocated to various occupations, industries, and geographic areas to yield the greatest output to society. The market sets the price or wages of labor that allocates labor to its most productive use. The labor market need not be a physically contiguous area, although it generally is; it can be widely separated spatially and linked only by information flows.

Employers perceive their labor market as being the geographic area containing people who either are in the labor force or are willing to enter the labor force if the firm offered an appropriate job (Lever 1980). A job's appropriateness can be judged in many ways, including skills required, wages, fringe benefits, opportunity for advancement, working conditions, and commuting distance to work.

The supply of labor is a function of the existing population base of the community and nearby area. It depends on the age, gender, education levels, skills possessed, and lifestyle of that population. Workers generally view the labor market as that spatial range of employment opportunities open to them without changing their residence.

In a perfectly competitive labor market, workers move among jobs, occupations, firms, and locations to eliminate differentials in wages and yield an efficient allocation of labor. Shortages of workers cause wages to be bid up, while a surplus of workers depresses wages. The wage rate becomes the market-clearing (equilibrating) signal.[3] The wage paid to the worker, however, depends on the firm's return from that labor service.

The textbook version of the perfectly competitive labor market assumes a nondifferentiated market with jobs being similar and workers being substitutable. As we saw in our regional neoclassical growth model in Chapter 2, jobs and labor are treated as commodities with no differentiation across communities. Labor markets, however, are segmented, and the boundaries between labor markets are often impenetrable. Segmented labor markets occur in a variety of ways. Three important ways to segment labor markets are spatial, occupational, and institutional (Clark 1983). Labor markets in California or southern Illinois are different from labor markets found in Alabama or northern Minnesota.

Spatial segmentation occurs when distance prevents the exchange of information about job opportunities and worker availability. *Occupational segmentation* occurs when workers possess different skills, jobs require different skills, and the workers are not very substitutable. The proverbial ditch digger rarely is substitutable for the lawyer or the chief executive officer. Substitution can occur, but it requires time. The *institutional segmentation* of labor markets can be very explicit or very subtle. Explicit segmentation occurs when accreditation or a license or membership in a union is required. Many occupations predominantly filled by either males or females display subtle labor market segmentation. Internal or external labor markets are another subtle form of labor market segmentation. An *internal labor market* refers to labor markets that exist within firms, usually larger firms. The *external labor market* represents individuals seeking employment who are not currently attached to the firm.

For the community practitioner to understand the local labor market, some boundaries—no matter how arbitrary—must be drawn to define the labor market. For individuals with comparable skills the labor market boundaries typically include the area where conditions of information transfer exist about employment opportunities and potential workers. The boundaries of a labor market are not fixed; they vary according to the worker's occupation, mobility, transportation systems, and spatial location of development. The unskilled worker's labor market boundary often extends a relatively short commuting distance (Clark and Whiteman 1983). Distance and transportation systems are playing an increasingly critical role as barriers to the supply of low-skilled labor (Blumenberg 2002) with notable race and gender linkages. In contrast, labor market boundaries for professionals extend to regional or even international boundaries. The geographic boundary also depends on the information transfer mechanism about job opportunities and potential workers (Chapple 2002).

An added dimension to defining labor market boundaries is a tendency for firms and laborers to be only partially integrated into the external labor market. Workers are more informed about job availability, job requirements, and the associated work conditions within the internal labor market. Likewise, firms know more about existing workers and their potential to fill new job openings. Thus, the tendency is for the labor market boundaries to extend no farther than the business premises.

Some important elements defining labor markets need to be re-emphasized. First, labor market or labor shed boundaries depend on a person's ability and willingness to travel from his place of residence to his place of employment.[4] Second, there is an emphasis on the matching of skills in the work force with job requirements (i.e., nonhomogeneous components). Third, the flow of information about job opportunities and availability of workers is an important determinant of the labor market boundary. Informal communication is a much more powerful explanatory factor in understanding the job/worker matching process than are formal institutions.

A well-functioning labor market is critical to community economic development. A labor market that allows mobility of workers from low- to high-value output jobs increases community production and per worker income. Furthermore, properly functioning labor markets improve the distribution of income by facilitating the transfer of workers from lower- to higher-paid jobs. As lower-wage jobs become less attractive to workers, those businesses offering low wages will be forced to bid up wages. Perhaps most importantly, a well-functioning labor market provides job opportunities to workers of all skill levels.

Demand for Labor

The demand for labor consists of existing employment opportunities, filled and unfilled, both locally and nonlocally as well as in both formal and informal labor markets.[5] Unlike our simple regional neoclassical growth model, there is no single demand for labor. The numerous demand functions for labor vary among occupations/skills, places, and sectors, and over time. The nonhomogeneity of demand must be recognized in any effort to analyze and influence the labor market.

The demand for labor is a derived demand. Employers do not desire workers because of their intrinsic value. An employer's demand for labor derives from labor's productivity and the price of the product labor produces. The demand for labor is really a function of several economic forces. First is the demand for the good or service that labor actually produces, expressed in terms of the price of the output. Second is the productivity of labor embodying the existence and efficient use of other resources. Third is the price of those other resources used with labor. Thus, if the price of a *substitute* resource falls, there will be a *decline* in the demand for labor. If the price of a *complementary* resource falls, there will be an *increase* in the demand for labor. Fourth, a shifter of the demand for labor is a change in the firm's goals, such as dropping profit maximization for maximizing market share, or maximizing the owner's leisure time (Bernstein and Hall 1997; McConnell, Brue, and MacPherson 2003).

The demand for labor is a function of labor productivity, the price of the output being produced by labor, the price of other factors of production, and the price of labor itself. But it is the price of labor that drives the labor market in terms of both demand and supply. For a representative firm, demand for labor (VMP_L) can be expressed as

$$w = MPP_L \times P_o = VMP_L \qquad (6.1)$$

The firm is willing to pay a worker wages (w) based on that worker's marginal physical product (MPP_L) and the price of the output (P_o) produced. The statement that labor has a derived demand becomes explicitly clear in equation (6.1). First, the marginal physical product of any given worker depends on the worker's inherent productivity (e.g., human capital). Second, the marginal physical product of any worker depends on the amount of physical capital and other fixed resources used by the worker, along with the technology embodied in the capital.

Third, the wage the firm would be willing to pay the worker depends on the price of output produced.

The demand for labor in equation (6.1) presumes competitive markets. Thus, the workers in a community are willing to provide all the labor needed at any given wage rate. In other words, for simplicity we assume for now that the supply of labor is perfectly elastic. At the same time, the demand the firm faces for its product is also perfectly elastic, or it can sell all the product it can produce and therefore the firm is willing to demand all the workers it can acquire as long as their productivity justifies it.

In Figure 6.1, the labor demand (VMP_L) curve represents the firm's short-run demand for labor. The position of VMP_L is governed by the productivity of labor (MPP_L) and the price of the output produced (P_o). The MPP_L is governed by the amount of fixed physical capital used with labor and the specific type of technology embodied in the physical capital.[6] The price of the firm's output (P_o) is governed by the competitive conditions in the product market and consumer preferences. Under constant or diminishing rate of returns, if more labor (L) is hired, the marginal productivity of labor (MPP_L) declines, decreasing the marginal value product (MVP_L). Hence, by equation (6.1), firms are willing to pay less (w) for labor. This inverse relationship is captured by the downward-sloping demand curve.

In a competitive labor market, the labor supply curve maps out the cost of labor from the perspective of the firm and can be interpreted as the firm's marginal cost of labor. The firm will hire L_c labor and pay w_c if it faces competitive product and factor markets. When the firm faces a less than perfectly competitive product market, its demand for labor will be altered. Regardless, in our simple model of the community, the sum of firms' demands for labor represents the community's demand for labor.

If a community has a less than perfectly competitive labor market, the equilibrium will differ from the competitive equilibrium (Kreps et al. 1980; McConnell, Brue, and MacPherson 2003). This occurs under situations of monopsony or oligopsony (when a firm is the only purchaser or one of a few purchasers of labor in the community). Noncompetitive labor supply can also occur as a consequence of unions, certification of qualified workers, and restrictions on in-migration or in-commuting of workers. In Figure 6.1, the marginal cost of labor (MRC_L), in the less-competitive labor market, lies above and to the left of the marginal cost of labor

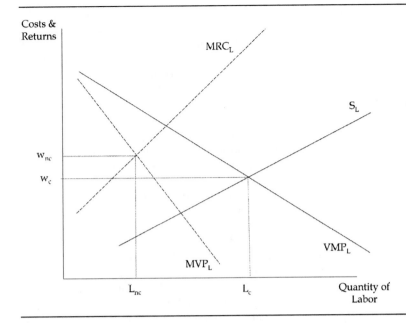

Costs & Returns

MRC$_L$

S$_L$

w$_{nc}$

w$_c$

VMP$_L$

MVP$_L$

L$_{nc}$ L$_c$ Quantity of Labor

Figure 6.1. Demand for labor under various competitive conditions.

curve (S_L) in a perfectly competitive market. In this market environment, the firm maximizes its profits when $MRC_L = MVP_L$ (assuming a less than competitive product market). The amount of labor used (L_{nc}) declines, while the wages paid (w_{mc}) increases.[7]

By equation (6.1), as output prices increase under noncompetitive output markets and the quantity of output declines, the amount of labor hired declines, resulting in an increase in the marginal physical product of labor. As output price and marginal physical product of labor increase, wages must increase. The less than competitive conditions in the labor market could be caused by a geographically isolated market or by specific occupational demands with little substitutability among jobs and workers (Madden 1977). The net impact of imperfect competition on total labor income in the community hinges on the interplay between higher wages and lower employment.

Beyond the competitiveness of local labor markets, the level of community demand for labor (the sum of individual firm demands) is a function of industrial mix, technology use, community priorities, and level of national economic activity. The composition of the local economy (industrial mix) affects the level of labor demand and composition of the demand for labor (e.g., skilled/semiskilled/unskilled).

The question of technology extends the discussion of industrial mix one more step (see Chapters 2 and 8). Obviously, different sectors use different technologies, but firms in the same sector of the economy may also use different technologies to produce the same product or service. In the most general sense, technology is either labor intensive or capital intensive. In a more specific sense, different technologies require different types of worker skills. For example, the change in the delivery of health care from hospitals to clinics alters the types of health care skills required. Again, the issue is that each and every job is not like every other job. They differ in terms of skills required, occupational classification and wages amongst other factors.

The product cycle theory outlined in Chapter 3 also can be used to help gain insights into the spatial demand for labor. *New products* require the maximum amount of flexibility by producers to adapt in getting the right production mix. In figuring out the right production mix, there is an extensive use of highly skilled technical labor. Furthermore, as new products and their production technologies evolve, there is a need for extremely close ties to management and advanced business services. More generally, new products need ready access to the types of inputs found in urban centers.

The second phase of product cycles is found in *maturing products,* where firms move production toward economies of scale. As the product's demand grows, standardized production methods are put into place and the location of production now depends on traditional shipment and labor costs. This phase starts the movement of production (and hence the demand for labor) away from technically advanced regions and loosens the firm's ties to urban centers.

The final phase of production focuses on *standardized products*. The product now exists in a situation of considerable price competition for both inputs and outputs and requires a significant amount of unskilled labor. The tendency for spatial location of mature production is a general movement of production to low-wage labor supplies. Traditionally, these have been located in rural, nonmetropolitan regions. With rapid globalization during the past 30 years, this spatial shift in labor demand is increasingly being absorbed by foreign workers in developing countries. The North American Free Trade Agreement, for example, has stimulated significant mature product manufacturing in northern Mexico, where labor costs are minimized.

An alternative way to think about the product cycle theory has been offered by Thompson (1987), who argued persuasively that five broad occupational-functional classes define local development paths which have explicit impacts on local labor markets:

Entrepreneurship
Central Administration
Research and Development
Precision Operations
    Blue-collar skill (tool and dye workers)
    White-collar occupations (auditing and
      advertising)
    White-coat, laboratory technologies
Routine Operations

Entrepreneurship emphasizes innovation, risk identification, and implementation of business ideas that can be independent of spatial location. *Central administration* and the associated business services tend to be clustered in urban centers. *Research and development* also tends to be clustered in urban areas but it also is tied to institutions of higher education. Each of these classes can be associated with the first stage of the product cycle theory.

Precision operations are places that emphasize skilled work, while *routine operations* are places where the major occupational category is low-skilled labor. These latter two functional classes cor-

respond to the final two stages of product cycle theory.

The occupational demand for labor will vary in each of the functional classes. One can think of this in terms of the spatial endogenous growth theory discussed in Chapter 2, as well as the structural change theory.

The demand for labor varies over time and space due to changes in technology, changes in consumer taste and preferences, changes in income levels, changes in the price of other factors of production, or changes in the amount of other factors available. The incidence of these changes affects communities and groups of laborers differently. The dynamic nature of the demand for labor means that workers must change jobs or residences, or acquire different skills. The ability of labor to respond depends on labor's mobility. The demand for labor in a community is influenced by the inherent skills and productivity of the resident labor force (i.e., human capital), competitive conditions in both product and input markets, and the type of technology and capital used as represented by the types of sectors present in the community, community priorities, and nonlocal economic activity.

Supply of Labor

The economic supply of labor refers to the various combinations of wage rates and quantities of labor offered. For the individual, the wage rates and quantities of labor offered represent a trade-off between time spent in the formal and informal labor markets. The *formal labor market* includes working in a factory, farm, or office for which the individual receives a monetary wage. The *informal labor market* (sometimes dissected into leisure and nonwage work) represents those activities individuals do to meet their own personal and psychic needs, such as home maintenance, child care, food preparation, sleep, and vacation (Becker 1965; Larrivee 2000). Since individuals face the constraint of having only 24 hours a day, they allocate those hours between work and informal work daily, weekly, annually, and over a lifetime.

The *community labor supply* is the sum of individual decisions that arise from the local population base, in-commuters, and in-migrants. The composition of the labor force in any given community derives from many factors, including (a) the participation rates of various components of the population (influenced by age, gender, race), (b) attitudes toward work for pay, (c) attitudes toward work after certain ages (retirement), (d) acceptability of teenagers joining the labor force, (e) labor legislation (e.g., antidiscrimination provisions, minimum wage laws), and (f) rules governing retirement and welfare benefits (Bernstein and Hall 1997; Hotchkiss and Kaufman 2002; Kreps et al. 1980; McConnell, Brue, and MacPherson 2003; Scott, Smith, and Rungeling 1977).[8]

The total supply of community labor offered to employers depends on the number of hours each worker is willing to work and the number of workers available. Of relevance is not the supply of labor to the economy as a whole, but the supply of a particular kind of labor to a well-defined group of employers requiring specific skills at a particular location. This specificity means that the supply of labor depends on the length of time involved and the skills required.

The time required to acquire skills, the worker's need to have nonmonetary aspects of life satisfied, time to move among occupations and places, and division of total hours in the family among work and informal work activities lead to the various shapes of the labor supply curve. The labor supply curve could be vertical or backward bending, but it most likely slopes upward to the right. To suggest that it could actually slope upward to the left (i.e., backward bending) appears to defy economic logic, but it does recognize the noncommodity aspects of labor. Typically, a backward-bending supply curve means that above certain wage rates, workers choose informal over formal work (e.g., workers rebelling against mandatory overtime). In a sense, wages are sufficiently high that workers can cut back hours and still maintain their desired standard of living.

The concept of reservation wage plays an important role in labor force participation. The *reservation wage* displays the relationship between income and leisure, and informal work. Essentially, this is a function of nonwork income (dividends, rents, welfare, gifts) and the indifference curve (utility function) displaying preferences between leisure and income needed to purchase the goods/services needed to maintain the desired lifestyle. The market wage must exceed the reservation wage for people to enter the labor force and offer time for paid work in order to maximize their utility.

Another dimension of the labor supply curve is *elasticity*. The long-run supply curve of labor is naturally more elastic (flatter) than the short-run curve because time permits greater changes in human capital and fixed capital, and greater mobility among occupations and locations. Long-run cultural changes, such as attitudes about the age of retirement, child labor, or the participation of women in the formal labor force, alter the supply of labor. Elasticity of the labor supply varies among skill levels also (Kreps et al. 1980; McConnell, Brue, and MacPherson 2003). The more specialized the skill, the more inelastic the supply of labor function because of the time it takes to acquire the skills. The geographic scope being analyzed also affects the elasticity of labor supply. The elasticity of labor supply for the entire economy is more limited than the labor supply in a particular locale because of the potential to transfer workers from one locale to another. Changes in wage rates affect labor force participation and the elasticity of labor supply. An increase in wages or in number of job openings attracts individuals only marginally attached to the labor force (e.g., youth, discouraged workers). These individuals essentially substitute formal for informal work (Larivee 2000).

It is important to distinguish between the supply of labor (as defined above) and the labor force as officially defined. The *labor force* is officially the sum of employed and unemployed workers (unemployed are those actively seeking work but unable to find it). The major difficulty with this definition is hidden unemployment, which takes two distinct forms. The first is underemployment, where individuals work in jobs that do not fully utilize their skills; these workers count as employed. By the same token, people may be overemployed (e.g., their job requirements exceed their skills).[9] The second form of hidden unemployment is represented by the group of discouraged individuals who are no longer actively seeking work because they perceive that their chance of locating a job is minimal. Since they are not looking for work, they are not unemployed or even in the labor force. These people enter the labor force when they perceive employment opportunities. The reason people may be discouraged could be entirely institutional, such as discrimination.

The employed worker represents an often overlooked source of community labor because employed individuals are presumed to be satisfied with their current jobs and not willing to change. Employed individuals, however, may find themselves holding jobs for which they are either overqualified or underqualified, or they may be dissatisfied with the working conditions or may desire higher wages. Regardless of their willingness to pursue new employment opportunities, the existing employment base typically represents the major portion of the local labor force in the community.

A hidden supply of potential labor is new entrants and returnees to the labor force. New entrants are individuals entering the labor force for the first time. They may be recent graduates from high schools, vocational schools, or universities, or they may be individuals leaving the informal labor market (e.g., housewives) to enter the formal labor market. Returnees, on the other hand, previously worked in the formal labor force and left because of unemployment or discouragement about employment opportunities, family responsibilities, or changes in lifestyle. These individuals now choose to re-enter the formal labor force. In many communities, new entrants and returnees represent a large reservoir of individuals available for employment if they perceive an opportunity to earn a wage.

The preceding discussion focuses on factors affecting movement along the supply curve. Shifters of the supply of labor include changes in nonlabor income, such as Social Security and investment income; changes in wages or earnings of other family members; changes in household productivity; changes in desire for wage versus nonwage work; and changes from the goal of utility maximization to a different goal. The local labor supply can shift because of in-commuters, in-migration, or new entrants or because of changes in the investment in human capital. In-commuters and in-migrants explicitly recognize the spatial dimension of local labor markets. In-commuters reside in nearby areas and commute to the community on a regular basis to fill local employment opportunities. In-migrants are new residents who move to the community for a variety of economic reasons, including employment opportunities, or noneconomic reasons, such as retirement. A less obvious source is labor mobility among jobs and among occupations. The human capital change is a long-run phenomenon that is discussed later.

Age, family responsibilities, and current skills influence labor mobility.[10] Lack of family responsibilities, lack of inertia, lack of job security, and lower wages reinforce job mobility of younger workers. The labor supply depends on the time allowed for adjustment to occur, the level of wages offered, the skills required, the preferences for wage/nonwage work activities, and the composition of the local labor force, as well as the extent to which people will either commute or move to the community. One must keep in mind that the decision to enter or leave the labor market is a household or family, versus a strictly individual, decision.

Institutions/Rules

Institutional rules move labor from a commodity to a product by affecting both the efficiency and equity outcomes of labor supply and demand. Labor market institutions provide the rules that govern the mobility of workers among jobs, the acquisition of skills and training, and the distribution of wages and other rewards obtained from working (Kalleberg and Sorensen 1979). Since labor is not a commodity, labor responds to market signals in a slightly different fashion than commodities would.[11] Rather than a single characteristic, jobs possess a bundle of qualities, such as working conditions, location, climate, company prestige, chances of promotion, seniority rules, and friendliness of co-workers, all of which alter the character of the job. These qualities combine to affect the worker's choice of where to work and for what wage (Clark and Whiteman 1983; Lester 1966). At the same time, firms do not perceive workers as homogeneous in skills, attitudes, motivation, stability, and health (Kreps et al. 1980).

A major societal rule is one that affects both the entry and exit into the labor force. For example, it generally is accepted that retirement age is 65, but recent federal legislation is moving the retirement age to 67 and 70. Child labor laws also affect entry into the labor market. There is also a wide range of rules and regulations dictating safe working conditions, unemployment insurance, and workers compensation.

Wages are the job's most obvious characteristic. Who has not asked questions about wage rates or annual salary associated with a particular occupation or job? Yet, fringe benefits are increasingly important. What type of health insurance, life insurance, vacation plan, and sick leave are linked with the job? Working conditions are also important. Conditions include whether the work environment is inside or outside, relatively clean or dirty, noisy or quiet, safe or dangerous, and the perceived nature of labor-management relations. Likewise, individuals often find co-workers instrumental in helping to form positive or negative perceptions about a job. The opportunities for advancement, either perceived or real, also shape opinions about the job. Some jobs offer no opportunity for advancement, while others represent stepping stones toward something better

(Dresser and Rogers 1997; Rogers and Dresser 1996). A final important dimension of a job is stability. How sensitive is the job to seasonal variation (e.g., construction work, farm work), to cyclical variation (e.g., automobiles, construction work, heavy manufacturing), or to secular trends (e.g., the replacement of many low-skilled jobs with jobs requiring different or higher skills because of changes in technology)?

In essence, the prior discussion condenses to characterizing jobs as good or bad jobs. The criteria for making this judgment typically includes wages, fringe benefits, irregularity of employment, working conditions, job security and opportunity for promotion (Dresser and Rogers 1997; Gordon 1972; Rogers and Dresser 1996). A good job provides adequate wages and fringe benefits, job security, decent working conditions, and an opportunity for advancement and control of one's work environment. The important point in defining good and bad jobs, however, is that people have different perceptions of good and bad. Workers place different values on the prestige, income and fringe benefits, authority and power, career opportunities, ability for self-direction, need to think about what they are doing, and interpersonal relationships associated with a job (Kalleberg and Sorensen 1979). This essentially says that communities in their quest for jobs must not discount a particular job because it is not a good job. But at the same time, the community can't just take any job. Communities need to recognize that workers desire jobs possessing various characteristics.

LABOR MARKET THEORIES

This section reviews the major labor market theories that can be grouped as supply oriented or demand oriented.[12] Some have focused principally on the characteristics that people bring to the market place (i.e., the supply side). Others have focused on the structural factors that affect the jobs available in local areas (i.e., the demand side). The theories do offer some insight for community economic analysis purposes.

Supply-oriented or Human Capital Theory

Human capital theory dominates theories emphasizing the supply side of the labor market. Human capital theory is rooted in the works of Schultz (1961), Becker (1962), and Mincer (1962). It builds on the neoclassical model of perfectly competitive labor markets and argues that human capital investments influence future monetary and psychic incomes by increasing the productive capacity of individuals (Beaulieu and Mulkey 1995; Becker 1985; Clark 1983). Examples of human capital investments include schooling, on-the-job training, and medical care. Such investment makes a worker more productive by augmenting her skills, knowledge, health, or other productive attributes. The improvements in human capital can have utility to a variety of firms or they can be firm specific and have applications only to the particular enterprises involved.

According to human capital theory, workers bring to the job skills and knowledge acquired through schooling, training, and work experience. In a perfectly competitive world no worker will be paid more than the value of his marginal product because a replacement worker is willing to work at a wage rate equaling the value of marginal productivity. Likewise, an employee paid less than the value of her marginal product will seek a job with a wage rate reflecting her productivity. A worker's productivity is a function of the skills and knowledge he acquires through schooling, training, or work experiences and also his health status and mobility. Workers increase their future earnings by investing in productivity-augmenting activities. Human capital theory describes how an individual chooses the "correct" amount and type of human capital investment.

Investment in schooling or training activities provides a worker with skills and knowledge to perform more-complex job tasks and, therefore, command a higher wage because she is more productive. Each worker weighs the discounted benefits of increased earnings against the cost of obtaining additional schooling or training. If the discounted value of the future benefits from a productivity-augmenting activity exceeds its cost, the investment will be undertaken.

At least five reasons account for differences in the amounts of human capital among individuals: (1) different expected annual returns, (2) different costs, (3) different amounts of time expected to be spent in the labor force, (4) different discount rates, (5) and different cognitive abilities. According to human capital theory, people receive different returns from investment in schooling or other training because of variations in wages paid or the kind of firm they work for, thus leading to variations in human capital investments (Becker 1985). Time remaining in the labor force also influences individuals considering additional schooling, training, or other productivity-augmenting investments. Women often spend a substantial amount of time out of the labor force or work reduced hours to care for children, which gives them an incentive to invest less in human capital because they have less time to recover their investment (Kahne and Kohen 1975). People differ in their time preference for income (i.e., different discount rate), which influences the amount of investment they will make in school or training. This discount rate also varies because of uncertainty about future returns.

Mainstream economists using the human capital approach insisted that passing through schools increased earning power of people independent of differences in family origins and inborn or acquired cognitive abilities (Blaug 1985). It has been suggested that an individual's abilities and level of schooling are significantly shaped by important family characteristics. This is the focus of research on *status attainment* (Beaulieu and Mulkey 1995). Status attainment research contends that a family's social economic status (SES) plays a substantial role in shaping a person's success in school and in influencing his early occupational choices. SES conditions the environment of support for aspirations and achievement, and individuals from higher SES families are often more socialized to place a high value on educational achievement. This view can also be used to help understand the cycle of poverty.

Bowles and Gintis (1976) argued that the human capital approach to education is flawed. The human capital approach says that people are rewarded for the cognitive skills that they gain in education. Bowles and Gintis saw that schools perform a socialization function which may be as important, or even more important, than the cognitive skills portion. In other words, for lower-level occupations to which unqualified school-leavers are largely condemned, the behavioral traits of punctuality, persistence, concentration, docility, compliance and the ability to work with others are important. The higher occupational levels (university graduates) call for a different set of personality traits, such as self-esteem, self-reliance, versatility, and the capacity to assume leadership roles (Blaug 1985). But even at the level of professional studies, the cognitive knowledge frequently consists of general communication skills and problem-solving abilities rather than occupation-specific competencies. This implies a combination of particular personality traits and certain cognitive achievements are involved (Blaug 1985).

To accumulate a productive labor force requires a substantial investment by the community or the worker. From the community's perspective, the investment takes the form of providing elementary, secondary, vocational, or university education. The community also bears a burden in the form of foregone output or poor-quality output during the training process. From the individual's perspective, the investment takes the form of foregone income during training, the cost of tuition, and other costs such as transportation back and forth to class. Both the community and the individual, however, perceive that future returns from these investments, both in increased wages and in output, will more than offset the cost.[13] Investment in human capital appears to conform to the law of diminishing returns.

Unlimited investment in human capital need not yield continuing increases in income and output both nationally and locally (Dresch 1977). The rational worker/community invests only to the point where the future benefits from the investment equal its cost.

The important insights from human capital theory can be summarized as follows:

- The amount an individual invests in human capital depends on the expected annual return from the investment, the cost of the investment, the expected time in the labor force to recoup the investment, and the discount rate.
- Persons who expect to spend a significant amount of time out of the labor force to raise children will invest in less human capital than workers planning to be in the labor force for several years.
- Compared with other members of the labor pool, youth possess less human capital because they have not completed their formal education and they lack work experience.
- Compared with individuals in the middle of their life cycle, individuals at the end of their labor force careers possess less human capital because some skills have become obsolete. Because their remaining time in the labor force is short, they are less likely to invest in updating old skills or learning new skills. In addition, they are more likely to have health problems.

The human capital theory provides a foundation for understanding how local labor markets function. This approach emphasizes efficiency in allocating labor resources among competing uses. Yet, the major problem in local labor markets is access by some individuals to full participation in the labor market. The neoclassical model does not provide very good insight into these equity questions. Another limitation of the neoclassical approach for deriving policy is that the worker carries the implicit burden for all the adjustment. The theory is silent on how adjusting labor demand (i.e., quantity and skill requirements) influences the use of labor and employment/unemployment levels.

While human capital provides an excellent starting point, it places too much reliance on the rational person understanding her investment return options as well as having control over her life and having equal access to job opportunities. People are subject to family and psychic ties as well as institutional constraints (e.g., discrimination) that prevent equal access. These shortcomings of the human capital theory have stimulated the creation of alternative explanations.

Demand/Rules-oriented Theories

According to neoclassical theory, the production skills and other worker characteristics demanded by a given employer depend on the firm's production technology. Specifically, the occupational requirements and the type of physical capital employed determine an employer's choice of workers from alternative groups in the labor pool because these factors define the type of production skills workers need to perform a job. In the neoclassical model, the match between workers and jobs is relatively uncomplicated. Workers with production skills most closely matching those required for a given job are hired.

The refined version of human capital theory, as outlined above, diverges substantially from the neoclassical theory of labor markets. First, the assumption of homogeneous workers and job requirements is relaxed. Second, restrictions on occupational mobility are recognized through the need to acquire additional training. Third, the uncertainty about educational investments is implicitly recognized in the variation of discount rates among individuals.

Critics argue that human capital theory masks the importance of behavioral parameters on the demand side of the labor market. Alternative labor market theories focus on the role of hiring rules established by companies when they set up their personnel policies, by trade unions, and by the actions of government (Cain 1976; Kerr 1982). These rules state which workers are preferred by employers in matching workers with jobs. Three major theories emphasize the role of hiring rules: signaling/screening theory, job competition theory, and segmented labor market theory.

Signaling/Screening or Credentialing Theory

The roots of signaling/screening theory lie in organizational behavior theory under conditions of uncertainty and imperfect knowledge (Arrow 1973; Spence 1974). Advocates of signaling/screening theory argue that the majority of job skills are acquired on the job, either by specific training or through *learning while doing* (Piore 1973; Spence 1974). Consequently, employers search for workers who are efficient in learning new job tasks and are dependable once they are trained. The firm needs quick and cheap techniques that enable it to improve its prospects for hiring dependable workers efficient in learning new skills (Blaug 1985). This exists as a critical information problem in many labor markets. Signaling/screening theory suggests hiring standards are not prerequisites for satisfactory job performance, rather they are proxies for personality characteristics such as ability, motivation, dependability, and willingness to learn (Spence 1974).

In the screening hypothesis, employers, lacking full knowledge of the productivity of potential employees, rely on certain devices that offer them a signal of the productivity level of individuals. The key signaling device that job candidates invest in is education. The screening advocates do not contend the education intrinsically enhances an individual's productivity (human capital); schooling simply certifies that those who have successfully completed a given level of education possess certain qualities (skills, ability, or family background) that should be rewarded.

The limited number of points of entry to the firm reinforces the employer's interest in hiring standards to screen potential employees. Many firms possess distinct families of jobs that are progressive as career ladders (Dresser and Rogers 1997; Hotchkiss and Kaufman 2002; McConnell, Brue, and MacPherson 2003; Reich, Gordon, and Edwards 1973). Within each family of jobs is a skill or responsibility hierarchy. Jobs at the bottom of each hierarchy require the least skills and are associated with the least responsibilities. Many firms customarily hire from outside the firm only to fill jobs at the bottom level of the hierarchy. Intermediate and upper-level positions are filled by promoting workers from a lower level in the hierarchy (Doeringer and Piore 1971). Lower-echelon hiring thus determines the future labor pool to fill higher-echelon positions. Hiring unskilled workers frequently determines the group from which key semiskilled workers and supervisors will eventually be selected. In hiring this year's salesmen, an employer may be selecting a future sales manager. For this reason, minimum education and experience standards may be set higher than required for the entry job (Kalachek 1973). Hiring standards reflect the desire to hire workers capable of ascending the skill and responsibility ladder.

Job Competition Theory

Job competition theory explicitly incorporates the concept of labor market screening into the model of firm profit maximization (Thurow 1975). Like screening theory, job competition theory assumes that most job skills are acquired either formally or informally on the job after the worker finds an entry-level position and the associated promotion ladder. Firms profit by minimizing the training cost associated with bringing the worker up to the standard level of job performance.

The bundle of personal characteristics (age, race, gender, work experience, and education) differentiates workers. From the employer's perspective, these personal characteristics indicate the cost of training workers for new job tasks. Based on past experiences in training workers, employers rank workers with different bundles of personal characteristics according to their perceived cost of training. Workers perceived by employers as the least costly to train occupy the front of the labor queue. Workers judged to be more expensive to train for new job tasks occupy positions further down the labor queue. There is also a queue of vacant jobs. Jobs paying the highest salary, with the greatest promotional opportunities and nonmonetary benefits, such as status and desirability, occupy the front of the queue. Jobs further down the queue possess less desirable monetary and nonmonetary benefits. The hiring process, then, matches the queue of job seekers to the queue of vacant jobs. The highest-placed person in the labor queue gets the job with the highest monetary and nonmonetary rewards in the job queue. Under this theory, the matching process effectively allocates employees to those slots that will minimize the training costs of the employer.

Segmented Labor Market Theory

The dual labor market theory divides the economy into the core and peripheral economies. The core is composed of large firms that have significant social,

economic, and political power and the periphery is composed of smaller firms without power (Althauser 1990; Bluestone, Murphy, and Stevenson 1973; Doeringer and Piore 1971; Gordon, Edwards, and Reich 1982; Osterman 1984; Reich, Gordon, and Edwards 1973). The *core sector* tends to be powerful. High productivity, high profits, intensive utilization of capital, and high incidence of monopoly elements characterize firms in the core economy. Core sector employees are likely to be very skilled, highly productive, and unionized. Because internal labor markets and on-the-job training are readily available, wages tend to be high and employee turnover low.

In contrast, small firm size, high labor intensity, low profits, low productivity, intense product market competition, and a lack of unionization characterize firms in the peripheral economy. The *periphery sector* is composed of firms that are in a constant struggle to survive. They lack the size, financial resources, or the political might to be major economic forces. Workers in this sector possess limited job skills, have limited access to internal labor markets, or have limited access to on-the-job training programs. Both worker productivity and wages tend to be low, and high worker turnover is common. Thus, worker compensation tends to be tied more to the sector one works in than the investment in training and education.

Large core economy firms enjoy economic advantages (Averitt 1968):

1. *Extensive assets,* which allow core firms to outspend smaller peripheral firms
2. Better geographic and product diversification
3. Vertical integration, which permits core firms to become their own suppliers and distributors, thus ensuring low-cost inputs and a distribution network
4. Favored access to finance when credit is scarce or restricted
5. Political advantages of favorable laws and administrative rulings on tariffs, taxes, and subsidies achieved through information preparation, public relations, and political lobbying
6. *Availability of extensive resources,* which allows spending greater amounts on research and development and early access to new technology

Along with dualism in the industrial structure is a corresponding dualism of working environments, wages, and mobility patterns (Doeringer and Piore 1971; Reich, Gordon, and Edwards 1973). Firms in the core economy, with more stable production and sales, offer jobs reflecting that stability. Peripheral firms, with unstable product demand, offer jobs characterized by instability. Because core firms are larger and more diverse than are peripheral firms, they can provide more promotional opportunities. Firms in the core economy are more capital intensive and earn higher profits than do peripheral firms. Because of their economic advantage, jobs in the core economy generally provide more monetary and nonmonetary benefits.

According to dual labor market theory, dualism in the economy leads to segmentation of the labor force. In essence, different subgroups of the labor pool fill different jobs slots (Gordon 1972). In particular, minority workers, women, the less educated, and youth tend to be concentrated in low-wage, unstable employment with few promotional opportunities in peripheral industries.

The reason for labor market segmentation, according to this theory, is due less to initial differences in human capital than to hiring standards established by employers and unions, as well as to custom and tradition. As in job competition theory, employers rank workers according to their preferences for personal characteristics, such as education, training, work experience, gender, age, and race. Because of their economic advantage, firms in the core economy pay a wage higher than the market-clearing price, thus permitting the firm to follow selective employment practices. Again, since the community is just the sum of firms present, this could have significant implications on community labor markets.

Dual labor market theory is concerned with the segmentation existing in labor markets, not with industry characteristics. The *dual economy theory* focuses on segmentation among firms, and the *dual labor market* focuses on the segmentation of jobs (good and bad jobs). In the primary sector, internal labor markets are pervasive, employment is stable and jobs are secure, wages are high, working conditions are good, workers are punctual and dependable, investment employee training is extensive, and worker turnover is low. In the secondary labor market, the workers have few opportunities for advancement given that internal labor markets are rarely present, employment is unstable and jobs are insecure, the requirements for gaining entry to these positions are virtually nonexistent, and wages and work conditions tend to be poor. These tend to be referred to as dead-end jobs. Moving between primary and secondary labor markets is difficult.

INFORMAL WORK AND LEISURE

The theories above explicitly assume you will be working for wages.[14] The individual implicitly faces the choice each day of whether to work for wages or become involved in nonwage work (informal work). Families view work as a major way to obtain resources to provide for the family. There are many types of activities that may function as alternatives to formal market work. Typically, informal work includes the broad household category of those activities traditionally considered to be family-related activities inherent in family life (child care, cooking, cleaning) and do-it-yourself work in which home production is intended to replace goods and services typically purchased in the market. This has become increasingly difficult as more and more traditional household work, such as child care or cooking, has become more readily available on the market (day care and preprepared meals) because of the changing roles of family members with time and social conventions. The real implication of informal work is the impact these changes have on job development, job training, and safety nets (Blair and Endres 1994).

Sources of income affect a person's choice of formal or nonformal work. In Figure 6.2, the individual has OH hours in the day, of which HH_w are used for work and OH_w for nonwork activities. This is labeled tangency point A. The family of indifference curves (u_i) represents the trade-off this individual makes between work and nonwork activities. The budget line ($w_0 w'_0$) represents the income individuals earn if they use all their time (24 hours daily) for work, leisure, or combinations of work and leisure. The slope of the budget line $w_0 w'_0$ represents the ratio between work income and nonwork income (*nonwork* is used here to represent the informal labor market). Nonwork income consists of nonmarket income, such as positive externalities, and other income sources, such as transfer payments, dividends, or interest. In the figure, total income (OY_t) is the sum of nonwork income (OY_{nw}) and work income ($Y_{nw} Y_t$). An increase in wage rates to w_1 (change of slope to $w_0 w_1$) increases hours worked from HH_w to HH'_w and the worker achieves a higher indifference curve (u_2). This is labeled tangency point B. If nonwork income increases to w_3, with no difference in wage rates (e.g., a parallel shift in the

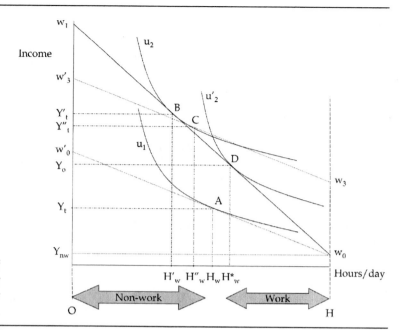

Figure 6.2. Influence of wage rates on labor offered to formal labor markets.

budget line to $w_3w'_3$), then the worker achieves u_2 with only HH''_w hours offered for work.

The movement from u_1 to u_2 is an *income effect* due to an increase in either work or nonwork income. The movement along u_2 because of the new ratio of work and nonwork income is a *substitution effect*. This is movement from tangency point B to tangency point C. When a worker receives a wage increase, his observed response is the sum of the income and substitution effects. The income effect occurs when the individual increases his demand for nonwork activities as wages increase. It means that he can maintain current income while reducing his hours worked. The substitution effect means that leisure hours now have a higher opportunity cost, thus creating an incentive to work more hours. The net effect on hours offered for work depends on the indifference curves of the individual and on whether the income or substitution effect dominates (Ehrenberg and Smith 1985). When the substitution effect dominates the income effect, increases in wages increase the amount of time supplied for work (a labor supply curve sloping upward to the right). If the income effect dominates, the labor supply curve becomes negatively sloped (upward to the left). In Figure 6.2, the workers utility curves are u_1 and u'_2, the increase in wages from w_0 to w_1 actually increases the worker's utility by reducing the time spent in the formal labor market (e.g., HH^*_w). The implica-

tion of Figure 6.2 is that the labor supply curve is backward bending beyond some wage. This is labeled tangency point D.

Although we assumed a formal labor market in this chapter, remember that each worker must decide daily where she will put her next hour. Also note that many labor decisions are family or household decisions, not an individual's decision. This may complicate the simple discussion presented here.

COMMUNITY LABOR MARKET ISSUES

If everything were to work correctly, labor would move among uses and places in response to differences in wages. Higher wages would draw labor, while lower wages would cause workers to seek opportunity elsewhere. As we saw in Chapter 2, per capita incomes would equalize and the economy would be achieving the maximum output from its given supply of labor. Some important and traditional assumptions associated with the "correct" functioning of the labor market follow:

1. Workers and employers have complete and adequate knowledge of job opportunities and wages.
2. Workers and employers are rational; they will maximize either their satisfaction or their profits.

3. Workers and employers are in sufficient numbers that neither can influence the wage rate, or there is no collusion among the parties concerning wage rates.

4. Labor and jobs are homogeneous and interchangeable within a particular market.

5. There are no barriers to occupational or geographic mobility of the worker.

6. Unemployment arises only when these assumptions are relaxed.

Well, not everything works correctly, so we will review some of the failures and discuss their ramifications for the community practitioner.

Unemployment

Any community labor market analysis must address the question of unemployment. While the initial reaction to unemployment is adverse, some unemployment can be considered normal or even beneficial.

> All unemployment is not socially disruptive or economically wasteful. Some unemployment is a natural concomitant of technical progress and free labor markets, and can be considered socially and economically beneficial. We need a framework for distinguishing the harmful from the beneficial or indifferent instances. (Kalachek 1973)

Unemployment is not a homogeneous phenomenon; it can be divided into components that respond to different market forces. This differentiation helps focus public policy response. While labels vary, typically unemployment divides into demand deficient unemployment, structural unemployment, and frictional unemployment (Ehrenberg and Smith 1985). *Demand deficient unemployment* occurs because the total number of unemployed workers exceeds the total number of job vacancies across all occupations and places. It really measures the vitality and strength of the national economy. *Structural unemployment* appears as a mismatch of the job requirements and the skills available in a particular locale.[15] This recognizes the nonhomogeneity of the labor force, which prevents any worker from filling any particular job. It also recognizes that differences in the location of jobs and the location of the workers cause unemployment. *Frictional unemployment* measures that irreducible minimum where unemployment and vacancies occur simultaneously in the same occupation because of imperfections in information about job availability and unemployed workers, the time required for workers to shift between jobs, or a spatial separation of vacancies and workers.

When examining local unemployment, it is important to look beyond these three unemployment categories. One hypothesis maintains that market forces will not reduce the imbalance between labor demand and supply. These situations of wage differences in a specific locale can occur even over long periods of time regardless of growth in aggregate demand and shifts in the supply of labor because of labor market segmentation. This imbalance is due to immobility of labor among uses and locations and labor's non-monetary attachment to an area and co-workers that resist wage differentials. Also, because of the dynamics of the economy, communities continually face a situation where demand changes and supply tries to adjust. The cost of information and immobility prevent instantaneous and complete adjustment of local employment to these shifts. The capacity to adjust also varies among geographic areas.

Job/Worker Matching

In the exchange or matching process between workers and firms, it is critical to examine the workers, the job, and the characteristics of the matching process itself because the creation of a labor force or workers and the creation of jobs in a modern society are almost independent processes (Ranney and Betancur 1992; Sorensen and Kalleberg 1974). Job/worker mismatches lead to reduced output, negative job attitudes, and low job satisfaction. Potential discrepancies exist in this exchange. The first is a discrepancy between the skills possessed and the skills required. Workers are not homogeneous. They possess different cognitive abilities, different personal interests, or different physical abilities. The second failure is the immobility of workers among occupations and places. Third, there are problems in information flows and possibly even in comprehending that information.

Gillis and Shaffer (1985, 1987) explored the link between specific industrial (job) attributes and the probability of hiring workers from the specified groups in the labor pool. Workers in each group are characterized by a specific sex, age, level of education, and other personal factors. The labor market theories briefly reviewed above indicate that the type of occupational demands, degree of capital usage, size of establishment, and wage level characterizing industries are linked to the probability of hiring workers from each group in the labor pool. Occupational mix defines the type of work skills demanded by employers. In general, what they found was not too surprising.

If the objective is to increase job opportunities for men, expanding employment in the manufacturing sectors is generally the most effective.

- Small, capital-intensive manufacturing firms are the most likely to provide job opportunities for disadvantaged men.
- Small, capital-intensive, high-wage manufacturing firms are the most likely to hire advantaged men.
- Among trade and service industries, employers demanding high proportions of craft occupations are the most likely to hire both advantaged and disadvantaged men.

If the objective is to provide jobs for women, trade and service firms are generally more effective than manufacturing firms.

- Trade and service businesses that demand relatively high proportions of less skilled occupations are the most likely to employ women.
- Small, capital-intensive firms with relatively high proportions of skilled occupations have the greatest tendency to hire mature advantaged women.
- Larger, labor-intensive manufacturing businesses are the most likely manufacturing firms to hire young advantaged women.

An advantaged worker tends to have higher levels of education, greater occupational skills, and more work experience and tends to be younger. With the findings above, communities can structure their development efforts to enhance the likelihood of hiring particular types of workers. Development efforts that create jobs in a variety of economic sectors will help to employ a wider range of workers than will more narrowly focused efforts.

Communities cannot stop their analysis once they have identified industries they wish to encourage to start, expand, or locate. Other factors, too, are involved, like the production processes used by the firm. Another important issue, not considered here, is that even if the workers of concern were hired, there is no guarantee that their economic situations will improve. That is, while it is important that the targeted workers be hired, the wage level and stability of jobs for which they are hired are of equal importance.[16]

Then there is the possibility of a spatial mismatch (Kain 1992; Preston and McLafferty 1999). Here the problem is the worker's residence and employment location do not match or transport between them is difficult. The worker certainly could move, but the cost of the move and psychic costs tied to friends and family are perceived as being too high. Likewise, using public transit or having undependable private transport may prove problematic.

There is also a strong potential in the long-term for jobs created at the local level to be taken by new migrants to the community. This reduces the likelihood that targeted job creation for specific local population segments will be successful. Often communities with high unemployment rates that target certain business growth strategies do not see significant changes in unemployment (Summers et al. 1976; Shields 1998).

Relaxing the assumptions of homogeneous workers does affect the ability of the individual to participate in the labor market and it can alter outcomes. First, there are differences in cognitive and/or physical (health) abilities among workers that may prevent them from fully engaging in the labor force. Second, there are differences in family considerations and the weight the worker may put on them. Third, the worker may pursue an objective function of satisficing rather than maximizing utility.

Information

For a labor market to work well, information on the location of jobs and the skills required must be widely available to members of the labor force (Kreps et al. 1980; Lever 1980; Ranney and Betancur 1992; Shaeffer 1985). Likewise, information on worker availability, skills, and willingness to work or to move to a new job location is necessary.

One problem is that the market may not give the appropriate signals to allocate the labor force optimally. Workers and employers are hampered by their own ignorance, including the ability to correctly comprehend or the uncertainty not only about current but also about future manpower needs.[17] Institutional wage rigidity from such things as union contracts may be higher than the equilibrium wage, sending a market signal of labor shortage in certain occupations even though those occupations are actually shrinking.

Widespread dissemination of labor market information depends on numerous forces, some of which are geographic size of the labor market, number of workers and employers, and relative market power possessed by either workers or employers. The larger the labor market in numbers of job opportunities, numbers of workers, and geography, the more difficult it is to dispense information to all parties. The greater the economic concentration or the more noncompetitive the market and the greater the homogeneity of economic activity, the fewer the sources of information needed by labor market parties.

Job information can be disseminated both formally and informally. Information is disseminated formally through want ads, job service bureaus, and private employment agencies. Informal channels are friends or relatives, who often are a major source of job and worker information (Chapple 2002; Weber and Theodore 2002). Often market signals about current and future job/occupation needs are particularly inadequate or ineffective in motivating some groups that are less sensitive or less responsive to market signals (e.g., racial minorities, youth). Much of the local hiring occurs because labor markets rely on relationships in the workforce for job referrals and new job opportunities. One of the negative effects of network hiring is the spatial distribution of job growth in the suburbs and job seekers in the inner city (Giloth 1995). Sometimes even though the market signals are there, the individuals are unable to comprehend the meaning and implications of those signals.

Labor Mobility

As we saw in the spatial model of the neoclassical growth model, labor is assumed to freely migrate from low-wage to high-wage communities to equalize real wages among the communities (Chapter 2). This presumes wage differentials are the only cause of migration among jobs and/or communities (Krumm 1983; Lester 1966). Yet, other forces also influence labor migration. Wage differentials no doubt affect the migration of labor most sensitive to financial incentives. These include the most mobile workers and those with the least social attachment to their existing residence (Ehrenberg and Smith 1985). Such workers value the monetary portion of their work more than the psychic or nonmonetary portion. Yet, other nonwage reasons accentuate or dampen the migration.

The distance of the move affects labor mobility (Krumm 1983). Distance becomes an obstacle to the movement of labor or economic activity and must be overcome by more than just a minor wage differential. As we saw in Chapter 3, the greater the distance, the greater the uncertainty involved in the move. Higher levels of uncertainty translate into high psychic costs.

The amenities associated with a community either sending or receiving the migrants affect the extent wage-stimulated migration occurs (Krumm 1983). Some individuals will accept lower wages or seasonal unemployment because they substitute such things as mountains, hunting, fishing, or access to the opera for monetary wages (Deller and Tsai 1998). The mobility of labor among communities sometimes depends on the psychological attraction of family and friends. Age, race, or gender discrimination can hamper worker mobility from eliminating the wage differential (Krumm 1983; Madden 1977).

Even with a large movement of labor among communities, wage differentials could continue to exist (Krumm 1983). One reason is because the people moving among the communities are the unemployed, wages are not reduced in the high-wage community nor are they driven up in the low-wage community. Likewise, differences in the production technologies and economies of scale associated with similar industries in different communities can yield a continuing community wage differential. Institutional rigidity, such as caused by union contracts, maintains wage differentials in the face of extensive labor mobility. Other reasons are that labor is heterogeneous, communities have different occupational structures (i.e., different jobs pay different rates), and communities produce different products (price of the output), thus altering the demand for labor.

Discrimination

Although numerous laws make it illegal, discrimination still occurs. It clearly prevents the labor market from fully functioning, so yields are either less than optimal output at the least or a waste of human resources at worst. This discrimination can be gender based (Chapple 2002; Dewar, Fitzgerald, and Green-Leigh 1994) or racial (Ranney and Betancur 1992).

Exploitation

While exploitation may appear to be a strange topic as a community labor market issue, we include it here for two reasons: living wages and economic exploitation. An issue that has gained increased importance during the 1990s and the beginning of the 21st century is the concept of a *living wage* (Mier, Vietoriz, and Harrison 1993). At the heart of living wage proposals is paying workers according to some need rather than on some skill. Typically, *need* is defined as the wage rate that will raise a person, in a family of four, above the federal poverty line. In particular, what communities need to do is encourage employers to pay living wages. Some communities have made those types of requests as part of gaining municipal contracts (see Chapter 10).

A second concern is labor market exploitation. Pure labor market exploitation occurs when labor is paid less than its *VMP* (Kreps et al. 1980). The firm maximizes profits by equating marginal revenue (*MR*) and marginal cost (*MC*). Since the firm's demand for labor is $w = VMP_L = MPP_L \times P_O$ in a competitive market, exploitation in a pure economic sense occurs whenever $w < VMP_L$. The most likely case is when the firm faces a less than competitive product market and $MR_O < P_O$. Thus,

$$w = MPP_L \times MR_O < MPP_L \times P_O$$
$$= VMP_L \qquad (6.2)$$
$$= w_c$$

The extent of labor exploitation is the degree $w_{nc} < VMP_L$. This does not deny the possibility that exploitation (in the popular connotation) can arise from other sources.

Contingent Labor Force

A recent phenomenon in labor markets is the movement of firms and workers to informal attachments to the job and/or employees (Belous 1989; duRivage 1992; Ofstead 1998). Firms are finding that contracting with personnel supply firms (temp agencies) is providing a heightened level of flexibility for firms. In more traditional labor markets, when a firm hires an individual, there is an implicit contract for full-time work plus benefits. Firms may not have sufficient work to justify hiring a permanent worker and find it too costly to search and hire workers. Most labor markets theory implicitly assumes the worker and firm engage in negotiations about employment. The temp agency intervenes between the firm and the employee and handles much of the negotiations, paperwork and potential part-time nature of the job. The temp agency tries to patch together several part-time jobs so the worker may end up working nearly full time rather than part time. Firms can have ready access to skilled labor for short periods, and the worker has a certain degree of flexibility in hours available for work. In addition, the temp agency incurs the cost of the job search borne by both the worker and the firm.

In summary, while the general perception is that labor mobility will eliminate geographic and occupational wage differentials, differences among community economies, workers, time, and such rules as

discrimination and contracts will probably prevent full equalization of wages.

POVERTY/WELFARE REFORM AND STRATEGIES TO ADDRESS LOW-INCOME LEVELS

The economic boom of the 1990s was also coupled with an increasing awareness that many people were not benefiting from the overall economic growth (Blank and Haskins 2001; Dresser and Rogers 2000). This is particularly true as the welfare reform of the 1990s plays itself out nationally and across the states (Blumenberg 2002; White and Geddes 2002). Poverty is a normative rather than an objective concept in that the standard/threshold varies over time, among places, and among groups.[18] Typically, we think in terms of some minimal standard of living that is monetary, culturally, politically, and socially determined and which no one should ever fall below. We must remember that many are in poverty only temporarily (e.g., job loss, age, health), not permanently. There is a tendency to define poverty in economic terms, but it includes cultural, racial, gender, and institutional aspects.

Although poverty reduction through income and wealth distribution is not the same policy goal as welfare reform, they are considered in a chapter on labor markets for two reasons. First, jobs and the wages associated with them are a major determinant of income distribution (Bartik 1994; Green-Leigh 1995). For a variety of political and social reasons welfare has changed its philosophy from entitlement to work for benefits (Brown 1997). A major reason is the emerging belief that there are two types of poor (Blank and Haskins 2001; Deavers and Hoppe 1991). The current debate on welfare reform has judged that there are *deserving poor* (e.g., invalid, aged, children) and *undeserving poor* (e.g., able bodied). Just as importantly, aid to the poor must not interfere with the functioning of the labor market. The concern is that welfare reduces people's willingness to engage in productive work. Specifically, if the value of the welfare package is equal to or greater than the reservation wage, people will not enter the labor force. Thus, with the new philosophy about welfare, the idea is that welfare should encourage people to engage in work.

Fundamentally, we are talking about targeted economic development or job-centered or employment-linked development (Giloth 1995), which refers to decent jobs and incomes for the disadvantaged.[19] *Targeted economic development* refers to the combining of employment training, human services, and enterprise development to enhance access to and creation of jobs, careers, and self-sufficiency for the disadvantaged. Targeted economic development is concerned with where the good jobs are, how to gain access to them, how to create jobs, how to support businesses likely to hire poor people, how to make bad jobs into good jobs, and how to support disadvantaged people as they enter the labor market.

The focus of targeted economic development is on specific tangible, short-term employment and income gains for disadvantaged people. The most effective strategies are market oriented, networked, empowering, integrative, and community based (Giloth 1995). Being market oriented means that job access and creation efforts recognize the economic dynamics affecting jobs and occupations, paying attention to what firms want in job skills and work preparation. *Networking* refers to those efforts that knit together an array of institutions, resources, and services to make job access work for job seekers and firms (Chapple 2002; Weber and Theodore 2002). Programs cannot be isolated if they are to be flexible, responsive, and entrepreneurial. *Empowering* means building on people's dignity, skills, and participation. Being *integrative* means that conventional boundaries of programs—problem definition, space, and authority—are overcome and that appropriately tailored partnerships and packaging of services are assembled to effectively promote job access and creation. Effective projects make a habit of crossing boundaries. Targeted economic development must be community based—connected to local institutions and people—so that jobs and opportunity reach the disadvantaged.

Several structural factors affect labor supply, especially for the poor and near poor. Housing (cost and location), child care (dependability, cost, access), and transportation (both public and private) are some factors that need to be considered because the goal is to link the jobs that are available with workers without causing them to have to change their places of residence.

We present some common strategies for promoting employment for low-income disadvantaged people below (Mueller and Schwartz 1998; Nightingale and Holcomb 1997).

- *JOB CREATION:* First, link the jobs generated through subsidies to those who need it. Local hiring ordinances and linking agreements are examples. Second, promote local development

shaped around the community's industrial structure and labor force. This could be targeted small-business development, industrial retention schemes, micro-enterprise programs, and business incubators focused on community-based enterprises.

- *DIRECT EMPLOYMENT STRATEGIES: Job assistance* can be in a group setting or through one-on-one counseling or coaching, sometimes through job clubs with workshops, access to phone banks, and peer support. *Self-directed job search* is where individuals search and apply for jobs on their own. *Job development and placement* counselors refer individuals to openings, often by using computerized job banks or developing relationships with specific firms, gaining knowledge of potential job openings or commitments to hire through the program.

- *JOB TRAINING STRATEGIES.* Three types of programs fall into this category. *Classroom occupational training* at training or educational institutions may include formal post-secondary programs leading to certification or licensing in a particular occupation. *On-the-job training* (OJT) can be with public or private sector employers, who usually receive a subsidy to cover a portion of the wages paid during the training period. *Soft skills/job readiness* helps workers understand the hiring process (appearance, resume preparation and interviewing). Soft skills include such things as being a team player, motivation, politeness, verbal skills, and neatness.

- *EDUCATION STRATEGIES. Remedial education* (such as preparation for the GED, basic skills instruction in reading and mathematics or English as a second language) is an example of an education strategy.

- *SUBSIDIZED PUBLIC EMPLOYMENT STRATEGIES. Work experience* can include unpaid workfare assignments, where recipients work in exchange for welfare benefits. Short-term unpaid work experience can be designed as basic exposure to the work environment.

- *WAGE SUBSIDIES.* The final approach is to *subsidize the wages* of classes of workers that you hope firms will hire and pay a living wage.

SUMMARY

Two subtle shifts in labor market analysis in recent years are a focus on inclusion and a focus on job quality. The first shift explicitly recognizes that some individuals, because of race, gender, age, or other factors, have been systematically excluded from full participation in the labor market and its associated rewards—that is, institutional barriers and heterogeneity in the labor force prevent certain segments of the population from earning a living. The second shift recognizes that a job is more than mere employment and that some jobs possess characteristics that workers appear to value more than others. Yet, not everyone desires jobs having the characteristics so highly valued by many workers. For example, some people do not want a job that requires them to solve problems or to work full time. Thus, it is important to maintain a mix of jobs in designing any community employment program.

To correct conditions of underemployment or unemployment, emphasis must be put on both the demand for and the supply of labor. Focusing solely on either demand or supply dimensions of the labor market presumes that only one force lends itself to policy manipulation or is the cause of the unemployment.

Often overlooked in community economic development is that an operating business yields two economic outputs. The first and most obvious is the product or service the firm produces. The second and less obvious is a worker with a set of skills and attachments to the labor force. This second dimension becomes an important interface between community economic development, the labor market, and business activities. If the community can structure a series of employment opportunities that match workers' differing skills and desires for work hours and conditions, then the community has created a full range of employment opportunities that allows people to meet their own individual needs.

Regarding poverty and labor, it is important to create opportunities to acquire and develop skills. Those without such opportunities will be left further and further behind and this is an enormous waste of human resources. The focus needs to be on more than just jobs (Reese and Fasenfest 1999). Communities should be asking these additional questions:

- What kinds of jobs are created in which economic sectors?
- Who gets the jobs?
- How do new jobs affect the distribution of income in a community?
- Do more jobs actually lead to other benefits, and are there concomitant unintended costs?
- How do the policies affect specific areas, neighborhoods, or groups, particularly the poor?

The importance of labor markets and the availability of livable jobs in a community cannot be overstated. Again, asking the appropriate economic development question provides for community dialogue and improved economic decision-making. Often, dialogue with respect to underlying labor market issues can be more useful than analytical tools that generate answers for the wrong questions.

Our previous discussion was primarily in reference to labor demand in the private sector, but a significant amount of labor demand also comes from public sources, such as government agencies, foundations, religious organization, and the like. Community or society priorities encompass the level of public and private sector production and the type of output. The level of public-sector production refers to the type of public goods desired (e.g., fire protection, defense, social workers) and whether the public sector or the private sector is perceived as the most appropriate employer.

The community practitioner should keep in mind that investment in human capital as an economic development strategy is people based, not necessarily place based. Because labor is mobile, the community that benefits from the investment may not be the community that made the initial investment. Given the mobility of the investment, some communities may underinvest in people-based programs. People-based strategies are harder to see because they are embodied in the people themselves, while place-based strategies, such as industrial parks, infrastructure investments, and downtown beautification programs, tend to be more visual.

STUDY QUESTIONS

1. What is a labor market and what economic purpose does it serve?
2. Is a labor market homogeneous? If not, how are labor markets differentiated from both the employee and employer sides?
3. What is meant when the demand for labor is referred to as a derived demand?
4. The demand for labor depends on what factors? How might these factors shift the community demand for labor?
5. What are some of the socioeconomic characteristics influencing the supply of labor?
6. The community's supply of labor is composed of the sum of several decisions: work versus leisure, number of hours to work, choice to work here or there. What forces influence these choices?

7. Why might labor mobility not achieve wage equality among communities?

8. Why is labor mobility so critical to community economic development?

9. How do the official and the economic definitions of the supply of labor in a community differ?

10. What are some characteristics associated with a job that distinguishes it from other jobs?

11. Human capital theory suggests a direct link between education and productivity that leads to differences in earnings. What are some additional assumptions needed to support this assertion?

12. What are the similarities and differences between the human capital and dual labor market theories of labor markets?

13. Why is it important for the community to create a range of employment opportunities?

14. Why distinguish between internal/external and core/periphery labor markets? Are they the same concepts?

15. What are the distinguishing characteristics between jobs in the primary and secondary sectors?

16. Thompson argued that "how you produce something" is more important than "what you produce." What does this mean for communities' choices for economic development paths?

17. How does relaxing the assumptions of homogeneous workers affect the ability of the individual to participate in the labor market?

18. Why are poverty reduction and welfare reform different policy goals?

NOTES

1. Upon reflection, in many other countries the possession of a job is even more important than it is in the United States. This is demonstrated by attempts by both public and private actors to ensure that full employment is achieved, rather than permitting it to be a coincidental outcome of private market forces.

2. A *commodity* in this context is a good that has little if any differentiable characteristics. Milk for example is a commodity, but organic milk is a product because it has a differentiable characteristic that can be identified. Labor is a product in that not all labor is the same and varies by gender, education, and work ethic.

3. Clark (1983) argued that contract theory, when applied to the market between laborer and employer, suggests far less volatility in wages and that the variability appears as employment changes. He contended that firms set a wage, and labor either agrees to work for that or seeks other work. Blaug (1985) also contended that labor markets adjust the quantity of people (unemployed) hired rather than wages.

4. This distinction between place of work and place of residence is important when examining secondary data on labor markets. Some secondary data reports employment on the basis of residence (e.g., Census of Population). While looking at employment by place of residence gives the analyst some perspective of the labor force skills present in that locale, it does not give an accurate picture of the economic structure of that locale. Likewise, using the employment by place of work gives the analyst a better picture of the economic structure of the local economy, but it may bias estimates of the profile of the local labor force.

5. Typically, formal is working for wages on which you pay taxes. Informal is housework, or noncash work (i.e., yardwork and housework), and typically you don't pay taxes (Larrivee 2000).

6. The short-run demand for labor is conditioned by a given quantity of capital and land. Changes in the price of these resources can alter the demand for labor. If the firm added more capital to the production process or workers became more productive, the demand curve for labor would shift up and to the right in Figure 6.1. The new equilibrium conditions would yield more labor hired and higher wages. In the long run, when the quantity of fixed resource is also allowed to vary, the demand for labor incorporates differences in amounts of other factors of production.

7. In this situation, public policy aimed at increasing employment and output would shift labor markets to a more competitive state.

8. The traditional model suggests the existing population base supply of labor depends on individuals' willingness to exchange informal work time for formal work time. This is far too simplistic because it does not capture the interactions within the family unit (Becker

1965) and ignores the implications of spatial and occupational frictions that reduce the capacity of workers to move between jobs and places or to trade-off work and informal work activities. Thus, to examine a community's true labor supply requires an analysis of not only the community's population characteristics, but also population characteristics of nearby surrounding areas and the ability of those people to move on a daily basis to jobs available in the community. It also requires recognition of family structures rather than just the socioeconomic characteristics of individuals (Beck and Goode 1981).

9. A variation on this idea is the *Peter principle,* where people are promoted till they reach their level of incompetence.

10. A major criticism of neoclassical theory is its failure to recognize how ageism, sexism, racial discrimination, and other social characteristics affect mobility. Job mobility, or "job hopping" can be viewed as being negative if the individual cannot become socialized into the work force or find a job to match skills, or as being positive if the individual uses this as an informal way of acquiring information about the quality of various jobs and working conditions.

11. Showing up for work on time, staying for the allotted time, and not consistently having blue Mondays and blue Fridays are examples of socialization skills that are typically acquired over time through family, cultural, and prior job experiences.

12. This section on labor market theory draws on Gillis (1983).

13. Labor mobility has high economic and social costs, however, not only to the individual but also to the firm (Lester 1966). The individual pays the traditional psychic costs, but the firm also experiences costs in the form of lost investment in training and on-the-job experience and worker knowledge of value to that particular firm and its production processes or clients. These costs tend to reduce the speed or rapidity of labor mobility under changing conditions.

14. This section draws on the work by Larrivee (2000).

15. Displaced workers are individuals with established job histories who have lost their jobs through no fault of their own and who are likely to encounter considerable difficulty finding comparable employment (Hamrick 2001). Displacement is considered structural unemployment, not employment due to economic cycles or due to the normal matching process between workers and employers, but instead unemployment due to skills or geographic demand/supply mismatches.

16. A disturbing finding is that women and disadvantaged workers tended to be hired by firms offering relatively low wages and less stable employment opportunities (Edin and Lien 1997; Hanson and Pratt 1995; Harris 1993 and 1996). Significantly improving the economic position of these workers may require upgrading their skills through training and education.

17. This is often referred to as *bounded rationality,* which essentially says that people are unable to comprehend all the options and rank them top to bottom. People are subject to information overload and cannot adequately process the information they possess. In other words, the all-knowing rational economic person is a figment of economists' imaginations.

18. Well-being is an interaction of economic, social, cultural, environmental, medical, attitudinal, and geographic factors. The poverty line is a measure of well-being that focuses on the individual's capacity to satisfy the basic needs of food, clothing, decent shelter, and good access to medical or social service. Such capacity is usually defined by income. It also depends on the stage of the life cycle of the person.

19. In the mid-1990s, a single mother who had two children and worked full time all year at a minimum wage job earned a little over $10,000 a year. The spendable income for this family, adjusted for payroll taxes, the earned income tax credit, food stamps and child care expenses was roughly $11,000, just below the federal government's official poverty line for a family of three, which was $12,273 in 1996 (Sawhill 1998).

7

Financial Capital Markets

COMMUNITY CAPITAL ASSETS AND MARKETS

Capital assets of communities and their markets provide critical components in the understanding of community economics. As such, they provide an important part of the resources node found in our guiding community economics star (Fig. 1.1). Broadly defined, community capital assets can take on many different forms. For instance, Flora (1997) defined capital as involving stock assets tied up in environmental, physical, intellectual/human, social, political, and financial aspects of communities. Although this broad approach to defining capital assets is important for understanding stock resources, its use presents a widely overlapping array of theoretical and empirical issues that do not neatly fall into a factor market approach to explaining community economics.

In this book, we have chosen to separate and more narrowly define capital assets into *private financial capital* (the focus of this chapter) and *public capital* (amenity resources in Chapter 9 and publicly provided goods and services in Chapter 10). In such an organization, we gloss over several elements that reflect a community's capital asset base here but incorporate discussion of these where they provide relevant additions to other chapters. For instance, in Chapter 12, Policy Modeling and Decision-making, we include a relevant community economics discussion of social capital. In Chapter 5, Land Markets, we discuss land-based environmental resources within the context of land markets.

Capital Stock Assets versus Flows of Funds

Before discussing financial capital markets, we must distinguish between capital stock assets of a community and the flow of funds. *Capital stock assets* refer to the total level of resources available within a

community at a given time. Capital stock levels can fluctuate up and down; rates of change are determined primarily by their relative rate of growth or productivity and the respective rate of removal. Removals from capital stock assets reflect short-term flows drawn from capital stock reserves. A *flow of funds* is a periodic removal or injection of a portion of the overall asset. The ability to draw flows from capital stock assets in the future is determined primarily by the rate of removal and the growth or productivity of the capital stock.

The Importance of Financial Capital Markets to Community Economics

A key element in community economic development activities is the availability and use of money within private financial capital markets. Indeed, money is the medium of exchange that sets the world in motion. The flow of money has sweeping implications for the sustained vibrancy of community economic activity. The rapid movement of money today in our modern financial markets, such as the stocks of Wall Street and the market in Tokyo, can mask the true investment in our economies and communities. We trade more than $100 worth of stock and bonds for every dollar raised for investment in a new plant and equipment, a ratio almost four times greater than 30 years ago (Institute for Local Self-Reliance 2002).

The importance of the structure and functioning of an economy's financial institutions and its capital markets to economic development cannot be overstated. This has been long understood and is well summarized by Baumol (1965, pp. 1–2):

> The allocation of its capital resources is among the most important decisions which must be made by any economy. In the long run, an appropriate allocation of real capital is absolutely indispensable to the implementation of consumer sovereignty, . . . for unless the flow of capital goods is responsive to the goals of the

members of the public, the community will only be able to exercise a very short run and temporary control over the composition of output and its activities. After all, capital is the economy's link with the future.

Baumol also described how the critical relationship between financial markets and real capital formation occurs:

> The selection of the physical forms which constitute the embodiment of our capital resources is largely controlled through the funds market—the market in which money capital is provided. Thus, the allocation process is heavily influenced by the decisions of the nation's financial institutions, its banks, its insurance companies, and a variety of other bodies. In addition, the government's monetary and fiscal policy obviously plays a highly important role in a variety of ways.

Financial capital is the mechanism permitting the community to purchase or develop labor and physical capital critical for community economic development. Private financial capital generally appears in two forms: debt and equity.

In the following sections, an overview of financial capital markets within a debt context is followed by a discussion of equity and venture capital. We deal with important *failures* that exist within private financial capital markets and conclude with public policy to address capital resources within a community economics framework.

AN OVERVIEW OF FINANCIAL CAPITAL MARKETS

The basic theoretical model of financial markets contends that capital is allocated among uses, regardless of place or type of use, in such a fashion that the return on capital in all uses and places is equal. This also maximizes returns to the lender/investor. Kieschink and Daniels (1979, p. 13) defined *capital market efficiency* as

> a web of institutions and mechanisms by which resources are saved, investment opportunities are identified, and savings channeled to enterprises for their productive and possible use. These markets are operating efficiently if the way funds are channeled meets every opportunity for improving the overall welfare of consumers.

In the circular flow model of the economy (Fig. 4.1), the suppliers of capital are households and businesses with more income than they currently need (i.e., savers). The demanders of capital are households and businesses with less income than they currently need (i.e., borrowers). These two parties meet and negotiate a transaction that mutually benefits one another. The *capital market* moves funds from the suppliers to the demanders of capital. The exchange of financial resources can be a relatively simple or a fairly complex exchange. It may involve a transaction between an individual demander of financial resources and an individual supplier of financial resources, a combination of a single demander/supplier facing a multiple supplier/demander, or a multiple supplier/multiple demander exchange. Because individuals who supply and demand capital have difficulty matching exactly the time, place, amount, or judgment about the soundness (risk) of capital market transactions, financial institutions have developed to facilitate the flow of capital among the parties in the market.

The scope of financial markets has changed with the passage of legislation (e.g., the Federal 1999 Financial Services Mobility Act) that broadens the structure of borrowing and lending and removes many former barriers (Kilkenny 2002). Examples of financial institutions now include commercial banks, savings and loan associations (thrifts), consumer credit companies, venture capital firms, stock exchanges, insurance companies, and investment bankers (Table 7.1). These institutions offer suppliers of financial capital a place to put their excess funds until they need them at some time in the future. Financial institutions accumulate funds from numerous savers/investors into amounts that meet the needs of those requiring funds in excess of their current income flow. Thus, a commercial bank collects deposits from numerous households and compensates them with interest or services (such as checking accounts). In turn, the commercial bank becomes the single place where a borrower can acquire necessary funds without approaching every saver.

The organization and structure of the capital market will determine its efficiency in clearing surpluses and deficits as well as the flexibility of its pricing adjustments. Under ideal conditions, if there are demand/supply inequities, financial markets, at any given time or space, create a set of interacting forces to bring the market into balance. As in any economic market, the equilibrating force is the price of money, or interest rates. Ideal performance in the capital market occurs when the market adjusts smoothly and quickly and with minimal economic costs. Straszheim (1971, p. 219) indicated "a perfect financial (capital) market exists when financial assets (loans and equities) of comparable maturity and risk yield the same return in all communities, not including a premium for location rent to cover the cost of transmitting information and funds to and from borrowers and lenders." This last restriction

Table 7.1. Sources of capital available to businesses

Internally generated funds (profits)
Personal savings of entrepreneurs (including credit cards)
FFA money or family, friends, and associates
Commercial bankers
Lease financing
Account receivable finance companies
Commercial finance companies
Public programs—federal, state, local government
Life insurance companies
Bonding companies
Investment banking
Employee stock ownership programs
Mortgage bankers
Venture capital industry (including angels[a])

[a]Angel networks are individuals of high net worth who are willing to make investments in business ideas.

indicates that if transaction costs are significant, borrowers located some distance away from the lender will pay more. This transaction cost is associated more with the movement (acquiring) of information than with the actual physical movement of capital. (The issue of risk is dealt with in Risk and Uncertainty below.)

Nonhomogeneity of Capital Markets

Capital markets are not homogeneous; rather, they segment into numerous submarkets (Brealey and Myers 1988; Hill and Shelley 1995). Some submarkets are defined by the type of capital, such as short-term, long-term, operating, real estate, commercial/industrial, and farm markets. Debt, equity, and venture capital represent other categories of submarkets. Geography also segments capital markets. Many financial institutions have service areas that they will seldom operate outside of. This is most obvious with community banks, savings and loans and credit unions.

Both the type of capital traded in the market and the providers of capital can be diverse (Kilkenny 2002). Examples of capital providers include households, businesses, government, young families, retired workers, pension funds, venture capital firms, and insurance companies. The demanders of capital also are diverse: young households, schools, farmers, retailers, and speculators, for example. Most financial institutions serve only a subset of all the various components of financial markets.

The products exchanged in financial markets can be highly differentiated in terms of risk, length of time to size, priced components of the pledged assets or collateral, maturity, location, type of busi-

ness, and type of borrower. The size of transaction encompasses the fact that small borrowers are usually limited to local commercial banks or local investors for capital sources, while large national corporations can utilize national capital markets.

Traditionally, the four reasons why financial institutions are active in only a few of the various segments of the capital market are information costs, types of liabilities, regulation, and conscious business decisions. Because of recent changes in federal legislation, regulation and types of liabilities are becoming less important, but they still exert some influence.

Information costs refer to the financial institution developing the capacity to evaluate the feasibility of various investments. For example, banks located in agricultural communities have developed a capacity to make relatively good judgments about various credit requests by farmers. Since such banks may service only a few commercial-industrial loans each year, however, their evaluation of requests often exhibits less art and more recipe. This can limit their support of new development opportunities in the community.

Types of liabilities refer to the fact that financial institutions must return cash to depositors when they ask for it. Money market funds, while yielding higher returns, must also be liquid and ready for customers who want their money back today.

Regulation can be a limitation on financial institutions because the regulations prevent some financial institutions from becoming involved in certain types of activities. In addition, because nearly all financial institutions are profit driven, many may make *conscious decisions* to stay away from certain types of markets, such as start-up businesses and farming.

DEBT VERSUS EQUITY IN CAPITAL MARKETS

Note the difference between debt capital and equity capital (Barkley et al. 2001a). *Debt capital* creates specific obligations for the borrower to repay the loan on a predetermined schedule. Failure by the borrower to meet the repayment terms typically allows the lender to attempt to recover the outstanding debt, even if the borrower is forced into bankruptcy. *Equity capital* conveys a share of ownership in the firm to the individual or institution that provides the funds. The equity investor gives up the right to a predetermined repayment schedule and a preferential claim on the assets of the firm in exchange for a share of future profits (or losses).

A distinction between equity capital and venture capital also needs to be made. *Equity capital* is taken here to mean owners' capital or the residual claimant on assets when a business is dissolved. The owners of a firm have the ultimate opportunity to reap the success or failure of the firm (Pulver and Hustedde 1988). The owners have the ultimate control of management decisions, often because they are one and the same. Equity capital is an owner's personal capital that is put into the neighborhood gas station, restaurant, clothing store, or the local agriculture custom-service operation. This can, and generally does, include the owner's own money invested into the business with no idea of ever liquidating, except maybe at retirement. Often it takes the form of "sweat equity."

Equity can also take the form of outside investors who become part owners of the business. Corporations are a form of business that allows a means for stockholders to invest in the company. For the largest of firms, this can take place on the New York Stock Exchange or any number of regional markets. Increasingly, many small companies are incorporating, which allows local investors to buy into or help capitalize the business. Many "family-owned" businesses are actually corporations and the ownership takes the form of stock ownership. The purchase of the stock represents a capitalization of the firm.

Venture capital, on the other hand, is typically money from individuals/institutions who see themselves liquidating their interest in the company within a three- to seven-year planning horizon. Most venture capitalists expect to earn a 20- to 30-percent

compounded rate of return on their money over the life of their investment.

DEBT: COMMUNITY CAPITAL MARKETS

The community's capital markets can be viewed in a resources/markets/rules context of the community economics star (Fig. 1.1) (Borts 1971; Hill and Shelley 1995; Moore and Hill 1982). There are two ways to think about debt/equity capital. The first is demand, supply, and institutional. Here *demand* comes from households, businesses, or government that needs financial resources to undertake some type of investment. This could be the purchase of a new home, the expansion of a warehouse, or the replacement of a bridge. The *supply* of the financial resource comes from savers, and again it can take the form of households, businesses, or government. This could be retirement savings, unspent profits, or government surpluses. The *institutional* element is purely the mechanism that brings demanders and suppliers together.

Second, a slightly different perspective of a community's financial markets appears in Figure 7.1, which demonstrates (1) the significance of community boundaries and the flow of capital across those boundaries and within the community among the various providers and users of capital, and (2) the variety of mechanisms available for a community to support either household or business borrowing. One could think of this as detail of the lower half of the circular flow model (Fig. 4.1).

The three major components in Figure 7.1 are households, financial institutions, and businesses (Mikesell and Davidson 1982; Straszheim 1971).[1] These components represent providers, facilitators, and users of capital, respectively. As providers of capital, *households* have several options. They can hoard idle cash in the back yard, they can deposit their financial assets in a local or nonlocal financial institution, such as a commercial bank, savings and loan, insurance, or they can buy and own stocks and bonds. Households can provide capital directly to the business sector of the community through an equity investment as an owner of a business or through some form of loan made to the business. Households can also export capital from the community by making either direct loans or equity investments in nonlocal businesses.

Financial institutions can choose between using community capital for local or nonlocal activity (Borts 1971; Cebula and Zaharoff 1975; Romans 1965). Local financial institutions can use the funds by (1)

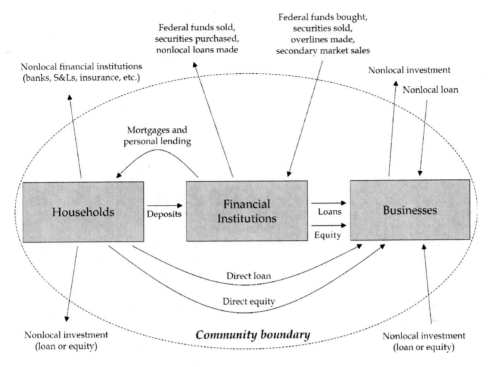

Figure 7.1. Community capital markets.

holding idle cash balances and/or reserve balances, (2) making local loans, or (3) exporting funds through the sale of federal funds, the purchase of securities, or participation in nonlocal loans. Local financial institutions obtain most of their funds via deposits by local households. These institutions also purchase federal funds, sell previously purchased securities, participate with outside financial institutions in locally generated loans, use overlines from correspondent banks, and sell local loans in secondary markets.

The local *business sector* is the major user of community capital. Local businesses acquire their local capital through direct loans from households, equity investments from households, or loans from local financial institutions (Kieschnick 1981a). They can acquire nonlocal capital through loans from nonlocal financial institutions and equity investments by nonlocal households.

It is important to recognize that a community's economy operates within a regional, national, and international financial capital market. In other words, space enters the decision. The capital market is relatively expeditious in facilitating the movement of capital among regions and communities. Today's

international financial markets mean that a retired couple in a small rural community could have part of their savings in a mutual fund that specializes in small, capitalized Pacific-rim companies.

Two important elements of the local/nonlocal capital market movement are secondary markets and correspondent banking.

Secondary Markets

Secondary financial markets are a very important and often little-understood mechanism by which capital moves among communities. A *secondary market* is created when a bank sells all or part of a loan into a regional or national money market. The bank then gets funds to be invested elsewhere. Generally, the key to selling in the market is the buyer's trust that the initial loaning institution has made a good loan. A simple example demonstrates the importance of secondary markets. Let Joe's Bar and Grill obtain a $10,000 loan from a local bank. The bank can sell $9,000 of the loan in the secondary market, then use the $9,000 received to make another local loan (i.e., import capital). The $9,000 sold in the secondary market carries with it a variety of assurances and guarantees that the loan is good. Often the person or business taking out the initial loan does not even realize that the loan has been sold on the secondary market.

Guarantees are a mechanism for reducing risks to financial intermediaries while reducing the demand or need for information about borrowers.[2] The reduced information needs lower information costs because there can be some economies of size in delivering that information and the guarantee provides some information (Litvak and Daniels 1979). Loan guarantees standardize financial instruments, thus facilitating the trading of such loans on secondary markets. The most common form of guarantee comes from governmental agencies (e.g., the Small Business Administration, the Rural Development Administration, the Economic Development Administration, and the Department of Housing and Urban Development); it is a widely popular economic development strategy. Such agencies have a relatively long tradition of guaranteeing loan risk. But all loans sold in the secondary market do not have guarantees; in these cases, the loan evaluation capacity of the initiating bank is crucial in providing assurances.

Correspondent Banking

Banks also can acquire funds through a correspondent bank (Mikesell and Davidson 1982). A *correspondent bank* provides funds by participating in loans made by the smaller bank (i.e., overlines), by trading securities on behalf of the smaller bank, by facilitating the trading of federal funds, and by borrowing on behalf of the smaller bank from the Federal Reserve Bank's discount window.[3] In return, the correspondent bank charges a fee, which may be a couple of points or compensating balances. Compensating balances are funds the smaller bank leaves in interest-free accounts at the correspondent bank (i.e., some free funds for the larger bank to invest).

Correspondent banks may not respond to community credit needs for several reasons, including lack of sensitivity to local borrower needs, higher cost of funds, or different documentation and evaluation standards.

Secondary markets and correspondent banks, along with government programs, are traditional sources of debt capital. Now we turn to some nontraditional sources.

Accessing New Sources of Capital

To meet the challenges ahead, communities and businesses must have access to the capital required to modernize and remain competitive (Markley and Shaffer 1993). Community economic development requires capital to support new business formation and expansion. While community bankers are the primary source of debt capital for local businesses, their ability to meet new demands for capital may be limited for two reasons. One, traditional financial institutions continue to be constrained from providing equity capital. For start-up enterprises and expanding industries, future capital needs may be for equity-like capital rather than debt.[4] (More on this point later in the chapter.) Two, regulated community bankers must be sensitive to the risk involved in lending activities. Lending to support community economic growth may be risky and involve loans with limited or untraditional collateral, loans to new enterprises with limited business experience, or loans to existing firms that want to expand into new markets.

To meet capital needs in support of economic development, community bankers must form partnerships with other private and public entities. These partnerships may be forged with public sector institutions, such as a state development finance program, or with private alternative financial institutions. These partnerships are necessary to pool limited resources and leverage funds to support economic growth. These relationships are also helpful in allowing banks to become more involved in financing local economic activities without incurring unacceptable levels of risk. Not only are funds pooled through these partnerships, but risk is shared as well.

The Role of Banks in Community Economic Development

Depository financial institutions can be the core of community economic development (Grzywinski and Marino 1981).[5] Banks cannot be expected to finance all the investment needs of the community, nor should financial institutions be expected to shoulder the entire burden. They must maintain some investment diversity (local, nonlocal; agricultural, commercial, personal) as a hedge against a major economic disaster in any one place or sector. Bank deposits are short term to medium term (monthly in duration), thus preventing the commitment of funds to long-term loans. Furthermore, the credit needs of some businesses may exceed the legal limit a single bank can loan to a single business or individual. Also, the banker may not feel sufficiently informed about the potential of a particular business line and decline to make a loan because of this uncertainty. Finally, many new small business ventures lack market potential or management abilities that capital cannot overcome. Despite these caveats, depository financial institutions, especially banks, are critical actors in community economic development.

Commercial banks can affect local income in a couple of ways: through the returns paid on deposits, and lending and investment policies. While the deposit interest rate approach is not as dramatic as bank lending policy, it may affect more people because of the greater number of savers, although each is affected marginally less. The advent of money market accounts means that individuals with small savings accounts can gain relatively higher-yielding returns on their savings.

Another way a bank can affect income levels in a community is through the investment decisions the bank makes. A simultaneous relationship exists among bank lending policies, local income, and local bank inputs (time and savings deposits and demand deposits). An analysis of commercial banks in Wisconsin found that if the banks increased their average loans outstanding over a four-year period by 1.0 percent, the change in county per capita income over that period would increase by 0.4 percent (Ho 1978). An example indicates the significance of this relationship. Let banks in a county have an average loan balance of $1,000,000 over a four-year period, and let the county's per capita income increase by $100 over that same period. If the average volume of loans increased by 10 percent (or to $1,100,000), then the change in per capita income would increase to $104. For a county with 1,000 people, the bank's additional $100,000 of loans would yield $4,000 additional personal income.

Some specific questions can be raised about bank performance and roles that banks can play in local community economic development. There are at least four ways that bank practices can be examined against community goals and objectives.

1. *Where* is the commercial bank using the dollars generated in the community? Are community funds being used in local or nonlocal investments?

   If the local bank is using a relatively large share of its available funds for nonlocal investments, it may be hampering the long-run economic health of its home community.

2. *How* are local banks using community funds? Are bank loans being used for commercial and industrial, agricultural, personal, or real estate purposes?

   Each of these local investments carries direct implications for future economic activities in the community. Real estate loans are typically very safe, but often fail to generate jobs and income for the vast majority of the people within the community. Housing loans can help a community two ways. The construction of a new house generates employment, and it also provides housing for another family. Commercial and industrial loans to main street merchants and manufacturers are likely to be the source of continued employment opportunities for the community.

   What types of commercial and industrial loans are being made? Are loans being made for inventory and working capital, or for real estate? Are long-term loans available? Are the terms of the loan conducive to smaller firms struggling through growth pressures? Is the bank actively involved in promoting the growth of existing businesses, the formation of new businesses within the community, or the attraction of new businesses to the community through its lending policies?

3. *What* types of *services* does the local bank offer that may facilitate economic activities within the community? Does the bank provide financial counseling to businesses within the community? How does the bank gain information and/or improve its capacity to judge unique loan requests? Is the bank willing to purchase industrial revenue bonds and water and sewer bonds in the municipality? Has the bank formed a development finance subsidiary that permits it to make equity investments?[6]

4. *What* type of *relations* does the local bank have with other financial institutions? (Often a single local bank cannot meet the credit needs of some of its customers.) Has the bank actively sought participation with other local financial institutions on major community projects? Has the bank used Rural Development Administration, Federal Housing Authority, or Small Business Administration lending assistance to help finance a major local project?

Assisting New Business Formation

Community bankers have more than capital to offer a potential business borrower (Markley and Shaffer 1993). The banker's financial expertise is an additional important resource to entrepreneurs since access to business assistance services in many communities can be difficult. Yet, several surveys suggest that many smaller firms feel they are not being served by their local bank or are unaware of services offered (Shaffer and Pulver 1990; Steinbrink 1992).

Small businesses represent a continuing experiment by individuals who think they have an idea that the market will support. Often these ideas require serious revisions. While experienced business managers anticipate those questions, new business managers or owners may not. The community bank can play a crucial role during the business formation process by increasing access to management counseling and support.

Banks could pursue several options for providing this support. The bank could support business management education programs for new and current small-business customers. Community banks can create quasi-separate organizations (e.g., community development corporations, small business advisory committees) that enable the bank to actively support small-business development while maintaining an arm's-length relationship with potential borrowers.

Linking Information

Community bankers are in a unique position within their communities. In most cases, the local bank is the first stop for a business needing capital for start-up, expansion, or modernization. To encourage economic growth, community bankers need to serve as an information link for these enterprises. Bankers can maintain information about state development finance programs, equity investors in the state or region, technical assistance providers, university or state industry modernization programs, and other relevant business assistance providers.

Providing Community Leadership

New economic realities signal that the days of passive community banking are over. Bankers need to be more aggressive in identifying entrepreneurs, encouraging and prodding community leadership in the pursuit of economic growth, and supporting economic development activities (Taft, Pulver, and Staniforth 1984). In other words, community bankers need to provide leadership to the community and help develop a vision for how the community can adapt to economic change.

Successful communities develop comprehensive strategies for guiding economic changes. Community bankers have an important role to play in formulating and implementing these strategies. Bankers can engage community groups in determining how the bank can respond to emerging community needs both as a community citizen as well as for Community Reinvestment Act (CRA) purposes. Bankers can lend needed financial expertise in support of economic development endeavors and make prudent lending decisions in support of community economic change. Community bankers must consider how to balance the potential returns from making short-term investments outside the community against the need for supporting the long-term investments identified as being necessary for growth within the community.

We saw that financial capital markets are extremely varied, and most financial institutions deal in only a few of them for a variety of legitimate reasons. In particular, we saw that capital markets consist of demanders, suppliers, intermediaries, and a local/nonlocal dimension. The behavioral changes suggested here do not require an abandonment of fiduciary responsibilities on the part of community bankers. These suggestions provide a way to meet fiduciary and community economic development objectives by sharing knowledge, spreading risk, encouraging innovation, and taking leadership.

EQUITY: THE TRADITIONAL VENTURE CAPITAL INDUSTRY

Traditional venture capitalists act as conduits between investors and entrepreneurs seeking equity capital.[7] Venture capital institutions control pools of money to be used for investments in companies that show considerable promise of high future profits. Venture capitalists provide cash to companies in return for a share of ownership and control. In addition to providing money, they generally exercise considerable control over the management of the companies (Sahlman 1990; Zider 1998).

In principle, venture capital funds attempt to invest in firms that have the potential to generate annual returns on investment in excess of 30 percent, and that can be readily sold within three to seven years from the time of the original investment. Venture capital funds then seek other innovative ideas and begin the cycle again. Although traditional venture capital funds demand a high rate of return, they can rapidly provide entrepreneurs with large amounts of money, offer critical advice in management and marketing, and with their investment, give the new firms credibility in the marketplace.

The high rates of return are from favorable growth in sales, assets, and net worth of the firm over the time that the venture capitalist participates. The reason a venture capitalist expects these returns is because they tend to invest in firms that have a great opportunity to grow due to having a technological advantage that can be exploited in a new/existing market, and/or have an experienced management team that will make a success. It is important to remember that venture capitalists are always looking for their exit strategy, specifically how they can liquidate their investment favorably and get it into something else. Venture capitalists supply money without collateral and servicing requirements, but become part owners of a new enterprise. As owners, venture capitalists have a greater role in the actual operations than does a debt-holder (Waddell 1995).

Venture capitalists have four principal tasks or responsibilities (Elango et al. 1995; MacIntosh 1997). They assemble pools of money from investors, identify companies in which to invest, and monitor the management of those companies after the investments are made; finally, they develop a strategy for exiting the investment. The exit strategy returns cash to the venture funds, allowing them to begin the cycle again. In recent years the preferred exit strategy has been an initial public offering (IPO) of stock. Other exit strategies include a private sale or the purchase of the firm by a larger business, both of which return money to the fund. In those cases where firms are not successful, venture capitalists may exit by selling their shares at a loss or by liquidating the companies.

High profit potential, a short time horizon, and a clear set of rules governing exit decisions are central parts of the venture capital process.[8] From this perspective, investing in a firm that makes only modest profits is only slightly better than investing in a firm that is failing. Unfortunately, the majority of investments fall into the modest profit or failing categories, thus venture capital funds have to achieve some large successes to offset inevitable losses (Sahlman 1990).

Impediments to the Venture Capital Investment Process

Traditional venture capital funds choose not to operate in rural places, in low-income areas, and in many smaller metropolitan areas because of

- *investment opportunities with profit potential below that sought by traditional venture capital funds*. Small market areas do not provide the investment environment venture capitalists prefer: a large number of firms with high projected growth rates and the likelihood of lucrative exits.
- *too few investments to provide adequate deal flow*. In sparsely populated areas, few firms need and qualify for venture capital investments. As a result, the cost of searching out promising entrepreneurs and identifying prospective deals is higher.
- *too great a physical distance between investment opportunities*. Venture capital investing is a hands-on process that often requires frequent contact with portfolio companies. If a traditional venture capital investor cannot travel easily and quickly to visit a firm, the investment is not likely to be made.
- *inadequate infrastructure to support venture capital investment*. Attorneys, accountants, bankers, and business consultants are often needed to help put a deal together and ensure the success of the investment. Such a business service infrastructure is limited outside larger urban places.
- *difficulty in defining a viable exit strategy*. Most businesses located in rural or small market areas are unlikely to provide rapid, lucrative exits. For example, IPOs are not a viable exit strategy for many traditional manufacturing enterprises and businesses found there.
- *limited interest by many small business owners in accepting the conditions set by the venture fund to get its money*. To many business owners, particularly in family-owned businesses, giving up any ownership stake in exchange for capital is unacceptable. Concerns about the intergenerational transfer of the firm may make a business owner unwilling to accept the terms of a venture investment.
- *difficulties in attracting venture capital staff to the community*. A venture capital institution is only as good as its management team. In more

isolated communities, venture institutions may have difficulty attracting and keeping the qualified staff needed to invest successfully. Recruiting management for portfolio companies when an injection of new leadership is required may also be more difficult.

If venture capital is to become a significant source of funding for rural and small city businesses, the impediments listed above must be addressed. One way of dealing with a portfolio of lower profit opportunities is to include other factors. Traditional venture capitalists focus strictly on the financial rate of return on their investment, thus only benefits that can be captured in the net worth of the portfolio firm are important. From a larger social context, however, some benefits of an investment are not reflected on a venture firm's balance sheet. Property values may rise in a community as a result of new business activity. Individuals with real estate investments may accept a lower rate of return on an equity investment in a local business if its presence leads to increased rates of return on their property holdings.

Many have argued that future prosperity for small towns, rural areas, and low-income areas will require a new economic base of small- and medium-sized firms. With roots in their communities, these firms would use local resources to produce goods and services to meet growing demand in the global economy. But creating this new economic base requires the support of entrepreneurs and their firms, a key component of which is access to equity finance. If potential entrepreneurs do not believe funds are available to bring their ideas to reality, they will not act on them. Thus, while a supply of venture capital may not create a demand for these funds, if the entrepreneurs are not present, the recognized absence of a supply may convince a potential entrepreneur not to try to start a business. Without this venture capital infrastructure, potential entrepreneurs cannot be identified and their ideas will not make it to the marketplace (Gaston 1990).

The Decision-making Process

The concern about financial capital has led to many initiatives to expand the supply of business capital. Most new programs provide loans rather than equity. Because both debt and equity capital are usually necessary for start-up and expanding firms, a critical financing gap may still exist even where ample debt capital is available. Thus, a core issue facing those seeking to implement a locally based economic development strategy is whether they can develop programs to provide better access to equity capital. What follows is an overview of the decision-making process followed in establishing venture capital institutions. Understanding this process will enable interested individuals to learn from the experiences of others and develop venture capital institutions that work best given differences in goals, market conditions, and institutional constraints among places.[9]

Decision-making is presented as a sequential process. There are clear feedback loops among the various stages of this process. Decisions made early in the process constrain later choices. In addition, if constraints are subsequently identified in the process, such as deal flow, modifications and adaptations to earlier decisions or goals may be needed. The recommended decision-making process has seven principal steps or stages.

First, the decision-makers must identify or recognize the impetus for creating a nontraditional venture capital institution. This may be motivated by a desire to promote overall regional economic growth and development, create jobs in distressed communities and for low-income individuals, or address inefficiencies and gaps in local venture capital markets.

The impetus for starting an institution will affect the geographic area served and the type of business targeted for investments. Thus, in step 2, market analysis of investment opportunities (deal flow) in the prospective service area is conducted. This information, in turn, helps determine whether an effective demand for a venture capital institution exists, what niche the institution might occupy, the potential size for the fund, and the managerial expertise required. If deal flow is inadequate to support a venture capital fund, fund organizers need to re-evaluate the reasons for establishing the institution, investigate alternatives for increasing deal flow, or end the process.

If market analysis demonstrates the need for a venture capital institution, the next step is to more specifically articulate the institution's goals (step 3). Maybe funds seek to maximize financial returns (for example, with a target internal rate of return [IRR] of 30 percent or more). Other nontraditional institutions also attempt to maximize financial returns but realize that such returns will be less than those expected by traditional venture capital funds. Social returns from investments such as jobs, income, retail sales, and housing starts, however, are considered as benefits of the investment process and are valued by the founders of the institution. Finally, a third type of funds seeks to maximize social benefits while earning an IRR sufficient to ensure the institution's long-term sustainability.

In step 4, the institution's size is selected based on funds available for investments. Institutional size also is related to deal flow (step 2) and the type of management selected to run the fund. About $10 million is the preferred minimum size for a self-standing nontraditional fund (Freshwater et al. 2001). A $10 million fund is large enough to support a professional management team, permit a diversified portfolio of investments, and provide a reserve of capital for follow-on investments. If deal flow or fund-raising prospects result in a potential fund much less than $10 million, fund organizers have three alternatives: (1) They can re-evaluate goals and area and industry focus to increase deal flow and potential investors. (2) The institution can operate a smaller fund and subsidize the costs of the smaller institution by partnering with a public or private organization. (3) The organizers can terminate their plan for a venture capital institution and investigate other ways to assist area businesses.

In step 5, funding sources for capitalization of the institution must be identified, though potential funding sources are a consideration throughout the process. Funding for capitalization and management can come from various sources. Examples include public funds (local, state and federal), private funds (individuals, commercial banks, and public utilities), private funds with public incentives (tax credits), and nonprofit foundations. Organizers need to understand any restrictions that are placed on funds invested in the institution. For example, if a foundation is identified as a lead investor, the foundation's goals will be instrumental in setting the goals of the venture capital institution. If state government is the lead investor, restrictions on investments to in-state firms will be likely. These constraints will influence the options available to institutional organizers in other stages of the decision-making process.

Organizers select the appropriate legal and organizational structure to manage the institution's investment activity in step 6. The institutional structure can include for-profit, nonprofit, and public enterprises with structures ranging from angel networks (Table 7.1) to limited liability partnerships to corporations. Each structure has advantages and disadvantages with respect to generating deal flow, raising capital, attracting qualified management, and attaining the institution's goals and objectives. In the last step, the venture capital firm will need to manage its investment portfolio.

The establishment of a venture capital institution is complicated, time consuming, and expensive. Decisions regarding goals, funding, management, and investments have different implications on later decisions. Thus, using a structured decision-making process can reduce the likelihood of inappropriate decisions and increase the probability of establishing a successful venture capital institution.

CAPITAL MARKET FAILURES

The discussion up to this point has implicitly assumed that the actors in the capital market and the signals given lead to market clearing performance, and that capital is used in its most productive manner. Yet capital markets often appear not to allocate capital to its most productive use. Litvak and Daniels (1979, p. 26) indicate that capital market failure occurs when

> specific enterprises cannot get the funds they need, even though either private or social returns justify it. Sometimes a failure manifests itself in a firm having to pay an unnecessary high price. ... Other times it shows up in the complete rationing away of credit at any price. In either case, capital is effectively unavailable.

We now discuss four causes of capital market failure—mobility, information, risk, and regulation—and some biases that apply more to venture capital.

Mobility

Capital typically is perceived as the most mobile factor resource in the economy. Examples of retired couples with a mutual fund specializing in small-cap Pacific-rim firms or companies drawing on investors from around the world through IPOs in Wall Street give the impression that all capital markets are global in scope. To a large extent, capital does not possess the perfect mobility it is presumed to have. Something less than perfect mobility prevents capital from moving among uses and places in response to minor changes in interest rates and rates of return.

In a purely competitive market, capital flows freely among production processes in a community or among communities until it yields an equal marginal product in each use and place (Richardson 1969b; Roberts and Fishkind 1979). In other words, the marginal product of capital in each use in every community is equal to the interest rate or profit rate after adjusting for the transportation cost of capital:

$$MP_{k\,i}^1 = MP_{k\,j}^1 = MP_{k\,i}^2 = MP_{k\,j}^2 \ldots = r \quad (7.1)$$

where i and j are different communities, 1 and 2 are different uses (commodities), and r is the interest rate determined by the national capital market. This is one of the equilibrium conditions in our spatial neoclassical growth model of Chapter 2.

Capital, in its financial form, comes as close as any factor of production to being a truly mobile resource, but its mobility is still far from perfect. Capital movement among communities in response to local interest rate differentials may or may not be sufficient to eliminate those differentials, for a variety of reasons (Moore and Hill 1982; Richardson 1973).

First, the vast majority of community capital is not in monetary form but in some physical form, such as housing, factories, office buildings, stores, trucks, inventories, and machinery, that defies quick adjustment to minor changes in interest rates and rates of return. These forms of capital do not or cannot move in response to minor market signals or within a short time. Changes occur only at the margin as replacement occurs.

A community's capital base and investments are indivisible and lumpy; for example, you purchase a complete machine, not just a portion of it. This prevents marginal shifts in response to minor or even major interest rate differentials.

Communities may have differences in taxing structures that either encourage or impede the movement of capital between communities. Tax structure differences are more likely to occur between states than within a state, but they can occur within a state if communities hold different operational philosophies about property taxes. Common examples of such differences in taxing structure are the state system of taxing capital gains and inheritances or the local assessor's valuation of property.

Differences in the private and social effects of investments can affect the movement of capital. Often the private investor makes a decision equating private return with private cost. The decision to cease investment is consistent for the individual but inconsistent with the community's desire for social benefits. For example, the private investor cannot capture the returns derived from investments in most forms of social overhead capital, such as education and waste disposal equipment, and will not invest.

Another reason for continuing interest rate differentials is uncertainty and differences of opinion about investment risk involved in various communities. This uncertainty will cause investors to require rates of return based on perceived risk. Finally, if investors make decisions based on average returns

and the average returns exceed marginal returns, the resulting overinvestment creates a continuing interest rate differential. The converse is true also.

Dynamic mobility also creates continuing interest rate differentials between communities. Investment in the community is a function of one community's investment opportunities and returns relative to those in other communities. A community will attract capital as long as its demand for capital equals or exceeds the demand for capital in other communities. This demand will be expressed as higher interest rates or higher return on investment and is measured by the marginal product of capital times the price of the output produced ($MP_k P_o$). As long as a community produces a higher priced product or uses the capital to produce more units of the same priced product, it will demand more capital than another community.

In a dynamic setting, shifts in community demand for capital can cause interest rate differentials. Community demand for capital shifts for a variety of reasons. Technological change resulting in new products or new production processes causes a community to demand more, less, or different types of capital. A shift in the demand for community output will affect the investment demand associated with the production of that output through the capital output ratio. Changes in expectations may shift a community's capital demand curve in or out, such as when speculators build residential housing in energy boomtowns. Finally, changes in other factor prices may make capital either more or less expensive relative to labor and cause a shift in the demand for capital. These shifts in the demand curve for capital, which do not occur uniformly over time or space, generate continuing interest rate differentials among communities.

In summary, capital is often not sufficiently mobile to eliminate interest rate differentials between communities. This situation occurs because of the form of capital, uncertainty, indivisibilities of investments, ignorance, and shifts in demand and technology over time. In most cases, these forces can be amended by community actions.

Information

The high cost of information and transactions can obstruct the smooth functioning of capital markets (Hill and Shelley 1995; Litvak and Daniels 1979; Moore and Hill 1982; Roberts and Fishkind 1979). It takes time and skill to arrange a financial transaction, and the greater the transaction and information costs are, the higher the rate of return on the investment must be before it will be funded.

The evaluation of a project is a costly and time-consuming process that need not be proportional to the size of the loan. Activities such as record inspections, site inspections, due diligence, and discussions with decision-makers can be discriminatory against smaller loans/equity investments or more distant locations (Mikesell and Davidson 1982). The high cost of information encourages lenders/investors to develop a conservative, skeptical attitude toward new or unfamiliar products, technologies, production processes, locations, and firms.

Some transaction costs incurred when putting together a financial package are governmentally imposed for public protection, such as a prospectus, accounting or auditing standards. Transaction costs also include the time required for the entrepreneur, financers, lawyers, and accountants to write business plans, prepare financial statements, negotiate the terms, and complete the investment agreement. These business costs (time and resources to arrange a financial transaction) do not necessarily represent a market failure, but these costs can be disproportionately high for some investments. Information costs increase for firms that are smaller, newer, involved in new technology or unusual markets, or located in an isolated place. In the end, investors tend to continue to invest in known firms, known markets, or known technologies even though the expected return may be lower (Kieschnick and Daniels 1979; Taft, Pulver, and Staniforth 1984).

Another information problem occurs when investors may be ignorant or may have imperfect information about investment opportunities and potential returns in different communities. Owners of capital must comprehend the opportunity for higher returns for capital to move to alternative uses and places. An important dimension of capital market information is inertia (failure to change investment patterns because of the cost and effort involved) and bias (belief that community X is economically depressed and does not offer any investment opportunities) about alternative uses and sites. Economically depressed communities must convey information about local investment possibilities to local and nonlocal owners of capital. Unfortunately, a high degree of inertia and bias prevent the full and complete movement of capital to equalize interest rates even with the information.

Returning to the role of local financial institutions in bring savers and borrowers/investors together

(Fig. 7.1), the banker plays a key role in smaller communities. The ability of the banker to collect, process, and use the relevant information can be a linchpin in the efficient function of local capital markets.

Risk and Uncertainty

Closely intertwined with risk is the information that supports an investment decision. With information in mind, it is useful to distinguish our concept of risk from uncertainty. *Risk* can be defined as a situation in which the decision-maker has enough information to establish a probability distribution of possible outcomes. Risk involves an understanding of the variability in income and net return: the greater the variability, the greater the risk. *Uncertainty*, on the other hand, is when a decision-maker cannot formulate a probability distribution for the outcomes and has difficulty forming a strategy to deal with the uncertainty. This could be due to the lack of quality information or the inability to process the information.

During the past 25 years, several economic approaches have been developed for addressing decision-making within a risk context. Most of these fall within the literature on game theory. These include (1) choice based totally on maximum gains devoid of the probability associated with the outcome (nonprobabilistic criteria), (2) application of a risk discount factor that addresses risk as a time preference, (3) the use of a "safety first" rule, where choice is based on a minimum level of chance relative to disaster, and (4) the use of the expected value of returns, where choice is based on the minimum and maximum outcomes defined by their associated probability levels. (A comprehensive discussion of risk is beyond the scope of this text. Interested readers are referred to the work of Robison and Barry (1987).)

Capital markets operate efficiently whenever capital is allocated to ventures based on the expected rate of return adjusted for risk or when the investor is not able to earn more than a competitive rate of return on assets. For individuals making an equity contribution, risk is part of the return to equity investors; riskier investments require higher rates of return. The investor must be compensated for taking on the higher risk. Because government securities are generally considered risk free, their rates of return are generally low.[10] For individuals and financial institutions making a debt contribution and expecting to be repaid in full, risk takes on a different dimension. Obligations to third parties (e.g., depositors) often cause financial institutions to be reluctant or demand risk premiums that appear exorbitant before they will participate in the activity.

Systematic risk results from factors that affect all firms and are not specific to an individual company or its products. Fluctuations in monetary supply or demand, inflation, and other economic, political, and sociological phenomena are the collective source of systematic risk.

Unsystematic risk results from factors that are specific to the individual firm or are directly related to its industry. These uncertainties are grounded in internal and external conditions, with their varying impacts on the effectiveness and efficiency of the firm's operations, under which the firm must operate. For instance, given current technologies, the structure of the firm's industry, and the competence of the firm's management, how likely is it that a proposed product can be developed over the time period and for the cost anticipated? How likely is it that the developed product can be manufactured at the expected level of fixed and variable costs? How likely is it that the manufactured product can be sold at the expected price? How likely is it that the market for the product will grow at the expected rate and the firm will capture the expected share?

Default risk and liquidity risk are two major types of risk (Hill and Shelley 1995). *Default risk* reflects the possibility that a borrower or equity user will fail to repay the loan completely or lose the value of the asset completely. This can occur in two ways. First is the way the firm finances its activities and the effect of this on its ability to remain solvent. Given the firm's current financial structure and level of operations, there are two operative questions: How likely is it that the firm can generate enough cash to meet its current debt-service obligations and sustain growth? Is required external financing likely to be available in the expected amounts and costs? Second, financial institutions face greater default risk in communities experiencing economic decline than in communities experiencing economic growth, again for two reasons: Declining sales imply reduced cash flow to pay off the loan or extract full value of assets, and declining asset values reduce the collateral pledged against the loan. The availability of fire or property insurance for a business may be a very subtle but key element in facilitating a lender's willingness to make loans.

Liquidity risk occurs when the lender experiences an unanticipated cash outflow. Depository financial institutions must always have on hand, or be able to acquire quickly, sufficient cash to meet surges of cash outflow caused by depositor withdrawals. These surges can occur for a variety of unanticipated reasons. Financial institutions must meet cash outflows through the cash inflow from new deposits, loan repayments, the sale of liquid assets such as government securities, purchase of federal funds, repurchase agreements, or discount windows (Fig. 7.1). This is a major reason why the liabilities a financial institution acquires (e.g., demand deposits, time and savings deposits, insurance benefits) become a major determinant of the type of loans the institution can make. If the deposits are all short-term deposits, then the institution will be vulnerable to a liquidity risk problem if it makes too many long-term loans.

Another form of liquidity risk from the perspective of the investor is the ability to convert the asset into cash in a timely manner. One of the powers of the New York Stock Exchange is that the volume of sales and institutional backstops ensure a ready market for all stocks. For smaller companies, investors who wish to sell their stock may not be able to find ready buyers in a cost-efficient and timely manner. In addition, fixed capital resources, such as land or buildings, may not be able to be converted into cash in a timely manner.

Perceptions of unduly high risk levels can cause some suppliers of capital to withhold their funds or require very high returns by demanding a large risk premium. Financial institutions can deal with risk through four general strategies.[11] First, the financial institution can diversify the type of investments made. This can include *diversification* by type of activity (e.g., farming versus manufacturing versus retail establishment loans), by type of loan (e.g., installment loans, real estate mortgages), by geographic area (e.g., some local loans, some nonlocal loans), or by type of risk. Diversification by risk means the lender uses portfolio analysis and attempts to bring new loans into the loan portfolio with a negative default risk correlation compared with loans currently in the portfolio. Thus, diversification reduces the default risk of the total portfolio even though the risk on any given loan may not have been altered.[12]

Second, a major mechanism for financial institutions to deal with risk is through *pooling* (Kieschnick and Daniels 1979). Pooling does not reduce risk per se, but it spreads the risk among enough other financial institutions that no single lender is overly exposed when a given loan defaults.

Third, a risk-reducing mechanism is to guarantee or insure against default. In this case, someone (usually the government or a co-signer on a loan) stands behind the borrower and will repay all or part of the lender's loss in case of default. Fourth, another mechanism to deal with risk is improved knowledge. The lender's ability to improve the evaluation of the loan request reduces the level of risk associated with lending judgments.

Implicit in all of the analysis of risk is that the financial institution is able to receive information and comprehend all the information that it receives. In other words, it is the rational economic person presumed in so many of our economic theories. Simon (1982) basically indicated this person does not exist, and we all operate in a world of *bounded rationality*. Thus, lenders/venture capitalists face the problems of identifying firm risk and their own capabilities to properly judge risk.

Regulation

Many forms of governmental regulation affect the flow of capital among uses and places (Daniels 1979; Kilkenny 2002; Litvak and Daniels 1979). These include various regulations and their interpretation by bank examiners, and the Community Reinvestment Act (Green and Cowell 1994).[13] Some suggest that one capital market problem is that financial institutions in the United States are over-regulated, and that many of these regulations, despite their obvious intent to protect depositors, have an adverse affect on small- to medium-sized businesses (Hansen 1981a; Hansen 1981b; Mikesell and Davidson 1982). The emphasis on protecting depositors' monies leads banks to try to reduce the risk of loss, and therefore lend away from the newer, smaller, more innovative, more dynamic, and riskier businesses and toward the larger, more established businesses with less growth potential. As a consequence, banks tend to favor loans for real estate, consumer products, low-risk business, and government lending. The problem with this is that the new small business, so critical to community economic development, becomes an unacceptable risk at the local neighborhood bank.

Government regulation of capital markets is usually initiated to correct some type of market failure. These regulations, however, can create some very perverse effects. For example, many regulations regarding the assets of banks, savings and loan associations, life insurance companies, and pension funds are designed to control the financial interme-diary's level of risk and thus protect the depositor or policyholder. This in essence prevents these institutions from actively supporting risky enterprises that may yield very high employment and income growth potential. Another form of unintended regulatory side effect is the cost of meeting security regulations and information requirements designed to protect the investor from fraud; such costs may effectively close off these capital sources to the smaller firm (Litvak and Daniels 1979). An unanticipated outcome of regulation or overly cautious responses to regulation is that lending by financial institutions does not yield the new productive capacity required for community economic development.

Although regulations generally are constrictive and try to get an economic entity to perform in a much more social fashion, some regulations, such as the Gramm-Leach-Bliley Financial Services Modernization Act of 1999, are expansive and open up entirely new opportunities to the regulated.

Discrimination

An implicit market failure is to recognize that people are discriminated against for reasons of gender, race, location, and age.[14] There is no reason that the market would discriminate against these people except that lenders have either explicitly or implicitly chosen not to make loans to them. While one would like to believe that capital flows to those opportunities that offer the best risk-adjusted return, evidence suggests that it is just as likely not to occur as it is to occur. In particular, numerous programs exist to help women and minority entrepreneurs get started and keep operating. While most of these are debt oriented, there is no reason to believe that the equity side is anymore enlightened.

Specific Venture/Equity Capital Market Failures

The preceding sources of market failure applied to both debt and equity/venture capital. What follows are some qualities that apply more to venture capital.

Bias against Business Life Cycle

Venture capitalists can invest at anytime, but to get the high rates of return, they tend to invest at start-up and expansion (early and accelerating growth) phases. Furthermore, businesses involving new technology are likely to require investors who specialize in evaluating the market potential for innovative products and are active monitors of their investments (Henderson 1989). These informational asymmetries are especially high when a venture's assets consist largely of firm-specific physical capital and intangibles such as an owner's stock of knowledge. Outsiders may be reluctant to invest in new ventures because of the high cost of monitoring management decisions. That is, the investor will not seek out those firms that may offer the best prospect of increasing the value of the venture capitalist's investment because of a natural tendency to feel more comfortable with an ongoing business.

Bias against Market

Investor prejudice against new markets and technological products makes the capital shortage facing rapidly growing, innovative producers even larger. In the end, money always passes between people, and it is the trust and confidence between investor and firm that determine whether a particular investment is made. Institutional investors often hesitate to finance new production techniques because of unfamiliarity with the complexities of the production process or because of unwillingness to accept highly specialized plant and equipment as security. Thus, the investor, because of information requirements, may choose the existing market or product or technology rather than the newer market or product or technology.

Bias against Giving Up Ownership

While it is presumed that both demanders and suppliers of venture capital will negotiate the best deal possible in their transactions, it is also presumed that there is no inherent bias against outside investors by those demanding venture capital. Fundamentally, venture capitalists are always looking for ways to liquidate their investments in three to seven years. This can be achieved by an initial public offering (IPO), by being acquired by someone else, or by the current owners buying the venture capitalist out. If the business owner is seeking an illiquid investment (i.e., an investment being made for succeeding generations), there is a natural conflict and capital may not flow toward certain investments. Another dimension of this is the firm that has a chance to grow but chooses not to do so for personal reasons of the owner.

Bias against Social Rate of Return (SRR) versus Internal Rate of Return (IRR)

Venture capitalists in the social investment tradition are sometimes referred to as *social purpose* equity investors (Waddell 1995). A more correct term would be *socially guided* equity investors since both

financial and social objectives are considered. These investors believe that in addition to creating profits, returns should be thought of as including factors such as reduced welfare costs, lower crime rates, and greater self-esteem. Some of these investors also see themselves as making particularly valuable contributions to society by developing particular types of enterprises, such as environmentally sensitive or employee- or minority-owned enterprises.

Policy Responses to Capital Market Failures

In general, there are two approaches to solving capital market problems (Bearse 1979; Kieschnick and Daniels 1979). One approach is to modify the behavior of private investors by giving subsidies that lower the cost of capital, by imposing taxes that raise capital costs, by guaranteeing investments, thereby reducing the risk, or by improving the flow of information. The second is that new public financial institutions can be introduced to act in ways consistent with the public interest. Either approach can be used to address both efficiency and equity problems.

The rate of return is the criterion lending/venture institutions use. For the community to cause capital to flow to underfinanced viable firms, it must enable the firm to pay those additional interest premiums (i.e., some type of interest subsidy) or it must assume some type of guarantee or co-lending position to reduce the lending institution's risk exposure and to reduce interest premiums.

We must remember there is no such thing as a perfect market, and governmental intervention into the capital market can only improve the working of the market, not perfect it (Bearse 1979). The dynamic character of the economy means there are shifting patterns of opportunity, technological innovation, and investment decisions that change market conditions. The public sector should address long-term market trends and conditions rather than try to respond to short-term symptoms.

SUMMARY

Financial capital by itself is not sufficient for community economic development, although it is a necessary component. Markets, management, and labor are equally if not more important to community economic development. Financial capital, however, is an important ingredient once the other factors are available. Furthermore, not all forms of financial capital are equally useful. For the newer, smaller firms with wide fluctuations in earnings, debt capital is less useful than patient or equity capital.

A review of the failures within the capital market identifies supply, demand, and institutional/rules gaps. A *supply* of capital gap could be a situation where a community's capital market cannot attract sufficient funds to support local investment. In this case, local financial institutions do not attract deposits or use regional money markets to support local investment. It could also occur when a community is unable to generate sufficient equity capital. A *demand* gap exists when an area lacks the management and entrepreneurs capable of transforming existing financial capital into physical capital required for production. In this case, the community needs to assist entrepreneurs and managers in the preparation of financial applications. An *institutional* gap exists when the area does not have the institutions to bring together providers and users of capital. Possibly as important as any concern about the shortage of capital is the shortage of talent to mobilize and utilize capital in a fashion both publicly and privately rewarding while protecting the parties against loss. This is particularly important in the public sector (Bearse 1979). The area may lack a bank, a credit union, a small business investment corporation, a community development corporation or a venture capital firm. These gaps are not mutually exclusive.

Of course, our idealized world of perfectly functional financial capital markets does not hold in the real world. Markets can fail in several ways, ranging from lack of information to ability to process information to discrimination. The community economics development literature abounds with studies that purport to document the shortcomings of the financial markets and offer solutions to market failures. In this chapter, we alluded to several strategies, including public guarantees to public venture capital programs.

The availability, rather than the cost, is critical when capital becomes an important location determinant. Capital cannot make a bad deal good, but the wrong type of capital can nullify a good deal or stop an otherwise promising enterprise from being born or growing. The right kind of capital on the right kind of terms can help a good venture start or expand and thus create vital economic activity in the community.

Capital is obviously crucial to the process of economic development. It is important in changing the productivity of labor because of its contribution to the quantity/quality of capital goods available to work with. However, although capital is necessary, it is not sufficient in itself for economic development.

STUDY QUESTIONS

1. When capital markets were defined, the clause "the return on capital in all uses and places is equal" was used. What is the significance of this statement to community economic analysis?
2. What role do financial institutions play in an economy? In particular, what role do they play between demanders and suppliers of capital?
3. How would you respond to a statement that capital markets are homogeneous?
4. What are some of the reasons for capital market failure? What is the significance of each to community economic development?
5. Why might capital not be mobile enough to cause the rates of return to equalize over all uses and places?
6. What is the significance of information costs in the functioning of capital markets?
7. Four questions were suggested for comparing commercial bank practices against community goals and objectives. What are they and why are they important to community economic development?
8. In what form is the community's capital base found? What is the significance of these forms to the mobility of capital?
9. How do the terms *stock* and *flow* relate to the community's capital base?
10. What are the implications of information and transaction costs, risk, mobility, liquidity, substitutability, and regulation to capital mobility?
11. Why might not equity/venture move to firms or places most in need?

NOTES

1. Note that households also can be users, and businesses also can be depositors.
2. The most elemental guarantee is co-signing a loan.
3. Each bank, whether it has a federal or state charter, is limited in the amount of loans that it can make to any one individual or business. This lending limit is typically 10 percent of the bank's capital base. Banks usually enter

into these markets to ensure that the 10 percent minimum cash reserve is met. Entering these markets to expand their pool of loanable funds is an extremely aggressive practice on the part of the bank.

4. Lenzi (1992) argued that bank Community Development Corporations (CDC) might be the answer.

5. Throughout, the term *banks* is used interchangeably with *depository financial institutions* because banks are the most common depository financial institutions. Other specific types of depository financial institutions include credit unions and savings and loans.

6. Lenzi (1992) argued that bank CDCs are another important mechanism for banks to use in their lending portfolio, and they can even take an equity position.

7. This section draws on Freshwater et al. 2001.

8. One could think of venture capitalists as seeing opportunities in a firm that will earn themselves short-term monopoly rents. As we have seen in our discussion of endogenous growth, these opportunities for monopoly rents can be a powerful economic force.

9. This section draws on Barkley et al. 2001b.

10. The difference between the rate of return paid on risk-free debt, such as treasury bills, and the rate of return actually paid is called the *risk premium.*

11. In many cases, smaller banks and rural banks, in particular, are at a disadvantage because the local capital market is undiversified and small, so they are highly exposed to cyclical changes in the local market. Furthermore, when they wish to sell or to pool risk by selling loans in a secondary or national money market, they are hampered because of uncertainty about their ability to judge risk, because of uncertainty about the quality of the collateral pledged and so on. The end result is an impediment in the flow of capital into those markets (Mikesell and Davidson 1982).

12. This is exactly the logic behind diversifying the economic base of a community in the name of stability (Wagner and Deller 1998).

13. Congress enacted the Community Reinvestment Act (CRA) in 1977. The law requires banks to meet credit needs, as well as depository needs, of the communities in which they are chartered. Previously, individuals and businesses in low-income areas were often denied credit because of the perceived high-risk nature of such loans. Under the CRA, a bank is evaluated approximately every two years on its community reinvestment performance by federal bank regulators (Haag 2002). If a bank's CRA record is poor, it can be denied a request to open a new branch or expand through mergers and acquisitions. Community groups can make written comments on a bank's performance at any time and make sure that the comments are put into the bank's CRA file so that regulators consider them. Not all regulations would necessarily be viewed as negative. Regulations that address predatory lending, ATM fees, lifeline accounts, and involvement in community development corporations may be viewed as very positive types of regulations.

14. Redlining is a form of spatial discrimination.

8
Technology and Innovation

Technology and, more importantly, innovation are major drivers of economic growth in endogenous growth theory (Chapter 2). While the neoclassical model of economic growth provides us with a framework to think about the growth process, it does not adequately explain or predict why firms would invest in research and development of new products or in technology. Although shifts in technology and innovation drive the growth of a modern economy, they are considered random and exogenous to the economy in most theories of growth and development. We concluded that this is clearly not acceptable either theoretically or practically when considering economic policies.

Current thinking acknowledges that firms benefit from internal and external economies of scale, or what regional economists call agglomeration economies. If a firm can make an economic profit on new products, technologies, or innovations, the firm has a short-run profit incentive to invest in research and development. The key is the institutional role of patents. Because firms or entrepreneurs can obtain patent protection to ensure monopoly profits over the life of the patent, there is an incentive to invest in developing new technologies. We also saw that this new way of thinking about the economic growth process—endogenous growth theory—can work against smaller communities that do not possess agglomeration characteristics.

Communities compete for limited resources, and in a spatial world, the competition occurs as communities try to position themselves as the location with a comparative advantage. Comparative advantage is a result of having some advantage in access to or costs of inputs and markets, but that is largely static. Dynamically, the key is technology and innovation. While economists firmly believe that prices are a market signal that will eventually lead to equilibrium, technology and innovation are constant disruptive forces. Prevailing utilization of technology, in the form of new inventions or imitations, and its rate of change clearly affect the respective competitive positions of firms and the resulting growth/decline of the community and its component industries (Erickson 1994). In addition, output and product composition changes in the innovative firms over time would be expected to require continuing changes in the set of communities from which inputs originate for such firms, or changes in the set of communities that are the destination of their products.

Three notions of technology and innovation must be clearly defined. The first hinges on the notion of the creation of new technologies or ways of doing business. These are patentable and can earn the creator profits in the short run. The second centers on the creation of new ideas that are nonrival in their nature. This could be a new business practice, such as just-in-time inventory. While certain business methods like implementing just-in-time inventory might be patentable, the overall concept cannot be patented. The third centers on adoption rates of the new technology or innovation. Early adopters can gain more from increased efficiencies than can later adopters. As more firms adopt the technology, the gains from early adoption fade away.

The neoclassical model essentially assumes that information about technology, a prerequisite for transfer or adoption, is easy to obtain, does not cost much, and is comprehended by the receiver. *Technology transfer* can be broadly defined as the transfer of a technology, technique, or knowledge that has been developed in one organization to another, where it is adopted and used (Melkers, Bugler, and Bozeman 1993). The emphasis on technology to spur economic development is not new. A widely accepted view is that new technology is an early link in a chain of activities that leads to a reworked industrial structure.[1] This view holds that technology is catalytic and can be expected to stimulate the rise of new industries. But the adoption of a technology by firms does not necessarily lead to the creation of new jobs and economic growth in a community, causing much concern in communities.

In Chapter 2, the neoclassical model assumes a continuous relationship linking the output of the area to the land, labor, and capital available in that area. This relationship, referred to as a production function, is defined as

$$Y = f(S, L, K, A) \qquad (8.1)$$

where

Y = level of real income or output
S = land resource inputs
L = labor
K = capital (Used in a generic sense here, it refers to natural and human-made capital as well as public, private, and social capital.)
A = technology (which is assumed to be constant across communities)

Recall that the neoclassical model suggests that sources of output or income growth for communities arise from any of the following economic actions: (1) an increased supply of land, (2) an increased supply of labor or population growth, (3) an increased supply of capital or capital accumulation, (4) some form of technological change, and (5) some form of increased economic efficiency by shifting resources from lower- to higher-productivity uses. (The land, labor, capital, and amenity portions of a community's production function are discussed in Chapters 5, 6, 7, and 9, respectively.)

The neoclassical model suggests innovation creates higher profits for adopters by reducing their average costs. These excess profits create an incentive for technological adoption and lead to diffusion. The spatial diffusion of technology or innovation significantly affects the rate of development among communities. The neoclassical model assumes distribution of technology instantly over space and uniformly over time. But technological change does not occur at a constant pace, nor is it uniformly adopted everywhere. The hindrances to the instantaneous spread of technology contribute to a market *imperfection,* preventing every community from equal or instantaneous access to the same technology. The

end result is communities develop at different rates because they use different technologies.

The community can become vulnerable to forces beyond its control. Technological change can shift the relative production advantages among communities (e.g., less of a local resource being needed or other local resources becoming more productive). The development of new cotton varieties, fertilizers, and irrigation methods, for example, lead to a shift of cotton production from southeastern to southwestern United States. Furthermore, over time, substitute resources may replace a key resource, such as artificial rubber replacing natural rubber. This gives us an idea of how technology generically fits into the picture, but what specifically is technology?

Technology and innovation are a bit like Pandora's box. Once you open it, it's very difficult to close it. Technology and innovation are critical in a dynamic economy, but invariably winners *and* losers are created when new technology or innovation is applied. This is directly related to Schumpeter's idea of creative destruction (Chapter 2). As new technologies and products are developed, they replace or destroy older ones; the losers are those that held the patents on the now old technologies. In addition, if the new technology is radical enough, early adopters can exert strong monopoly powers and force later adopters from the market.

Technology tends to be sector specific, although there are excellent examples throughout history when that was not the case (e.g., small electric motors, computers, and the Internet). Technology also tends to be fairly spatial in its impact because the industries are located within space and the inputs are located within space. The losers in the technology game are those who see their standard of living, wealth, and job prospects and occupational desires decline. On the other hand, those places and workers who naturally can capture the benefits from the application of new technology will see their wealth increase.

Knowledge is a precondition for all forms of production (Andersson 1985). This involves not only the results from research and development, but also adoption of new technologies or business ideas. Firms must have knowledge of the technology itself, as well as knowledge of how to build the technology into their own firms.

From one perspective of technology, we are in the resources node of Figure 1.1, and from another perspective, we are in the decision-making node. These distinctions will become clear as we discuss what technology and innovation are, and why they are important. We also turn to the innovation process or how technology moves spatially, then to economic theories that explicitly consider technology/innovation.

It is important to remember what we are talking about with regard to community technology. First, the technology in the community is the sum of the technology of individual firms, the sectors they belong to, and the technology employed by the local public sector (e.g., municipal government, schools, water and waste treatment, fire protection). Second, much of the community's technology is process technology (i.e., how we deal with each other). This involves interaction between constituents and governing bodies, how teachers teach, or how local businesses network to build synergies. Process technology emphasizes management technology or management information technology; within the business world, it focuses on operation research.

Community growth and decline are largely due to differential growth rates based on the particular composition of industries a community hosts and the competitiveness of those industries in relation to those of other regions, both of which are deeply influenced by the prevailing utilization of technology and its rate of change (Erickson 1994). Schumpeter's entrepreneur is vital because she possesses the skills required to bring new ideas and technologies to the market.

WHAT ARE TECHNOLOGY AND INNOVATION?

For the most part, people tend to associate technology and innovation with high tech and advanced technology firms and industries, but it is much broader than that. In the simplest sense, technology is how land, capital, and labor are brought together to produce an output. Technological advances are changes in how the inputs of land, capital and labor are brought together to produce goods and services. When a community or industry experiences technological change, the inputs required in the production process are likely to change. When these inputs change, other areas may become competitive, by gaining a comparative advantage, in the delivery of those inputs.

Another way to look at technology is in the products being produced. Here again, as the products being produced change, the communities and industries that produce them probably gain some type of competitive advantage.

The final form of technology is process technology, which focuses on how we do things or on different forms of organization to accomplish the same thing, or how we deal with each other.

New technology can be in (1) an *embodied form* (new products, capital equipment for production), (2) an *altered process* (the light bulb, the microchip, and software), or (3) a *disembodied form,* which focuses on new managerial or organizational changes or processes (just-in-time inventory control, decentralized organizational structure, new uses of existing products, and the Internet) (Erickson 1994; Mathur 1999).

For our purpose, *technology* is defined as "a formal and systemic entity of knowledge and skills in order to realize and control complex production techniques/processes" (Stöhr 1986). Technology is not a single item; it is a spectrum of activity, including the advancement of scientific knowledge and the incorporation of this knowledge into products, processes, and management improvements. Technology also includes the flash of insight that provides new understanding of basic questions. Invention can be either autonomous, a largely a random event, or induced, the result of research and development efforts. *Invention* is conceiving and presenting new combinations of preexisting ideas, or it could be the result of an inventive act or some discovery of something previously only imagined (Friedmann 1972). It need not be used immediately and it can remain dormant for a considerable time.

Technology also ranges from basic to applied technology (Thwaites and Oakey 1985). *Basic research* is conducted with little appreciation for what its commercial possibilities may be. It generally is conducted at universities, think tanks, or research labs. *Applied research,* on the other hand, is research conducted with either a practical question or a commercial application already in mind. The recent growth in the number of patents flowing from basic research, such as gene splicing methods, has seen a significant acceleration of private firms investing in basic research. Indeed, the much more rapid application of patents to what was once viewed as basic research has significantly blurred the distinction between basic and applied research.

Innovation is the successful introduction of ideas perceived as new into a community or as the first commercial or genuine application of some new advancement outside of experimentation (Friedmann 1972). From a community economic development perspective, innovations may be based on ideas that are inventive, borrowed, or imitated. Again, from the community's perspective, what is already established in one place may, by borrowing or intimation, become an innovation in another. It is something new in this organization or place, which is the decisive criterion for innovation. Innovation is the adoption of technology and its application to production, marketing, or organizational problems. While also dependent on flashes of insight, innovation is more likely to be influenced by entrepreneurial attitudes typified by a willingness to seek different solutions to problems, economic conditions, and availability of investment funds. Innovation and the organization into which it is introduced must be structurally compatible. Innovation requires individuals or institutions to organize the necessary resources and assume the risks of failure. At this stage, technological change begins to have considerable impact on the structure of economic activity, changing the demand for capital, materials, labor, and skills and introducing the possibility of substitution as consumers make choices among competing goods within their budget constraints (Stöhr 1986; Thwaites and Oakey 1985).

Technology is an ongoing process that is somewhat independent of macroeconomic conditions at any one time and is more dependent on the inherent curiosity of the population. In endogenous growth theory, one could argue that the inherent curiosity of the population is driven by monopoly profit motivations. Innovation is much more dependent on economic conditions (e.g., growth rates and competitiveness in the market) and the ability to generate investment funds to implement the new technology.

Technological change can be divided into two parts: *things* and *ideas*. Things, it is argued, are *rival* in that things, such as a new robot technology, can be patented. Ideas broadly defined, on the other hand, are *nonrival* and easily transferable (Mathur 1999).[2]

Because nonrival ideas can be copied and communicated, their value increases in proportion to the size of the market in which they can be used. A nonrival good is that good whose use by a firm does not preclude or diminish its use by others (Mathur 1999). It is non-excludable when a firm or an economic agent cannot prevent its use by others without incurring substantial cost relative to the value of the good. Its use with other inputs like land, capital, and labor overcomes the limitations imposed by diminishing returns to these inputs in production.

We tie all these pieces together with the use of Figure 8.1 and an example of the Japanese firm that first introduced just-in-time inventory. The idea of just-in-time inventory is a way of doing things and, in our context, a process technology. The broad idea is characterized as nonrival and easily transferable. The means by which the Japanese firm implements the idea of just-in-time inventory, however, can be patented and is the thing or product technology.[3] In essence, the specific business plan and computer software developed by the innovating Japanese firm is rival property of the firm. Here the business plan can include methods for interacting and contracting with input suppliers and shipping firms. The Schumpeterian idea of innovation or entrepreneurship is

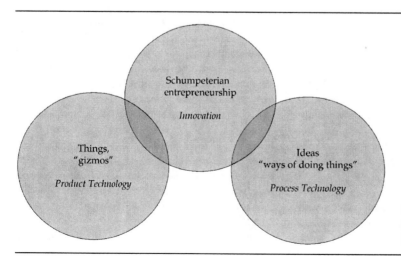

Figure 8.1. Relationship between things, ideas, and innovation.

the human capital required to pull all the pieces together and earn those monopoly profits. The Japanese firm is in a win-win situation because it has developed and adopted a profit-enhancing technology, or way of doing business, and is now in a position to sell or, more specifically, to license the specific business plan. From the community's perspective, innovation is fostered by local firms adopting or transferring these new technologies.

INNOVATION OR TECHNOLOGICAL ADOPTION

Work on the innovation process has indicated that the process is not linear (Davelaar 1991; Melkers, Bugler, and Bozeman 1993). The relationship of science to technology and technology to economic growth is complex, interactive, and iterative. The transformation of an innovative success to a market success occurs at the level of the firm. Information about the new technological opportunities and new market demands is filtered through the firm (Oerlemans, Meeus, and Boekema 2001). The firm's interpretation of demand, along with internal firm factors, determines which markets the firm seeks to exploit with an innovation. The entire process is characterized by uncertainty, trial, and error. Adopting an innovation does not mean the firm can capitalize on it; it depends on the firm's technical capacity, organizational ability, and knowledge of the markets. The type of technology transfer, whether it is the transfer of scientific knowledge, physical technology, technological design, or process, affects the nature of the transfer.

Another important component in assessing technology transfer is the time line. Time is needed to find out about the technology, to study it for appropriateness, to acquire the capital needed to purchase it, and finally to incorporate it into the production process.

Andersson (1985) talked about creative regions that directly apply to communities, for regions are composed of communities. The spatial dimension is just smaller. His idea of creativity can be viewed as a series of steps up a pyramid, with each building on the lower one. He claimed *data,* the most elementary concept, can be disaggregated and aggregated without loss. The next higher step is *information,* which is structurally ordered data. An analogy is viewing data as variables in a system of equations, and information is the set of equations containing these variables. Concepts, ideas, and patterns are subsets of information. The third step, *knowledge,* can be seen as embodied information. This means that knowledge is information that is embodied in man's relations to other humans, machines, and the environment. *Creativity* is at the peak of the pyramid. Creativity presumes a capacity to order and reorder data with the aid of an information system. We assume that the creative process is synergistic and dynamic. This implies that data, information, and knowledge are brought into an intensive interaction with each other in order to fashion new knowledge. Andersson argued that creative communities are the pinnacle to which every community should aspire. Note that Andersson's ideas apply equally well to product and process technology (Figure 8.2).

Andersson (1985) also argued that the creative process at the community level must include three basic concepts: knowledge (and its relationship to information and data), synergisms, and structural instability. We have already discussed knowledge. The concept of *synergism* denotes that the "effect obtained from the combined action of two distinct chemical substances is greater than that obtained from their independent action added together" (Goldstein and Luger 1993, p. 156). In community economic development, this idea suggests not only the presence of specific agents and institutions within a community, but also that their mutual dynamic interactions are a prerequisite for optimizing creativity and innovation under conditions of structural instability (Stöhr 1986). Where you have synergism

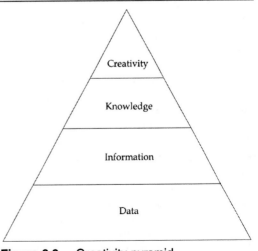

Figure 8.2. Creativity pyramid.

combined with knowledge, you can transform an unstable situation into a creative situation (Andersson 1985). A completely new approach to an old problem can emerge because there is no stabilization to prevent the force of synergism.

In terms of understanding the conditions of research, development, and creativity, however, the importance of structural instability cannot be minimized (Andersson 1985). Uncertainty is fundamental to creativity. Uncertainty of this kind is not an obstacle but a precondition for a creative state. The idea of instability might be thought of as nonroutine decisions made over time (Thomas 1985). The nonroutine decision involves unlearned acts of insight and learned acts of skill by the firms or communities. When making nonroutine decisions regarding their future well-being, firms or communities synthesize past and present flows of information generated within the firm and community with information from external sources concerning the present and future decision environments of the firm or community.

An innovative firm or community has established a flexible approach to decision-making (Thomas 1985). These have also been referred to as entrepreneurial communities (Chapter 12). The organizational routine behaviors probably would not deal with unprecedented situations or new and changing conditions in the decision environment of the firm or community. In such new or unexpected situations, the firm or community's well-being will depend on how effectively it can adapt or replace current organizational routines. It would be expected that firms or communities would initiate organizational search and judgment activities that might lead to modification, drastic change, or replacement of current routines. The appearance of the need to apply nonroutine decision-making is largely of a stochastic nature.

These three notions—decision environments, organizational routine, and search—provide a foundation for understanding the behavior of a firm or community when it deals with nonroutine decisions, such as those related to innovative activities or unexpected internal or external events (Thomas 1985).

Communication is key in understanding the creativity process. The idea of horizontal or internal and vertical or hierarchical communications is fundamental to the creativity process. (These ideas are similar to bonding and bridging social capital in Chapter 12.) This critical role of communication tends to occur in metropolitan areas and favors them

in the creative process. Andersson (1985) concluded that creativity is a social phenomenon that primarily develops in communities characterized by high levels of competence, many fields of academic and cultural activity, excellent possibilities for internal and external communications, widely shared perceptions of unsatisfied needs, and a general situation of structural instability facilitating a synergetic development.

Characteristics of the environment are important because they represent the economic and infrastructural preconditions necessary for the circulation of information (Camagni 1985). The environment provides the base for those psychological, cultural, and social variables that define the taste for risk, the capacity for extensive organizational change, and attitudes toward technology. The adoption of an innovation also depends on its appropriateness to the firm or community. A different rate of innovation adoption through time can be caused by the accumulation of information by its previous adopters. For the individual firm or community, the adoption process is based on the possibility of access to information, the estimation of profitability or usability, and the evaluation of adjustment costs. Subsequent adoptions of an innovation imply active imitation characterized by minor marginal improvements.

Costs, but more importantly profits, represent an important consideration in adopting any new technology. First there is the cost of acquiring information about the new technology, then there is the investment cost of acquiring the technology. But an important cost, often not considered *ex ante*, is adjustment cost. *Adjustment costs* are the time and resources to acquire a set of necessary skills and know-how (basic and advanced understanding of the technology) in order to fully utilize the new technology and realize its maximum potential productivity gain (Ahn 1999). Furthermore, it includes lost production and declines in productivity as the new technology is incorporated into the production process.

Consider, for example, a small company that elects to upgrade from a collection of freestanding personal computers to a network of workstations with a central server. The company must not only buy the hardware and software to make the system work but also must install the wiring to network each workstation. After the initial effort, there are costs associated with all users becoming familiar with one common set of software. During this frustrating transition period, initial system crashes may cause work to be lost. Because of these costs associated with the adoption of new technology, its implementation tends to reduce productivity in the short term, even though the potential productivity gain in the long run outweighs this short-term loss.

At the organization level, just as with individual learning, it takes time and effort for an organization to learn how to fully utilize a new technology and realize any productivity growth. For example, to most effectively implement a new technology, a new organizational structure might be needed. Of course, restructuring an organization requires tremendous time and effort. This restructuring can also be regarded as a part of organizational learning.

In particular, the ability of a firm or community to realize and ably exploit the potential of a technological or commercial idea, its receptiveness toward information, and its flexibility in a turbulent environment depend on the adoption of an adaptive internal organization rather than a mechanical, noncreative one (Camagni 1985). The former is characterized by the existence of limited regulations or procedures, workloads that are notable for the absence of routine, wide areas of responsibility, and free communication among people rather than their positions. Essentially, it is typified by the continual redefinition of individual responsibilities, a weak hierarchical control structure, a more limited sense of loyalty toward superiors, and a consultation network of horizontal communications that are not geared to the reporting line and control structure. Stöhr (1986) argues that this can be viewed as organizational flexibility. This is required so the firm, organization, or community can reshape itself to capture the benefits from the new technology.

New technology and innovative economic activities display a tendency toward agglomeration and concentration (Malecki and Nijkamp 1988; Thwaites and Oakey 1985). Given our current understanding of nonroutine economic activities, there appear to be sound reasons for this agglomeration. The information-intensive nature of technological activities and the resultant need for face-to-face communication favor those places that offer (a) high levels of confidence in the information received, (b) many fields of cultural and economic activity, (c) excellent possibilities for internal-external communication, (d) widely shared perceptions of unsatisfied needs by customers, and (e) a general situation of structural instability facilitating a synergistic development. The linkage with the concept of local knowledge spillover is provided by the frequent claim that knowledge transmission is mostly a matter of face-to-face contacts and labor mobility. In other words, the most important knowledge carriers are people, in particular, people who know and possibly trust each other, meet frequently, and trade job offers very often. More formal means of transmission, such as scientific articles or technology licenses, are seen as playing a lesser role (Breschi and Lissoni 2001).

Simply put, *technology transfer* is a complicated process involving numerous factors (Melkers, Bugler, and Bozeman 1993). Technology transfer varies according to differences in characteristics of the transfer type (product/process), the mechanism used for transfer, attributes of the transfer recipient, and market contingencies and whether the recipient is closer to being a developer or a user.

MODELS EXPLICITLY INCORPORATING TECHNOLOGY

While some authors suggested that "technological change [was] the *terra incognita* of modern economics" (Erickson 1994; Schmookler 1966), others stated that "technological change is arguably one of the single most important (and most overlooked) influences on regional (community) change" (Malecki 1983, p. 89). As we have seen, Schumpeter's (1942) work made the role of technology and technological change the engine of economic growth. He used the idea of creative destruction as an explanation for the dynamism of industrial change focused substantially on entrepreneurship and the disequilibrium effects of new product and process technologies. The theories that explicitly incorporate technology/innovation can generally be classified as product/innovation life cycle theory, growth center/propulsive industries theory, disequilibrium theories, and the new economic geography (Krugman 1991a, 1991b, 1999), the latter of which we discussed in Chapter 2. The new economic geography has consumed the current thinking on the other theories, but it is important to review the prior theoretical frameworks because each brings a slightly different perspective to the larger problem at hand.

Product/Innovation Life Cycle Theory

The *product life cycle theory* articulated by Vernon (1966) and Hirsch (1967) can be traced back to the 1930s and the idea of heritage of regional and industry life cycles. This approach conceptualizes growth of product output in phases of early development, rapid growth, maturity, and decline (Erickson 1994). Decline may be staved off for a time when new products or innovations are introduced. *Product life cycles* are based on the premise that the commercial life of a specific new product is finite. While product enhancement and changes may very well extend a product's commercial life, eventually it ends.

Technology plays a key role in the product cycle framework (Blair 1973; Davelaar 1991; Erickson 1994; Mack and Schaeffer 1993; Malecki 1983). Innovation is critical in the early phase of the cycle, in which the product is initially developed and adapted for commercial use. In the growth phase, production technology is pivotal as firms move the product to the mass market. The individual supplier uses a combination of the characteristics of its products as its most important measure to attract customers to it and away from its competitors (Karlsson and Larson 1990). The important characteristics are not only its physical and aesthetic characteristics, but also the way the product is delivered, including the services provided, before, during, and after the delivery. Price is still important, but it is not the supplier's primary competition parameter.

A firm introducing a new type of good or service can influence the market and exert monopoly powers. Such a firm is said to gain a head start in the market that lasts until existing or emerging competitors find efficient imitations or to develop competitive substitutes. A firm gains this head start because of the following:

- It has close contacts with its customers.
- It produces after receiving orders.
- It can influence the price of its products.
- It has well-established and functioning information channels to discover what's new in their markets and their products.
- It views product development and knowledge handling as important competition parameters.

By the mature phase, the product and its production processes become relatively standardized and the competition from new entrants is more likely than in the previous phases, so achieving lower production costs is clearly a paramount goal of the firms. Even in the decline phase, improved technology may enable a community's producers to capture an increasing share of the stagnant industry's sales and therefore contribute to an expansion of the community's economic growth.

The importance of innovation and rapidly changing technologies in the early and growth phases of the product cycle means that the firm's own activities and many linked activities need to be performed in more technologically sophisticated communities (Erickson 1994).[4] These areas tend to be large metropolitan places where a range of skilled engineers and other professionals and inter-firm mobility as well as appropriate subcontractors are available. Agglomeration in research and development results from the preferences of technical personnel for large city locations and pleasant environmental conditions as well as the attraction of existing research and development labor pools for additional corporate research activities (Malecki 1981). A related process is the spin-off, the formation of new firms by entrepreneurs formerly employed by other firms. Since spin-offs generally originate in the same area as the founder's former employer, a cumulative process of new firm creation is often characteristic of such research and development complexes. The creation of new firms, in turn, increases the local agglomeration of research and development and attracts further corporate activities.

The market becomes saturated because other firms attempt to capture those monopoly profits earned by the initiating firm by imitating the product or developing alternatives (Karlsson and Larson 1990). To find new customers, the imitating firms must lower the price of their products. The way they achieve this is to standardize the product, to automate the manufacturing process, and to make the deliveries less customized (i.e., mass market). Such a market has the characteristics of price competition, cost hunting, and increased firm survival problems. The struggle is over market share within markets that unfailingly shrink in the long term. One crucial cause for the emergence of market saturation is that every product will sooner or later be the victim of substitution competition from new products that give rise to new product cycles.

During the later phases of the product cycle, capital and unskilled labor take relatively more importance in bringing unit costs into line with those of imitating producers. There is considerable evidence of the spatial division of labor, including that mass production activities are increasingly performed at branch plants that have been shifted to nonmetropolitan areas or urban cores. Here, firms are likely to find relatively docile, lower-wage, and less-skilled labor.

In the context of community change, technology leads to the creation of new products, new firms, and even new industries (Malecki 1981). Research and development activity, however, tends to concentrate in relatively few urban areas. This tends to perpetuate the agglomeration of technical change to the same few core communities. The variation in labor skills needed for research and development versus routine product manufacture is a major factor in the spatial division of labor. Through the process of "de-skilling" the work at certain plants, the proportion of more-skilled, higher-wage jobs is reduced at those locations. Changes in entry-level skill requirements can be called *skill-training life cycle*—an extension of product, process, and technology life cycle models—to directly address labor markets, firms, and individual workers (Malecki 1989). Products, production processes, and technologies are seen as dynamic phenomena in which skill and training requirements change as they evolve. Communities and other political jurisdictions worry that they may end up in a downward spiral. The de-skilling process makes them increasingly unattractive to those firms that use skills and pay higher wages.

The product life cycle theory focuses on the emergence of new products and their eventual decline over time and space. Technology is the key variable in new product development. For the community, this theory holds implications depending on whether the local economy is concentrated in firms in their early development, rapid growth, or maturity stages.

Growth Center/Propulsive Industries Theory

Growth center theory stresses the role of *propulsive sectors* in shaping the economic growth fortunes of communities (Darwent 1969; Erickson 1994; Friedman 1972; Hansen 1975; Perroux 1950, 1988; Thomas 1972). These propulsive industries are comprised of technological innovators (or lead firms) with growth-stimulating effects on both backward- and forward-linked sectors through which positive growth (spread) effects or negative (backwash) effects are propagated. In short, growth center theory suggests there is a dynamic disequilibrium where cost-reducing and growth-inducing actions feed on themselves and pull other areas along.

The tendency for high technology to agglomerate is due equally to traditional organizational needs and agglomeration advantages and the locational preferences of professional workers (Malecki 1989). One agglomeration advantage is the mutual relationship among firms and workers because an area with many firms of all kinds, especially firms in the same industry, becomes an attractive location for technical workers. They have a variety of potential employers should they decide to leave their current jobs. A second agglomeration advantage is the contact networks of technical workers. Firms rely on the heterogeneous contacts and networks of individuals to acquire information and technical knowledge from other organizations. Growth center theory argues that innovative/propulsive firms are the sources of growth and change. These firms tend to agglomerate and, through inter-industry links, transmit growth and change throughout the economy.

As argued by Krugman (1991a, 1991b), Barkley, Henry, and Bao (1994, 1996), and Henry, Barkley, and Bao (1997), the spread and backwash effects can occur through several broad economic forces.[5] These include investment, spending for goods and services, migration, knowledge and technology, and political influence and government spending. Each of these can be shown to have both spread and backwash dimensions. For example, urban funds are invested in rural areas to take advantage of low labor and land costs, resulting in a spread effect. Simultaneously, rural funds are invested in urban areas to take advantage of relatively rapid growing goods and services markets. Another example focusing on knowledge and technology flows: Urban centers are the generators and diffusers of information and innovation for the surrounding hinterland, resulting in spread effects. Rural to urban migration is selective of the better-educated and more highly skilled rural residents. Here, the urban center will draw in the human capital of the surrounding hinterland.

The balance between countervailing spread and backwash effects will vary from community to community. Henry, Barkley, and Bao (1997) concluded perceived quality of life differences between the urban core and the surrounding hinterland play a primary role. If the urban center is perceived to be a better place to live and conduct business than the surrounding area, backwash is likely to be present. Conversely, if the urban center is perceived to be crime-ridden and run-down, while the surrounding hinterland is blessed with high levels of natural amenities, strong spread effects are likely to occur (see Chapter 9).

Disequilibrium Theories

Another group of theories includes entrepreneurship, seedbed, and regional creativity theories in which environmental factors and nonspatial synergies contribute to an area's overall receptivity to technology and innovation (Davelaar 1991; Goldstein and Luger 1993). Entrepreneurship, seedbed, or creativity theorists ask: Why have some locations been economically more dynamic than others, and what role can public policy play in creating and fostering these conditions?

The important point is the idea that new technology and innovation induce people to start or spin off new firms. This causes a dynamic sense of disequilibrium in that many things in both input and product markets and production processes are changing. As in the neoclassical model, shifts in technology and the introduction of innovations cause the growth path of the economy to shift, resulting in short-term disequilibrium. The structural changes that arise from technology have implications for capital and labor, particularly from a perspective of multiplant business organizations and strategic decision-making and the conflicting interests of both workers and firms in the process of facing changes (Erickson 1994).

Correspondingly, a shift toward a behavioralist paradigm since the 1970s has been caused by the increasing emphasis of multilocational firms. The behavioralist paradigm emphasizes strategic decision-making with respect to technological and market changes, location of research and development or other innovative activities within the organizational structure of the firm, and differences between small and large firms. It also includes the effects of mergers and acquisitions on the rationalization of productive capacity, the role of capital availability for venturing, and foreign direct investment and international trade in advanced technology processes or products.

Theories of innovative and creative communities focus on the factors that make some communities dynamic and synergistic (Goldstein and Luger 1993). These theories emphasize the innovativeness and creativity of human agents in the development process. The agents are typically entrepreneurs in the broad sense as we have defined them (see Chapter 12), or include powerful or insightful politicians, business leaders, and university officials. These creative agents both stimulate disequilibrium (new technology leading to spin-offs) and see it as an opportunity for exploitation.

POLICY

Technological change often alters traditional comparative advantages, creating new locations for economic production (Thwaites and Oakey 1985). The end result is that governments frequently intervene in the transmission of technology because of their concern about relative comparative advantage. Communities lagging in technological development are likely to see their economic base eroded by

external competitive forces, leading to more decline in output, incomes, and employment. If these communities continue to find themselves disadvantaged by technological change, they will seek to have some offsetting governmental intervention.

A variety of forces retards the transmission of technology over space (Malecki 1983). The most important reason is differences in the rate management accepts and adopts technology. Mere receipt of information about an invention does not ensure adoption. Rather, management requires repeated and redundant messages about the invention, coupled with new information about how the invention has performed for similar firms. Management reduces risk by requiring additional information. The transmission costs of technology are not zero. These costs include becoming aware of the new technology, figuring out how the new technology can be applied, and disrupting production and training workers in the use of the new technology.

It takes time to incorporate new technology into the capital stock. In some cases, new technology requires a completely new production process, which retards adoption of the technology until the present capital stock is fully depreciated. New technology requiring only minor changes is adapted more quickly.

Since communities are composed of different industrial sectors, the rate of technological progress among sectors varies. Communities with sectors experiencing rapid technological change also experience more rapid technological change than do other communities. Patent agreements and secrecy in the use of the technology hampers the instantaneous transmission of technology among communities. Thus, some of the reasons for different adoption rates of technology include the risk attitude of management, ignorance about the new technology, time to implement, and restrictive agreements.

Backward regions are unlikely to spawn the entrepreneurs who are needed to bring new technologies to market. It appears that only rather densely populated regions in the vicinity of large urban centers will produce sufficient entrepreneurship. In the context of small peripheral regions, innovative development is not likely. The human capital of small regions is too limited and too mobile; many of the talented people will simply leave. Community and regional policies can only do so much, for as Malecki and Nijkamp (1988) note:

- Regional policies cannot create entrepreneurs.
- Policies are unable to substitute for a critical mass.

- Regional policies can have indirect effects.
- Policies cannot drastically alter the position of lagging regions.
- Regional policies may have a limited impact on spatial mobility.

Government policies designed to address technology transfer must work or be aimed at the organizational level (Melkers, Bugler, and Bozeman 1993). Successful technology transfer requires transfer of knowledge across disciplines, professions, industry sectors, regions, and communities. Thus, it is an organizational and cultural process as well as a knowledge-transfer process. The idea of *recipient organization* describes an organization, community, or firm that is open to new technologies defined in the broadest context. This goes back to the idea of the entrepreneurial spirit of the community and the businesses that help make up that community.

Policy can take different forms depending on the level of government involved. At the local level, policy can be aimed at aiding technology transfers. But at higher levels of government, the spectrum of policy options broadens. The government has a role in supporting basic research and historically this has been done by supporting research universities and research labs that cater to the Department of Defense and NASA-related programs.

Another key element as technology becomes more important to the economy is the ability of the labor force to take advantage of and use technologies and innovations. This translates into the increasing role of educational opportunities beginning primarily with K–12 education, institutions of higher education, and continuous retraining opportunities.

We describe this spectrum of policy involvement by using Figure 8.1. At the local level, the community can craft policies to help foster the right two-thirds of the technology spectrum, focusing on technology transfers, fostering entrepreneurship, or most broadly process technologies. Higher levels of government can influence all three elements of the figure but have a comparative advantage in fostering the product and innovation or left two-thirds of the figure. Indeed, one could reasonably argue that many elements of new technologies have a public or nonrival nature, such as the outcomes of basic research, that government has a central role in investing in research. A classic historical example is the agricultural experiment stations in the system of land-grant universities. Unfortunately, because of public budgetary issues, an increasing level of funding for basic research is coming from the private sector. The hope is to capture short-term monopoly profits from any patents flowing from the research.

SUMMARY

Historically, technology has been viewed as invariant among communities where all firms (assuming complete information) have access to the same knowledge. But it is more likely that technology and innovations in the form of new products, new kinds of processes, new forms of organization, and new management systems are not equal everywhere. Heterogeneity across communities provides entrepreneurs and innovative firms with advantages, enhancing their effectiveness in various forms of *behavioral competition* or *creative destruction* behavior (Thomas 1985). Innovations provide the firm with a *competitive advantage* over firms in the same industry and same market. The concept of behavioral competition helped to direct attention toward the unlearned acts of insight performed by innovators. Innovation, investment, and location decisions were viewed as significant nonroutine decisions.

In a world with profit motivation, firms have an incentive to invest in new products and processes. This investment can be in the form of research and development of new ideas, products, or processes with the hope of capturing profits from patents or adopting new technologies before competitors do. From a community perspective, most local firms are likely adopters of new technologies developed by others. The new idea, product, or process in and of itself is not sufficient to foster economic growth and/or development. The notion of the Schumpeterian entrepreneur who can bring the pieces together and introduce the whole to the market plays a vital role.

Innovation and its diffusion throughout the economy is a continuous process that exerts a major influence on growth and development (Romer 1984, 1986; Malecki 1989). Many innovations target the development of new products or different production methods within specific sectors. Other innovations come about by the development of new general-purpose technologies that give rise to changes across a wide range of sectors and affect production methods, inter-industry relationships, organization of work, and skill requirements. Commonly, it takes a long time before general-purpose technologies are implemented and used to their full potential. An example is the steam engine replacing waterpower, and eventually being replaced by diesel and electric motors. This transition led to organizing work and skills differently, such as moving from craftsman to assembly line. The growth of output or income in a community comes from an increase in the quantity of labor and/or capital used, adoption of new technology, and reallocation of capital or labor to more productive uses within the community.

Innovation is a complex phenomenon that requires technological, institutional, and social change. The existence or provision of single factors to promote innovation is usually not a sufficient condition for the actual emergence of innovation. Additional characteristics of innovative communities include coexistence of vigorous competition and efficient knowledge and information networks, which in turn depend on cooperation, trust, and reciprocity (Hansen 1992).

Returning to the Shaffer Star (Fig. 1.1), we see that technology in its broadest sense includes both product technologies, which act as resources in the economy, and process technologies, which describe the interaction of economic agents and the ability to make decisions and move forward. From a purely microeconomic perspective, the profit motivation of the firm drives the economy. From a community perspective, however, the motivation can be broader than just profits. The notions of welfare maximization, income distribution, quality of life, and Rawlsian's perspective of how a society treats its worse-off members come into play. What distinguishes successful communities is their willingness to engage in new process technologies, that is, they are communities that act entrepreneurially in their decision-making and policy implementation.

STUDY QUESTIONS

1. What is technology?
2. What is innovation?
3. What is the difference between product and process technology?
4. How does Schumpeter's view of entrepreneurship play a role in technology?
5. What is meant by the recipient organization must be ready for or compatible to the new technology?

6. What does it mean when we say technology is a source of dynamic disequilibrium?
7. How might technology lead to changes in spatial comparative advantage?
8. Is information about technology costless?
9. What is the distinction between basic and applied research and development?
10. Why is agglomeration such an important force in research and development?
11. What do we mean by nonroutine decisions when discussing technology or innovation?
12. What do creative communities mean to you?
13. Why is instability so important in creative communities?
14. Compare and contrast the role of technology/innovation in the four major theoretical perspectives outlined in this chapter.
15. Why might larger urban areas have a comparative advantage over small rural areas?

NOTES

1. Changing factor prices and changing demand structures also contribute to new or altered products, production processes, or changes in industrial structure.
2. These concepts have also been with us for several years. They are comparable to the ideas of product and process technology (Perroux 1988).
3. Another example of a business practice that can be patented and licensed is ISO2000 quality control methods and certification.
4. The resultant spatial reorganization of industry leads to a spatial division of labor. The spatial division of labor is the division of high-skill, high-wage administrative and innovation jobs into specific locales, generally separate spatially from low-wage, low-skill jobs in standardized production (Malecki 1983).
5. In Krugman's terminology, backwash effects are called *centripetal* and spread effects are referred to as *centrifugal*.

9

Nonmarket Goods and Services: Amenities

Resources are a primary foundation of community economic activity and represent important determinants of future economic vibrancy. Resources are a critical component of the community economics star (Fig. 1.1). Communities across North America vary widely in both the level of initial resource endowment and in the manner in which these resources are utilized, or their current productivity. Up to this point in our discussion, factor resources have included land, labor, capital, and technology. In Chapters 5 through 8, we have only partially dealt with key public aspects of these factor resources that result from community decision-making. The role of the public as a stakeholder in providing, maintaining, and improving resource endowments helps characterize the economic condition of communities. This public component can take on two alternative forms that help direct the organization of Chapters 9 and 10. First, in this chapter we provide useful background that helps us understand naturally occurring amenity assets. Specifically, we deal with the nonmarket complexities associated with public land-based goods and services. Chapter 10 covers the role of local governments in providing publicly available infrastructure as another form of regional capital assets. Necessarily, we include the fiscal aspects of local government operations and taxation as a mechanism for the development of public capital resources.

This chapter is organized to present the relevant issues and concepts associated with amenity resources important to community economics. We begin with the spectrum of goods and services, from private to public. The valuation of nonprivate goods (or nonmarket goods) is seen from the standpoint of economic value. This represents the set of alternative demands we place on the natural environment and outlines the array of goods and services that are not well represented in operating markets. This is

followed by a discussion of naturally occurring amenities that differentiate them from other important factors of regional production. In particular, we focus on amenities as nonmarket economic goods and services. Next is a summary of the relevant techniques developed to date that have been used to help quantify the value and demand structure of nonmarket goods and services. Finally, we discuss the current literature and contemporary regional models with respect to how they attempt to incorporate amenities and other nonmarket goods and services.

THE CONCEPT OF NONMARKET GOODS AND SERVICES

Amenities and publicly provided goods and services viewed as part of a community's set of economic assets extend the traditional notion of primary factors of production into a rather nebulous, less tangible situation. In addition to the traditional market-based aspects of land, labor, and capital, we now have begun to mix market-based assets with those that are fundamentally nonmarket in nature. To discuss this, it is valuable to present a broader approach to how economists view the spectrum of economic goods and services. This extended view accounts for problems of excludability and the diminishing aspects associated with additional users.

The Array of Economic Goods and Services

An overview of amenity resource values and publicly provided infrastructure would not be complete without a brief notation that distinguishes, in a very crude way, a significant problem associated with these types of assets. This dilemma arises because of the rather ill-defined nature of property rights with respect to the array of goods and services demanded. Certainly, property rights with respect to the land itself are relatively straightforward as a result of standard land surveying techniques, record keeping, and the historical development of real estate institutions. Much less clear, though, are the rights that people have over how land is used and how the public owns publicly provided infrastructure and resource endowments.

Property ownership has often been described as encompassing a bundle of rights (a more complete discussion of institutions and rules can be found in Chapter 11). This bundle can extend to include mineral rights for endowments found beneath the land, air rights above the land, and the surface use rights for agriculture, forestry, or alternative uses such as residential, commercial, and industrial purposes. Contemporary land use tools (discussed briefly at the end of Chapter 5) use this notion of a bundle of rights. Especially important is the idea that regulations, policies, and programs can affect various aspects of this bundle and separate the bundle into smaller packets of rights. For instance, land use zoning specifies the nature of surface rights with respect to land development. With zoning, the freedom to decide about these development rights rests with the publicly elected decisions-makers of the jurisdiction within which the land falls. This notion of separable rights builds on the unique characteristics that we will discuss as *rivalry* and *exclusion*.

The economic values we place on natural amenity resources and publicly provided infrastructure are, in large part, determined by their characteristics as goods. Property rights affect the type of good provided by the resource because property rights deal with ownership. Ownership of a resource deals with the ability of an owner to exclude use to those who are not willing to pay. Thus, excludability is an aspect that often presents problems with valuing natural amenity resources and publicly provided infrastructure. Another aspect that affects a resource's consideration as a good has to do with how additional users diminish the utility of others who use the resource. We sometimes refer to this as the level of *rivalness* that a good exhibits. A rival good is one in which the addition of another user acts to diminish the original user's value. A nonrival good is one in which additional use does not diminish other users' value. Once again, in characterizing resource value, it is important to bear in mind that land-based amenity resources and publicly provided infrastructure vary in their level of rivalness. In other words, the *congestability* of a resource will vary depending on the resource being examined.

Nonmarket goods and services can be arrayed with respect to their levels of exclusivity and rivalry. The type of good is largely determined by the extent to which the good is excludable and rival (Fig. 9.1). Many natural amenities and publicly provided pieces of infrastructure are sought competitively (characterized as rival in nature). This is due to the very essence of increased demands on a limited supply of resources. In other words, for many uses we put to land, additional use has a diminishing effect on others' value. For example, an additional recreationist in a forest is likely to diminish the value of others in using the forest for recreation. In general, people regard more-solitary experiences with outdoor amenities highly, thus values for fewer users are often higher than that with crowded use. As shown in Figure 9.1, most land-based recreational uses view the amenity resources as a *common prop-*

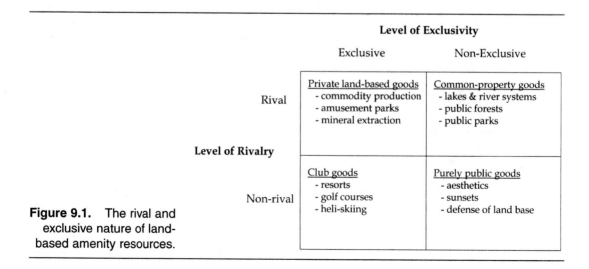

		Level of Exclusivity	
		Exclusive	Non-Exclusive
Level of Rivalry	Rival	Private land-based goods - commodity production - amusement parks - mineral extraction	Common-property goods - lakes & river systems - public forests - public parks
	Non-rival	Club goods - resorts - golf courses - heli-skiing	Purely public goods - aesthetics - sunsets - defense of land base

Figure 9.1. The rival and exclusive nature of land-based amenity resources.

erty resource; namely that they exist as nonexcludable and rival goods.

Certain final products rely on exclusion within a competitive environment. For instance, timber harvested from a forest is considered a *private good*. The value we place on this type of good, however, is typically a function of the costs of processing, delivery, and marketing, which are not so directly related to the land itself. *Club goods* are sometimes applicable with land resources but, again, relate to the packaging of an experience to an individual. An example of this type of good might be the value an individual places with an exclusive lease for hunting. Finally, some values primarily relate to what is discussed in the next section as *non-use* values. These typically are referred to as purely *public goods,* where exclusion is not possible and use is not competitive. Examples are the values people place on the simple existence of untouched wilderness or in certain endangered species. Someone else's value for this simple existence will not affect an individual's own value (nonrival). Also, an individual's existence value is not something that the individual can be excluded from.

Many uses associated with land-based amenity resources view the amenity itself as a *common property resource*. Publicly owned parks and forests can be used to illustrate the characteristics of a common property resource. Because these public lands are open to the public, no one can be excluded from using them (exception of user fees noted). Eventual-

ly, however, as more and more people use these lands, congestion becomes a problem. Consequently, each individual's enjoyment of the property will be reduced. This is in stark contrast to the items that we purchase in the marketplace.

Common property resources exist in between purely public goods and private goods. The nonexcludable nature of many land-based amenity resources makes it nearly impossible for market forces (the purchase and sale) to operate due to the presence of what are commonly referred to as *free riders*. Free riders create a situation where no one is willing to pay for the use of the resource if other people are also able to have access to it without paying. This is only partially the case with the problem of common property resources. Indeed, people may be willing and able to pay, but no market mechanism within which payment can be extracted exists.

The Value of Nonmarket Goods and Services

Again, community economic problems associated with nonmarket goods and services deal with the inability of private markets to provide signals that direct efficient allocation. This lack of signaling is based on the general absence of prices that lead to an inability to directly identify the value of nonmarket goods and services. In an effort to more clearly articulate the array of alternative values, a more comprehensive economic approach is found in Figure 9.2.

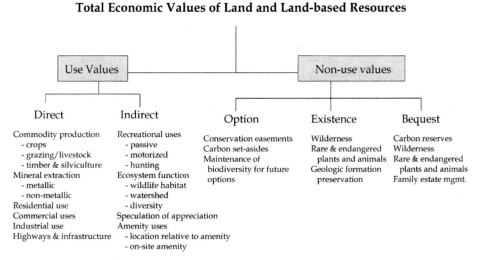

Total Economic Values of Land and Land-based Resources

Use Values		Non-use values		
Direct	Indirect	Option	Existence	Bequest
Commodity production - crops - grazing/livestock - timber & silviculture Mineral extraction - metallic - non-metallic Residential use Commercial uses Industrial use Highways & infrastructure	Recreational uses - passive - motorized - hunting Ecosystem function - wildlife habitat - watershed - diversity Speculation of appreciation Amenity uses - location relative to amenity - on-site amenity	Conservation easements Carbon set-asides Maintenance of biodiversity for future options	Wilderness Rare & endangered plants and animals Geologic formation preservation	Carbon reserves Wilderness Rare & endangered plants and animals Family estate mgmt.

Figure 9.2. The broad spectrum that makes up the total economic value of land and land-based resources.

Use Values

Direct

In Figure 9.2 what might be termed *productive* uses (i.e., those uses that rely on site productivity, endowments of minerals, or location as a basis for activity) are categorized at the far left as *direct use values*. These include commodity production for row crops, livestock, and timber. Extraction of mineral endowments is another example. Estimating direct use values associated with production of these commodities is relatively straightforward and involves the quantity of commodity produced multiplied by the market price of the commodity at the time of sale. The value of land in this case is a function of its ability to produce commodities.

Managers of commodities falling within this category attempt to maximize returns and minimize costs of production given site productivity and some predetermined time frame associated with sustained production. For example, forest managers interested in a sustainable harvest of timber into perpetuity will account for a forest stand's rotation age (largely determined by site productivity) and some objectively based preference for the time value of money to determine both the type and level of timber extraction. From an economic perspective, the profitability of using land for timber production is based on initial growing stock levels, wood prices, and the direct costs of extraction (including land rent largely determined by stand accessibility and distance to market). From a strictly private decision framework, noncommodity outputs from forests are not included.

Another example of direct use value related to land has to do with the consumption of land for residential, commercial, industrial, and infrastructural uses. Indeed, these are direct applications of land value, but in most circumstances, the portion of total value related to land is relatively small and generally is not related to the productive capacity of soil or the endowment of mineral wealth located beneath the site. The key element of land associated with these types of direct use values has to do with land location relative to urban or recreation-oriented development. These locations are often wrapped up in trade-offs between land parcels located close to market centers, infrastructure, and amenities or quality-of-life measures (distance-based land rent also is discussed in Chapter 5). Once again, the estimation of value in this circumstance is reflective of the amount of land involved and the market-based price of land with respect to the rivalness of avail-

able land and the demand for land. In regions surrounding urban centers, stiff competition for land and a limited supply regularly lead to scarcity and resulting prices that often greatly exceed those values derived from using the land for commodity production. This is particularly so with respect to prices for land that contains or is proximate to high levels of amenities. The *highest and best use* of land in urban, suburban, and amenity-rich exurban areas is for residential and commercial purposes. These purposes supercede the use of land for production of commodities.

Indirect

In addition to these direct use values, however, less direct forms of land exploitation lead to what is termed *indirect use value* associated with resource use. Indirect use valuation is more complex given the various intervening mechanisms that relate to economic activity within local communities. Included within this category are recreational values of using land for leisure purposes. Estimating the value of land use for recreation has followed several alternative methods. A straightforward approach to valuation of recreational land use involves estimation of traveler expenditures for items purchased in local markets. Additional approaches include the development of stated and revealed preference models.

Another indirect use value associated with land falls within a category broadly referred to as *amenity uses*. Closely related to recreation, this form of land use relates to the hedonic, or pleasurable, aspects of land and land-based resources. Scenic resources associated with the outdoors are key elements of amenity use and include several different types that can be further broken down into sociocultural amenities and natural amenities. Sociocultural amenities include historic sites and special events such as festivals. Natural amenities include regional features such as lakes, forests, mountains, and other recreational sites. The presence of sociocultural and natural amenities and their impact on regional growth have been shown to have significant positive relationships with regional economic growth (Deller et al. 2001; English et al. 2000). This is consistent with our definition of community wealth in Chapter 1.

Another indirect use value is for the maintenance of operable ecosystems (typically referred to as ecosystem function). Specific examples of this type of land use include the tacit incorporation of wildlife habitat or watershed concerns within silvicultural prescriptions. Increasingly, these ecosystem-function roles of land and land-based resources take on important land management considerations. They have provided the primary objective of public (local, state, and federal) agencies and private not-for-profit institutions in situations where critical ecosystems need to be maintained.

Non-use Values (Option, Existence, and Bequest)

Perhaps the most challenging aspects of both management for and valuation of land-based resources are for the relatively less tangible aspects of economic value associated with providing resource service functions for current existence, future options, and transference of resources among generations. Classified in Figure 9.2 as *non-use values,* they represent an important aspect of land and land-based resources. From a land use practice perspective, examples include the basis for preservation efforts to set aside tracts of land for wilderness, bioreserves, and carbon sequestration efforts.

From a valuation method perspective, the stated preference models discussed in the following section can provide estimates of people's demand for these services. Note that estimates of non-use value can, and often do, greatly surpass levels of use value. There are many explanations for this, but one important point of reasoning comes from the notion that non-use values are often societal in nature and reflect the overall value placed on resources by society as a whole, while use values are more specific to individuals, private concerns, and focused interest groups.

THE ROLE OF AMENITIES IN COMMUNITY ECONOMIC DEVELOPMENT

Naturally occurring land-based resources abound and present both raw material inputs to production and an amenity base that serves as a backdrop for regional quality of life. We turn to the amenity aspects of natural resources. Here, we examine the issues surrounding amenities and argue that a community's amenity base is an important determinant of economic development. While many communities in the United States are experiencing depopulation and economic decline, others are experiencing rapid in-migration and its resulting demands for residential, retail, and service sector development. Increasingly, arguments examining the course of community vibrancy focus on the importance of amenities (Johnson and Beale 1998; Nelson 1997). Again, the status of community change is dependent on available resource endowments and the manner in which these resources are utilized.

Particularly in regions richly endowed with natural resources such as lakes, mountains, and forests, amenities act as factor inputs into the economic system. Indeed, amenities as a separate factor resource are more elusive to measure and more latent in terms of production. This said, they are becoming widely accepted as a primary motivator behind a community's quality of life and increasingly they are driving individual consumption patterns throughout the developed world (McGranahan 1999). Much of this demand stimulus for amenity resources can be attributed to changing consumption patterns led by increasing disposable incomes, rising demands for quality-of-life attributes associated with residential living, and technological innovations. Amenity-led community development presents a new breed of economic, social, and environmental issues and helps explain rapidly changing land use patterns, high-end residential developments, and rapid retail and service sector growth. A vibrant literature has developed on amenity-based development and the nonmarket goods and services on which this development is based.[1]

For our purposes, the term *amenities* refers to the hedonic, or pleasurable, aspects associated with natural and human-made features found within and surrounding communities. These features include wilderness, agricultural landscapes, historic structures, and cultural traditions (OECD 1999). *Natural amenities* refer to the more specific community characteristics directly associated with land and water resource endowments.[2] Typically, the values of these as amenities are driven by human perceptions of esthetics associated with trees, forests, open space, water (lakes, rivers, coastline), and topography (mountains, canyons, hills).

Recreational demands also bring about amenity-based developments that extend definitional complexities. Take, for instance, an alpine ski resort or a golf course development. These types of artificial amenity enhancements extend naturally occurring endowments of land and land-based resources into alternative forms that more closely tie natural endowments to the amenities specific to a given recreational pursuit. In many regions, these artificial amenity enhancements have significantly altered the original amenity base. As demands for outdoor recreation grow, the natural amenity base is placed under increased developmental pressure, thus creating substantial adaptation of the natural ecosystems through the design and construction of recreational amenities.

It is becoming increasingly apparent that both natural and recreation-related amenities do indeed matter to the economic, social, and environmental development of communities and their surrounding rural regions (Beyers and Nelson 2000; Booth 1999; Deller et al. 2001). From a spatial perspective, regions can be characterized by their amenity endowments. Work that more closely defines natural amenity types remains a growing area of academic interest. Previous studies have tended to identify natural amenities in an ad-hoc manner and rely heavily on climatic characteristics (Nord and Cromartie 1997). More recently, efforts to systematize regional natural amenity-based endowments and recreational developments into standardized categories, or typologies, act to more clearly capture environmental and recreational attributes of rural landscapes (Beale and Johnson 1998; English et al. 2000).

Research suggests that regions endowed with a significant amenity base have attracted in-migrants at disproportionately high rates (Johnson and Beale 1998; Johnson and Fuguitt 2000; McGranahan 1999; Rudzitis 1999; Shumway and Davis 1996). In-migration of amenity seekers to rural regions raises concerns about the impact of development on economic, social, and environmental characteristics of rural communities. Amenity migrants, those who make locational residence and travel decisions based on the availability of amenities, create much of the demand for development in amenity-rich areas. They include recreational homeowners, teleworkers, retirees, and tourists.[3] Amenity-driven settlement has been shown to follow an evolutionary pattern, proceeding as people discover a high-amenity area as tourists or recreationists purchase developed or undeveloped parcels for occasional use, build permanent structures, and later settle permanently or semipermanently in later working years or retirement (Bennett 1996; Godbey and Bevins 1987; Shumway and Davis 1996).

Telecommunication is another recent phenomenon identified to be an interesting and important explanatory factor leading to disbursive locational effects, exurbanization, and rural in-migration closely tied to amenity-based migration (Gillespie and Richardson 2000; Helling and Mokhtarian 2001). The ability to telecommute has the potential to allow significant work to be accomplished in remote locations. Teleworkers who are able to take advantage of advanced telecommunications technology to work in places ever distant from urban centers are a small but growing segment of the nation's workers. The magnitude of this phenomenon for the future, however, is still unclear (Graham and Marvin 2000).

The Temporal Aspect of Amenities

Prior to discussing specific amenity characteristics, it is valuable to briefly discuss the dynamic context of amenities with respect to stages of economic development. Natural resources, such as forests, prairies, lakes, and rivers, and indeed, much of the North American landscape, have been transformed from their original natural state to their present condition by human activity. Until recently, this activity of transformation was driven first by subsistence and then by market-based production of tangible commodities for consumption with primary motivating factors centered on the generation of income. What exists today in terms of a community's natural resource endowment is largely the product of what's left over from previous productive activities.

Amenity values build on this current natural resource base and exist as a demand component of later, more mature stages of economic development. These values, driven largely by changing demands, can be stimulated by efforts to conserve natural resource endowments, conscious planning to develop recreational sites, and attempts to capture perceptions through marketing. Furthermore, our future value for amenities is equally dynamic and can be expected to continue to change over time. What we value today as an amenity, we valued differently yesterday, and we can expect to value differently tomorrow.

The dynamic nature of amenities can be linked to several developmental attributes. Over time, infrastructure has allowed us to become much more mobile. Furthermore, time has allowed us to progress into more mature economic conditions that shift the relative importance of natural resource dependence away from use of natural resources as production inputs (raw materials) to more of an amenity basis. These changing natural resource dependencies follow the accumulation of wealth and disposable income and represent a progression in developmental stages (Mäler 1998; also see the discussion of the stages of development in Chapter 2).

Examples of this transition can be found throughout rural and urban North America. Take, for instance, the Lake States forests of Minnesota, Wisconsin, and Michigan. During the late 1800s and early 1900s, these forests were cut down, processed, and sold to build the great midwestern cities of Chicago and St. Louis. Lands once rich in virgin timber were cut and burned, with the most productive sites reverting to agricultural production.[4] The value of these forests in an early stage of economic development was easily measured in production-oriented price-quantity terms (volume of timber multiplied by market-determined price). At the time, amenity values held for these forests were, at best, modest relative to the direct use values associated with trees for timber. At worst, the amenity value of these forests was nonexistent. Today, however, there has been a dramatic shift in value types. Although a vibrant wood products industry remains and continues to draw on significant timber volumes from regrown forests, the indirect use values of forested landscapes now supports a vibrant recreational home market and tourism industry that dominate many communities throughout this region. Also, these forest-based amenity values now play a significant role in determining how forest management practices are applied to forested lands. This natural landscape exists today in a highly transformed state originally driven by production-oriented human activities.

In addition to the stage of economic development, transitions to consumption of natural amenities often require an initial input of some productive factor that allows an awareness of the resource. There is also a temporal aspect to these inputs that relates to the use of resources as a production input. For instance, the development of infrastructure (highways) for travel to amenity-rich regions and recreational-site developments that facilitate amenity resource use plays an important role in determining overall amenity value. Without infrastructure, the amenity's overall value is diminished as few people are aware of and decide to utilize amenity-based resources.[5] A continued public investment in infrastructure serves both the purpose of production (access to markets and raw materials) and amenity access.

Characteristics of Amenities

Amenities are unique from other regional factors of production. Their uniqueness can be summarized along four basic lines that represent fundamental characteristics of amenities. These are the notion that amenities tend to be (1) nonproducible, (2) irreversible, (3) subject to high-income elasticity of demand, and (4) regionally nontradable.

Nonproducibility

Amenities, particularly natural amenities, are difficult to produce. The supply of natural amenities tends to be usually restricted in an absolute sense. It is very difficult to recreate events that lead up to natural amenity change in the short term. Thus, it is typically not feasible to produce natural amenities. Attempts to produce natural amenities are often limited to gradual, or incremental, transformations of the existing resource endowment.

Nevertheless, there are mechanisms that can be used to increase the regional capture of amenity values. Public investments in infrastructure allow regions to more fully utilize natural amenities. Public and private expenditures to develop recreational facilities and the forward-linked hospitality sector (retail and service sector businesses that cater to visitors) can also serve the purpose of more fully utilizing regional amenity endowments. Also, resource management practices that are sensitive to the effects of resource use on amenity values have the opportunity to affect amenity outcomes.

Irreversibility

Changes in natural systems occur over relatively long time frames that are best measured in decades and centuries rather than months or years. Consequences of natural resource management decisions are difficult to ameliorate in the short term. For example, a forest that is clear-cut will take several decades to regrow to a point of comparable size and density where the characteristics important to amenity value are restored. Minerals mined from the ground are, in effect, nonrenewable within any reasonable human-based time frame. This said, attempts to reverse amenity-diminishing resource decisions in the short-term are possible but only at very high costs. An example is replanting clear-cut areas with large-diameter trees—in effect, a very expensive endeavor. Another example comes from the mining industry. Mine reclamation is a common approach to restoring the function of land for ecosystem and/or amenity uses. This also is a costly endeavor and clearly identifies the short-term irreversible nature of amenities with respect to resource management decisions.

The level of irreversibility in natural resource decisions depends on the temporal aspects of resource renewability. Certainly, different types of natural resources can rejuvenate themselves at varying rates. In general, natural resources that rely on geomorphology (e.g., plate tectonics, volcanism, soil building) as a regenerating mechanism are extremely slow. Here, the temporal frames are measured in millennia or longer. On the other hand, natural resources that rely on biomorphology (e.g., tree growth, wildlife production, prairie restoration) are relatively faster. Temporal units here are on the order of decades or centuries. Human-created development of amenities would be the fastest, but the most costly.

High Income Elasticity of Demand

Does the consumption of natural resources for amenity value depend on the relative wealth (or income) of individuals who make up the demand base? In other words, are amenity values representative of *luxury* goods? These questions would raise important public policy issues of both efficiency and equity. It is generally assumed that amenities can be characterized by income elasticities of demand (the percentage change in the quantity demanded of a good in response to a 1 percent change in income) that are greater than unity (McFadden and Leonard 1992). Empirical research has confirmed this theoretical basis.[6] Thus, the demand for environmental goods as amenities increases more rapidly as income increases.

If demand for amenities is positively and strongly correlated with income level, then equity issues become important. From an equity perspective, arguments that focus on distributive aspects associated with trade-offs are compelling. Given appropriate safeguards against environmental degradation, is it fair to trade off production of extractive marketed commodities that generate income to local residents regardless of income level for amenity-based outputs that are nonmarketed? These nonmarketed outputs disproportionately provide benefits to people of higher incomes who often reside outside of the local community. Indeed, questioning who benefits and who pays is needed to extend aggregate cost-benefit analysis into a distributional realm. However, introducing trade-offs without a more critical assessment is naive. A growing literature identifies compatibility of alternative land use as a primary empirical research target (Clawson 1974; Marcouiller and Mace 1999; Van Kooten 1993). The literature suggests that different land uses will have varying levels of inter-use compatibility and that the nature of management practice can have a significant effect on the outcome of land use trade-offs.

Nontradability

Much like land itself, natural amenities exist as fixed assets of regions. For our purposes, this is primarily important from the standpoint of the mobility of amenities as a primary factor input, which is a supply component of a region's production capability. Amenities as fixed regional assets cannot be traded among regions. A consumer's amenity value is linked to the region in which the amenities lie. Unlike capital or labor resources, a community is isolated from the amenity inputs of other regions but is in direct competition with other regions for people attracted to similar types of amenity resources. What exists in terms of regional amenity value can be considered fixed in the short term. This supply characteristic of immobility holds for amenities as regional factors of production.

This immobility aspect of amenities is a supply characteristic. It breaks down as we consider demand characteristics of amenities. Certainly, a region can enhance use of its amenity assets through marketing itself to the outside world or through affecting demand from the outside for regional amenity-based assets. Like non-amenity natural resource outputs (such as agricultural commodities or timber products), the level of demand for amenities can and often is affected through marketing to individuals and firms beyond the boundaries (or outside) of the region. Thus, we can view amenity demand in a similar fashion to commodities and raise the specter of natural amenities as export-based (or basic) goods. The trades that take place with amenities are now in the form of traveler demands and demands for recreational housing. It is important to note, though, that it is the demand for the amenity that is affected through trade, not the supply of the amenity.

Approaches to Estimating the Value of an Amenity

As with private goods, the market determines who gets to use a resource and who does not by granting use to the highest bidder. But for many land-based amenity resources, this only works when a market allows trading to take place. Without trading, the market is not able to place a value on uses that are not producing any sort of saleable good. Thus, we have to rely on a combination of market and nonmarket valuations of land use to determine which uses are the most valuable to society. Nonmarket valuation techniques have been used for many years to determine the value of alternative uses of land and land-based resources. These techniques can be broken into two primary categories: *stated preference models* and *revealed preference models*.

Stated Preference Approaches

This aptly named category of methods for estimating the demand for nonmarket goods and services relies on surveys of people who are involved in some aspect of the nonmarket good. It allows people to directly state their preference for a nonmarket good or service. The key elements of interest in these types of approaches typically involve a two-step process of developing a reasonable market for trading a nontraded good, then querying people to estimate what they would be willing to pay for this good or service. Contingent valuation and its variants are specific methods that fall into this category of estimation techniques.

To be sure, many possible problems and biases are associated with this type of approach (Boyle and Bishop 1988). Those who have been applying this method have worked to minimize several key types of bias that crop up. Examples of problems with these types of surveys include corrections for *strategic bias* (or the ability of a respondent to strategically effect the outcome of the research), *hypothetical bias* (or the inability on the part of the respondent to understand the contingent market being set up), and *starting point bias* (or the inability of the respondent to respond accurately given the absence of any reasonable benchmark on which to conceive of value). Although beyond the scope of this chapter, specific correction mechanisms have been instituted to allow use of stated preference methods as a reasonable approach for assessing the demand structure of nonmarket goods.

Revealed Preference Approaches

Throughout this chapter, we have argued that amenities are indeed critical to the understanding of regional economic activity. This argument is easily substantiated by a more closely focused assessment of economic relationships surrounding high-amenity sites. Conceptually, the economic importance people place on amenities can be proxied by the amount people are willing to pay for a market good closely related to the amenity. In this way, revealed preference approaches can be thought of as *indirect* valuation techniques. An example of this indirect approach is embodied in what's referred to as the *hedonic modeling approach*. Consider the premium paid for residential developments surrounding lakes, rivers, and coastlines. The affect of amenities on land value can be thought of as analogous to our previous discussion of bid-rent surfaces for land rent (Chapter 5). The effect of a lake or river on property values for residential lands directly proximate to the amenity is outlined in Figure 9.3. The amenity premium paid for land is highest for sites that directly border the water but fall away rapidly as distance to the lake increases.

These nonmarket goods that are capitalized into the price of land are affected by such issues as the size of an adjoining water body, quality of the water body, and distance to public lands such as parks. In essence, the nonmarketed amenity value is capitalized into the price paid for the proxy good. This is summarized in Figure 9.3 where the vertical axis

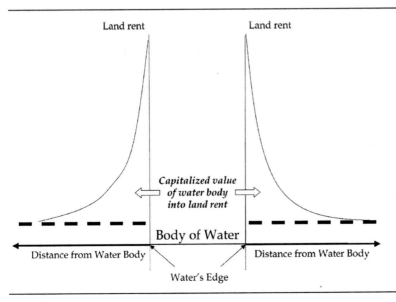

Figure 9.3. An example of amenity-driven land rent gradients that represent the hedonic model's proxy for value of the amenity.

measures land value (or rent) and the horizontal axis measures distance. In our example, if we were to model the nonmarket amenity value of a water body, we would be interested in the premium paid for waterfront property. Necessarily, we would need to control for factors involved in real estate purchases to assert the notion that water bodies cause people to pay a premium for real estate in close proximity to water bodies. This control would focus on assessing real estate transactions in a *with* against a *without* amenity context. In this way, the baseline value of the premium can be isolated (the heavy dashed line in Figure 9.3).

As one moves away from the amenity, the importance of the amenity as an economic factor declines because the amenity is found in a fixed location (represented by the erosion of the amenity-based distant premium in Fig. 9.3). To state the obvious, the demand for leisure-based visits of high-use recreational sites is most evident at the high-amenity site itself. In other words, the supply of an amenity is fixed in space while people are attracted to the amenity from outside.

Increasingly, applied research has shown that amenities are a significant explanatory component behind land-based economic activity in close proximity to amenity sites (Benson et al. 1998; Doss and Taff 1996; Lansford and Jones 1995, 1996). In the Lake States, work has been done to apply hedonic models to lakefront development, specifically dealing with the value of alternative land use zoning tools in regulating waterfront residential development (Spalatro and Provencher 2001). The practical significance of hedonic models has to do with inherent believability of analysis that focuses on an active real estate market and the logic of how nonmarket goods are capitalized into the trades that take place within such markets.

Revealed preference represents an aptly named category of research methods that attempts to uncover evidence of people's value for nonmarket goods through how they spend money in proxy markets. In other words, these approaches try to develop an indirect estimate of a person's *revealed preference* for a nonmarket good or service by examining how they spend money in the marketplace for a closely related market-based good or service. Examples of specific techniques within this category include hedonic price models and travel cost models.[7]

In closing, it is important to step back from the alternative techniques to consider an essential element of community decision-making that represents believability. Often, stated preference approaches suffer from the lack of a believable market mechanism on which we base the estimation of nonmarketed goods value. On the other hand, revealed preference models, with their reliance on actual real estate markets, have stronger inherent ties to how most lay people (and their representative policy-makers) view the world. Policy-makers tend to have less reservation in relating to and accepting results of revealed preference models as a basis for decision-making.

INCORPORATING AMENITIES INTO THE REALM OF COMMUNITY ECONOMIC ANALYSIS

We now turn our attention to a discussion that places valuation of land-based resources within the realm of economic modeling. Given the unfolding nature of environmental economics as a theoretical subdiscipline within economics and the rather incomplete nature of contemporary empirical models that attempt to capture latent nonmarketed production inputs, this discussion necessarily remains more conceptual than prescriptive.

In this section, we make the case both for an approach to incorporate amenities into economic consideration and for the contemporary relevance of doing so. We outline the incomplete nature of how we have traditionally measured a community's economy (Chapters 14, 15, and 16). Furthermore, quantifying the value of amenities is important to understanding contemporary issues of community change, such as the increasingly rapid exurbanization phenomenon as we begin the 21st century.

The Incomplete Nature of Traditional Regional Accounts

The challenge of contemporary community economic modeling focuses on more fully integrating causal elements associated with economic growth. Thus, regional modeling of amenities is attempting to inject public inputs into regional production systems to more fully explain the regional economic output and its growth. The traditional view of regional production held that regional output was a function of the wise combination of inputs as follows:

$$X = f(S, L, K, T) \qquad (9.1)$$

In the traditional view, output (X) is a function of land (S), labor (L), capital (K), and technology (T).

Given our previous discussion, it is easy to state that this traditional approach to modeling is overly simplistic and overlooks important public inputs. During the recent past, regional economic models have looked at a rather straightforward specification of public inputs that have included variables that track a region's infrastructure and educational status, both of which are important public goods to the production process. This more complete view of regional production now specifies output to include the traditional inputs of land, labor, and capital plus the public inputs of infrastructure and education as follows:

$$X = f(S, L, K, T, H, E) \qquad (9.2)$$

Output now is a function of land, labor, capital, and the public inputs of highway expenditures (H) and educational expenditures (E). This more complete specification has been found to generate significant and more useful empirical results in explanatory models of regional growth (Johnson and Stallman 1994), which are more fully described in Chapter 10. It is now widely accepted that publicly provided inputs have a significant and positive effect on regional economic growth.

Our interest here, though, is to focus on the incorporation of amenities into community economic analysis. This could be extended into the amenity realm by considering the regional production in rural amenity-rich regions to include the value of important land-based amenity resources:

$$X = f(S, L, K, T, A, H, E) \qquad (9.3)$$

where output in an amenity-rich region now includes land, labor, and capital and also the public expenditures to maintain or improve an amenity (A), highway expenditures (H), and educational expenses (E).

Relationship to Our Conceptual Basis

In tying this discussion of amenities into our conceptual basis for community economics, it is important to briefly outline the role of amenities as a factor of production. In our conceptualization, amenity values act as a latent factor of production into regional economic activity (see Fig. 9.4). In an analogous fashion to land, labor, and financial capital, amenities provide an important component of regional production.

The relevant industries that rely on amenities as a latent factor of production include those that would traditionally be termed the tourism sector as well as those comprising real estate development and residential construction. Certainly, natural amenities are critical in locating firms reliant on outdoor recreation and are directly relevant to those businesses that cater to seasonal homeowners and those new residents that relocated to rural regions as amenity-based migrants. As shown in Figure 9.4, the production process of these sectors can utilize the natural amenity input in a flexible fashion to produce alternative forms of output, each valued as common output levels as represented by the iso-surface (Marcouiller 1998; Smith 1987, 1994).

Our current thinking of how amenities relate to community economic development are represented in Figure 9.5. We really are faced with four interdependent issues, all of which have been discussed so far in this chapter. At the center of our thinking is the theoretical paradigm relating amenities to economic growth and development. This is in essence the problem of the incomplete nature of regional accounts outlined above.

But there are three important dimensions that are more practical for the community practitioner. How do we go about estimating and measuring the value of amenities (A in Fig. 9.5)? How do we collect data on the value of amenities? Is it the miles of white sand beaches or mountain peaks or the pleasure derived from viewing or using the amenity? What is the relevant spatial unit to think about with amenities (B in Fig. 9.5)? Some amenities have a very localized effect, while other amenities have a large regional effect. For example, the Chicago Art Institute draws from a much larger market than, say, the Elvehjem Museum of Art in Madison, Wisconsin. Similarly, the white-water rapids of the Colorado River running through the Grand Canyon have a much different market and, hence, role in regional growth than do the white-water rapids associated with the Snake River in the Ozarks of southern Missouri. At issue are not only the range of the amenity but also the relevant geographic unit of measurement. The final issue at the forefront of amenity research has to do with empirical modeling. How are amenities best captured in the empirical model estimation procedures (C in Fig. 9.5)? Each of these three dimensions provides ample opportunities for future research to more clearly address the role of amenities in community economic development.

At the local level, the real policy question is one of responsibility. Who should be vested with the responsibility of maintaining and adding value

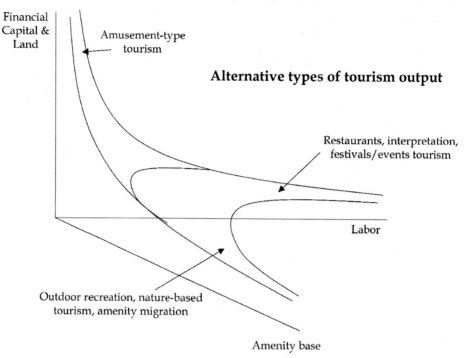

Figure 9.4. The conceptual basis that views amenities as a latent primary factor of production relies on the notion that amenities are an important aspect of an overall regional production function for tourism and residential development.

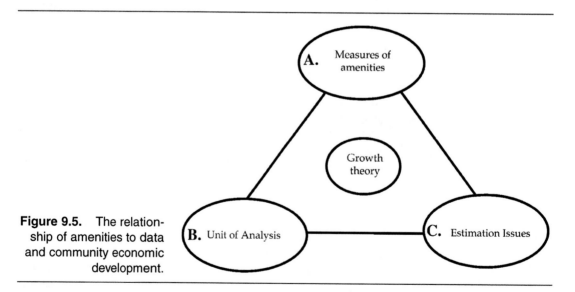

Figure 9.5. The relationship of amenities to data and community economic development.

to the amenity? Is it the responsibility of a local community to maintain a national treasure such as the Twelve Apostles in Victoria, Australia? Should a local community with a vibrant wood-processing industry forgo that industry for the sake of ecosystem management policies?

In the end, we have only begun to scratch the surface of our understanding of the interplay between amenities, quality of life, and community economic development.

SUMMARY AND POLICY IMPLICATIONS

Many of the drivers behind community economic vibrancy are the result of nonmarket goods and services. These take the form of both land-based amenities and publicly provided infrastructure. This chapter focused on the former; the next chapter will deal with the latter. Important fundamental aspects that relate to both include characteristics of resource ownership and the effects of additional use on utility of current users. We formalized these concepts as *resource exclusivity* and *rivalry*. These characteristics allowed us to distinguish alternative types of economic goods and services. Given the general lack of operating market mechanisms that signal prices for many of these economic goods and services, alternative types of value and alternative approaches to valuation were discussed.

Amenities serve as important latent inputs to production in amenity-rich communities throughout North America. They exist as latent inputs and present a complex mixture of market-based and nonmarket goods and services into the analysis of community economic development. Rapid change experienced within amenity-rich communities continues to point to the importance of amenities as key factors. This change is driven by the demands of short-term visitors, in-migrating newcomers, and long-term residents. The latter two categories drive change through residential developments and individual decisions about locating living space in proximity to amenities. The economic, social, and environmental changes brought about by residential developments are typically not fully understood by decision-makers within affected communities. Although rapid residential development and its planning complexities have been recognized since the 1970s (ASPO 1976; Coppock 1977; Ploch 1978; Greason 1989), unfettered growth has persisted throughout many rural American communities. Several factors have contributed to this, including a general lack of planning resources, desperation for economic growth in hopes of alleviating persistent rural poverty, and a more conservative political environment.

Communities often have very little control over amenity-based residential developments in their midst, at least in the short term. Land use controls are often weak in rural communities, whereas planning resources are typically meager. Nevertheless, new housing development in rural areas is popularly viewed as having significant net benefits for small towns and rural county governments, with little or no analysis to verify this belief. There is no question that such housing affects local land values and respective property tax bases. However, rarely is there any regard for or understanding of the comprehensive economic, fiscal, environmental, and social impacts that amenity-driven residential development creates within rural communities.

STUDY QUESTIONS

1. How do rivalry and exclusivity help us characterize the goods and services available within communities?
2. Many components that make up total economic value are not well represented in markets. What does this mean for our ability to allocate scarce resources to the production of these goods and services?
3. How does the fact that amenities exist as nonmarketed goods and services diminish their value from the standpoint of free-market economics?
4. What examples can you give that represent production of amenities?
5. How do the temporal characteristics of natural amenities affect the ability of society to produce these nonmarket goods and services?
6. Are amenities a luxury good? If so, how does this affect societal interest in producing these goods?
7. What distributional considerations could be raised with respect to a community's reliance on amenities for future economic growth?

NOTES

1. We attempt to capture the academic literature that relates to this specific phenomenon of the late 20th century. It builds on previous assessments of the literature compiled by Marcouiller, Clendenning, and Kedzior (2002), Dissart and Deller (2000), Jepson et al. (1997), and Marcouiller (1995) and focuses on natural amenity-led rural residential development with specific interests in its implications for development planning, land use change, environmental impacts, and social aspects of community life.
2. For purposes of clarity, we interchange the terms *community* and *region*. Particularly in

reference to natural amenities, community can also extend well beyond the limits of city and town borders to include rural areas. Thus, our use of the term *community* with respect to amenities implies less about market centers than about groupings of people living within a region.

3. For clarification, we have included tourist as an amenity migrant when there is often a lack of association between tourism and migration. We argue, however, that tourists are indeed short-term migrants to regions for the sole purpose of pursuing leisure-based activities.

4. Indeed, forested lands converted to agricultural production experienced a checkered history of success in the North woods. Given marginal soil fertility levels and shorter growing seasons, many of these northern farms were unsuccessful and reverted back to forests. Interestingly, these more-productive sites made up the bulk of tax-reverted lands during the 1920s and 1930s and now are largely owned and managed by public agencies such as the U.S.D.A. Forest Service and state/county-level units of government. Readers interested in this pattern of changing land ownership are referred to Stier, Kim, and Marcouiller 1999.

5. The development paradox of wilderness is a useful concept to reinforce this notion of accessibility. Wilderness, by definition, connotes a general dearth of infrastructure. The awareness of natural amenities found in wilderness areas originates from those who actually access wilderness areas. If our interest were to develop amenity-based market-driven economic values of wilderness areas, we would need to increase their use by developing access and recreational sites within the wilderness, a pursuit inimical to the very essence of wilderness.

6. This said, there is some disagreement on the extent of the empirical relationship. In examining the growing empirical literature, some have had difficulty in substantiating a claim that the income elasticity of demand for amenity values exceeds 1.0. For instance, Kriström and Riera (1996) examined several European-based contingent valuation studies for the income elasticity of demand and found widely varying results. Indeed, many of the studies they examined suggested that income elasticity of demand was less than unity.

7. A complete discussion of these approaches remains beyond our scope. Interested readers can obtain a more thorough detailing of nonmarket goods valuation from benchmark texts in natural resource economics (Garrod and Willis 1999; Kahn 1998; Lesser, Dodds, and Zerbe 1997; Tietenberg 2002).

10
Local Government and Public Goods

Governments at the community level play a vital role at all levels of economic activity. Local government policies related to taxation, spending, and borrowing impact all actors in the economy in their various roles of consumers, workers, employers, and producers. Governments provide transportation infrastructure, police and fire protection, and parks and recreational services to name a few. The topics we discuss in this chapter are in the portion of the Shaffer Star (Fig. 1.1) in which factor inputs are addressed; they focus on the importance of government as a key provider of inputs into production activities that characterize community economics. Governments also monitor, regulate, and stimulate economic actors. As we will see in Chapter 11, part of this set of responsibilities is the government's role in directly managing and regulating natural resources. Regulation and stimulus strategies are the focus for Chapter 11, Institutions and Society.

For the community, local governments are not only important partners in the economic development and growth process but also are responsible for the public provision of goods and services; they make the fundamental decisions related to what services to offer and at what level these services should be provided. *Public goods* include components that fall within the overall infrastructure of a community, such as the construction and maintenance of streets, roads, public utilities (sewer/water, energy, and others), and public buildings. Publicly provided services include fire and police protection, sanitation services, and education. Decisions about where to make investments in infrastructure can have a significant impact on the location decisions of people and firms. Equally important are decisions about taxation policies and how to pay for those services. Should taxes be placed on consumers or businesses, on what employees earn, or on the wealth of individuals? How government answers these questions influences how the economy will function, where economic activity will take place, and how it will change over time.

One could generalize and argue that the economy is composed of two parts: the public and private sectors. In most discussions of economics, we tend to focus on the private sector, which is composed of businesses and consumers interacting in the market. Market forces affecting both the supply and demand sides drive the economy to determine what goods to produce and what prices to charge (Chapter 1). We have seen how this plays out at the community level in our discussions of land (Chapter 5), labor (Chapter 6), and private capital (Chapter 7) markets. What often is overlooked in these economic discussions is the role of the public sector. As we will see, private entrepreneurs in a market economy do not normally produce some goods and services that are vital to the economy and its growth and development. If it were not for the public sector of the economy ensuring that investments are made in infrastructure and public services, our economy would grind to a halt. Firms would not be able to ship their goods to customers, consumers would not be able to travel to shopping centers, and crime could become so rampant that citizens would not feel safe. Ultimately, if governments fail to provide sufficient goods and services or do so with excessive cost, firms often elect to relocate to another community. As we will see, the public sector is a vital and integral player in the growth and development of the market economy.

This chapter provides basic insights into the underlying economic principles that influence the decisions made by government. We begin with a review of how markets can fail and why government has a legitimate place and role in our market economy. But if markets fail, how does government go about deciding what public goods and services to produce and at what level and how to pay for them? To answer this, we present the logic of Tiebout and *voting with one's feet* within a broader context of simple supply and demand. Finally, we discuss the notion of a fair tax structure, along with the characteristics of what makes a good tax system. By becoming more versed in the public sector and how it plays out in the market economy, the community practitioner has a more complete picture of the complex puzzle of community economics.

WHEN PRIVATE MARKETS FAIL

The purpose of a market economy is to allocate scarce resources of society in a manner that maximizes the well-being of citizens within that economy. Given limited resources of production, land, labor, and capital, how should these resources be combined? What goods and services should be produced and at what level? How are these goods and services to be allocated among the citizens of the community? Within a market economy, these questions are answered by the interplay of buyers and sellers through the forces of market demand and supply (Chapter 1).

In an idealized world of perfect competition, utility and profit-maximizing behavior, and rational decision-making, examining local markets is a relatively straightforward proposition. Pareto optimality is achieved and the *first fundamental theorem of welfare economics* holds. Here, the interaction of well-behaved demand and supply relationships yields stable equilibrium and economic well-being is maximized. All is right with the world.

When we move away from this idealized world and introduce complexities of the real world, life for the community economic development practitioner becomes difficult very rapidly. For example, when one considers the possibilities of externalities in either consumption or production, the simplicity of Pareto optimality breaks down. On the consumption side, market failures result in demand that is not reflective of the true underlying preferences of consumers. Because people or firms do not consider the presence of externalities in the calculus of their decision-making, market equilibrium is not Pareto efficient and our world becomes messy.

But what exactly is an *externality?* With reference to our discussion in Chapters 1 and 9, externalities can take on four specific types: ownership, technical, public goods, and pecuniary externalities. All of these externality types characterize imperfections in

how individuals and/or firms interact to create misplaced benefits and costs. In the community setting, these can be both negative and positive. A common example of a negative externality is pollution as a by-product of the production process (an example of an ownership externality). When a firm makes its production decisions, it is most often looking for the lowest cost means of production and has no incentive in a market economy to include the costs of pollution in its decision-making process. Pollution is essentially a negative by-product that is discarded because pollution itself is not normally a privately owned good. The firm does not consider the negative impact that pollution has on other components of the economy, such as the value of neighboring properties. Without some type of corrective mechanism put in place, such as explicit regulations or the attachment of some enforceable pricing mechanism, such as a tax that acts to assign ownership of the offending pollutant, the level of negative externalities produced will be too high from a societal perspective.

But not all externalities are bad for society or the economy. For instance, painting one's house or investing in landscaping creates a positive externality since one's well-being is increased by having a neighbor's property well maintained. This is an example of an ownership externality where the positive amenity has an associated impact that is not owned by the maintainer. Indeed, one's property value can be directly affected by the positive amenity values resulting from maintenance practices of neighbors (Chapter 9).

As we discussed in Chapter 9 in our idealized world, goods and services can be characterized as *excludable* and *rival* (see Fig. 9.1). Here, potential consumers can be excluded from consuming the good or service, most commonly through pricing mechanisms. For example, most people are effectively removed, or excluded, from the market for Ferraris due to the pricing structure. Goods and services exhibit the trait of being rival if consumption by one consumer precludes consumption by another. By some rich entrepreneur owning and driving that Ferrari, someone else is precluded from consuming the benefits of owning and driving that car.

Some goods and services, however, can be neither rival nor excludable. Take, for example, protective services offered by the local police and sheriff departments. Ideally, an effective police and/or sheriff department should deter crime within the community. Crime prevention or deterrence, the output or product offered by these protective services, has the characteristics of a positive externality. If your neighbors are consuming the benefits of crime prevention, then so are you. I effectively cannot be precluded or prevented from consuming the same benefits as my neighbors. Due to the nonrival and nonexcludability of crime prevention, private businesses are not able to capture the market and charge prices. Hence, in the real world, private businesses will not attempt to protect the public at large from crime through the provision of police protection. If, as a society, this is a service that we demand, government effectively has to step up and provide the service.

When considering public goods, life becomes difficult for the community economic development practitioner. Samuelson (1954) demonstrated that when public goods are present, there is no market mechanism to ensure that a Pareto optimal allocation of goods and services will be achieved; Adam Smith's *Invisible Hand* fails us. In short, because of the nonexcludable and nonrival nature of public goods and services and the resulting problem of *free ridership,* the nature of the aggregate demand curve is not clearly defined. In the case of private goods and services, individual demand sets are aggregated in a horizontal manner (summing individual demand curves to the right). But in the case of public goods, individual demand sets are aggregated vertically (summing individual demand curves upward). The key with public goods is that any number of individuals can consume a fixed level of the good or service, or there are economies of scale in consumption that are not present with private goods. Samuelson referred to this situation as the opposite of the first fundamental theorem of welfare economics derived in Chapter 1 and identified it as the *second fundamental theorem of welfare economics.*

In addition, with private goods, people will only pay in proportion to benefits received. If the price is higher than benefits received, the consumer will not purchase the good (Chapter 1). But with the case of nonexcludable public goods, if prices (taxes) are attached to benefits received, then there exists an economic motivation for consumers not to reveal their true preferences (benefits) and act as a free rider. With an unwillingness to reveal their true preferences, individuals present government officials with an informational problem, since in order to know how much of the good or service to provide, local government officials are left guessing as to what residents actually want. In short, the demand

side of our economic way of thinking about markets becomes tricky for understanding how the local economy actually functions.

To complicate the demand side one step further, the excludability–rival spectrum just described (Fig. 9.1) is not bimodal; it is a spectrum of degrees. For example, few public goods can be jointly consumed perpetually. As more and more consumers of the public good or service are added, the phenomenon of congestion becomes very real. A police department that does an excellent job of deterring crime for a community of 10,000 persons may be stretched to its limits if the population of the community doubles, and it may completely breakdown if the population doubles again. A public swimming pool designed to comfortably handle 200 swimmers experiences no congestion with only 50 swimmers in the pool. At 175 swimmers, again the users of the pool are not suffering any loss of benefits. But at 225 swimmers, congestion becomes an issue and the benefits to all swimmers decline with each additional swimmer entering the pool. At some point, the pool becomes so congested that safety becomes a concern and all benefits are lost and real negative costs may enter into the picture.

Another factor that adds to the challenge is the production structure of most public goods and services. Numerous studies have documented that the production structures are subject to what economists call nonconvexities, specifically economies of scale and size (see, for example, Deller, Chicoine, and Walzer 1988; Deller and Nelson 1991; Deller and Halstead 1994; Doeksen and Peterson 1987; Fox 1980). For most communities, particularly the more numerous smaller communities, the relevant part of the supply curve slopes downward rather than upward. In other words, the level of service provision is in the downward-sloping portion of the long-run average cost curve. Also, by their very nature, public goods and services have positive spillover (or technical externality) effects that are often not considered when communities make allocation decisions (Deller 1990). For example, if a community invests in a high-quality park and recreation system, nothing prevents residents from neighboring communities from using and benefiting from those services. Again, because of the notion of free ridership, the true demand for services is unknown. Theory dictates and available empirical evidence suggests that the level of the public good will be Pareto inefficient—specifically, too little of the public good or service will be supplied. In the end, by the very nature of public goods and services, there does not exist a clean market mechanism to ensure that any level of the good or service will be provided (i.e., the Samuelson result embodied in the second fundamental theorem of welfare economics).

As we have seen here and in Chapter 9, the broad phrase "public goods and services" covers many items, from clean air and scenic beauty to public swimming pools and police and fire protection. In this chapter, we limit our discussions to those goods and services directly provided by local governments, including police and fire protection, transportation services, such as road construction and maintenance, public schools, water and wastewater services, garbage collection and disposal, and street lighting. The public services provided through land conservation, parks and recreational services, and clean water and air are discussed in detail in Chapter 9.

Tiebout's Solution

One of the dominant areas of study in public economics is the provision of public goods and services. Public finance economists have spent considerable time and energy attempting to model individual preferences for public goods and services and the efficient or Pareto optimal level of the good to provide. While this literature is insightful and can help us conceptualize the problem, it is limited in helping us think about the public sector within community economics. What distinguishes the study of local public good provision from the general theory of public goods is the notion that people migrate between communities.

As we have seen throughout our discussion of community economics, the role of space is of utmost importance. Tiebout (1956c) refuted Samuelson's conclusion that people will not reveal their true preferences, by acting as a free rider, for public goods and services. Tiebout noted that, at the local level, there does exist a market mechanism in which consumers will reveal their true preferences. If consumers are mobile among communities, they can shop for the community that offers a mix of local public goods and services best suited to their tastes and preference. By voting with their feet, consumers reveal their true preferences for local public goods. Because people will self-select communities with certain bundles, in equilibrium communities will move to homogenous groupings: People within any given community will have similar tastes and preferences for public goods. Tiebout claimed that the

problem of preference revelation could be solved and there would be no obstacle to the efficient provision of public goods, provided there are a sufficient number of communities offering a wide-enough variety of public goods.

Perhaps the most famous empirical test of the Tiebout hypothesis was a study of local capitalization levels of local taxes and expenditure levels into property values by Oates (1969). Oates argued that if the quantity and quality of local public services and property taxes were determinants of residential choices, they should influence local property values. If public goods are a normal good, specifically as income increases demand increases, then higher levels of the good should attract more people, all other things remaining equal. The increased demand to locate in these higher-service communities should put upward pressure on local land prices. Equilibrium would be established in which part or all of the value of the public good or service was capitalized into property values. The opposite logic applied to taxes. Using data for New Jersey municipalities, Oates found statistically significant evidence that higher levels of public education expenditures positively influenced the median value of housing, while higher tax levels negatively influenced housing prices. Oates interpreted these results as evidence supporting the Tiebout hypothesis.

Because of the intuitive attractiveness of the Tiebout view of the world and these early empirical findings by Oates, the Tiebout approach dominated much of our thinking about local public finance and how markets for public goods and services work for years (e.g., see Rubinfeld 1987). These theoretical and empirical results can be very helpful for the student of community economics. First, it provides a framework to link public goods and services with human and firm migration patterns and local land/housing prices, two important elements of community economics. Second, the wealth of empirical work lends practical insights into how local community factors influence local government decision-making. For example, how does an aging population affect local government, or how does the growth of one type of industry over another type place different demands on local public goods and services? This work also has allowed us to gain insights into the types of public goods and services that need to be in place and at what level to promote certain types of economic development and growth. Much of the empirical work on the role of public infrastructure on growth and development that has flowed from the Tiebout world allows us to conclude that certain levels of infrastructure must be in place, but its presence does not guarantee growth or development.

Unfortunately, the Tiebout approach to the problem provides only a partial foundation. More recently, the Tiebout approach has fallen from grace within the public finance literature. Some have challenged the logic of Oates, suggesting that if the system were in equilibrium, public goods and tax levels would be fully capitalized into property values and the positive/negative coefficients reported by Oates and others are actually evidence of disequilibrium, hence causing Tiebout to fall short. Others (Bewley 1981; Stiglitz 1977) have attempted to develop more stylized statements of Tiebout's verbal arguments with no success. Studies looking at property value maximization models of local public goods (Brueckner 1982, 1983; Deller 1990) suggested that under this alternative paradigm, evidence can be found of Pareto efficiency in the provision of local public goods.

This latter line of theoretical thinking is particularly exciting because it helps local officials better grasp the challenge before them. Tiebout does not provide local officials with any type of decision-making rules or guidelines for service provision levels or taxation policy. In the world of Tiebout, local officials simply offer a bundle of goods and services at some tax level, then sit back and let people and firms react through migration. The Brueckner-type world of property value maximization, on the other hand, provides local officials with some guidance in terms of what that bundle of goods and services should look like. Specifically, local officials should keep an eye on local property values and monitor how values react to various service and taxation levels. Officials can increase service levels, and correspondingly taxes, up to the point that local land and housing markets respond unfavorably. If local property values start to stagnate or decline, then service and tax levels are too high.

A Stylized Model of Local Public Markets

Parallel to the capitalization work following the spirit of Tiebout, there is a rich empirical and normative analysis of local government fiscal behavior focusing largely on the demand for local public goods and services.[1] Based on the early voter theory work of Black (1958), the *median voter model* has evolved as a framework frequently used to specify fiscal behavior under situation of majority rule. By focusing attention on the preference of the median voter rather than of the whole heterogeneous population, our problem is greatly simplified. The logic of the median voter exactly parallels the classic Hotelling beach vendor location problem. Recall Hotelling's example from Chapter 3, where customers are evenly distributed over a finite beach. Two vendors are seeking the optimal location for their establishments. Beach goers will walk to the closest vendor. After several iterations, the vendors will cluster at the center of the beach.

In Black's political world, voters are evenly distributed over a political spectrum. Voters will vote for the candidate or political party that is as close to their preferences as possible. In this scenario, much like in the real world, candidates at the extremes of the political spectrum will never win a majority of the voters. The candidate or political party that comes as close as possible to matching the preferences of the median voter tends to carry the election. This simple logic explains why those at the extreme ends of the political spectrum view the two major U.S. political parties as being too similar in their views: Both parties are vying for the political mainstream, or the median voter.

Originally developed by Barr and Davis (1966), Borcherding and Deacon (1972), and Bergstrom and Goodman (1973) and expanded on by Deacon (1978), Pommerehne (1978), Inman (1979), Bahl, Johnson and Wasylenko (1980), and Bergstrom, Rubinfeld, and Shapiro (1982), the median voter approach has provided the community practitioner with a rich insight into local decision-making with respect to public goods and services. Jurisdictions are modeled as purchasing factors at market prices, and individual demands are assumed to be aggregated through a competitive political process, specifically Black's (1958) voting rules. Within this framework, service demands are assumed to depend on both income levels and tax prices, and any public officials spending far from the median will be driven from office by an opposition that proposes an expenditure level closer to the demands of the median voter.

Determinants of Supply

Despite the complicating characteristics of public goods and services, we can gain insights into the widely used median voter framework by integrating the supply and demand side of the local goods and services market by first focusing attention on the supply side. Assume that the local government production function follows a well-behaved Cobb-Douglas technology and that the local government is assumed to be a price taker in terms of factor inputs. Output can be expressed as

$$X = \alpha L^{\beta} K^{1-\beta} \qquad 0 < \beta < 1 \qquad (10.1)$$

where X is defined as physical output and L and K are labor and capital inputs, and α and β are a constant level of production and a factor productivity parameter, respectively. In this idealized world, public managers charged with the production of public goods and services are given a budget by elected officials, and those managers are assumed to maximize output/outcomes given the budget constraint:

$$B = TC = wL + rK \qquad (10.2)$$

where r is the interest rate and w is the prevailing wage rate and the budget (B) is exactly equal to total costs (TC). The optimization problem facing the local public manager is the maximization of equation (10.1) given a limited amount of money to spend on labor and capital. Identical to the problem of the producer discussed in Chapter 1, the standard optimization solution can be expressed as

$$\frac{\partial X / \partial L}{\partial X / \partial K} = MRS_{L,K} = \frac{w}{r} \qquad (10.3)$$

where $MRS_{L,K}$ is the marginal rate of technical substitution between labor and capital and is just equal to the ratio of wages and rents.

As officials make production decisions, they will pay utmost attention to costs, so what we now need is a way of determining the value of each factor price in relation to total marginal cost. The theoretical solution to this problem states that the local government, in an effort to optimize the use of its resources, will consume that quantity of a factor input up to a point where the value of the marginal product equals the input factor cost:

$$\partial X / \partial L * MCX = w; \qquad \partial X / \partial K * MCX = r \quad (10.4)$$

where MC_X is the marginal cost of producing an additional unit of the public good or service. Finally, using the well-behaved properties of the Cobb-Douglas production technology, factor demands can be expressed as

$$L = \frac{\beta MC_x}{w}; K = \frac{(1 - \beta)MC_x}{r} \qquad (10.5)$$

By substituting the factor demand equations in (10.5) into the production function (10.1), we can solve for the marginal cost function, which is needed for solution of the demand side of the model:

$$MC_X = (1/\alpha)(w/\beta)^{\beta}[r/(1 - \beta)]^{1-\beta} \qquad (10.6)$$

The marginal cost for production of local government goods and services is expressed as a power function with two arguments: municipal employee wages and the rental rate of capital.

Keep in mind that local government officials act just as a private firm would in its production decisions. Local officials face market prices for labor and capital, they are given a budget in which to operate, and their job is to maximize the level of the service offered to the public. The key here is that local officials will pay particular attention to the cost function, or equation (10.6).

Determinants of Demand

Several problems are unique to the market solution to the supply and demand for local public services. The two that are most troublesome for this simple stylized model are (a) the description and quantification of public goods and services and (b) the decision-making process that aggregates individual demands for public goods and services into a single bundle over the effective period of the collective decision. This latter process, often referred to as the political decision-making process, is discussed in Chapter 12.

In the discussion of production, the nondescript X measures the public good. The true output is eminently linked to the value of the good and is derived from the flow of services coming from the good or service itself. For example, a stretch of road has no intrinsic value itself other than the land the road is located on and the gravel and concrete making up the road; its value is determined by the transportation services it provides. Road mileage can be counted or traffic times and potholes can be counted, but these barely begin to capture the value of the local road system to individuals or to the economy. Alternatively, the output of police services is not hours of officer patrols, tickets issued, or even crimes solved, but rather the effectiveness in terms of crime prevention. At issue is the difficulty of measurement (Ostrom 1977). There are strong distinctions between inputs, outputs, and outcomes (Chapter 12). With the case of public goods and services, inputs can be measured easily, outputs can be approximated, but outcomes are very difficult to quantify. How do we know, let alone measure, the level of crime prevention for a given level of public expenditures?

In the approach adopted here, we assume that higher-level services are more expensive, hence expenditures are a proxy measure of outcomes. But this is not without several inherent limitations. The production technology for many public goods exhibits economies of scale in such a way that as service levels or outcomes increase per unit, expenditures decline. Conversely, for some public services, costs are actually inversely related to the quality of the service. Transportation service is a classic example. The cost of maintaining a high-quality road (high flow of service or outcomes) is lower from an engineering perspective than the cost of maintaining a road in a high state of disrepair (low flow of service). Contrary to common production economic logic, higher costs (expenditures) can be associated with lower service levels. A similar argument can be made for educational and police services: Costs of producing a certain output level are higher in poor, depressed areas than in wealthy, affluent communities. In our simple stylized model, the constant term in equation (10.1) varies from community to community in a way that economists do not fully understand.

Despite the serious limitations to the use of expenditures as a proxy for service levels, one must keep in mind that the local policy debate on setting service levels is a budgetary debate over expenditures and taxation levels. For the community practitioner, the questions posed are in the vein of "How much should we spend on roads?" and/or "How will local taxes affect the economy?" Insights into these questions provide the community practitioner with insights into how the local economy functions and reacts to changes in local government spending and taxing decisions.

As in Chapter 1, let us assume marginal cost pricing, which allows us to define per capita expenditures directly as price times quantity consumed:

$$e = sq \qquad (10.7)$$

Here the quantity consumed by an individual is specified as his or her relative share of total production of the public good:

$$q = X/N^\rho \qquad (10.8)$$

where X is the level of the public good as defined above, N is the population of the local government where the good is consumed, and ρ, a measure of

congestion, ranges between 0 (zero) and 1 (one). As the congestion coefficient approaches 0, the public good X exhibits characteristics of a *pure public good,* where congestion is never a factor. As ρ approaches 1, the level of congestion in the consumption of the public good becomes an issue. Put another way, the congestion coefficient has also been called the degree of *privateness* coefficient and as it approaches 1, the good takes on more characteristics associated with a private good.

Before continuing with the demand side of our simple model of local governments, we need to more clearly and consistently define price. Again, in a perfectly competitive market, firms will charge prices where marginal cost (MC_X) equals price (s) (Chapter 1). But in the public sector, local governments do not enter or leave the market if marginal cost does not equal price ($MC_X \neq s$). Nor is there a quid pro quo between buyers and sellers where this good is delivered for that amount of money. In the local public setting, payment for the good or service is through some form of a general tax, most commonly the property tax, the level of which may or may not be directly tied to the locale of the good consumed.

As argued by Black (1958), the relevant taxpayer to worry about is the median voter. In the simplest framework, price can be defined as

$$s = \tau MC_X \qquad (10.9)$$

where τ is the tax share of the median voter. Given equations (10.8) and (10.9), we are now in a position to more fully explore the demand side of the model.

An individual within the community is assumed to maximize utility over a public good (X) and a composite private good (Q) facing money income (y) and prices for the public good (s) and the composite private good (p), or to maximize

$$\text{Max } \kappa Q^\lambda X^{1-\lambda} \text{ subject to } y = pQ + \tau MC_X X \quad (10.10)$$

Analogous to the production side of the model, we assume that the utility function is well behaved, permitting the use of traditional optimization methods. Using equation (10.8) to capture the level of congestion in the public good, the budget constraint in equation (10.10) can be revised as

$$y = pQ + \tau MC_X N^\rho q \qquad 10.11)$$

with price now defined as

$$s = \tau MC_x N^\rho \qquad (10.12)$$

If we assume a Cobb-Douglas specification of the individual's utility function, we can derive a demand equation for the public good in which per capita demand is a function of the parameters of the utility function (κ and λ), the tax share (τ), the marginal cost function, population of the community (N), and the congestion parameter (ρ). But from equation (10.6), we know that marginal cost is a function of the wage rate (w), interest rate (r), and the technical coefficients of the production function (α and β).

We can express per capita demand or consumption of the public good in general terms as

$$q = (w, r, N, y \mid \alpha, \beta, \rho, \tau, \kappa, \lambda) \qquad (10.13)$$

In the simplest of terms, quantity demand of a public good is a function of wages (w), interest rates (r), the population of the community (N), and money income (y). The parameters of the demand function include the parameters defining the marginal cost function (α and β) and the utility function (κ and λ), taxes (τ), and our measure of congestion (ρ).

What does this stylized model provide the community practitioner? First, it helps us think through the problem of producing public goods and services. In many ways local governments are just like private firms; they are attempting to maximize output given certain constraints and factor prices. Production decisions are highly dependent on the underlying cost structure of the production function and the characteristics of demand. In short, as in much of economics, our stylized models help us to get a handle on complex situations.

Second, the structure of this stylized model provides a framework for thinking through how local governments may react to changes in the economy. For example, if the local economy is growing with increasing population and wages, what will be the impact on local governments? Increasing income and population will raise the demand for public services, perhaps leading to increased congestion. But the local government will be faced with higher wages that it must pay for labor. At the same time, however, increased income yields more tax revenues, shifting the budget constraint the government faces outward. How will all this play out? It depends on the values of the parameters of the system (α, β, ρ, τ, κ, λ), and these will vary by the type of public good and community.

Limitations and Additional Insights

While this stylized model provides the community practitioner with a foundation on which to begin, this approach is not without some significant limitations. Some of the difficulties already highlighted include the explicit use of expenditures as a proxy for the level and quality of the public good and service, and the inability of the model to capture the lumpiness of the production technology underlying the marginal cost curve. Publicly provided goods often suffer from lumpiness in the sense of excess capacity and the need for large fixed cost investments to increase capacity beyond a certain point. In short, to simplify the modeling process, several limiting assumptions have been made that may confound the empirical implantation of the model.

On the supply side of the model, there is a real potential problem with the assumption of marginal cost pricing. While this makes intuitive sense in the private sector, a strong case can be made that in the public sector local governments often work off the average cost rather than the marginal cost function. Two reasons are advanced. First, in many states, local governments are statutorily limited in how charges and user fees for services can be structured. By law, local governments must focus attention on average cost. This attempt to follow the average cost function in pricing is clearly demonstrated in the block rate pricing schemes for public utilities. Second, because many local public services exhibit economies of scale in production, the marginal cost of a unit of the public good will fall below the average cost of that same unit. Despite the fact that for many public goods marginal cost is lower than average costs, average cost is easier to measure and think about in the public setting. Indeed, marginal cost pricing would lower prices to consumers.

The weaker side of the model is on demand. Because of the nonrival and nonexcludable nature of public goods and services, the resulting problem of free ridership is at the foundation of the problems facing the model. To operationalize the median voter model as used above, several restrictive assumptions are required. These are nicely summarized by Bailey and Connolly (1998, p. 349):

> The explicit and implicit assumptions of the median voter model are voter sovereignty, single issue decisions, single-peaked preferences, majority voting, all voters vote, the cost to the local electorate of producing alternative levels of services is known, the full marginal cost of changes in service provision is reflected in the median voter's payment of local government

taxes, all voters benefit equally from the provision of local government services, . . . a perfectly functioning local democracy exists where local governments have full autonomy regarding local tax levels and service provision, . . . utility functions are separable between private and public sector expenditures so that preferences expressed through voting reflect only the marginal costs and benefits to be derived from local government services, individual voters vote for their most preferred alternative, the political process is competitive . . . and politicians seek to maximize voters allocated to them, bureaucrats serve politician's instructions who in turn reflect voter preferences, the median voter is the pivotal voter whose preferences are decisive, the median voter's identity is unaltered when grants are received or incomes change and, finally, a menu of distinct alternative policies and programmes is offered to voters.

Not only can each of these assumptions be questioned on an individual basis, but when taken in aggregate, it becomes readily clear that the demand side of our stylized model is perhaps at best an extreme oversimplification of the real world. Elections are seldom single dimensional, such as a direct referendum vote; rather, candidates have stated and unstated positions on a range of issues. Individual budgetary decisions are not made in isolation but within the framework of a comprehensive budget. Seldom if ever do voter participation rates even come close to approaching 100 percent. The restrictions that such a strong set of assumptions place on the demand side of this simply stylized model of the local public sector has been long acknowledged.

Efficiency, Effectiveness, and Evaluation

The challenge to local government officials to provide optimal levels of public goods and services at a reasonable tax level can be viewed conceptually as a simple demand-and-supply problem. But economists are not alone in the pursuit to provide insights into this problem. Professional public administrators and political scientists have devoted years of study to finding ways and means of guiding local public officials in making important investment and taxation decisions. While economists think in terms of demand and supply, public administrators and political scientists offer an alternative centered on the notions of *effectiveness* and *efficiency*.

Demand considerations are focused on the ability of local decision-makers to effectively match public service levels with the desires of local residents. As local public resources become increasingly scarce, the question of effectiveness centers on the ability of local officials to match resources to the demands expressed by local residents. Supply considerations, on the other hand, are predominately concerned with the ability of local officials to efficiently produce the public good or service in a manner that minimizes costs to local taxpayers.

Another way of thinking about the problem is to use Inman's (1979) two-step model of local government operation. When one thinks about how local governments operate, decisions usually follow a simple sequence. First, locally elected officials are faced with the responsibility of determining what public goods and services should be provided and how to pay for them. This is the demand side of the problem and local officials must take a read of what local residents want or demand. Once these decisions are made, responsibility for production decisions are turned over to the various department heads within the local government, such as the chief of police. They, in turn, are charged with supplying as high a level of service given the budget afforded to them. Effectiveness relates to the first step or stage, and efficiency relates to the second. The importance of the distinction between effectiveness and efficiency cannot be overstated. *Effectiveness* of the local government refers to collective choices that determine what goods and services to provide, how to raise the necessary revenue, and how to arrange for production. *Efficiency* refers to the purely technical process of transforming inputs into outputs.

While residents, through the local government, decide which goods and services to provide (demand), the local government need not actually produce (supply) the goods and services. For example, if local officials are faced with allocating limited resources between road maintenance, police and fire protection, and park and recreational services, the effectiveness of the decision-making process hinges on the local decision-makers' abilities to match resources to the values of local residents. If local residents place greater value on protective services over, say, recreational services, effective local decision-makers should devote more resources to protective services than to recreation. The efficiency of the local government then is dependent on the ability of the police and fire departments to produce the desired level of protection at the lowest possible cost to the taxpayer. It may be that the lowest cost alternative may not be local public employees performing the task; contracting with another unit of government or contracting with a private firm may be an alternative.

A simple framework to think about the effectiveness and efficiency of local governments is illustrated in Figure 10.1. Clearly, category I is the ideal position, where local demands are clearly satisfied and produced at the lowest cost to the taxpayer, and category IV is the worst position. Categories II and III deserve some further comment. Category II suggests that the local government is efficient in the production of public goods and services but produces the wrong mix or level. An example would be a very efficiently operated public golf course that has very few users. In category III, the government is effective in that it has correctly identified the priorities of the local residents, but it is not efficient in the actual production or supply of the public goods and services. Unfortunately, this is the category in which government is most often believed to be operating: Contrary to widely held perceptions, much of the evidence suggests there seldom is significant fat and waste. Local government inefficiency is most often the result of the lack of scale economies imposed on the government due to its jurisdictional size. This is a result of many local governments operating on the downward portion of the marginal and/or average cost curve.

The identification of the effectiveness and efficiency of local governments is a difficult task and requires rather precise empirical measures (Ammons 1996; Hatry 1999). It is necessary to make a number of decisions as to the type of the public good or service in question as well as the behavior of public officials involved.

When evaluating effectiveness, public administrators often turn to measures of citizen satisfaction and perceptions, service accomplishments, or community conditions, such as keeping an eye on property values as in the Brueckner-type world of property value maximization. In addition, families of performance standards are adopted and implemented, such as fire insurance ratings or engineering standards for road surface conditions.

Efficiency is directly related to costs of government and the level of the public service produced. Here output/input ratios based on engineering work standards are constructed to assess government efficiency. Measures such as dollars or employee hours per unit of output are common methods of discussing government efficiency. Economists have offered more advanced methods to develop statistical relationships between measures of output and input levels (Deller, Nelson and Walzer 1992). The lack of flexibility and oversimplification of these measures, however, are widely recognized by both public administrators and economists.

There are two primary complicating factors in evaluating the effectiveness and efficiency of local governments. The first (and perhaps the foremost) difficulty is to develop a meaningful and practical empirical measure of the public good or service. Again, performance measures for evaluation purposes struggle with the distinctions between inputs, outputs, and outcomes. For some services, well-defined engineering standards and monitoring methods are in place, such as road surface conditions. Other services, such as police protection, are much

Levels of Effectiveness

Demand Matched   Demand Mismatched

	Demand Matched	Demand Mismatched
Efficient	I	II
Not Efficient	III	IV

Levels of Efficiency

Figure 10.1. Trade-offs between notions of effectiveness and efficiency.

more difficult to measure. Designers of these methods caution against the serious potential of distorting the policy-making process if improper performance standards are employed. For example, if the police department is evaluated based on the number of arrests or tickets issued, law enforcement officers have a perverse incentive to behave aggressively with local residents.

The second complicating factor is the need for benchmark comparisons of the selected indicators. An indicator, no matter how well constructed, is meaningless unless there is a reference point on which to make comparisons. The reference point is what makes for a reasonable benchmark comparison (Chapters 14, 15, and 16). Typically, local governments compare themselves to other similar communities or to statewide averages, but this requires that the data for the performance measures be widely available. In addition, given the complex structure of local governments, no two local units are directly comparable. One community may contract with the county sheriff, while another may have its own police department or a joint department across many communities, which is common with fire departments. Also, residents in one community may demand a higher level of service than residents in another community.

The pressure to develop and implement performance standards is increasing. Important local decisions will be made based on the interpretation of these evaluations. While this is an extremely useful exercise, care must be taken in designing, implementing, and interpreting the standards. In the absence of performance measures, it is difficult for local officials to discuss and evaluate whether the services they provide meet resident preferences and whether the services are efficiently provided.

TAXATION

Local government officials are charged not only with the decision-making responsibility of choosing what public goods and services to provide and at what levels, but also how to pay for those services. In a market economy, there are generally three ways in which governments can generate funds to pay for public services: the levying of various taxes, issuing public debt, and creating money. Unfortunately for local government officials, the creation of money is reserved for the federal or central government and its regulator authority over the financial markets. Local officials are then left with the two options of imposing more forms of taxes or issuing debt. No single issue creates more conflict at the local level than decisions over local tax policy. While the decision to provide higher levels of a service seldom generates conflict, the correspondingly higher tax levels to pay for that decision is a focal point of heated debate. Seldom in community economic development and growth discussions does one hear concerns about public service levels being too high, but inevitably arguments about taxes being too high are at the forefront.

Sources of Revenue for Government

Throughout the world, the two general sources of revenue that governments use to pay for public goods and services are taxes and fees/charges. Taxes are the predominate source of revenues for nearly all levels of government; they include individual and corporate income taxes, property taxes, sales taxes, excise taxes, and dedicated payroll taxes such as social security contributions. The income tax is the predominate source of revenue for the federal and many state governments. The sales tax is a significant source of revenue for most states and many local governments. The property tax is generally a local tax and is often the primary source of revenue for local governments. Excise taxes are targeted taxes on the consumption of specific goods, such as alcoholic beverages and tobacco products—the so-called sin taxes—and the motor fuel tax. Other less predominate forms of excise taxes include targeted taxes on airport and telephone usage and the purchase of firearms.

User fees and charges are not a tax per se; they act more as a direct price because the fee or charge is attached to a good or service that is being directly sold to the consumer.[2] For some public goods and services, the flow of benefits to the consumer can be more readily measured. For example, the users of a recreational facility such as a public swimming pool can be directly observed. The free-rider problem is not as distinct, and direct pricing schemes can be put in place. As the debate over high tax burden is coupled with growing demand for public services, local officials are increasingly turning to user fees and charges. It is common for people to be charged for services (e.g., ambulances) that were once provided as part of the mix of public services. Library cards that were once free to any local resident of the community now are available for a fee.

But many government services cannot be directly sold because of their public nature. Because of the nonexcludability nature of many public goods and services, a charge or fee structure cannot be reasonably enforced. Others offer positive externalities, and charging an amount sufficient to cover all costs would defeat the purpose of government undertaking the activities. With other services, such as education, the distributional effects of offering the good dictate against charging. Strictly distributional activities, by their nature, cannot be sold to the users. Accordingly, charges and fees do not represent a viable alternative for financing most governmental activities. In addition, the revenue-generating potential is often so limited that the cost of administering the fee cannot justify the fee or charge.

A detailed discussion of the pros and cons of each type of tax used is beyond the scope of our discussion of community economic development. Specific examples of such pros and cons are given in our discussion of the characteristics of what makes a fair and workable tax system below. Particular attention to the impact of taxes on economic development at the local level is provided at the close of this chapter.

For local governments, a third source of revenue that warrants attention comes in the form of intergovernmental aid. These flows of dollars from higher units of government to local units can be significant. These monies, however, generally come with two conditions. First, the local unit of government has no influence or decision-making ability as to the size of the flow. Second, they often come with "strings attached" that dictate how these monies are to be used. This latter characteristic can cause distortions in decision-making at the local level, such as communities chasing the intergovernmental flow rather than addressing community needs. Just because the money is available for a certain activity does not mean that the activity is what the community needs or wants.

Qualities of a Good Tax System

When thinking about taxation policy to pay for public goods and services, certain attributes that taxes exhibit can be used for evaluation of alternatives: (1) revenue potential, (2) efficiency, (3) equity, and (4) accountability. While these attributes do not lend insight into the question of whether or not local taxes are too high, they provide some guidelines for thinking about the issue of how local tax policy should be formulated.

Revenue Potential

One important criterion for evaluating alternative taxes is their revenue potential. All other attributes

aside, a good tax must be able to generate enough revenue for the purpose it is imposed. User fees, for example, may be fair because they are directly tied to the use of a particular service, but generally they fail to generate significant revenue volume. For example, some communities charge a nominal fee for the right to utilize the community public swimming pool. While a $10 or even $20 fee for a pool pass helps defer the costs to provide the services represented by the pool, this fee structure will not generate sufficient revenues to pay for all the costs associated with operating and maintaining a public pool.

Further, revenue from the tax should grow in proportion to the growth of the local economy. As a community grows the revenue base must be able to keep pace with increased demand for government services. As we saw with our simple stylized model above, as the community grows, the demand for local public goods as services will increase in proportion to population and income growth. To meet those high levels of demand, the budget constraint also must expand proportionately.

Finally, the revenue flowing from the tax should be as resistant to economic downturns as possible. The sales tax, for example, may be able to generate significant revenues and grow in proportion to the community, but revenues decline with downturns in the economy. Thus, the tax must not only reflect the growth of the economy but also afford local officials with some degree of stability for budgeting and planning purposes. Because of the large expense associated with numerous public goods and services such as construction of new infrastructure or the specialty of certain equipment, such as fire trucks, many local governments have long-term capital improvement plans that require stable and predictable sources of revenue.

Tax Efficiency

When local decision-makers consider how a particular tax might affect economic activity, they are considering tax efficiency. The taxes that disturb market decisions the least are more efficient taxes. This issue is most often at the front of the local government economic growth and development debate. The commonly held perception that governments are inherently wasteful (inefficient) drives taxes unnecessarily higher and, in turn, harms the local economy. This view is too simplistic. In the broadest sense, in diverting resources to the public sector, taxes of different sorts impose varying degrees of distortions on the economy. These distortions impose *welfare losses,* or what has been called *deadweight losses,* on the economy by causing a departure from a Pareto optimal allocation. These losses to welfare are also referred to as the *excess burden* or *welfare cost* of taxation.

Unfortunately, there is no completely efficient tax, since all taxes distort the economy to some extent. Income taxes may cause some people to work less than otherwise. Excise taxes may cause some people to buy less of the taxed good or service. Property taxes can affect where people and businesses locate their homes and establishments. The issue when designing a tax system or modifying an existing one is to be aware of potential distortions a tax may have on economic decisions. In the public finance literature, this problem of the design of the most efficient tax system is called the *optimal tax problem.*

A second dimension to tax efficiency concerns the costs associated with administering and enforcing the tax. Clearly, imposing and collecting a tax places an administrative cost on government and compliance costs imposed on the taxpayer. An efficient tax is one that minimizes those costs. Some argue today that the federal income tax, which was once an efficient tax to administer, has become excessively cumbersome, hence inefficient, and that the tax needs to be simplified. The complexity of the U.S. federal income tax code introduces a tax-induced distortion. A user fee or charge is argued to be the easiest to administer because it is collected at the time the public service is used. Unfortunately, user fees and charges often fail the criteria of revenue generation potential. Taxes must also be designed in a manner that minimizes the ability of individuals to evade the tax.

Tax Equity

Tax equity asks, what is fair? While fairness is in the eye of the beholder, there are some guidelines to help think through tax equity. Tax equity often is judged on the basis of the *benefits principle* and the *ability-to-pay principle*. The benefits principle suggests that who benefits most from a government good or service should bear most of the tax burden. A user fee, such as a toll road, is an example of a tax that follows the benefit principle because the users of the road system are directly paying for it, as long as revenues from the toll are devoted to the road system.

Many public goods and services, such as police protection, cannot be readily supported by a user fee–type tax because there is no reliable or efficient way to measure the benefits each resident receives. Many people believe that individuals with more income or more property and other wealth can afford to pay more to support government services and therefore should pay higher taxes. This concept of tax equity is known as the ability-to-pay principle. For example, it is argued that a person with a large home with many improvements receives more benefits from police and fire protection than does a person with a modest home. Based on the benefit principle, the former homeowner should pay more in support of those protective services.

Two generally accepted notions of equity associated with the ability-to-pay principle are horizontal equity and vertical equity. Horizontal equity essentially means that equals should be treated equally. That is, if two people have the same level of welfare before the tax, they have the equal welfare after the tax. In practice this translates into individuals with equal income should have equivalent tax burdens. Vertical equity means that people with different levels of welfare should be treated differently. To judge the degree to which a tax system is vertically equitable, one must be prepared to make a judgment about the appropriate way to treat people at different welfare levels. In practice this is a particularly difficult concept to apply; while the idea of equal is clear, the notion of being treated differently is not as clear and requires subjective judgment. Should an individual with twice the welfare of another pay exactly twice the taxes, one-third more, or five times as much? To answer that question requires interpersonal welfare comparisons that are value judgments (see Chapter 12).

The concept of vertical equity can be reflected in progressive, proportional, or regressive taxes. A *progressive tax* is one that takes a higher percentage of income in taxes at each level of income. The federal income tax system is progressive by the construction of increasing marginal taxation rates. With a *proportional tax,* the same percentage of income is paid in taxes for all income levels. Finally, a *regressive tax* is the least defensible on the basis of ability-to-pay or benefit principles. A tax is regressive if the percentage of income taken in tax declines with higher income levels. Fixed dollar fees, such as licenses and some user fees, are regressive.

Most people would judge a tax to be equitable if it followed at least one of these equity principles, and most taxes do. Besides the toll road example, most excise taxes (e.g., for stadium seats and airline tickets) follow the benefit principle. Federal and state income taxes usually have progressive rates, thus adhering to the ability-to-pay and benefit principles. Property taxes follow both equity principles indirectly. People with more property pay more taxes for two reasons. First, owners of more property receive greater benefits from local services such as police and fire protection. Second, those with more property usually have a higher income, with the exception of some landowners.[3]

Accountability

A tax should be visible and have a clear link between the taxing unit and the services provided. That is, people should know when they are paying taxes and how much they are paying. If local residents become separated from the public policy process, they may become unaware of how local tax dollars are raised and spent. This is not a difficult scenario to imagine in light of the many special districts for certain purposes, the array of public charges for services, the state collection and redistribution of revenues (state aids to local governments and special districts), and the many boards involved in the decision-making process and their unclear relationships to one another.

In some instances, taxes become excessively cumbersome and complex. Here individuals may become confused about what they are being taxed on and at what level. Income tax systems that treat different types of income differently, such as wage and interest income, or offer a variety of special tax credits or deductions can distort the market. The tax code may be altered to encourage, or discourage, certain types of activity, and people become confused about how they are being taxed. At the local level, tax incentives used in the name of economic development may not only fail the notion of horizontal equity and benefits principle but also create unintended distortions in the market.

When individuals become confused about the taxes they are paying and to whom they are paying them, economists point to what is referred to as *fiscal illusion*. In the extreme, public officials may find that they are not held accountable for their actions because residents become disconnected to the decision-making process through fiscal illusion. Although in extensive reviews of the fiscal illusion literature both Oates (1988) and Dollery and Worthington (1996) concluded it was nearly impossible to document and measure the extent of fiscal illusion and its impacts, this literature has

profound implications on the community economic development practitioner.

TAXES, SPENDING, AND ECONOMIC GROWTH

What role does the provision of local public goods and services have on the performance of the local economy? In the name of tax relief should local governments cut back on services provided? Should more of the limited tax dollars be spent on transportation infrastructure or public education or parks and recreational services? Specific answers to these questions hinge on the unique characteristics of the community. But the available literature overlapping public finance and economic growth and development provides us with insights into the general answers to these types of questions.

The volume of public finance work that has flown from the Tiebout debate outlined above provides us with partial answers. Nearly all of the empirical tests of Tiebout have found that public goods and services play an integral part of the local economy. The logic of these models is that property values capture much of what is happening in the local economy. If the economy is strong and growing, property values will increase. If the economy is weak and stagnating, property values will stagnate and perhaps decline. Studies that have examined the relationship between government service levels, proxied most often by expenditure levels, and local property values have found strong positive relationships. Simultaneously, many of these same studies have found that higher local taxes have a negative impact.

But the real question concerns the balance between local government spending and taxation policy.[4] While the literature focusing on the Tiebout debate has helped, there is a second pool of literature looking directly at the impact of taxation on firm location. According to Ladd (1998), five benchmark studies document the progression in the thinking about local taxes and their impact on economic growth: Due (1961), Oakland (1978), Wasylenko (1980, 1981), Newman and Sullivan (1988), and Bartik (1991). In his survey of the literature on firm location, Due (1961, p. 170) concluded that "while the statistical analysis and study of location factors are by no means conclusive, they suggest very strongly that the tax effects cannot be of major importance." Due based his conclusion that taxes are inconsequential on firm location decisions on the pretext that taxes account for such a small percentage of operating costs. He concluded that other costs associated with labor, land, and transportation dominated the effects of any small variation in taxes across locales. In his update of Due's earlier work, Oakland (1978) accepted without question the conventional wisdom founded by Due that taxes have little effect on interstate or interregional location decisions.

Wasylenko (1980, 1981) expanded the discussion of taxation and local economic growth by explicitly examining the notion of intraregional competition for firms. While the interpretation of the literature by Due and later by Oakland concluded that taxes account for little in a firm's decision to locate in one state or a metro area over another, they did not address the role of taxes in the selection of one locale within a metro area, for example, over another within the same metro area. Citing a limited number of statistical studies, Wasylenko concluded that statistical evidence identifying a marginal role taxes play in intraregional firm location is outweighed by other more relevant factors. Wasylenko suggested that the limited role taxes may play is due to the limited variation in taxes across regions. While Wasylenko attempted to address this latter issue, he concluded that our thinking about and measuring of relative tax burdens need to be refined.

Ladd (1998) noted that the 1980s witnessed a proliferation of statistical studies challenging the conventional wisdom advanced by Due and reaffirmed by Oakland. Newman and Sullivan's (1998) attempt to summarize this newer work found three distinct approaches: general equilibrium, partial equilibrium adjustments, and dynamic adjustment models. Because of the escalation of studies and approaches, Newman and Sullivan concluded that the impact of local fiscal policy, taxes in particular, on economic growth through firm location "should be treated as an open rather than a settled question" (Newman and Sullivan 1988, p. 232). They were encouraged by the introduction of new theoretical approaches and empirical data and the sophistication of econometric methods.

Perhaps the most influential review of this literature was conducted by Bartik (1991, 1992). Using a modified delphi method summarizing 57 empirical interregional and 25 intraregional studies conducted since 1979, Bartik provided compelling evidence that taxes do matter in economic growth. While previous reviews of the literature discussed individual studies, Bartik's use of delphi methods allowed systematic averaging of results across studies. While individual studies may have limitations, there would have to be serious systematic errors cutting across all studies for the consensus results to be invalid. In striking contrast to the previous reviews of the literature, Bartik concluded that taxes have quite large and significant effects on economic activity. Of the 57 interregional studies reviewed, 70 percent reported at least one statistically significant negative effect of taxes on one or more measures of economic activity such as employment, output, or business capital. Ladd (1998, p. 92) argued, "This observation alone suggests that the conventional wisdom that taxes do not matter deserves to be questioned."

White (1998) suggested that Due's conventional wisdom and Bartik's challenge may both be right. She argued that the idea of firms becoming more sensitive to taxes over the past 30 or 40 years is intuitively appealing. According to White, first-order effects, such as labor, land, and transportation costs, vary less across regions now than they did in the past. Because firms have become more footloose, second-order effects, such as taxes, probably have become more important. Thus, both Due and Bartik may be correct. Perhaps more important is the increased incident of tax incentives at the local level to influence firm location. Municipalities are more willing today to "go to war" to attract, retain, and promote economic growth with tax incentives as a primary tool of war.

But some have argued that such wars may result in inefficiently low taxes and an underprovision of key public services (Wilson 1999; Brueckner and Saavedra 2001). One should keep in mind, as argued by Bartik, that taxes are only part of the local fiscal policy and economic growth debate. Taxes pay for public goods and services that are vital to the functioning of the local economy. In their now classic study of local economic growth, Carlino and Mills (1987) found that local government services could have a larger impact on local growth than taxes. Similar results have been uncovered by more recent studies, including those of Wagner and Deller (1998), Deller et al. (2001), Henry, Barkley, and Bao (1997), and Barkley, Henry, and Bao (1998).

Fisher's (1997) survey of the literature found that spending on public education has a positive impact on business activity in 12 of 19 studies reviewed. Likewise, spending on public safety also had a positive impact in five of nine studies. As noted in Chapter 9, the literature is reporting that not only the traditional public services but also secondary services such as parks and recreational services are important to economic growth (Dissart and Deller

2000). Clearly, the trade-off between lower taxes at the local level and adequate funding of key public services is still not clearly understood.

SUMMARY

Local governments play an integral role in the economic development and growth process. One of their primary roles is in the provision and production of local public goods and services. Because of the problem of free ridership, markets will fail to provide key goods and services that are vital to the functioning of the economy. These goods and services, widely referred to as public in their nature, include roads and bridges, police and fire protection, garbage collection and wastewater treatment, and education to name but a few. Because of market failures, local governments are vested with the responsibility of making sure adequate levels of these goods and services are offered.

The challenge facing local public officials is two pronged: First, what goods and services will be provided and at what level? Second, how will those services be paid for? Part of the answer to this complex question can be seen through simple demand and supply analysis. Notions of effectiveness and efficiency speak directly to the economist's concept of market behavior within the demand and supply framework. But with much of community economics, the answers to these questions are not always clear nor is the science fully developed. What public finance economics provides local decision-makers is a framework to think about the problem and make simple comparisons of alternative options.

While the growth and development literature is clear that certain levels of key public goods and services must be in place before the economy can be expected to grow, what that level is and the additional benefits of higher levels are not well understood. We know, for example, that a road system must be in place for an economy to grow, but the necessary expanse of that road system is not clear. A two-lane highway may be a minimum, but is a four-lane highway too much? The answer varies on the specifics of the community asking the question.

At the forefront of the government economic growth and development decision is the role of taxes. Clearly, for government to pay for these goods and services, it must impose taxes, but there is a wide range of taxing mechanisms, some better than others for a given situation. There is give and take between service levels and the underlying taxes

that are imposed to pay for that service. While the literature suggests that more of a public good or service is generally a good thing for the economy, higher levels of taxes place a drain on the economy. The challenge facing local officials is to find the best match between service delivery levels and taxes.

STUDY QUESTIONS

1. How does the local government determine whether a service should be provided and the level of service required?
2. How does the local government provide the service?
3. Why should community economic development practitioners worry about public goods and services and the mechanisms to pay for them?
4. How is the service paid for? Does that affect its efficiency or effectiveness?
5. How might a local government evaluate its performance?
6. How do local governments reach a balance between taxation and spending levels?
7. Does Tiebout's theory of voting with your feet solve the market failure problem associated with public goods and services?
8. What characteristics of a tax should local officials worry about in their decision-making?
9. In a democratic society, whose preferences tend to determine the outcome of public decisions?
10. How do the presence of public goods affect the Pareto optimality of competitive markets?
11. How do you define business climate?

NOTES

1. The discussion presented in this section follows the derivation of Beaton (1983).
2. The attempt of charges and fees is to try to get the individual to internalize the cost of his or her action. That can be done by using what is referred to as a Pigouvian tax. A Pigouvian tax is a tax or subsidy on an externality that brings about an equality of private and social marginal costs. (See Chapter 9 for a discussion of private and social costs.)
3. One of the most common critiques of the property tax is that it places excess burden on retirees, who find that they are at times unable to afford the tax on a family home.

This critique points to a clear distinction between income flows and property wealth. Individuals who may be property rich (wealthy) may have limited incomes with which to pay the tax. A somewhat rigid and perhaps uncaring economic argument is that such individuals are living beyond their means.

4. Often this balance is discussed in terms of the *business climate* of the locale. Frequently the discussion is solely focused on the taxation side of the equation. The importance of public goods and services paid for with the tax revenues and the relations of local government with the business community are seldom discussed.

Section III
Institutions and the Art of
Community Economics

Have you ever wondered why people who forecast the weather are only right about half the time? The science of meteorology is extremely complex with benchmark metrics, satellite imagery, and highly sophisticated models that forecast future weather events. To top it off, this is all orchestrated by some quite well educated and highly skilled professionals backed up with complex and expensive equipment. And yet no matter how sophisticated the models are and how skilled the meteorologists are, they are still wrong much of the time; sometimes more often than they are right. Community economic development is somewhat analogous to the ability to forecast the weather. We have very sophisticated models to predict change, but the actual paths with which communities proceed are much less predictable.

We make this analogy not to disparage meteorologists or community economists but to simply point out that the reality of the here and now with respect to community economic development is inherently unpredictable. The scientific skill of the forecaster often needs to be tempered by the art of hunches, social networks, politics, regulatory frameworks, and strategic actions of individuals. We need to realize that the manner in which economic decisions are made incorporates much more than the economic forecast.

Indeed, the reality of community economic development is considerably less predictable than theo-retical economics would lead us to believe. So much of our difficulty in moving smaller communities forward has to do with key aspects of how these communities are organized and the manner in which public policies that affect economic activity in these communities are made. This section is written to provide the reader with our sense of how community economic development plays itself out within the real world.

We begin with a chapter on decision-making that presents alternative philosophical perspectives important in how we distinguish just from unjust or right from wrong with respect to the impact of decisions on societal welfare. This philosophical background within the context of entrepreneurial communities provides a framework for the next chapter, which focuses on institutions and rules involved in community economic development. We conclude this section with a chapter that outlines alternative strategic approaches that practitioners can adopt to improve their chances of success.

Throughout this section, it is important to reinforce the notion that although we acknowledge the importance of a scientific approach to community economic development, the successful practitioner relies on a variety of skills, many of which are people oriented, to actually exhibit success in the art of community economic development practice.

11
Institutions and Society

Institutions and culture are basic to any form of social interaction. Their importance, however, is usually not recognized except when changes are proposed or when these institutions are performing unsatisfactorily. The interest here is limited to institutions facilitating or impeding community economic development. We are now discussing the rules and culture nodes of the development star found in Figure 1.1.

Institutions and culture are mutually interdependent. Social norms and interpersonal relations within a society drive culture. The entrepreneurial spirit or infrastructure of a community is a product of the culture of the community. Institutions are designed and put into place by the people of the community, and the culture of that community directly affects institutions. Once institutions are in place, however, their feedback can affect the culture of the community. When one compares developed and developing economies, the economic forces that are at play are the same, but the culture of the people and the institutions that are in place can be vastly different. Hence, an economic policy that may work in one community may be wrong for another community because of cultural and institutional differences.

The chapter begins with what we mean by culture and institutions. We then discuss the role of institutions in community economic development, including specific types of institutions and the notion of institutional change through time. Finally, we look at how government interacts with a private market system to bring about desired changes. Throughout the chapter, the terms *rules* and *institutions* will be used interchangeably. Likewise, *society* and *culture* will be used interchangeably. We deal with both institutions and society together because they tend to be in the background and because of their mutual dependence.

The manner in which private and public interests are balanced within institutions and culture is constantly debated. It seems as though the current market-oriented libertarian viewpoints common in contemporary U.S. society equate societal interest with individual and privately held freedoms to the detriment of a collective realm. On the other hand, modern market-based socialism is alive and well in Northern Europe and Canada. Personal ideology determines the proper balance between public and private interests. In essence, the culture of a community is driven by the aggregation of individual ideologies.

Institutions and society could be thought of as two sides of the same coin, but we are choosing to say institutions and rules are formal while society and culture are informal. Jones, Stallman, and Infanger (2000) also referred to three factors (two of which are forms of institutions/culture) affecting the structure of an economy and how that economy changes over time. The first factor is legal. What is prohibited or what is permitted? Are the economy and society built on the idea that, unless some stricture is written, people are permitted to undertake an activity? That is a much different legal structure than one that presumes every activity is prohibited unless explicitly granted. For example, is the economy decentralized and market based, or is it more centrally planned and societally driven?

The second factor is cultural. Does society take an individual or a communal perspective? An individualistic society would essentially place the individual at the highest and most dominant role in evaluating goals and values. This type of society might choose not to have any type of community economic development except as you would as an individual. This would be a libertarian view of the world. The other extreme is the communal society that essentially places itself into a more socialistic approach that says all values need to be considered and local decisions are made when everyone is satisfied. The key is to find a balance between these two extremes.

The third factor involves economic objectives. Does society value equity or efficiency as the primal concern in any type of transaction?

Defining terms will allow us to more clearly identify key elements that relate to community economic development. *Society and culture* largely are the unwritten and informal rules.[1] *Culture* consists of that collective system of societal characteristics, relationships, and processes that are slow to adjust because they are highly durable (i.e., almost constant) on a time scale that is pertinent to long-term (intergenerational) decision-making. Culture defines what behaviors are acceptable, which are often referred to as *social norms*.

Institutions and rules place into law those social norms. *Institutions and rules* define what are legitimate benefits and costs that accrue to private citizens versus those that remain public responsibilities. In some communities, starting small businesses is viewed as a positive occurrence and is supported by local institutions and culture. In other communities and societies, people who start small are viewed with some skepticism about their motivations, and the rules and culture in place hinder small business start-ups.

THE ROLE OF SOCIETY AND CULTURE IN COMMUNITY ECONOMIC DEVELOPMENT

The society and culture prong (Fig. 1.1) is frequently not recognized because it is so pervasive, tends to be background in local communities, and often is difficult to express.[2] The culture of the community can vary among communities.[3] Some communities have a culture that makes them feel they are victims to external forces, other communities take the external forces and amplify or deflect them to their advantage. The latter communities are more *entrepreneurial* in their outlook (McDowell 1995).

Swanson (1996) called culture the pattern of daily life strategies. It represents strategies for daily tasks and the rules for acceptable behavior. Culture (traditions, customs, attitudes) and governmental and legal arrangements set the framework in which economic units (households, businesses) make consumption and production decisions (Behrman and

Rondinelli 1992). Culture can be said to represent informal institutions. Swanson (1996) also gave three dimensions to culture:

1. *Ideas* or the knowledge gained by a society that are reflective of judgments about what should be
2. *Social norms* derived from values that reflect accepted ways of doing things and specific guidelines to appropriate behavior (how people act toward one another)
3. *Values* that reflect shared ideas about what is right or wrong

Flora et al. (1992a) contends that culture, at the most basic level, is the shared product of society. Some of the products are material (e.g., clothing, pottery). Some of the products are intangible, such as ideas, governments (one element of our concept of institutions), patterns of acceptable behavior, values, and myths. *Socialization* is the process by which we learn a culture, the process of becoming a member of society. Socialization takes place within the social organizations closest to us: family, school, church, and civic groups.

Culture defines community and values, which establish how a community views and defines legitimate problems and solutions to those problems, determines what is right and wrong, and determines what is desirable and undesirable. In other words, culture helps a community frame the big questions that, in turn, can vary from community to community and from region to region. A community that is ripe for community economic development has a strong level of dissatisfaction with the status quo (wants things to change), is willing to experiment, engages in a high level of discussion, debate, and networking to define the issues, and has a history of getting things done (Shaffer 1990). Each of these characteristics is an important part of a community's culture/unwritten rules and norms.

An example of changing cultural aspects important to labor markets involves women's participation in the labor force. After World War II and through the early 1950s, few women pursued permanent careers outside the home. Women working outside the home were simply not acceptable by social norms. Since the mid-1960s, the opposite has become common. Now more than half of the women aged 16 to 60 participate in the formal work force. As a result of this change, the labor supply and the economy's productive capacity has increased dramatically. In essence, the social norms about women working outside the home changed and laws, or rules, were changed to reflect the change in what was acceptable behavior.

Some authors (e.g., Flora et al. 1992b) refer to culture as social infrastructure. Three elements of social infrastructure are (a) local social institutions, (b) human resources, and (c) quality of social networks. *Local social institutions* are social organizations that provide services essential to the maintenance of physical, social, and cultural needs of the community. These include police and fire protection, churches, health care, community kitchens, and homeless shelters. The community's *human resource* base conceptually captures the ability of community members to accomplish their individual goals and vocational tasks. *Social networks* are webs of relationships that link individuals within a community. The quality of the social network affects the community's ability to effectively gain control of their social and economic development. Communities that have high levels of these three qualities are called *entrepreneurial communities* (Flora and Flora 1993). They include depersonalization of politics, development of extra community linkages, diverse community leadership, and a broad definition of community.

Other authors, including Blakely (1983), have even created typologies of community culture and how it relates to community economic development. Blakeley suggested five categories of communities:

1. *Entrepreneurial communities* have aggressive leadership, are pro-growth, try a lot of things, probably have lots of resources, and tend to be action oriented.
2. *Analytical communities* are more cautious, are resistant to change, probably have a somewhat limited resource base, and tend to be debate versus action oriented.
3. *Defender communities* are opposed to change.
4. *Destroyer communities* actively work to block change regardless of its beneficial effects.
5. *Desperate communities* have experienced major disruption and loss, have few resources, fear the loss of further resources, are often vulnerable to quick fixes, and tend to be symptom oriented rather than problem oriented.

For many communities, economic decision-making is hindered to a large extent by local culture and the tradition of doing things a certain way. Thus, members of a community that are seeking to change the culture of the community must challenge the status quo. This means that the community, or individuals within the community, must ask perceptive questions about why things have always been done a certain way. It also means that the askers of the questions must be willing to offer alternatives. The scope of a market-based economy (the primary emphasis of this book) is determined by tradition, a key element of the society node (see Fig. 1.1). The structure of a market economy and the rules that regulate the functioning of the market are products of attitudes of the society or culture.[4]

The divergent interests and pursuits of actors in the market do not automatically mesh to form a harmonious whole. Specific mechanisms are needed to keep competition productive and/or creative. The starting point of community economic analysis is that the economy is a subsystem of the societal system. Often, what occurs within the economy can be explained by society or culture. For example, the scope and level of innovation is in part determined collectively. The lower the culture of a community ranks economic goals, or productivity and efficiency gains, the lower the scope and level of innovation.

Culture is largely unwritten rules about how people deal with each other, deal with conflict, define fairness, and identify acceptable behavior. As we saw in Chapter 1, community development is concerned about nurturing the culture of the community such that the community is in a position to identify problems, make decisions, and act on those decisions. Since community economic development implicitly assumes such a culture is already in place, communities that do not have this culture in place are not ready for community economic development (Chapter 13).

THE ROLE OF INSTITUTIONS IN COMMUNITY ECONOMIC DEVELOPMENT

Institutions are the rights and obligations or social, political, and legal rules that determine what factors need to be taken into account in the use of a community's resources (production), exchange (markets), and the distribution of rewards (Davis and North 1971). Kraybill and Weber (1995) said institutions are the laws and norms that society lives by and the mechanisms created to enforce rules and norms. The role institutions play in community economic development possibly can be overlooked in the short run but cannot be ignored in the long run (Beauregard 1993; Bromley 1985; Schmid 1972; Wolman and Spitzley 1996). Rules represent the rights and obligations that must be considered in all social interaction. As society and the community move through time, a sense of stability has to be present, but it must be flexible enough to handle the new contingencies and the new opportunities that appear in a dynamic environment.[5] Rules need to be in place to help resolve conflict. Conflict can be about which set of values governs what goals the community pursues, and it can be viewed in light of the conflicting philosophies of libertarianism, utilitarianism, and contractarianism (more fully described in Chapter 12). Consequently, conflict resolution is essential for effective rule setting. Rules need to be in place to help guide the process of aggregating individual preferences into a community set of ideas and preferences. Rules help guide what behaviors are acceptable and provide a framework for community decisions by establishing what is politically, legally, or economically possible. Understanding current rules and how they may be modified is crucial to successful community economic development. Understanding the current rules allows communities to identify and possibly remove barriers to successful implementation of community strategies.

Specific Institutions

Discussion of institutions and community economic development leads to a series of questions about institutions and rules that influence resource use; the incentives and aspirations of individuals; orderly change through time; entrepreneurship; capital accumulation; technological change; labor supply; geographic and occupational mobility of labor; the resolution of conflicts; the negotiation of debts; the specification of rights, duties, liberties, and immunities; property rights and contracts; and the organization of businesses (Beauregard 1993; Behrman and Rondinelli 1992; Parsons 1964; Schultz 1968; Wolman and Spitzley 1996). Let us discuss each of these in turn.

A change affecting labor supply has been the changes in the retirement age and the availability of valuable human capital and experience. During the 1990s, mandatory retirement became unacceptable in the eyes of society and laws were passed to effectively prevent businesses from age discrimination. Society's changing views on when one is expected to retire is also reflected in the changing eligibility rules for social security benefits.

Institutions also affect economic markets because they set the framework for the bargaining process and the resolution of inevitable conflicts. Institutions need to be in place to ensure the enforcement of agreements, such as contracts. The proper functioning of markets, especially over great distances among unfamiliar actors, requires that the rules of the market be enforced. Likewise, institutions related to the method of settling transactions (e.g., barter, cash, credit cards) influence markets. Trade agreements help define acceptable markets. For example, the North American Free Trade Agreement (NAFTA) has clear implications for communities in northern Mexico, the United States, and Canada. Trade agreements can also focus on restrictions (such as tariffs and trade embargos) to the free flow of goods and services. For example, the United States has trade restrictions for the flow of goods and services to Cuba, Iran, and North Korea and has had restrictions on the People's Republic of China. While under restrictions, these markets are simply unavailable to U.S. companies.

Institutions affect income distribution directly by defining rates of reimbursement, by whom and how resources are owned, and by how the returns from the use of resources are distributed. Laws governing minimum wages and slavery are important labor market rules. The distribution of income directly affects development through its impact on incentives, the willingness to save, the willingness to invest, and the aggregate demand of society.

Institutions related to the ownership of resources and capital accumulation affect community economic development by defining the mechanisms to acquire and control capital. The form of business ownership, for example, can influence the access to certain financial instruments. Sole proprietor businesses have no access to regional or national stock markets, while publicly owned corporations do. Other institutions influencing capital accumulation are income and inheritance taxes. Income tax rules affect consumption and investment decisions as well as work incentives. Definitions of which forms of income are taxable, such as wages and salaries, or nontaxable, such as fringe benefits, affect the economy. The tax system encourages various forms of capital accumulation. Tax-free interest on municipal bonds encourages public capital formation, while depreciation rules and tax credits encourage investment in new buildings or the rehabilitation of existing buildings. Indeed, much of fiscal economic policy is aimed at changing behavior by altering the tax code.

The rules that define how a modern economy functions are extensive and can be extremely complex. Attempts to grasp the full range of all the rules affecting community economic development is daunting and beyond the scope of this text. The community economic development practitioner must be sensitive to the relevant rules affecting the problems the community is addressing. If the current institutional structure fails to support community economic development, it becomes necessary to implement institutional change. An important aspect of institutions/rules is that since they are created by people, they can be changed by people.

Institutional Change

Institutional change means changes in how individuals and organizations interact among themselves and with their economic-social-political environment. It may mean changes in rules. New institutional arrangements evolve in the face of new scarcities, new knowledge, new technology, and new tastes and preferences. Changes in relative factor or product prices, in resource endowments, or changing social norms creates social-political strains on existing institutions, making institutional change necessary (North and Thomas 1970). Inevitably, the creation of institutions has unanticipated outcomes that lead to a need to adjust or modify those institutions. Economic development can occur within the existing institutional framework, but frequently institutional change must come first.

One or more strategies are usually used to effect institutional change. The first is to make a selective modification of customary social, economic, and political practices within the existing framework. The purpose here is to strengthen those practices beneficial to economic development. For example, the Community Reinvestment Act requires federally chartered financial institutions to meet not only the deposit needs but also the credit needs of their community. Prior to the Reinvestment Act, some federally chartered banks would accept deposits from local residents of the community they serviced but would not seriously attempt to financially support local projects.

The second strategy is to adopt institutions from elsewhere. Here, the idea is to borrow institutions from a different cultural or economic context and superimpose them on an ongoing customary relationship. General Motors and the United Auto Workers union are experimenting with the German idea of co-determination, giving labor a formal role in the policy-making of the business.

The third strategy substitutes completely new institutional arrangements for the existing institutions. Here, the purpose is to superimpose an idealized system for which there may be no prior experience. Examples include *employee stock ownership plans* (ESOP) and *community cooperatives* that are alternative business ownership mechanisms designed to return control to community residents and individuals most affected by the employment decisions of the firm.

Institutional change typically is gradual and subtle rather than sudden and dramatic. Economic decisions related to investment, pricing, and employment are made within the existing institutional framework. A change in that framework yields both gainers and losers (Davis and North 1971). Who gains and who loses is often difficult to anticipate; almost every policy decision has some unanticipated outcome. Institutions generate around them economic, political, and social power that reinforces the existing institutional structure. These powers and forces will resist change and must be contended with when change is proposed or when it actually occurs. The deregulation of the trucking industry provides an example of institutional change and its ramifications. In some locales, service will increase and price will decline, while other locales will experience the opposite. The gainers will support deregulation, while losers will be indifferent at best. The gainers include those who can now more easily enter the trucking business or can alter services. The losers are those who lose economic power. In the end, changing the *rules of the game* is often the simplest strategy, but it can be the most difficult to implement.

GOVERNMENT AS AN INSTITUTION

As we have seen, institutions within the community can take many forms, ranging from laws and rules to organizations, including chambers of commerce, churches, and government.[6] In a market economy, the public institution of government is fundamental to community economic development. But government here is not a provider of key public goods and services that are vital to the functioning of the economy, such as transportation services or police and fire protection (as in Chapter 10); *government* is the institutional mechanism in which rules are determined, written, and enforced. These rules affect the operation of the economy. Our goal now is to explore how government can be a participant in economic development via its role in creating and enforcing rules.

The scope of government authority is wide in terms of geography and function. State government's taxing authority, regulatory power, and agenda set-ting ability allow the state to lead communities in broad economic development efforts. If the unit of analysis is the state, it is important to recognize the role of smaller communities in community economic development because local actors usually implement strategies at the local level. Therefore, municipalities, nongovernmental organizations, quasi-governmental organizations, and individuals are also important actors in community economic development. The general justifications for public sector involvement in economic development are market failure and/or unintended market distortions created by rules and laws (Bartik 1990).

There are several basic approaches to public sec-tor intervention in community economic develop-ment (Eisinger 1988; Sternberg 1987; Wolman and Spitzley 1996).[7] These approaches are not mutually exclusive, and some are more appropriate for nation-al than for local governmental units. The most ele-mental national government policy for local economic development is the encouragement of vitality and stability in the national economy.[8] Beyond the general encouragement of national eco-nomic activity, there are six basic categories of gov-ernmental economic development programs.[9] They are incentives, regulations, joint ownership, market expansion, governmental purchases/employment, and capacity-building.

Incentives and regulations are closely related. Incentives or subsidies seek to direct economic activ-ities covertly through market signals, while regula-tions seek to direct economic activity overtly through extra market limitations. Allen, Hull, and Yuill (1979) enumerate five types of incentives: (1) capital grants, (2) soft loans, (3) accelerated depreciation, (4) tax concessions, and (5) labor subsidies. The first three are capital incentives, the last is a labor incentive, and the fourth, tax concessions, could be both. Capital incentives influence both the public and private capi-tal market. Loan guarantees, subsidized interest rates, favorable repayment schedules, and the creation of secondary credit markets are programs to influence the private capital market. Grants and loans for indus-trial parks, water, sewer, and roads related to eco-nomic development needs influence the amount and type of public capital. Labor market incentives take the form of employment information, mobility assis-tance programs, and low-cost training.[10] The end result of labor incentives is either increased mobility of labor among places and occupations or reduced real cost of labor.

Some local government regulations attempt to direct private economic development decisions through outright refusal to permit an activity through formal regulation, such as environmental regulations on production practices, or through cost increases via tax fees, or regulatory delays. Regula-tions cause private decision-makers to incorporate public dimensions into the private decision calculus. Requiring social, economic, and/or environmental impact statements with agency approval creates awareness of the nonmonetary dimensions and com-munity implications of many investment projects. Banking regulations that require federally chartered financial institutions to meet both the credit and the deposit needs of their communities encourage using local financial capital in the community rather than exporting it to other areas. State regulations such as lifeline accounts are a mechanism for delivering cer-tain financial institution services to specific popula-tion subgroups, such as the elderly and people with low incomes. Regulations can be used in isolation, such as attempts to minimize a specific negative externality by limiting the choices available. The public costs are relatively low (just administrative) and are often paid by the firms through licenses and fees.[11]

The joint ownership approach, while not common in the United States, occurs with an explicit partner-ship between private investors and the government in selected economic activities. Typically, joint own-ership ventures pose great risk, require substantial capital, and target a basic industry needed to support further economic development. In a joint ownership program, the government takes an equity position in the company and it is represented on the manage-ment team. The purpose is to provide a broader pub-lic perspective in the decision-making. In the United States, local governments achieve joint ownership through a community development corporation (CDC), local development groups, or socially moti-vated venture capital firms (Chapter 7). CDCs, quasi-public corporations that can take an equity position in a business, often are initially financed with federal or local government dollars. The CDC can invest in businesses that provide employment for low-income or minority families, and can build the business management skills of the local popula-tion. A variant of the CDC concept is direct public sector involvement in enterprise and/or expanded public employment. Here, we mean such things as the Tennessee Valley Authority, the Wisconsin Con-servation Corps, and public service employment.

Market expansion programs offer considerable opportunity for local government involvement in economic development.[12] The setting aside of minimal amounts of governmental contracts for minority-owned businesses or small businesses is a form of market expansion for specified business segments. Likewise, local government attempts to buy locally are market expansion strategies. Conscious differentiation of freight rates for specific commodities, directions of travel, or locations are also forms of market expansion programs. A form of market expansion for isolated communities is to support business owners' attendance at trade fairs or conventions. This gives the businessperson an opportunity to make his or her product known to potential users who otherwise might not have been aware of that business. Some local governments attempt to stimulate tourism through their financial support of marketing efforts of local hospitality businesses.

The location of governmental investments in physical infrastructure is another mechanism to stimulate the geographic redistribution of economic activity (Halstead and Deller 1997). Other equally important mechanisms, however, are not as obvious. The first is income transfer programs to disabled, retired, or economically disadvantaged individuals and families (Deller 1995; Fagan and Longino 1993; Harmston 1979; Hirschl and Summers 1982). These transfers direct the flow of government expenditure to people located in economically depressed areas. The second is attempts by governments to purchase a minimum proportion of their materials and supplies from specified categories of businesses, such as those that are minority owned, small, or located in a depressed area. A third economic stimulant is the spending of workers employed at government facilities and office buildings.

Capacity-building is a conscious effort to increase the technical, financial, and managerial expertise of public and private decision-makers and their range of options. Plans for industrial parks, downtown renewal plans, and trade area surveys require technical expertise. Financial expertise includes the art of grant writing and the ability to package development projects so as to maximize the leverage of public funds. An understanding of how key business and political forces integrate and organize to achieve local economic development objectives is part of public managerial expertise. Counseling for small business managers on legal responsibilities, accounting methods, personnel management, and financial management is one way to create informed private decision-makers.

In summarizing what local governments in Western Europe did to promote local economic development, Keating (1989) said there were three general approaches: facilitative, interventionist, and directive (Table 11.1). In a *facilitative* role, the local government encourages communication between economic agents or enhances economic opportunities by investing in capital, both public, such as transportation systems, and potential private, such as an industrial park. In an *interventionist* role, government takes on a much more aggressive role by directly injecting itself into the operation of the economy through government purchasing rules, credit availability, and direct subsidies, among others. The *directive* approach is similar to the facilitative in that the government offers a plan or road map for community economic development. As we will see in the discussion of specific economic development strategies in Chapter 12, the approaches range from the simplistic to complex.

A final classification scheme that has seen much discussion in community economic development is about waves. As we explore the differences between the first, second, and third waves of economic development strategies, we can see a maturing of the science and art of community economic development (Bradshaw and Blakely 1999; Eisinger 1988, 1995; Schweke 1990). The first-wave strategies included industrial recruitment and offers of low-cost produc-

Table 11.1. Local government policy instruments

Facilitative
    Low tax policy
    Infrastructure development
    Land and buildings
    Lobbying other governments
    Education
    Promotion and publicity
Interventionist
    Advice and research help
    Science parks, technology transfer
    Municipal trading
    Local purchasing
    Credit guarantees and loans
    Equity investment
    Subsidies
    Contract compliance
Directive
    Planning agreement

tion sites. Such things as tax abatements, low-cost sites, low-interest loans, and training funds are the emphases of these types of strategies. The second wave of economic development strategies emphasized homegrown economic activity (expansion and retention). Activities associated with these types of strategies were the increase of investment capital for local firms, development of incubators, technical assistance for local firms, revolving loan funds, and tax increment financing.

Third-wave economic development strategies emphasize public–private partnerships. Here the emphasis is on increasing competitiveness. In third-wave strategies, you worry about creating networks to leverage capital, investing in human resources and high-skill and good-paying jobs. You may also be worried about creating strategically linked firms (clusters). Reasons for emphasizing third-wave development strategies are that service was fragmented and the customers were expected to integrate. Third-wave strategies attempt to integrate different programs for the customer. Some reasons focus on differentiation instead of an approach whereby *one size fits all* (endemic to first- and second-wave strategies). Other reasons for third-wave strategies include issues of scale, such as agglomeration effects favor larger over smaller communities (Chapters 2 and 3). First- and second-wave programs also often lacked accountability and seldom had measures of effectiveness or success. Finally, first- and second-wave programs tended not to have a link back to the economic and social goals that the community was pursuing.

An interesting argument offered by Eisinger (1995) is the internal conflict government officials face with second- and third-wave approaches. Community economic development envisioned in the second- and third-wave approaches is a long-term process where small steps today may not have significant payoffs for years to come. Given a democratic form of government, local officials face short-term election cycles. Local government officials looking ahead to the next election often do not have the patience to wait for long-term efforts. Short election cycles force local officials to look for immediate payoffs in economic development efforts. It is almost human nature to want and expect immediate rewards.

Another way for the community to become involved in development is to mandate specified outcomes (Goetz 1990). Some possible ways local governments can mandate desired outcomes are pre-sented in Table 11.2. The approaches or strategies outlined by Goetz follow the interventionist view as described by Keating (1989). (It is important to note that the strategies laid out here are fundamentally different from those strategies described in Chapter 12; the strategies we describe as the Shaffer approach are more comprehensive in that they include both private and public actions.) The strategies offered here are tied to governmental action via specific programs or rules. These government strategies can be much more powerful than private strategies because they have the full power (taxation, regulatory, program) of government behind them.

Government agencies and programs thus play a multifaceted and powerful role in local economic development. They are vital in introducing purposive sociocultural and technical change into a community. But the mere existence of government programs or a vibrant national economy does not ensure that opportunities will be utilized. While regulations play a direct role in economic outcomes, one of the indirect roles government can play is to send signals through the market place via incentives and subsidies. Because of the wide use of incentives and subsidies, they warrant special consideration.

Incentives and Subsidies

Incentives and subsidies are justified on two broad economic grounds: to lubricate and catalyze market processes (Eisinger 1988; Kieschnick 1981b; Laird and Rinehart 1967).[13] The lubrication argument contends that incentives speed eventual free market adjustments by reducing historical and institutional barriers, by reducing risk to the entrepreneur, by attempting to compensate for the lack of agglomer-

Table 11.2. Mandating economic development outcomes

Shared-equity partnerships with developers

Developer contributions for housing, parks, parking, other social infrastructure

Developer contributions to transportation mitigation

Hiring or training of local residents

Provision of social services by developers

Set-asides for minority- and female-owned businesses

Technical or financial assistance from developers to community-based organizations

ation economies, and by neutralizing resource mis-allocation caused by immobile labor, minimum wages, or other factors. The catalytic argument contends that incentives stimulate the development of specific skills in the labor force or stimulate investment in critical businesses or services that are precursors to other development efforts. Subsidies that operate at the margin, such as training workers in new skills not currently available in the community or encouraging investment in new technology or new types of business, alter the structure of the local economy. But there is no reason to believe that the changes that subsidies bring about will improve the long-term economic structure of the community (Buck and Atkins 1976).

The lubrication and catalytic arguments are typically stated in specific terms to justify the need for government incentives in community economic development. An obvious example would be the self-interested group that visualizes community economic development as using resources they have to offer, ranging from an industrial building or site for a retail store with excess sales capacity to some form of labor skill. The concern about local economic activity voiced by such a group is a perfectly legitimate economic response: It wishes to maximize its returns from the use of its resource.

Another example is that government may seek to overcome the problem of capital immobility through public involvement to alter inertia in the capital markets. This public effort reduces external/internal investors' imperfect information about opportunities in the community; it attempts to reduce risk to a level that stimulates investment or it attempts to overcome the "herd instinct" of always investing in other projects and/or other areas because that is the way it has always been done (Kieschnick 1981b). A variation of this argument is that the community is depressed because its existing economic structure does not respond to changing economic conditions; in particular, the products produced are not income elastic or are subject to wide cyclical variations. Conscious public efforts can alter the existing economic structure and link the community to the growing elements of the national/regional economy.[14]

Governmental involvement is also justified because human resources are often immobile among places and uses (Moes 1961; Rinehart and Laird 1972). Inadequate market signals and/or personal characteristics hinder the movement of labor into productive activities. Society loses the potential output of these irreplaceable human resources. The small public subsidies in training, health, job information, or mobility assistance that facilitate labor's adjustment can yield large returns to the community.

A final argument for subsidies parallels the infant industry argument in international trade (Alyea 1967). Subsidies allow a new local firm to begin operations under conditions of reduced competition (i.e., tariffs) or create cost savings that other more established industries enjoy through economies of size and agglomeration. The new business will eventually grow and become sufficiently competitive so that support is no longer needed. In the meantime, the community reaps the rewards of the intervening employment and income generated as a result of the program.

The theoretical support for local subsidies derives from relaxing some of the basic assumptions of neoclassical growth theory. The most important abandoned assumption is that wages are flexible both up and down as workers and management negotiate (Laird and Rinehart 1979; Moes 1962). In fact, union contracts, minimum-wage legislation, associated labor costs (e.g., social security and health insurance), and labor market regulations prevent full flexibility of wages. The second relaxed assumption is that labor is homogeneous among uses and places and is equally productive. The third relaxed assumption is that of complete mobility of labor and capital among uses and places.

The inflexible wages case for subsidies assumes that labor is wage immobile. In other words, labor will not move among places or uses even though there is a wage differential (Gray 1964). This inelastic supply of labor appears to exist in many rural and urban communities because labor mobility is not solely a function of wage differentials among communities (for additional details, see the Labor Mobility section in Chapter 6). In Figure 11.1, the current demand for labor (D_L) is incapable of employing all the workers at the fixed wage (w_1). The wage may be fixed by minimum-wage legislation or union contract. The amount of unemployment is full employment minus actual employment $(N_0 - N_1)$. The demand for labor is inadequate to use all the existing labor because the cost of labor (w_1) exceeds the value marginal product of labor (VMP_L) or the marginal physical product of labor times the price of the output $(MPP_L \times P_o)$. Labor can be fully employed if it is used in more productive processes (higher MPP_L), it yields a higher-valued output (P_o), or a community offers a subsidy of up to $w_1 - w_0$ to encourage hiring more labor.[15] This subsidy lowers the real cost of labor to the firm to w_0.

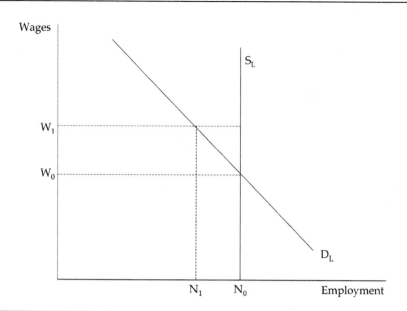

Figure 11.1.
Fixed wages,
inelastic labor,
unemployment,
and subsidies.

In the second case, again allow the free substitution of capital and labor in production processes but allow relatively free mobility of both capital and labor among uses and communities. In Figure 11.2, the demand and supply of labor curves are not perfectly elastic. In this case, the question becomes what size subsidy overcomes the less-than-complete mobility of labor. Again, let w_1 be some externally determined wage (e.g., national minimum wages or union contract). Assume for simplicity that movement along the S_L curve represents the movement of labor into or out of the community through migration or commuting.[16] A wage of w_1 will cause the supply of labor offered to be equal to N_2 but the demand for labor is only N_1, resulting in unemployment of $N_2 - N_1$. Excess labor moves to the community in hopes of getting one of the jobs paying relatively higher wages and labor is not discouraged by the unemployment.

Again, a subsidy can bring about an equality of labor supply and demand. The subsidy becomes a mechanism to reduce the cost of labor relative to capital and increase employment. But since labor continues to migrate into the community in response to the higher wages (w_1), the unemployment problem continues unless the demand for labor shifts to D_L'. The community's capacity to shift demand is limited. The only way to discourage continuing in-migration of

labor is to lower the wage labor receives. Since the wage (w_1) is externally determined, the remaining mechanism is to tax labor's wages for the cost of the subsidy (Buchanan and Moes 1960). If labor pays the full cost of the subsidy, wages to employees decline from w_1 to w_0 and N_0 employment results. The economy achieves full employment (i.e., demand for labor equals supply of labor) and continued in-migration will cease because labor will receive an effective wage of only w_0, not the visible wage of w_1. If labor pays none of its subsidy cost, the effective wage required to hire N_2 workers is w_2. In this case, a subsidy of $w_1 - w_2$ is required to equate labor demand and supply. This yields an inefficient use of labor. There are $N_2 - N_0$ redundant jobs in the community, and although everyone is working, many are underemployed. There is no wage incentive for underemployed workers to move, because they receive a full wage of w_1.

While the preceding analysis provides some insights into how subsidies can work, remember the analysis has some limitations (Cumberland and Van Beek 1967; Morss 1966). First, capital and labor are not completely and freely substitutable, thus preventing the marginal changes required in labor and capital use. Labor may have imperfect information about wage differentials and job opportunities elsewhere, reducing migration rates. More importantly, although the arguments for subsidies recognize

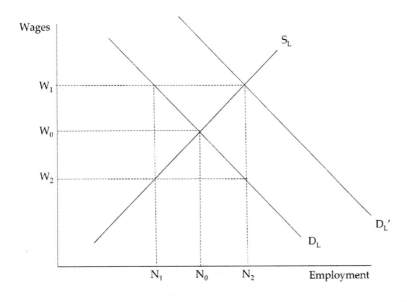

Figure 11.2.
Amount of labor subsidy required to create wage-mobile labor.

imperfections in the labor market, they do not recognize that similar imperfections occur in the capital market and the market for subsidies. Imperfect information exists about what subsidies are available or the subsidy required to overcome the excess wage. If the subsidy is not in the form of cash but occurs as public infrastructure or industrial buildings, the subsidy market suffers from the same capital immobility problems mentioned in the section Capital Market Failures in Chapter 7. Finally, the equity question remains: Why should labor pay the full cost of the subsidy or adjustment (Kieschnick 1981b)?

Types of Subsidies

Economic development subsidy schemes can be divided into labor, capital, and price subsidies. *Labor subsidies* generally reimburse the firm some proportion of the hourly or annual wage payments made, or they reduce the cost of labor by lowering the cost of training. Labor subsidies propose to increase the use of labor relative to capital. *Capital subsidies* provide capital to firms at below-market rates. This cost reduction occurs through an outright grant on some portion of the firm's investment or through favorable interest rates or repayment terms on loans. Capital subsidies increase the use of capital relative to labor, with the implicit assumption

that labor is immobile and unemployed. Neither labor nor capital subsidies are factor neutral. *Price subsidies* are factor neutral since they do not overtly encourage the use of either labor or capital relative to the other. Price subsidies take the form of raising the price the firm receives for its product by providing some form of market information or transportation cost reduction, or by artificially attempting to mimic agglomeration economies. Public infrastructure investments (e.g., fully serviced industrial parks) that artificially create agglomeration economies are a form of price subsidy.

Haveman and Christiansen (1978) identified two broad labor subsidy schemes: public service employment and wage subsidies. *Public service employment programs* offer the advantage of targeting jobs to specific categories of workers, such as minorities, youth, unskilled workers, veterans, and refugees. Public service employment programs hire workers for ongoing public services or public works projects. *Wage subsidies* can take the form of training and sheltered employment. By the latter, we mean public or private sector jobs for workers who are typically unemployable due to some type of handicap. Training and sheltered employment programs create new human capital, with sheltered employment typically offering long-term permanent jobs. Wage subsidy programs can also take the form of government paying some portion of the workers' hourly wage rate or wage income. The subsidy can be paid to the employing firms or to the workers directly; it can apply to all employees, to just newly hired employees, to employees above some previous workforce level, or to particular categories of employees.

Capital subsidies occur in numerous forms; some of the more obvious are grants, soft loans, accelerated depreciation, and some tax concessions (Allen, Hull, and Yuill 1979). National, state or community public agencies offer capital grants. The public agency, depending on legislative authority, conditions the rate of award by discriminating on the basis of size, physical location, type of economic activity, and project type. An example of these capital grant programs might be one targeted at promoting small tourism-based business start-ups along a deteriorated riverfront. A key element of the expected impact from capital grants is their effective rate. The *effective rate* is a function of the proportion of investment supported, taxability of the award, and the payment schedule. If the grants become taxable income or if payment occurs after the firm has made the expenditure, the effective rate of the grants is reduced. The higher the effective rate, the more impact the award has on capital investment.

A soft loan carries a below-market interest rate or has an extended repayment period, principal repayment and/or interest-free holidays, or reduced collateral requirements. An extended repayment period allows delays before any repayment is due and/or lengthened maturity to aid the firm's cash flow. The reduced collateral requirements occur as guarantees on the unpaid principal and interest, higher collateralization rates on assets, or acceptance of a wider range of assets as collateral. Soft loans offer the advantage of using the private financial market for screening applicants and administering the loans but the use of an intermediary reduces the discretion of the public agency in using funds for public purposes. The effective rate of soft loans is much lower than grants because the loan must be repaid and reduced interest costs increase the firm's taxable income.

Accelerated depreciation, income tax forgiveness, and property tax relief are capital subsidies. The first two subsidies require that the firm make a taxable profit before they have much impact; the lag before taxes come due reduces the effective rate of the subsidy. Accelerated depreciation permits the firm to postpone tax liabilities and improve its

cash flow. This subsidy can be linked to specific assets, such as new buildings and equipment and geographic areas. Income tax forgiveness programs are credits against income tax liabilities or exemption of a proportion of taxable income if a firm locates in selected areas. The tax credit will not occur until a profit is made, while the income exemption occurs with the first sale. Property tax concessions granted by a municipality are of varying durations up to 20 years, and do not require that the firm make a profit, and can apply to nondepreciable assets such as land.

An underlying premise of subsidies in a spatial economy is that cost minimization is the driving force behind the location of economic activity (Chapter 3). No consideration is given to the demand side of the market. The effects of any community economic development incentive or subsidy program depends on

- the type of subsidy offered.
- the importance of that inducement in the final location and/or production-level decision.
- what alternative production is lost by diverting resources into either the subsidy or the activity supported.
- the type of activity supported.
- who pays for the cost of the subsidy.
- the importance of costs to the firm.

Critiques of Government Subsidies

Several critiques of government subsidies have been put forward in the literature (Bartik 1991; Cumberland and Van Beek 1967; Eisinger 1988; FRB-Boston 1997; Harrison and Kanter 1978; Kieschnick 1981b; Thompson 1962; Wolkoff 1992; Wolman and Spitzley 1996).[17] These can be broken into seven major categories:

1. Political decision-making has no self-limiting mechanism that is similar to the forces of the market.[18]
2. Capital, labor, and subsidy markets have imperfections.
3. Subsidy programs have a tendency to be duplicated and become common among most communities rather than being unique to a few communities.
4. The diversion of growth from one locale to another is zero sum.
5. Subsidies are not critical in location decisions.

6. Firms are more sensitive to demand factors than to cost factors in their location decisions.
7. Subsidies are given for the wrong factors of production and to the wrong firms.

The response to the first criticism, lack of a self-regulating mechanism, is that communities seek to acquire benefits or jobs, and each community must pay some price for those jobs (Moes 1962). Viewed in this context, the offering of subsidies for new jobs is an economic transaction, although a somewhat unusual one, in which the new jobs go to the highest bidder (Morss 1966). The question becomes what is to prevent communities from becoming caught up in an endless cycle of increasingly larger subsidies?[19] Indeed, in the political process of making subsidy decisions, decision-makers can get caught up in a game of trying to outdo someone else. In doing so, the community can lose sight of the broader economic goal the subsidies are intended to achieve.

Rinehart and Laird (1972), on the other hand, offered three reasons why subsidies will be self-limiting. First, some communities will not participate because they already have all the jobs they need (full employment). Furthermore, communities will lose interest if they must pay all or most of the cost of the subsidy. Second, communities with unemployment will only bid the value of the new jobs to the citizenry. As the communities with unemployed workers experience success in creating jobs, the community reduces its subsidy efforts as its unemployment problem declines. Third, some communities with unemployed workers will not offer incentive programs because of the lack of interest, community beliefs, or expected out-migration and out-commuting of unemployed workers. These behavioral responses prevent unfettered subsidies from draining the community treasury.

The beneficial results from an incentive program are based on market imperfections in either the capital or labor markets. Theoretically, the decision to offer an incentive is made with full knowledge that the community must bear the full cost of the subsidy (Morss 1966; Shaffer and Tweeten 1974). In this case, there are no state or federal programs that lower the cost of the subsidy to the community. Full knowledge means the community knows what the benefits and costs of the incentive program are and the benefits are at least equal to the costs. Ideally, if some individuals in the community are made worse off by the subsidy program, then intra-community compensation is a legitimate cost of the program. Advocates for incentives postulate that communities freely bid for jobs, but only up to the point where the benefits equal the costs (Morss 1966).

The above assumptions are often violated in the real world. Subsidy programs are seldom completely financed by local resources; they depend heavily on state and federal funding. The full benefits and costs are not known and generally do not fall on the same individuals and/or communities (Morss 1966). Not every person in the community has equal access to the decision to offer a subsidy. The distribution of income among communities is not acceptable, as evidenced by the numerous state and federal programs to equalize income, employment, and tax burdens. Furthermore, the information about how much and what type of minimal subsidy is required may not be available to the community. The community is at the mercy of the business in terms of information about what is required for the firm to alter its investment or location.

Regarding the question of program duplication, there is nothing wrong with duplication except if every community has an identical program, one can legitimately ask how the situation has been altered from before the program was first created (Reese and Fasenfest 1996). One must remember that the implementation of an incentive program is to make marginal changes to the economic landscape. If every community makes the same marginal change, they in effect cancel each other out.

The commonly held belief that the diversion of growth from one locale to another is zero sum results from local economic development efforts. The net effects of subsidies depend on who receives them. Thus, changes in employment stimulated by the subsidy must be the net of the jobs lost by existing firms forced to leave or reduce their growth. The section A Zero Sum Approach in Chapter 12 addresses the question directly. Suffice it to say here that zero sum outcomes are not a foregone conclusion to local economic development efforts.

The importance of incentives to the location of economic activity continues as an empirical debate. Anecdotal evidence suggests great significance. The almost universal response in surveys about location decisions is that incentives were not critical in the decision about place or size of investment. As we discussed in Taxes, Spending, and Economic Growth in Chapter 10, the services that government offers through investments in transportation services,

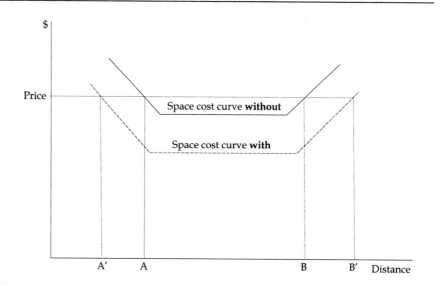

Figure 11.3
Space cost
curves with and
without subsidies.

police and fire protection and recreational opportunities, among others, can have a much larger, although indirect, impact on the growth and development of the economy.

Arguments that incentives reduce total costs of operation cannot be denied (Due 1961). But the important question is whether the firm is more sensitive to the total revenue or total cost portion of the profit equation (Harrison and Kanter 1978). To argue that costs drive the decision-making of the firm presumes the firm faces a competitive market and is unable to influence prices. If the firm is facing a competitive market, the dual labor market theory (Chapter 6) suggests that the types of jobs created are marginal and have low wages, low skill requirements, and no upward occupational mobility and provide limited fringe benefits. In other words, firms that are most likely to depend on incentives tend to be highly competitive and may not offer the best opportunities for economic development.

The jobs in the primary labor market (i.e., firms having some monopoly power) possess contrasting characteristics. If the firm has market power, it can control prices and output. There is no guarantee that the reduced costs from the incentive will be passed on to consumers as lower prices rather than becoming additional profits (Harrison and Kanter 1978). The increase in profits is a direct transfer of income

from the general taxpayer to the business owner. In addition, firms possessing some monopoly power tend to be more likely to seek incentives and receive the largest share of incentives.

Allow the firm in Figure 11.3 to be somewhat indifferent to its specific location within *AB*, since the sites yield identical profits. A subsidy to the *AB* area increases the profits of firms or leads to lower product prices. If profits increase or prices do not drop enough to eliminate the profit increase, there is a direct transfer from those who financed the subsidy to the firm to support an already profitable decision. Figure 11.3 also displays the significance of the unequal amount of information held by the firm and community and their relative power in the incentive bargaining process (Morss 1966). If the firm implies that it needs the subsidy to make the location profitable, and the community responds, an already profitable site yields a greater profit.

This returns us to the concern about asymmetrical information, where the community is dependent on the firm for truthful information. This exemplifies the negative connotation associated with subsidy programs (Thompson 1962).[20] A carefully structured subsidy program, however, can expand the range of profitable locations. In Figure 11.3, limiting the subsidies to *A'A* and *B'B* expands the range of profitable locations. If the additional profitable areas permit output to expand, then national welfare probably increases. If no increase in national output occurs but firms relocate into *A'A* or *B'B,* then just a transfer of activities occurs. This would represent a zero sum outcome unless *A'A* and *B'B* are the target of national concern and unequal welfare weights are used.

SUMMARY

Rules and culture are perhaps the most difficult parts of the Shaffer Star (rules, culture, markets, decision-making, resources, and space) to consider, but they can provide a sense of stability in the market in terms of investment. Yet it is important that rules remain flexible in a dynamic environment and the community's culture is adaptable to the changing economic situation.

Rules are a key component of community economic development. In the most basic sense, they govern how we bring markets and resources together through decision-making across space. Rules are the social, political, and legal tenets that must be accounted for in the use of resources, in exchange, and in the distribution of rewards. They define what is possible and what is illegal, which behaviors are acceptable and which are unacceptable, which benefits and/or costs are legitimate public issues and which are private. Rules and institutions define who receives the income generated from the use of resources, thus facilitating or hampering economic development. Economic institutions provide decision rules for adjusting and accommodating conflicting demands among different interest groups within society.

Economic theory typically assumes that the necessary institutions either do exist or will develop. The creation of an institutional framework supportive of community economic development, however, is not automatic and may be the critical element in a community's economic development efforts. Institutional arrangements evolve in the face of new scarcities, new knowledge, new technology, and new tastes and preferences. Any community economic development effort must be conscious of institutions and culture.

Rules are important in economic development because if we understand the rules, we may have some insight as to potential barriers to economic development. Also, rules can help ensure more equitable results (respond to/minimize market performance failure). They provide a means of conflict resolution and help us figure out how to sum individual preferences into aggregate community well-being. Some even suggest that public policy, which is a conscious attempt to set rules, is really a need or desire to redistribute advantages among members of society (Bromley 1985). The interpretation of eligibility criteria for various incentive and subsidy programs is an excellent example of a reinterpretation of rules strategy. For example, if the eligibility criteria for industrial revenue bonds were broadened from manufacturing only to include all export potential industries, then agriculture and many service and retail sector establishments become eligible. The broadening of the interpretation of eligibility is clearly an attempt at redistributing advantages among sectors.

An important institution in community economic development is government that has primary responsibilities to regulate and stimulate. In a mixed private/public economy, like in the United States, the majority of decisions about what type of economic activity and where it occurs remain in the private sector. The public sector and local, state, and federal governments, however, influence the type and location of economic activities. These public efforts can be broadly classified as having either a demand or a supply focus.

An important element to keep in mind is that the economic development policy tools available to government are not uniformly appropriate. Subsidies and moving allowances are financial incentives that work at the margin to cause capital and labor to migrate. Yet they are of minimal use in creating changes in local rules and/or attitudes about opportunities and perceptions of future courses of action.

An often-used phrase associated with local government and community economic development is *business climate*.[21] The phrase, much like culture, has no clear definition although everyone knows what it means. The typical connotation of business climate is low taxes. Yet to limit the public influence on business climate to low taxes ignores much of the involvement of government in community economic development. *Business climate really refers to the quality of the relationship between the private and public sectors in pursuing community economic development.* What is the nature of the partnership? Is it harmonious or turbulent? While there is rarely complete agreement between business and government, high-quality business climates foster the situation where businesses accept their social responsibilities and government supports legitimate business needs. In essence, the business climate is but one small part of the overall culture of the community. Unfortunately, the fullness of this relationship is often collapsed into the simplistic tax burden issue.

Culture is a persuasive force that largely is not recognized. It tends to be background to how the community goes about making decisions. Because culture defines what is acceptable or unacceptable, it sometimes inadvertently limits the options considered. Culture performs the critical functions of judging what is desired currently and in the future, opinions about entrepreneurship, and views about change. Society and culture define values that establish how a community views and defines legitimate problems and legitimate solutions to those problems, and determine what is desirable and what is undesirable. A culture that values traditionalism hampers economic development, while a culture that is willing to accept change and technological innovation facilitates economic development.

STUDY QUESTIONS

1. Why is culture largely unrecognized at the local level?
2. What is meant when we say the economy is a subset of society and culture?
3. What is meant by institutions and rules, and why are they so crucial to community economic development?
4. What is meant by institutions and rules define options on what you can and cannot do?
5. Why should institutions and rules of the economic game be considered in community economic development?
6. What is culture? How does it affect community economic development?
7. What community qualities or aspects of community culture would you associate with community economic development?
8. How might values and norms play an important role in community economic development?
9. What are some economic reasons to justify community/government intervention in economic development activities?
10. There were several different classification systems for government involvement in the market. What major difference is there among them?
11. What economic justifications can be made for the use of subsidies/incentives?
12. What is the economic rationale for why communities will not continue to offer ever-increasing subsidies to new firms?

NOTES

1. While not culture per se, a 1992 national poll by the Roper Organization found that a majority of people questioned (most of whom were urbanites) felt that rural communities had friendlier people, better personal values, a stronger sense of family, better quality of life, better community spirit, and greater honesty in business (Rowley and Freshwater 2001).
2. To a large extent in most economic analysis, rules or institutions or society/culture are the ceteris paribus (all else being held equal) in the equations.
3. Salamon (1989) had one of the more interesting studies of culture and community. She found two communities that she called German and Yankee. The two different cultures yielded different attitudes toward community, education, civic engagement, and commerce. They are supported by the fact that the spirit of cooperation is prevalent in rural areas. In another statement about culture, Castle (1995, p. 367) stated "cooperation, both formal and informal, has always characterized group behavior in the countryside."
4. This book accepts democratic capitalism as the fundamental institution or culture. However, at least two other forms are available: socialism and Marxism (Holupka and Shlay 1993; Molotch 1976; Wiewel, Tietz, and Giloth 1993).
5. One problem developing economies face is the temptation to constantly alter the rules defining the functioning parameters of the economy. As we discussed in Chapter 2, business investment is vital to the growth of the economy. If the rules are unstable, unacceptable levels of uncertainty are introduced into the economy and businesses will become reluctant to make investments.
6. Rules can be formal and informal. *Formal rules* are those that are implemented and enforced by government, such as zoning of land use. *Informal rules* follow more from social norms and the culture of society, such as wearing brown shoes with a tuxedo.
7. Several of these approaches require the federal government to create the program. Local government, however, often performs the critical role either of making use of the program itself or of encouraging the local private sector to do so.

8. *Stability* is used here to mean the future is somewhat known or there is minimal uncertainty about the future, not that the future of the community reflects the status quo in the name of stability.

9. An always present option that perhaps should be exercised more often is not to intervene at all. The market may be functioning adequately, government may be unable to do anything effective or its actions, if taken, may be too late to be of importance. For the purposes of this discussion, it is assumed that government action is felt to be appropriate and potentially useful.

10. All training programs must be sensitive to creating the skills required (to the demand for labor) rather than just working on the labor supply function. See Chapter 6 for more detail.

11. The private costs to the firm also may be insignificant, but one of the purposes of regulations is to make the firm internalize some of its externalities. An example is the disposal of an effluent discharge.

12. A variant of market expansion is the promotion of a community image or branding (Haidler 1992).

13. Traditional economic subsidies have been questioned. First there's a tendency of these subsidy types to shift the cost and risk of development to the public sector and leave the rewards in the private sector. Second these subsidies typically assume that the benefits will trickle down to the poor.

14. It is assumed in this argument that the cause of the economic depression within the community is a stagnated industry, not a lack of resources or limited access to markets or resources. If the latter is the case, public efforts of the sort discussed here will not alleviate the situation.

15. The increase in MPP_L could be achieved through capital investment, shifting the demand for labor outward. For further discussion, see Which Type of Subsidy? below and Demand for Labor in Chapter 6.

16. As we saw in Chapter 2, in-migration of labor causes a shift of the labor supply curve to the right. For simplicity in this discussion, we hold the supply curve fixed. The logic is the same if you allow the supply curve to shift but the demonstration becomes unnecessarily cumbersome.

17. *Clawbacks* are a method being used to reduce the risk of giving an incentive, under expectations of so many jobs or wages or tax base (Ledebur and Woodward 1990; Peters 1993). With a clawback, the firm agrees to pay back part of the incentive if it does not meet projections or promises. It is not clear how widespread the use of clawbacks is.

18. An adjunct concern is that the subsidization program will yield an uneconomic distribution of economic activity in terms of both use and space. Such a distribution, however, can occur as a result of factors other than the subsidy program, and that concern should not dominate the decision about their use.

19. One long-run effect of subsidy programs is the tendency for firms to bid up the cost of the factors used with the subsidized factor, thereby eliminating the cost advantage created by the subsidy. Often the policy response is to increase the original subsidy, setting off another round of linked factor price increases, or to expand the subsidy to other factors.

20. A critical element, from the community's perspective, is the danger of oversubsidization, which occurs when the community provides more incentives than required to influence a decision or when it gives inappropriate or noncritical subsidies.

21. While the level of competition among communities for new economic development probably has gone beyond the optimal, communities are essentially in a prisoner's dilemma. Unless every community believes every other community will also stop, none can unilaterally withdraw.

12
Policy Modeling and Decision-Making

In a typical economic model, men and women are considered independent, rational, and all knowing. *Independent* means that each individual makes his or her decisions regardless of the actions of someone else. *Rational* means that the individual orders all possibilities from best to worst and picks the best by using some objective criteria. *All knowing* means that the individual is aware of all of the options and can comprehend all of them.

In reality, most men and women do not approach decision-making in such an ideal fashion. In particular, community economics is concerned with decisions in which individuals are not independent and often times are not all knowing in their decisions. Before we proceed, it is important to point out that although decisions often boil down to a shoot-from-the-hip political position, our role as community development practitioners (economists, planners, and technical analysts) is to provide the best information about possible consequences that result from some policy decision. This topic of decision-making and the role played by community development practitioners provides the focus of this chapter. In essence, we are discussing the decision-making prong of Figure 1.1.

Decision-making translates into how the community goes about setting and implementing policies that affect development. Underlying these policies or programs are community values. When it comes to economics, how are values translated into decision-making? Does the community integrate sound and objective analysis with community perspectives and desires? Does the community involve a broad spectrum of interests or just a select few? Does the community really address the underlying problem(s) or is it just treating the symptoms? Indeed, can the community make a distinction between the underlying problem and symptoms that appear on the surface? What types of strategies can a community

implement to attain economic growth? Why are some communities better equipped for making economic decisions when conditions change? These are the questions we set out to answer in this chapter.

We will focus on formulating community economic development strategies that incorporate an awareness of the community's goals and objectives. The emphasis is not on group or political processes in policy formulation (Chapters 11 and 13) rather than on the economic objectives of policy. We begin our discussion with communities' capacities to make decisions related to economic development policies and social capital. Our interests focus on how values relate to the selection of community development policies. Given this basis for identifying the character of social utility, we then outline an objective and positivist approach to modeling policy effects. A review of generic community economic development strategies leads into a discussion of zero sum outcomes. The chapter closes with a discussion of entrepreneurial communities.

OUR CAPACITY TO MAKE DECISIONS (SOCIAL CAPITAL)

When we build decision-making capacity, we are essentially increasing the ability of people and organizations to do what is required of them to grow and prosper within the setting of the community.[1] In other words, in the context of community economic development, we are encouraging local people to manage their own affairs and to increase their level of self-reliance. Improvement in the quality of community economic development decision-making either through the efforts of governments or community groups requires full recognition of certain fundamental elements. There must be a group of local people who are committed to spending the time, energy, and the other resources it will take to make positive local change a reality. The presence of assertive, forward-thinking people in both the private and public sector may be the single most critical variable in community economic vitality (Pulver 1988).

The quality of local decision-making is strongly influenced by community residents' abilities to distinguish between problems and symptoms, and to build coherent responses to real issues (Honadle 1986). The continued pursuit of solutions to symptoms rather than real problems seldom results in community satisfaction. There remains no real substitute for well-informed local leadership. Successful development requires local leaders with strong analytical skills that have access to the information, data, tools, and techniques necessary for the full redress of root problems.[2] Inadequate access to current and accurate information is a serious stumbling block to local decision-making (see Chapters 13 through 16 for more discussion).

Widespread citizen participation and support is critical to most decisions affecting a large segment of the community (Gaunt 1998; Sharp and Bath 1993). Local residents control an array of resources that affect community life. These are controlled either as individual residents and businesses or through collective institutions, including local governments and citizen groups.[3] Individuals have financial and time resources to share on community projects. Businesses make decisions on the manner in which they treat their employees, the type of technology they employ, and the way they make local investments. Governments and community groups invest in schools, transportation systems, and other infrastructure, and take other collective actions that influence change. With widespread community support, many local resources may be brought to bear on local economic problems. Without this critical support, either the public or private sector can accomplish little to address community needs.

We have found that local solutions to local problems are almost a prerequisite to sustainable policies. Communities that rely exclusively on outside assistance for problem identification and solutions often do not implement long-run sustainable policies. It is almost human nature that people grasp and support self-generated ideas firmer than those externally generated. The trick for the community development practitioner is to help steer discussion in such a way that the community comes to its own conclusions about the underlying problems and policy solutions (see Chapter 13).

An item that needs to be considered in community decision-making capacity is a concept that has gained much popularity since the mid-1990s. This item is social capital, an important element in community decision-making. The intellectual foundation for social capital can be traced to Adam Smith (1759, p. 3), who recognized the interdependence of preferences when he wrote,

> However, selfish so ever man may be supposed, there are evidently some principals in his nature, which interests him in the fortune of others, and render their happiness necessary to him, though he derives nothing from it, except the pleasure of seeing it.

In the perfect neoclassical marketplace, anonymous transactions (in which trust and reciprocity are not considered or needed) occur among self-interested, rational buyers and sellers in a world of complete information. According to Robison and Hanson (1995), however, the underlying assumption of social capital theory is that personal relationships matter. They represent the degree to which a person's well-being is influenced by the well-being of another person, place, or thing (i.e., they influence economic choices, they are not independent). Accordingly, self-interested individuals—a typical assumption of neoclassical economics as we saw in Chapter 1—are capable of vicariously sensing the well-being of others. While these are not equal across individuals, they depend on those with whom they have significant emotional and social ties.

What we are talking about in decision-making is civic engagement derived through activities such as community organizing, citizen participation, and community-based decision-making (Turner 1999). Community-based organizations can bring residents together to learn decision skills that reflect a more entrepreneurial style which is conducive to strategic planning for local economic revitalization, including social capital.

The difference among communities in terms of civic engagement is due largely to the level of social capital (Coleman 1988; Putnam 1995; Turner 1999). *Social capital* refers to features of social organization such as networks, norms, and social trust that facilitate coordination and cooperation for mutual benefit. Networks of civic engagement foster norms of general reciprocity and encourage the emergence of social trust. Social capital consists of the social networks in a community, the level of trust between community members, and local norms. These networks, norms, and trusts help local people work together for their mutual benefit (Cavaye 2000a; Flora 1998). To say that a person is trustworthy means that the probability that she will perform an action that is beneficial or at least not detrimental to us is high enough for us to consider engaging in some form of cooperation with her (Hansen 1992).

Reciprocity is a mutual expectation or understanding that a given action will be returned in kind. Networks facilitate coordination and communication, and thus allow dilemmas of collective action to be resolved (Putnam 1995). At the same time, networks of civic engagement embody past successes and collaboration that can serve as a cultural template for future collaboration. There is some difficulty in addressing social capital due to its inherently abstract nature, and the fact that it is a public good. Social capital is inherent to the social relationships and interaction between people in a community (Cavaye 2000a).

Swanson (1996) talked about *social infrastructure* as the capacity and will of individuals and communities to provide or take advantage of opportunities that enhance their economic and social well-being, a form of social capital. There are several community characteristics associated with social infrastructure. First, the role of decision-making by individuals and communities, including the capacity of the community to make choices, is a primary characteristic. To a large extent this depends on the quality of information available and the ability to decipher that information, specifically the capacity to learn. This characteristic addresses the ability to make informed decisions from an often-confusing array of choices. Making a decision is the ability to act on your own volition—or free will.

Second is the negotiation on process within the community to determine the collective good, including the capacity to work together. What is the distribution of political and economic power? Is it widely distributed or narrowly held? What is the quality and focus of interaction among participants in the economic development decision-making? To have the capacity for innovative behavior, the community almost needs an atmosphere of immunity or indifference to permit individuals to experiment with ways of doing things differently. Without the presence of this type of atmosphere, tradition becomes the standard and change remains virtually nonexistent.

Flora and Flora (1993) talked about *entrepreneurial infrastructure,* rather than social capital, and how that might relate to community economic develop-

ment. Essentially they offered three dimensions that define entrepreneurial infrastructure. The first is *symbolic diversity,* or the community-level orientation toward inclusiveness rather than exclusiveness. Furthermore, rather than raucous conflict or superficial harmony, the community engages in constructive controversy to arrive at workable community decisions by focusing on community processes, depersonalizing politics, and broadening community boundaries. There is an acceptance of controversy in the community so that disagreements are not perceived as conflicts. People accept controversy and deal with it in a positive fashion. One way of looking at this is to think in terms of win-win outcomes versus win-lose outcomes. This is a comprehensive means of thinking of civic engagement.

The depersonalization of politics means disagreements are seen as honest differences to alternative solutions, and those offering alternatives are not seen as inherently evil. What to do in this type of situation is to focus on the process rather than winning. An analogy might be of parenting. What you are trying to do as a parent is not necessarily a win-lose type of outcome but to get the child to gain some understanding and perspective. The community also thinks in terms of a broad definition of who should be involved in the decision and there are permeable boundaries as to whom should be involved. In other words, you draw your boundaries as inclusively as possible and reduce the perception of us versus them.

Second, *resource mobilization* is when the community learns to rely on its own resources first rather than looking elsewhere. Looking elsewhere might include senior levels of government, nonlocal businesses, or foundations. But again, sustainability of local policies almost mandates that those policies be determined by local dialogue and decisions.

The third quality of entrepreneurial infrastructure is *quality of networks.* This includes such things as network diversity including, but not limited to, ethnicity, class, and gender. We worry about what the Floras call horizontal networks because we learn best from people like ourselves. But at the same time, there has to be an appreciation of vertical networks because a community needs to be linked to the world outside. For example, a rural community may learn from the experiences of similar nearby communities (horizontal), but it also is willing to draw on the experiences of larger and distant places (vertical).

It is not what you know, but whom you know that really represents the definition of social capital

(Schmid and Robison 1995). Relationships do count in decision-making, as do trust and networks. This awareness goes by several terms: social capital, social infrastructure, or entrepreneurial social infrastructure. Whatever it is called, it recognizes all decisions are not motivated by just self-interest or greed. Like other forms of capital, social capital is a productive resource in communities. Unlike other forms, social capital is not invested in physical things; it is a resource that exists collectively within and between individuals.[4] Social capital differs from other forms of capital in that it increases as people use it and it decreases if they don't use it (Cavaye 2000a). Social capital belongs to the community as a whole, not to individuals. The time and effort that an individual invests in civic organizations and social interaction benefits the whole community. A minimum level of social capital is a prerequisite to successful community development initiatives.

SETTING COLLECTIVE OBJECTIVES

The goals and values of a community's residents set the context for the choices perceived as needed and acceptable. For example, if the community believes it is important to keep the next generation of workers productive, residents are likely to insist their schools be of a high quality. If legalized gambling is proposed as a means of generating financial support for education, community values may judge that as an unacceptable mechanism for providing needed funds, thus requiring identification of another potential funding source. In short, local goals and values define what is considered and what is deemed acceptable.

We saw in Chapter 1 that a market economy, given some fairly restrictive assumptions, will yield a Pareto optimal allocation of resources, a state of affairs where no one can be made better off without at the same time making at least one other person worse off. Adam Smith's *Invisible Hand* of market forces works. We also saw that the presence of externalities results in market failures (Chapters 1, 9, and 10); competitive markets may not result in a Pareto efficient allocation. Also at issue is that Pareto optimality says nothing about the total welfare of that allocation. The final allocation, while efficient, may not be deemed fair by society. Much of the economic development process hinges on the notion that the final allocation achieved by a market economy can be improved on in the name of fairness or "justness."

One common goal expressed in the community economic development process is that the benefits of economic growth should be reasonably distributed across all residents of the community, or that the welfare of the community as a whole should be maximized. Traditionally this is accomplished by assigning monetary measures to the utility levels reached by an individual. But is it possible to derive an aggregate measure of social welfare so that alternative allocations might be compared, including allocations that are not comparable by using the Pareto criterion? Such an aggregate measure is called a *social welfare function*. A social welfare function is just some function of the individual utility functions: $W[u_1(x), \ldots, u_n(x)]$.[5] It gives a way to rank different allocations that depend only on the individual preferences and is an increasing function of each individual's utility.

Economic decisions are inextricably tied to one of several alternative value structures. These alternative value structures can be summarized philosophically as falling within three historical streams that represent alternative definitions of social justice. These alternative perspectives have corresponding economic decision criteria that act to guide the translation of philosophy into public policy. In a tacit manner, these alternative economic philosophies act to provide criteria reflective of a decision's economic justness. The titles people attach to these alternative economic philosophies vary widely. For our purposes, we will summarize them as within three alternative ways of thinking: (1) utilitarianism, (2) libertarianism, and (3) contractarianism. For these philosophies, social welfare functions can be visually compared and contrasted as shown in Figure 12.1.

Utilitarianism

Utilitarianism, a classical economic philosophy, relies on early concepts as described by Jeremy Bentham in England during the 1700s.[6] The classical utilitarian or Benthamite welfare function is a simple sum of the individual utility functions:

$$W(u_1, \ldots, u_n) = \sum_{i=1 \ldots n} u_i \qquad (12.1)$$

Or a slightly more generalized form, the weighted-sum-of-utilities social welfare function:

$$W(u_1, \ldots, u_n) = \sum_{i=1 \ldots n} \alpha_i u_i \qquad (12.2)$$

Here the weights $\alpha_1, \ldots, \alpha_n$ are weights indicating how important each individual's utility is to the overall social welfare. The Benthamian concept centers on the rightness or wrongness of actions that are determined by the goodness or badness of their consequences for

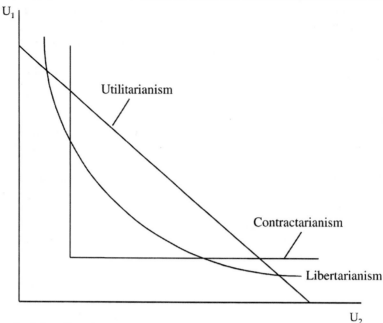

Figure 12.1.   Social welfare functions.

society as a whole. Utilitarians consider maximization of social utility as the basic criterion of morality. In essence, the phrase "greatest good for the greatest number in the long run" represents the basic decision-making criteria as espoused by the utilitarianist perspective. In utilitarianism, *social utility* is defined as the sum or average of the utility of all individuals in society. It rests on the concept forwarded initially by John Locke that all individuals be counted equally. Each person in society receives an equal weight and voices his (more recently, her) presence as a vote. At the local level, the notion of social capital is intimately linked to utilitarianism. It is important to note that utilitarianism often does not coincide with optimal decisions based on Pareto criterion because of differing levels of initial endowment. With utilitarianism, the one man, one vote criteria explicitly assumes a more equal initial endowment.

Libertarianism

Libertarianism, an alternative classical liberal philosophy, advocates an individual's entitlement to freedom from the interference from others.[7] With respect to justness of an economic decision, libertarian thinking believes that holding, using, and transferring resource endowments is just if it is beneficial

to an individual and if it doesn't impinge on the liberty of others. Thus, libertarians hold to Pareto optimality as a key decision rule. Namely, only programs that satisfy Pareto criterion as defined in Chapter 1 are acceptable. Recall that the Pareto criteria involve decisions whereby at least one person gains while making no one worse off. Libertarians hold fast to competitive free markets as "morally free" zones. Indeed, libertarians believe in a limited role of government intervention in freely operating competitive markets. The primary role of governments, from a libertarian perspective, is to ensure the institution of democratic decisions about rules of the game and, once instituted, full enforcement of these rules (Chapter 11). In practice at the local level, libertarians believe that the only public goods and services local governments should be providing are transportation services and police and fire protection. Government has little if any role in providing park and recreational opportunities, for example.

Contractarianism

Contractarianism, an extension of utilitarianism, rests on the assumption that individuals are willing to accept constraints on their actions because they agree that if everyone accepts, all will benefit.[8] It differs from utilitarianism in that the greatest good for the greatest number is often unjust because it comes at the expense of the weakest. This contemporary philosophy is often identified as Rawlsian, named after John Rawls, a Harvard philosopher who wrote extensively on liberalism. Rawls forwarded a philosophy of social justice based on two principles: (1) Individuals have a right to the most extensive system of basic liberties, and inequalities are based on differences in initial endowments. (2) Inequalities are prioritized to provide the greatest benefit to the least advantaged. As a decision rule, contractarianists rely on the *maxi-min* criterion, which holds that decisions should attempt to maximize the welfare of the minimum group in society (Fig. 12.1). This rule can be formalized as

$$W(u_1, \ldots, u_n) = \min\{u_1, \ldots, u_n\} \quad (12.3)$$

At the local level, this philosophy emphasizes policies that address people living in poverty first and foremost.

Arrow's Impossibility Theorem

While in theory the ability to design a social welfare function to assess the alternative policy options is appealing, can it be put into practice? The classic work that addresses this question, and on which most of the literature is built, is by Arrow (1951). Arrow set down certain properties or value judgments that a social welfare function should possess:

- *Pareto principle:* If all individuals prefer one allocation to another, society should do the same.
- *Unrestricted domain:* The social welfare function should allow all possible ordering of individual preferences. In other words, the same welfare function should apply regardless of the particular preference orderings of individuals.
- *Independence of irrelevant alternatives:* The social ordering of any two allocations should not depend on the ordering of any other two allocations. That is, the social welfare function reflects all possible pair-wise comparisons.

Arrow found that, indeed, a social welfare function could be constructed that satisfied these minimal characteristics, but the only admissible one would be unattractive, specifically a dictatorship, where all social rankings are the ranking of one individual. Perhaps the most surprising part of *Arrow's impossibility theorem* is that three plausible and desirable features of a social decision mechanism

are inconsistent with democracy. The implication here is that there is no perfect way to make social decisions. There is no perfect way to aggregate individual preferences to form a social welfare function.

For the community economic development practitioner the implication of the impossibility theorem is that our ability to measure social welfare is far short of our ability to theorize and think about social welfare.

In the end, these philosophies are just that, ways to think about and approach economic development. In the following section we discuss practical approaches to setting community economic development policy. Each of the approaches discussed can be viewed within the framework of the above economic social welfare philosophies.

MOVING FROM SOCIAL UTILITY TO POLICY ACTION

The full realization of local development opportunities requires consideration of a wide range of local and external resources that may be brought to bear on community concerns. These resources include time, effort, and knowledge as well as all forms of physical and financial capital. Finally, no amount of decision-making is of value unless it is accompanied by action. The willingness of a broad spectrum of local citizens to take risks and commit resources is essential. Community economic development is not a one-time event. It is only achievable through continuing analyses, decisions, and actions over an extended period. Limitations of time and money permit only a few concerns to be addressed at once. Furthermore, conditions change and prior decisions must be modified. If local leaders are to make progress in influencing the rate of development in their community and to assure continuing economic vitality through effective decision-making, all of these elements must be in place.

Jeep (1993) contended the question hinges on a broader issue: What is the public purpose of development? Contrary to the view of utilitarianists, community economic development has to be more than aggregation of individual private returns. Jeep suggested four traps we must be alert to in local decisions about community economic development:

1. Failure to develop a vision that is shared with the larger community and failure to achieve consensus on core values connected with that vision (see Chapter 13). He noted that community economic development is a continuous

process and does not have an end point. It requires work and cultivation. It entails imagining a future that transforms people and relationships, specifically moving from being controlled by external forces—victimhood—to community self-determination—freedom. It means gaining confidence to re-invent the community's economic base not on fantasy and wishful thinking, but on real possibilities.

2. *Misunderstanding by community leadership of what form of leadership is required.* The leadership required is not control, nor top-down, but is best understood and defined as a group learning process in which no single person has the answer but all have something to contribute. It is commitment and enrollment in group learning. The participants do not cling to conflicting positions or try to finesse or overpower each other. Conflict is not avoided but is used in a constructive manner as a source of energy to accomplish change (transformation) in people and relationships.

3. *Failure to appreciate the process, not events, as an important part of the substance.* It is the temptation to see situations as series of events rather than the expression of a whole dynamic human system (i.e., responding to individual events, not to the system as a whole). It is a failure not to respond to underlying factors driving events and not to see the human system with both public and private values, implications, and responsibilities.

4. *Failure to go to the roots of a problem and address it.* This appears as an unwillingness to address problems at their roots and to shift focus to symptoms, which can be more safely addressed.

Mier and Giloth (1993) used the term *leaders* rather than *decision-making*. They asserted that community leaders need to build common agendas from among diverse sectors, organizations, and populations. This leadership need not be associated with authority but is derived from collaborative efforts to create solutions to difficult and controversial problems. These leaders must be able to bring diverse interests to the table to engage in forming cooperative partnerships to create joint solutions to complex problems. This involves giving an active role and voice to the previously disenfranchised. Mier and Giloth claimed this new civic leadership requires

more than traditional leadership skills of vision, strategy, motivation, and communication to include building bridges to diverse groups. How widespread, representative, and interconnected community leadership is affects community's capacity to engage in collaborative problem solving. These notions are just the exact opposite of the good ole' boy network that frequently dominates community decisions.

AN OBJECTIVE APPROACH TO EXAMINING POLICY

During the past 50 years, economists and policy analysts have spent a great deal of effort in refining the process of policy formulation from what could best be described as ad-hoc and shoot-from-the-hip approaches. Much of our understanding of contemporary economic policy modeling traces its roots back to Jan Tinbergen.[9] According to Tinbergen (1967), a policy model needs to fully describe the context in which the community sets its economic development policy.

This structure of analysis can be likewise formulated as a deterministic model. In essence, the Tinbergen model can be represented by

$$W = f(O, P, X, I, Z, S, L, T) \qquad (12.4)$$

The focus of equation (12.4) is the explanation of what constitutes improvement in attaining some predetermined measure of social welfare (W). Change in W results from implementation of policy. Typically, W represents the social and economic welfare of community citizens, and it is based on values presented by alternative economic philosophies. As we have seen above, W cannot be quantified in some numerical manner as stated by Arrow's impossibility theorem. Here W represents some outcome of the community economic development process rather than a specific output, such as reducing poverty rates by some reasonable amount.

The policy levers (P), sometimes referred to as instrument variables, represent controllable independent policy tools the community uses to guide and encourage economic development activities. Examples include retraining programs or downtown beautification programs. The equation suggests that the relationship between community welfare (W) and policy choices (P) is well understood.

Exogenous forces (X) beyond the control of the community exert a definite influence on the community. Exogenous forces could be interest rates,

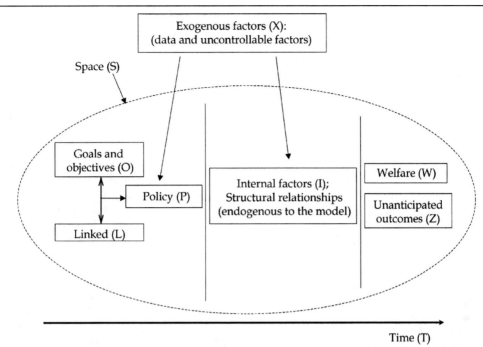

Figure 12.2.   A variant of the Tinbergen approach that provides a foundation for objective and rational community economic policy modeling.

consumer prices, changes in consumer tastes, or the price of energy. Internal variables (*I*) describe the community and its economy through a set of structural relationships. These might include such measures as local employment, income sources, and types of businesses.

Inadvertent and unanticipated side effects (*Z*) of the selection of a policy/program (*P*) are the externalities of the community's economic development policies. For example, the community is successful at promoting a computer software development industry, and the local high school may begin to emphasize computer skills in its curriculum. An unintended outcome is that these skills make local high school graduates attractive to businesses in other locales.

The all-important notion of space in Figure 1.1 is captured by the variable *S*. Within the Tinbergen framework, the effects (benefits and costs) of policy are likely to spread across the spatial landscape. Objectives and goals (*O*), while often expressed independently of each other, are implicitly linked to themselves (conflicting and complementary) as well as inadvertent and unanticipated side effects (*Z*). These relationships are expressed as *L*. *T* captures

time. It will take time to recognize the need for policy, implement it, and have an effect.

In equation 12.4, f(•) represents the structure of the community economy the policy-maker is trying to influence. The function f(•) demonstrates how the endogenous, exogenous, and instrumental variables interact with each other; f(•) is really a restatement of the community economic development theories reviewed elsewhere in this book.

The function f(•) in equation 12.4 can also be thought of visually (Fig. 12.2). Viewing the figure from left to right, the vertical lines in the figure distinguish the policy model from its required inputs and resulting outcomes. The local economy is described by exogenous factors (*X*) representative of data and/or uncontrollable elements. These can include macroeconomic conditions and the empirical basis of secondary data elements. Policy instruments are directed by goals and objectives of community residents as voiced by policy-makers.

The community sets its own goals and objectives (*O*) and some of these are linked (*L*) to other community goals and objectives. For example, the desire to increase employment opportunities for women may be linked to increases in the availability and affordability of quality child care. Then the community is in a position to try to manipulate whatever policy levers (*P*) are within their realm of influence. Policy levers (or instruments) must be specified for analysis; they reflect tangible economic characteristics that are imbedded within the set of structural relationships found in the model (identified in Fig. 12.2 as internal factors, *I*). An example of this distinction is the relationship between the export and nonexport bases in input-output analysis. What drives the internal factors in input-output analysis are demand changes. These demand changes specified outside the modeling framework by the analyst reflect the change brought about by the policy instrument under examination. The structural relationships embedded within the model then estimate changes in community welfare (*W*) and some unanticipated outcomes (*Z*). A possible unanticipated outcome in our example above is that unskilled women may now be hired to provide child care, but all unanticipated outcomes may not have desirable characteristics. This all occurs within a given geographic area (*S*) over time (*T*).

It is important to note that there is a logical progression from understanding how the community is structured to community outcomes or welfare. One of the issues that is not explicit in Figure 12.2 is the feedback loop from welfare (*W*) back to internal factors (*I*). As we will see in later discussions, the community economic development process is continuous and ever evolving.

The outcome of community economic development policy is community well-being or social welfare. Well-being includes the effect on individuals and among individuals as well as the monetary and nonmonetary effects. While the conceptual or theoretical aspects of community well-being and social welfare are known, the measures used are only partial and often are indirect (Reese and Fasenfest 1997).

ADDRESSING COMMUNITY NEEDS: THE RELEVANCE OF PUBLIC POLICY

What qualities need to be in place in a community before it may even consider thinking about community economic development (Shaffer 1990)?

1. *A slight level of dissatisfaction in the current community situation.* This means that the community is not complacent about its circumstances but sees that it can be improved. At the same time, the community does not view its situation as being hopeless.

2. *A willingness to experiment in the community.* By this we mean that they are willing to try something new, at the same time remembering what has worked for them in the past. Thus, they are attempting to blend new ideas (experiment) with existing ideas.

3. *A high level of discussion* among all the parties in *the community* about its future, its processes, and its strategies about what the community desires.

4. *A history of implementation in the community.* By this we mean the community does not just talk about doing something, it has a history of *doing* something. It's important that communities think in terms of what they are going to do in an actual policy/strategy implementation.

Any time a community attempts to alter its economic situation, it is setting public policy. Before the public or private sector embarks on any community economic development policy, however, it should ask these basic questions:

1. *Will intervention help?* If intervention will not help, for example when external factors are so dominant they overwhelm any local policy choice, then another avenue needs to be pursued.

2. *How will intervention help?* By this we mean how the community will be changed and what policies or internal factors are causing that change.

3. *What are the costs of intervention?* The costs can be monetized resource costs or nonmonetized costs in terms of community values and what might be any unanticipated consequences.

4. *When will the effects of the intervention actually occur?*

These questions should always be posed before a community embarks upon any form of collective action to alter its economic situation.

IDENTIFYING COMMUNITY STRATEGIES, GOALS, AND OBJECTIVES

A community consciously attempting to alter its economic situation can pursue a comprehensive strategy, or it can simply implement a collection of programs that may or may not be cumulative or even effective in achieving stated objectives. A collection of programs does not make a policy or strategy. Vaughan and Bearse (1981, p. 308) suggest strategic choices bear a positive integral and consistent relationship to the attainment of some objective. The set of actions must possess synergy: that is, have positive, cumulative, and mutually reinforcing affects on the attainment of objectives. Thus, if two actions are both chosen as part of a strategic set, each will reinforce, or at least not diminish, the effects of the other. Timing and phasing of actions are also important. Strategy is not merely the choice of actions but the ordering of each action to best achieve one's aim.

Any strategy must be based on a theory or model about how the entity functions and will respond to stimulus, such as the framework that the Tinbergen model provides. A theory of community economic development, multitudes of which are discussed throughout the book, must be in place to link internal and external forces before one can think about policy choices or specific strategies.[10]

Vaughan and Bearse (1981, p. 308) define an economic development strategy as

the set of programs which individual design and interrelationship correspond to an appropriate dynamic model of the economic development process. The choice of development programs must be consistent with how the economy actually develops, otherwise the programs may fail. ... There are important judgments to be made, not only with respect to the choice of objectives and instruments but with respect to the relative weights and timing to be assigned to those elements. All elements are not equally important, and their relative importance changes over time.

Using similar words, Bryson (1990) saw strategy as a pattern of purposes, policies, programs, actions, decisions, or resource allocations that define what an organization (community) is, what it does, and why it does it. A pattern means that there is some consistency and it is not just a one-shot event and that it has an action component. Recall that community economic development is a continuous process. Identifying the who and what of community economic development essentially says that strategies define the organization or community. Why a community does what it does is really a statement about the values the community or organization possesses.

There are numerous reasons for a community to create an *economic development strategy* (Bramley, Steward, and Underwood 1979; Townroe 1979). Perhaps the most obvious reason is to clarify the direction of economic development in the community (Chapter 13). A strategy statement includes the community's goals and objectives, the role of local and nonlocal actors/agencies, explicit recognition of the trade-offs among community goals, and sources and uses of political and technical/ professional input. Another

reason is that a community economic development strategy helps delineate priorities, such as identifying key issues and target groups. A strategy for community economic development assists in the allocation of resources by eliminating conflicts and linking or sequencing apparently unrelated efforts (e.g., female unemployment and day care centers), and it should increase the effectiveness of the community's effort. A local economic development strategy yields an organizational structure and response system that ensures knowledge of who is doing what and when, and it links the community with nonlocal public and private resources. Another reason is that a community economic development strategy provides a framework to guide the small, incremental events that accumulate into a major event or effort. Thus, a strategy links separate efforts within the context of a conceptual understanding of community economic development, and permits the community to organize actions and resources to achieve objectives more efficiently. Finally, developing strategies can help a community separate symptoms from the true underlying problems.

Often people are more interested in short-term projects than in long-term strategies. Both are important. The long-term strategies provide people the overarching direction for the community. The short-term projects provide tangible feedback that local people need to stick with the long-term strategy. Without a long-term strategy within which short-term projects fit, the projects might conflict with each other or not add up to effective development progress.[11] One could think of the long-term strategy as the envelope that collects the short-term projects. For example, a long-term goal or outcome of the community might be downtown revitalization. This goal is achieved through a collection of short-term projects that all fall within a longer-term strategy. One strategy might be to enhance downtown business profitability; short-term projects might include forming a business association and a business improvement district, using tax incremental financing (TIF) districts for capital improvement and/or focused business counseling and training opportunities.

A critical step in formulating strategy is drawing the distinction between inputs, goals/objectives, outputs, and outcomes (Reese and Fasenfest 1997). In the former example, inputs are the time and efforts of local citizens in undertaking specific projects. The goals and objectives are linked to specific projects, such as the creation of a downtown improvement authority, more retail shops, or brochure production, and the output is enhanced business profitability. The outcome of all this effort is a revitalized downtown or, in the extreme, the downtown becomes the heart of the community, including but not limited to business activity.

It is important that the strategy focus on the underlying root causes of the problem the community wishes to address. What is the strategic issue and why is it a concern? Generally this means that the issue is the root cause of several symptoms and other problems. A *strategic issue* is one that is fundamental and opens possible solutions to many other issues. The consequences of not dealing with the issue are serious. What are the barriers to solving the strategic issue? What are some of the major ways to resolve the issue? The strategic issue may be that the community has a continuing problem with low income. The causes could be that existing firms have low pay scales or that people in the community have minimal skills. One strategy might be that the workers need to improve their stock of human capital. Once the community has decided what the strategic issue is, then two questions emerge: What are major ways to resolve the issue? What are the barriers to resolving the issue?

Identifying strategic issues in communities often boils down to a relatively simple iterative process. When someone proposes a solution to a local problem, they need to be asked a simple question: Why do you want to do that? Every time a response is given, the question is asked again and then again until the fundamental issue is revealed. For example, if the proposal of adopting an aggressive tax abatement policy in our town is forwarded, the promoter needs to be asked the question "Why?" The response is likely to be "so we can attract new manufacturing firms in our town." "But why?" The response is likely to be "because we need more high-paying jobs than are currently available." Now the essence of a strategy begins to take shape. If higher-paying jobs are the ultimate goal, there are several ways to proceed. Some possibilities:

- Helping existing firms become more efficient so they can pay higher wages.
- Helping existing firms improve their marketing so they can pay higher wages.
- Helping labor learn new skills and become more productive so local firms can afford to pay higher wages.
- Improving the quantity and quality of the local labor force to attract new firms.

The point of this discussion is that the strategic issue involves improving the number of high-quality jobs. It may, or may not, have any relation to the initial proposal. Once the underlying strategic issue is recognized, then a variety of specific means for getting there can be explored. It does not bode well for a community to immediately rush to the conclusion that they need to attract manufacturing by subsidizing their relocation. There are many ways that a community can improve or change its economic circumstances beyond the first-blush response.

Any community economic development strategy begins with a discussion of community economic development goals (outcomes) and objectives (output) (Reese and Fasenfest 1997). *Goals* are general qualitative statements that suggest some attempted achievement that is largely immeasurable. Community economic development goals are often intangible and abstract, and they definitely are value laden.[12] *Objectives* are more specific statements that typically can be quantified and measured, such as downtown business profitability. Goals and objectives are intimately related. Objectives are derived from goals and are indicators of goal achievement. Goals and objectives represent the juncture of facts, values, and hierarchy of importance.

What is the goal of community economic development? There are a variety of goals for community economic development to pursue. One that we prefer is that *community economic development is about producing wealth and facilitating change/adjustment.* It is important that we recognize that we include both monetized wealth (income) and nonmonetized wealth. Nonmonetized wealth includes increasing choices or opportunities for individuals. It also includes the community's appreciation to amenities such as the view of a beautiful sunset. In the broadest sense, nonmonetized wealth includes many qualitative aspects of the quality of life. Facilitating change/adjustment means that it is a dynamic process in which new resources, new markets, and new rules are appearing both inside and outside the community. The change/adjustment means that the community is able to absorb those new factors through increased economic resilience.

Goals and objectives should possess the following qualities:

- They should be consistent with respect to expectations.
- They should be comprehensive and include *all* major dimensions of the problem(s).

- They should be precise to assure effective action and response.
- They should be internally consistent and not redundant.
- They should recognize resource constraints relative to any action program to be initiated.

For example, the community has a problem with youth getting into trouble. This could be a symptom of youth not having things to occupy their time. One potential cause of this problem is the lack of employment opportunities for younger persons. Our expectation might be to create jobs for youth. Our objective in this example is to create 150 jobs for people between the ages of 16 and 21. We can achieve this through a comprehensive approach by incorporating part-time after-school employment opportunities with focused training. In addition, the community needs to encourage businesses to tap into this unused pool of resources. These attempts to reduce youth unemployment should not emphasize adult job creation. The community may wish to use federal and/or state grants to support their efforts so that limited school resources are not further stretched. Or, the community could examine its high school curriculum to see if it is really addressing the vocational needs of these youths. To balance out the community's strategy might be to encourage more extracurricular activities within the school system.

These seemingly obvious qualities are often discounted because "everyone knows what we are all after and there is no real need to waste valuable time, energy, and enthusiasm belaboring the obvious." Communities and groups skipping this important step, however, frequently reduce their potential accomplishment because efforts are disjointed and nonsynergistic.

Different Perspectives of Community Economic Development Policy Choices

Implicit in the discussion of community economic development goals and objectives is the belief that once the goals and objectives are identified and the relationships (links) among goals are understood, there will be agreement on how to achieve those goals and objectives.[13] The wide range of policy choices available to achieve the goals a community sets for itself almost assure that unanimity on means will not occur. Earlier in this chapter, we presented three conceptual philosophies from an economist's perspective: libertarianism, utilitarianism, and contractarianism. Each of these three

philosophies can provide insights into each of the following perspectives. There are no right or wrong ways to answer the questions raised by the different philosophies.

These philosophies could be described as *macro* in perspective, worrying about the big picture. The practice of community economic development, however, requires us to worry about *micro* details. While keeping the forest in perspective, we need to worry about the details of the trees. For example, should community economic development policy be people driven or place driven? Should it be focused on economic efficiency or equity? What is the appropriate time frame for goals to be achieved? Is this an issue that should be in the public domain? Decisions about community economic development policies depend on value judgments of the policymaker and the individuals involved (Mier and McGary 1993; Reese and Fasenfest 1997; Richardson 1969a; Richardson 1978b). The assumption made in the discussion that follows is that economic development policies represent local initiatives rather than externally imposed choices.

The question of place versus people programs is more a matter of degree than a clear distinction (Bolton 1992). *Place programs* attempt to alter the geography of the economy in favor of some economically depressed area. Examples are water and sewer systems, and industrial parks. Place programs or place prosperity is really a means to an end—people prosperity. *People programs* attempt to alter the economic circumstances of particular categories of people, such as the disabled, unemployed, and elderly. Examples of people programs include sheltered employment, training, dissemination of job information, and Medicaid. Place programs, which usually offer a quick and visible payoff, are often chosen over people programs because the immediate value exceeds the present value of future returns from a people program. Place programs are clumsy tools to redistribute economic activity to population subgroups.

Ladd (1994) talked about three basic policy approaches for dealing with pockets of distress. The first, a pure people-oriented strategy, focuses on helping people, with little or no attention paid to revitalizing the area in which they live. The second, a place-based people strategy, uses a variety of place-specific strategies to increase the economic well-being of people living in a distressed area. The third, a pure place-based strategy, focuses on improving the physical and economic vitality of a geographically defined area without explicit attention to the people who live there.

A *pure people-oriented economic development strategy* would assist people regardless of where they live and would focus on increasing their human capital and mobility. A more direct approach to dealing with pockets of distress involves using people-specific assistance to help residents, especially disadvantaged residents, of distressed areas. Central to this approach is the view that the community plays an important role in its residents' well-being. Thus, to help people, one must help build or revitalize their communities.

Thus, a *place-based people strategy* starts from the view that in a very meaningful sense people cannot be separated from place, and that antipoverty strategies need to treat individuals in the context of their communities. A place-based people strategy aims to preserve and strengthen community institutions and ultimately to generate more jobs and a higher income and a higher standard of living for residents.

Pure place-based strategies involve either improvements to the physical landscape of the area or its economic revitalization, defined as new investment and new jobs within the area. Implicit in this strategy is the view that pockets of blight are detrimental to the economic viability or vitality of a larger jurisdiction. Externalities are sometimes used to justify geographically targeted governmental intervention.

Another philosophy affecting the choice of community economic development policy is the question of equity and efficiency (Hill 1998; Ledebur 1977; Mier and McGary 1993; Richardson 1969a; Stöhr and Todtling 1977). Creating the greatest return from the resources used characterizes a policy priority of *economic efficiency*. Limited resources available to support community economic development efforts, at both the local and national levels, require maximum efficiency. A public policy focusing on *economic equity* emphasizes the distribution of rewards and burdens of economic development among individuals and groups inside and outside the community. Consistent with the philosophy of contractarianism, this type of policy emphasizes the need to consciously include those segments of the community excluded from the mainstream of the economy. An equity program might also compensate groups who bear a disproportionate share of the burden of change.

It is important to remember that equity is not equality. *Equality* means dividing economic output evenly among all people in the community. *Equity*, as we saw in our discussion of utilitarianism, means distributing access to opportunity or avoidance of arbitrary and external constraints. Equity implies access to opportunity with some recognition of different attributes. People are treated in similar fashion regardless of the attribute chosen. An equity policy objective requires giving priority to areas of greatest distress (i.e., worst-first policy) to reduce their economic disparities from the norm (Ledebur 1977). Equity requires that the community give priority attention to the goals of promoting a wider range of choices for those residents who have few, if any, choices (Mier and McGary 1993). Compare the above equity objective with an efficiency objective, where the priority would be given to areas with greatest potential for high productivity. The policy dilemma is that areas with potentially high productivity typically are not the areas of greatest distress.

Under what conditions might we argue that local economic development efforts create efficiency or equity gains? The following conditions represent the most likely situations in which *efficiency gains* occur. It could be immobility of resources or a reduction in risk, or it could be overcoming externalities/spillovers or overcoming inadequate information flows. *Equity gains* probably occur when the aid is targeted to specific groups and when there are unequal welfare weights for people and places.

Some believe equity and efficiency are the major economic development trade-offs confronting communities and policy-makers (see A Zero Sum Approach below). This may not, however, be the major trade-off usually perceived (Vaughan and Bearse 1981). The major trade-off occurs when equity, defined as a static redistribution of economic welfare, conflicts with activities to improve efficiency. Development is a dynamic process concerned with creating new products, mobilizing new resources, improving the quality of existing resources, and altering structural and institutional arrangements that impede the effective utilization of resources. Thus, the pursuit of dynamic efficiency also may yield the equity results desired. The need for trade-offs between goals and objectives may remain. Short-run decisions can create some efficiency/equity trade-off problems, and more importantly, may conflict with the long-run dynamic efficiency concept. These can be thought of as unanticipated outcomes in our Tinbergen approach.

An often overlooked dimension of making community economic development policy decision-making is time. The politics of the problem place a premium on solutions achieved quickly—by the next election in two to four years (Eisinger 1995). One could think of time in Figure 12.2 as the election cycle. Development, however, is a long-run concept and the internal factors defining the well-being of a community are independent of the election cycle. The lack of development reflects a long-term, cumulative process. Short-run attempts to reduce distress, if viewed as single-shot, turn-the-corner attempts, are doomed to failure. Alleviating the causes of distress rather than the symptoms requires a long-run transformation of the economy.

Time also affects policy choices among alternative forms of economic development. The choice is seldom recognized because the development opportunities (choices) often do not occur at the same time. For example, a community may have an opportunity to help Industry A or Industry B but does not have the resources to help both. The chance to choose either A or B seldom occurs simultaneously, so the coincidence of time precludes some choices. Another set of values guiding the selection of policy is the degree of public or private sector involvement. A libertarian may argue that if the public sector becomes involved in the private sector or market, it is stealing from another location or sector. If the markets yield a Pareto efficient allocation of resources, then public sector involvement moving the economy away from that efficient allocation comes at the cost to at least one person (firm, locale). The libertarian-based policy choice is that the private sector will be responsible for achieving the community's employment goals and that there is no role for the public sector. A less severe position might be a contractarian- or utilitarian-based policy choice that would allow for the community to generate incentives to encourage private decision-makers to alter their decisions and actions to achieve community goals. Alternatively, there could be heavy public involvement in the decisions affecting employment in the community, which could take the form of a full partnership between the public and private sectors or a completely public sector project.

A final value influencing policy choice is that the community could take no actions by either public or private actors to achieve the desired goal. This policy choice is always available to a community, yet frequently it is not explicitly considered.

Beyond these values, several factors restrain the selection of policy. First, some policies affecting local economic development are beyond the control of a community. These would be the external factors in Figure 12.2, such as interest rates or federal tax policy. Political feasibility limits policy selection. Some activities simply may not be politically or socially viable for a given community (e.g., the operation of a casino). The chosen policies may be disconnected from the target population. If a community desires to increase youth employment opportunities, it could choose to increase manufacturing jobs, anticipating that some new employment opportunities will go to youth. A more "connected" policy option could be that the community encourages specific types of economic development that directly affect youth employment, such as fast-food restaurants. Finally, as we discussed with linked policies in Figure 12.2, other goals pursued affect the selection of policy. If a policy increases employment opportunities while adversely affecting air quality, an unanticipated outcome, that policy may need to be revisited.

STRATEGIES FOR COMMUNITY ECONOMIC DEVELOPMENT

There are numerous strategies a community can pursue to accomplish its goals concerning the economic future. The standard strategy is to attract a manufacturing plant with the associated secondary development.[14] A second common approach emphasizes the expansion of existing businesses. These are valid strategies, but a community that limits its economic development efforts to just the above needlessly restricts its opportunities.

The Shaffer Approach

Shaffer (1990, 2002) offers seven ways that an economist would look at the economic choices that affect the economic viability of an area.[15] The best way to look at these would be to turn to the logic of the star diagram (Fig. 1.1).

Referring to the *market* node of the star, two broad approaches become readily evident. The first is to increase the flow of dollars into the community. The second is to increase the recirculation of dollars within the community.

If you look at the *resources* node, there are really two approaches. The first is to increase the amount of resources available. The second is to use resources differently, including new uses.

When examining *rules*, the idea is to change or reinterpret the rules to an advantage for your community. In terms of *decision-making,* essentially all one needs to do is act smarter. And finally, just get lucky.

The *society* node of the star suggests that the community could encourage a stronger entrepreneurial spirit within the community, both in terms of businesses and of attitudes toward policy experimentation. Now let's turn to these in more detail and outline some examples of specific actions the community can undertake.

Increasing the Flow of Dollars into the Community

Increasing the flow of dollars means that the community is essentially injecting new money into the local economy by attracting new basic employers, by existing basic employers increasing their sales outside the community, by the community increasing its visitors, or by the community increasing its intergovernmental aids. In each case the community is bringing more money into the community. But this broad approach is only part of the answer. These are examples of specific activities a community can undertake to increase the inflow of dollars:

1. Develop local industrial sites, public services, and potential employee information.
2. Develop community and regional facilities necessary to attract new employers in these areas:
   a. Transportation (e.g., airports, railways, highways)
   b. Recreational facilities (e.g., parks, hunting grounds, restaurants, hotels, convention centers)
   c. Communications (e.g., newspaper, telephone)
   d. Business services (e.g., banking, computers, legal assistance, accounting)
3. Expand purchases by nonlocal people (e.g., tourists, neighboring citizens) through appropriate advertising and promotions.
4. Ensure that key public services (e.g., fire and police, water and sewer, general administration) are more than satisfactory.
5. Recognize the important role of transfers, such as retirement benefits, and unemployment compensation as a flow of funds into the community.

Increasing the Recirculation of Dollars in the Community

People and businesses must have some place to spend the money. They spend it either locally or elsewhere. Too often, communities think all they have to do is attract a basic employer, such as a manufacturing plant, and all the economic woes of the community are solved. Increasing the recirculation of dollars means that the community is plugging leakages of money out of the local community's economy. In other words, the community is actively seeking ways to get people and businesses to spend more locally. It could be altering store hours, encouraging new or different store types, physically renovating downtowns, or even talking to businesses about buying inputs locally rather than from nonlocal sources. The end result is that dollars recirculate at least one more time locally. Additonal things the community could do:

1. Identify market potential of retail outlets through surveys of consumer needs and buying habits.
2. Improve share of retail market captured through downtown analysis and renewal through
   a. using consumer and merchant surveys.
   b. providing convenient parking or public transit.
   c. reviewing store hours and merchandising.
3. Aid businesses in developing employee-training programs to improve quality of service.
4. Encourage local citizens and businesses to buy locally by providing information programs.
5. Encourage collective action through the formation of organizations such as Chamber of Commerce or Merchants Association.

Increasing the Amount of Resources Available

Increasing available resources simply means that the community has increased the amount of labor and capital, including both public and social capital, available for producing output. This could be local financial institutions making more loans available locally, or an outside business making a local investment, forming a credit union, or establishing a local branch of a community college. It could be people moving or commuting into the community or working more hours. This can be done by

1. Organizing community capital resources to assist new business formation or to assist in attracting new business.
   a. Encourage investment of private funds locally through formation of capital groups.

b. Encourage the use of secondary capital markets and public financing programs.

c. Encourage the use of industrial revenue bonding, bank loans.

2. Organizing training programs for youth, in-migrants, and resident population.
3. Encouraging population in-migration.
4. Providing the same services to start-up businesses as provided to businesses being sought from outside the community.
5. Creating an encouraging community attitude toward entrepreneurship.

Using Existing Resources Differently

Using existing resources differently generally means that you have applied new technologies. You have found new ways to combine existing capital and labor to produce greater output per worker. It could also mean that you have used existing capital and labor to produce a new good or service that previously had not been produced locally. It could also mean that you now have local jobs for workers who previously commuted out for work. Or it could mean that workers have received training and now are able to do different tasks than before. Examples of specific community actions follow:

1. Strengthening management capacities of existing firms through educational programs (e.g., personnel, finance, organization).
2. Encouraging business growth through identification of equity and loan capital sources.
3. Developing training programs for workers using new and different technology.
4. Increasing knowledge of new technology through educational programs in science and engineering.
5. Aiding employers in improving workforce quality through educational programs, employment counseling, and social services (e.g., day care, health services).
6. Developing community and regional facilities that improve local business efficiency and access to nonlocal markets (e.g., transportation, services, communications).

Changing the Rules

Changing the rules means that the community seeks a change in rules that would benefit the community or seeks a change in interpretation of rules. For example, we all know the rules to a typical Scrabble game. They are that you maximize your own score and hide your tiles. Think what the game would be like if you maximized the score on the board and laid your tiles face up and people could play your tiles also. Another example is if the community gets the state to re-interpret eligibility rules on some type of manpower training fund, thus making some community residents eligible. Remember that rules are societal constraints that govern how we either use resources or exploit markets. These activities include

1. Ensuring correct use of public assistance programs for the elderly, handicapped, and others who cannot work.
2. Supporting political activities to ensure fair treatment of community concerns by broader governmental units.
3. Reviewing how retirees and handicapped people might find services, access, housing, volunteer organizations, and community attitudes.
4. Minimizing contradictory regulations and regulatory barriers, including uncertainty.

Acting Smarter

Acting smarter translates into how the community goes about making decisions and sets up and implements strategies. Does it involve a broad spectrum of interests or just a select few? Does the community really get at the problems or just treat symptoms? Does the community integrate sound analysis with community perspectives and desires? This is what it means when a community is acting smarter, almost in an entrepreneurial manner.

1. Identify market potential for new retail, wholesale, and input-providing businesses.
2. Organize to provide individual counsel and intensive education for those interested in forming a new business.
3. Utilize aids from broader government whenever possible (e.g., streets, parks, lake improvements, emergency employment) through active monitoring and support of the activities of local officials.
4. Identify specific public programs, projects, offices, and/or services that could be located in the community and organize politically to secure them.
5. Encourage collective action through formation of organizations such as economic/industrial development corporations.
6. Ensure that quality and access and appropriateness of local school systems, including vocational-technical.

7. Identify through research the most desired type of basic employer with greatest potential.
8. Organize business-networking forums.
9. Sponsor business appreciation events.
10. Create organizations (including high school programs) to stimulate entrepreneurial thinking and action.

Getting Lucky

Getting lucky may seem like an unusual item, but think about it for a second. A small rural community could be located within the commuting shed of a growing metro area or 50 years ago it could have been the birthplace of a budding entrepreneur. While we like to think more than luck is involved, and it is, it also explains a lot of current economic activity. What a community could do:

1. Examine old high school yearbooks for previous graduates who might like to return to the community.
2. Promote locally available natural resources and amenities to outside visitors.
3. Design vacant residential sites for development.

One will notice significant overlap in many of the specific strategies offered. The important thing is to be comprehensive in your approach to community economic development. Also note that these strategies range from the simplistic to the complex. Many communities may find that pursuing a collection of simple projects to get started helps build social capital within the community. Short-term successes with simple projects can help build a foundation for more comprehensive and complex long-term strategies.

COMMUNITY ENTREPRENEURSHIP

As we think about community decision-making, it is appropriate to think in terms of entrepreneurial communities (Flora and Flora 1990, 1993; Johannisson 1990; Johannisson and Nilsson 1989; Malecki 1994; McDowell 1995; Stöhr 1989; Swanson 1996; Turner 1999). *Entrepreneurship* has been defined as the purposeful activity (including integrating sequence of decisions) of an individual or group of individuals to undertake, to initiate, to maintain, or to aggrandize profit-oriented business for production and distribution of economic goods and services (Hoy 1996; Malecki 1994; O'Farrell 1986; Schum-

peter 1961). If we think of a community as a multi-product firm seeking to maximize profits over different markets, then the analogy between a firm and a community becomes much more obvious. The community (firm) must know what resources (land, labor, public and private capital, technology) are to be used across various outputs that are destined for nearby and distant consumers. At the same time, the firm brings together its resources and markets over space and time within the constraints of the rules that society has imposed on its operations.

Community entrepreneurs can essentially be viewed as mobilizers of networks. They can mobilize internal networks within the community as well as external networks. Networks help individuals to know what, how, when, who, and why (Johannisson and Nilsson 1989; Malecki 1994). This includes the ability to tap into the right people to help solve a community problem. These people can be inside or outside of the community itself.

Entrepreneurial communities probably have a greater capacity to adjust to constantly changing economic conditions surrounding them. They learn from their own and other's mistakes (Johannisson 1990), and they rely on successful colleagues/communities as role models. They recognize that there is no cookbook because of the uniqueness of each and every community and situation. They also realize that the community economic development process is the long-term envelope of short-term projects and is in need of constant re-evaluation.

Community entrepreneurs create an acceptance of rebirth (Cavaye 2000b). They help create the visions and convincingly communicate that vision to local and nonlocal networks. These are visions of new comparative advantages. Finally, they integrate community resources, people, and networks. Community entrepreneurs are engaged in managing turbulent organizational change and deciphering complexity.

Several qualities appear to be associated with entrepreneurial communities but four need particular attention (Pryde 1981): (1) The ability to perceive opportunities and devise a strategy to exploit them, (2) the capacity to identify either the availability or shortage of a resource that's needed and to acquire it if necessary, (3) community entrepreneurs who can manage political relationships both within and outside the community, and (4) community entrepreneurs who can manage interpersonal relationships by working with other people and motivating them. Entrepreneurship is the adaptive process of replacing or changing social patterns that have become obsolete. The increasingly turbulent times around us require people who are able to make sense out of the overwhelming complexity.

An entrepreneur can be characterized as an innovator. A community needs to build a cultural environment that encourages entrepreneurial activity. Writers such as Hoy (1996), Malecki (1994), O'Farrell (1986), Vaughan (1985), Thompson (1965) and Schumpeter (1961) recognize the importance of innovative strength and entrepreneurial skills in economic development. The question becomes, how can a community create conditions that foster the attitude that opportunities exist and attempt to generate new jobs, income, and products be successful? Or stated differently, what can a community do to promote entrepreneurship? The community can promote an attitude of immunity and indifference that permits people to experiment. Social institutions in the community are differentiated and there is no single way of doing things (i.e., diversity is common). There is a diffusion of social-political-economic power within the community, such as weak vested interests. Finally, the avenues of social-economic mobility are widely available.

One common misperception of entrepreneurs, either as individuals or as communities, is that they enjoy taking risks. Within the community, they may be incorrectly viewed as promoting radical ideas or policy options. The success of entrepreneurs, however, is not in the love of taking risks but in identifying risks and implementing strategies to minimize those risks. To some, thinking outside the box entails risk, but to successful entrepreneurs, thinking outside the box opens alternatives that may not be otherwise seen.

A ZERO SUM APPROACH

Community economic development programs are often accused of being nothing more than a zero sum exercise because of the apparent shifting of economic activity from one location to another or from one group to another. *Zero sum* is defined as the increase of economic activity for one locality or population subgroup only at the expense of another locale or group. In a general sense, zero sum is a foregone conclusion in a perfectly competitive market, but it need not be in an imperfectly competitive market. The following examines the justification for these conclusions.

If the market is perfectly competitive or behaving in that fashion, public intervention cannot improve the outcomes derived from private decisions. Perfect competition and Adam Smith's *Invisible Hand* yield a Pareto efficient allocation. The result of Pareto efficiency tends to have many people believing that community economic development activities are zero sum. Economic markets yield efficient allocation of resources; attempts to intervene in that allocation must come at some cost. One person's gain is another person's loss. Yet, even if the perfectly competitive market conditions exist, this argument ignores the equity issues and the dynamics of economic development.

The equity perspective recognizes that all workers are not intrinsically the same, not all forms of capital yield equally profitable returns, and not all communities are blessed with the same sets of endowments, such as location or mineral deposits. Society can seek equity goals that offset the loss of efficiency. The argument against zero sum contends all economic actors (e.g., workers, businessmen, communities) are not created equal and public policy is justified in aiding those people, businesses, and places through difficult transitions. A libertarian might support the idea that community economic development is a zero sum proposition. A contractarian or utilitarian, however, would reach a different conclusion.

If the economy is not functioning in a perfectly competitive fashion or there are externalities present, the prospects for a zero sum outcome from community economic development efforts still exist but appear much more unlikely. There are several reasons why a zero sum outcome would not result and most are linked to a failure to meet the fundamental assumptions about a perfectly competitive market. Capital and labor resources in the economy are not perfectly mobile among places and uses. Without the mobility implicit in the perfect market analysis, community actions promote the employment of underused resources. Even if the economy might yield a higher return if resources were used elsewhere, any nonzero return from immobile unemployed resources is a net addition to both the local and national economy.

Information is not perfect through time and over space. Workers and investors do not have all of the information about future trends, and they may even make different choices given the same information. For example, firms may not be trying to maximize profits but rather market share. Many community economic development activities disseminate information on job opportunities, advertise community attributes to nonlocal investors, or announce interest subsidies to attract capital to depressed areas. Activities that improve the flow and quality of information do not yield a zero sum outcome.

The perfectly competitive model assumes constant returns to scale. Yet, many activities exhibit increasing returns to scale. Community economic development activities that permit or create increasing returns to scale reduce the probability of zero sum outcomes.

Externalities are not present in the perfectly competitive model, but in reality they are a significant component of any investment in either physical or human capital. In this situation, even if the direct returns from economic development activities among communities appear equal (indicating zero sum), if the spin-offs in one locale exceed those in another, a zero sum outcome is less likely to result.

The existence of perfect market conditions in all input and output markets is implicit in the perfectly competitive model. Nonoptimal market conditions in one market, however, can yield suboptimal conditions in another market. Community economic development activities that help markets approach a perfectly competitive mode of operation do not yield a zero sum outcome.

Unequal welfare weights (such as in equation 12.2) among groups or places, mean that society has chosen to emphasize these as a particularly important part of society's well-being, and that intervention in favor of those groups and places are more important than potential losses in other groups and places.[16] This would also be the position argued by Rawls (equation 12.3).

While it appears that the idea of market failure becomes the major justification for community economic development activities, that conclusion is insufficient. Market failure is a quasi-static concept and it ignores the dynamic functioning associated with community economic development.

COMPLEX INTERRELATIONSHIPS AND PERSISTENCE

Setting community economic development goals can be a frustrating experience. Goals can be simple or complex, complementary or conflicting. The setting of a community's goals and objectives for economic development is a very critical but difficult component of any development effort. It is important to specify the objectives clearly so that the community can effectively set a policy and a strategy. Without this, the inevitable result is a disjointed

nonstrategy, an eclectic collection of efforts. A community may not be able to satisfy all of the goals and objectives that it sets for itself because of one or more of the following factors: inadequate political support, inadequate resources, conflict with other goals and objectives, the need to conform with higher-level goals and objectives, and the inability to control the means by which the goals can be reached. There may be continuing tension in the community caused by differences of opinion about goals and objectives. Opinions vary about the nature of the problem and its priority, about the capacity of suggested programs to achieve the desired goal, and about the consequences of the actions proposed to correct the perceived problem. Even if all parties agree on these matters, differences of opinion about the proper political, economic, or social solution may continue to cause tension and fragmentation in the community.

Development is an inherently uneven process in terms of its sectoral distribution, its timing, and its geographic distribution. However, disruption (disequilibrium) is a necessity in a dynamic economy and is part of the dynamic process associated with development (Vaughan and Bearse 1981).[17] Schumpeter referred to this dynamic process as *creative destruction*, where old resources become less valuable, new resources become more valuable, and gaps and spillovers of externalities are generated among markets as some institutions, industries, or resources lead and some lag the development process. In reality, the economy does not adjust instantaneously and painlessly to the disruptions caused by development. There are inherent structural barriers, inconsistencies in the market that make this process less than smooth. One goal of development strategy is to facilitate this development adjustment. Remember that *the goal of development policy should not be to avoid disruption and adjustments but to facilitate the adjustment process* (Vaughan and Bearse 1981).

Economic development exists in an interconnected web of community stakeholders, business managers, nonprofit agencies, government departments, and multinational corporations. Managing contemporary local economic development requires the ability to facilitate interaction, to mobilize stakeholders, and to reconcile different goals and values among key development actors. Economic development takes teamwork. We joke that you need both collaborators and unwilling conspirators. While willing collaborators may be easy to find, many people are unwilling to get involved. To engage unwilling conspirators in building a long-term economic development strategy takes skill on your part. It means that often people are not interested in building long-term strategies, they are more interested in worrying about short-term projects.

There are several characteristics in communities that seem to be pursuing successful local economic development strategies versus those that are not. The strategy formers recognize they are the ones who will make it happen. They focus on what can be accomplished within a few months or years without forgetting their long-term objectives. They do not get to the solution before they know the problem. In other words, many "strategies" are nothing more than a collection of programs that treat symptoms rather than community problems. The strategy is specific in outcomes and actions. This means that who, what, where, and how are explicitly recognized. They tend to break actions into smaller parts to recognize the achievement of intervening goals and objectives. They remember that most others know very little about the effort, and they seek to find and cultivate allies. The final two characteristics appear to be contradictory. First, they discipline themselves to stay with their strategy over time. Second, they continually revise as they achieve some intervening goals and objectives.

SUMMARY

Community developers build the capacity of people when they encourage them to create their own dreams to learn new skills and knowledge (Hustedde 2001). Capacity-building occurs when practitioners assist or initiate community reflection on the lessons they have learned through their actions. Community economic development is about building the capacity to understand, to create and act, and to reflect.

Community economic development decision-making capacity is the ability of a community to initiate and sustain activities that promote local economic and social welfare. Why is decision-making capacity so important to community economic analysis? It means that the community thinks holistically or systemically. The community thinks in terms of problems rather than just symptoms. The community is able to create and implement appropriate responses. Community economic development encourages communities to think in the long-run and to understand how they should function. It encourages communities to think strategically. The community is able to reflect and learn from past decisions and it has a broad spectrum of people involved in decisions.

Community economic development argues for a holistic approach to problem solving that notes explicitly interlocked economic and ecological systems that won't change, so the policies and institutions concerned must change.[18] Of prime concern to us is the recognition that development is not a fixed state of harmony; it is a process of change in which exploitation of resources, direction of investments, orientation of technological development, and institutional change are made consistent with future as well as present needs. Development is driven by balanced views of present and future needs. Furthermore, this definition implicitly includes a system that secures effective participation in decision-making, provides for solutions that arise from disharmonious development, and is flexible and has the capacity for self-correction. In other words, we have returned to Figure 1.1.

The overall purpose of community economic development policy is to reduce and/or abolish the barriers in product and factor markets that prevent the positive culmination of economic development processes. Community economic development policy must be viewed in the long-run dynamic context of those markets and the surrounding socio-politico-physical environment. The fundamental justification for community economic development policy is to ameliorate failures in product and factor markets.

Four basic tenets underlie any attempt to set community economic development policy and generate community economic development action programs: (1) A community is a logical economic unit that can exert some control over its economic future. (2) Intervention in the form of conscious group decisions and actions will affect local economic welfare more than the sum of individual actions. (3) The action or policy must be comprehensive and cannot focus just on economic activity but also must include noneconomic dimensions. (4) The resources needed will be available to implement the policy.

Decision-making is important because several qualities associated with decision-making also have implications for community economic development. Does the community think holistically/systemwide? (Do they view decision-making as small incremental types of choices versus systemwide choices?) Does the community think in terms of problems rather than symptoms? Can the community create and implement an appropriate response? Does the community reflect and learn from prior decisions so it can make better decisions in the future? Who makes the decisions for the community?

Communities have a power structure. If the power structure is all that is involved in the decision-making, the end is not likely to be sustainable because the community at large will not embrace the decision. If, on the other hand, the community wants to get something done, they must engage the power structure. Too often the community has many actors with little power trying to bring about change. Community economic development must achieve balance between engaging the diverse interests and the power structure.

STUDY QUESTIONS

1. What are some basic elements of a community economic development strategy?
2. What are the elements of the community economic development policy model and how does each relate to creating community economic development policies?
3. Is it possible to achieve all worthy community economic development goals simultaneously? If not, why not?
4. Compare and contrast the three perspectives that an economist might use to drive the choice of community economic development policies.
5. Why is a perfectly competitive market almost a necessary condition before a zero sum outcome occurs?
6. What do you think is the goal of community economic development and why?
7. How do economic development goals, outputs, and outcomes interact?
8. What basic questions need to be addressed prior to initiating any community economic development policy?
9. Why create a community economic development strategy?
10. Why is it so important to understand the theoretical basis when formulating community economic development policy and strategy?
11. Why is the clear delineation of the cause of community economic change so important in designing community economic policy?
12. How can one justify government involvement /collective involvement in community economic development policy, especially in a decentralized market economy?

13. Why do economists worry about equity/efficiency with respect to government intervention? What is the implication of static/dynamic analysis to this?
14. Why is decision-making capacity so important to community economic analysis?
15. What is social capital and why is it so important to decision-making capacity?
16. Why does a high level of community entrepreneurship imply a greater capacity to adjust to changing economic conditions?
17. Are entrepreneurial communities characterized by deep thoughts or action?

NOTES

1. This essentially assumes that communities are free of outside constraints, specifically local and nonlocal decisions that are compatible. There is clearly a different view that local community decisions are constrained by external decisions such as local and nonlocal decisions are conflicting.
2. If it is not in the community, the community must search for this technical assistance elsewhere. The technical assistance must be tailored to fit local needs and provided by a reliable source as well as readily available, flexible in approach, and available over time.
3. This is very similar to asset-based community development as enumerated by Kretzman and McKnight (1993).
4. "Within" an individual means that a person holds a personal value set that sees relationships as a positive life event. "Between" individuals is the actual relationship itself.
5. This presumes that you can measure utility in a cardinal fashion. We do know that in reality the only measures are ordinal rankings. Most of the following discussion moves forward on the assumption of ordinal rankings.
6. John Locke (1632-1704) was an early English advocate for social justness of decisions based on one man, one vote. Utilitarianism rests on "the greatest good for the greatest number in the long run" as a basis for determining just economic decisions.
7. Milton Friedman (1912-) is a leading proponent of the libertarian approach to economic philosophy. Libertarianism, which is characterized by much of the contemporary economic thought as espoused by the Chicago

School, rests on the notion of Pareto optimality as a key decision rule for just economic decisions.

8. John Rawls (1921-2002) advanced a perspective that advocates a contractarianist economic philosophy in which individuals have a right to the most extensive system of basic liberties and inequalities are based on differences in initial endowments. As a decision rule, contractarianists rely on the maxi-min criterion, which holds that decisions should attempt to maximize the welfare of the minimum group in society.

9. A rational approach to economic policy modeling can be attributed to the work of Danish mathematician Jan Tinbergen (1903-1994) who advocated objectivity and a positivist approach to policy analysis. Tinbergen was awarded the 1969 Nobel Prize in economics for development and application of dynamic models to analyze economic processes.

10. In the last chapter of the book, we discuss the fact that we still lack a single comprehensive general equilibrium theory that ties all these pieces together into an easily understood framework. We do have a sufficient number of partial equilibrium theories to help understand and predict many elements of the phenomenon of community economic development.

11. In planning theory literature, there are two conflicting approaches to decision-making. The first, *radical planning*, tries to achieve long-term goals and objectives in a rapid fashion through dramatic sweeping changes. The alternative *incremental approach* moves toward long-term goals and objectives using short-term marginal changes. In the real world where people are often leery of change, the incremental approach is typically the most politically palatable.

12. In fact, community values go a long way in determining community goals. For example, if the community sees the marketplace as the way the world works, then issues such as unemployment, poverty, and affordable housing are not even appropriate topics for conversation.

13. There are different types of relationships among community goals. They can be complementary, independent, or conflicting goals at different levels of achievement of either goal through time. The two goals can be seen to be complementary (i.e., can attain more of both simultaneously) or independent (i.e., the level of goal 1 attained does not affect the level of goal 2 attained), or conflicting (i.e., can only achieve more of goal 1 at the expense of goal 2). This example is the most likely situation in setting community economic development policy. The extremely frustrating dimension is the uncertainty about the nature of the relationship. The relationship may change at different levels of achievement, with variations in resource base or variations in external (national) conditions, or through time.

14. A common misconception in community economic development efforts is that secondary effects will flow naturally from the primary effect. With the increased mobility and intra-organizational linkages in today's economy, there is no reason to believe that the hoped-for secondary effects will unquestionably occur.

15. Another widely referenced approach was presented by Glen Pulver (1979). Pulver's approach itemized five strategies that, in layman's terms, outline how a community can change the economic conditions of the community:
    1. Attract new basic employers.
    2. Improve efficiency of existing firms.
    3. Improve ability to capture dollars.
    4. Encourage new business formation.
    5. Increase aids/transfers received from broader governmental levels or units.

16. One of the more interesting cases of equal or unequal welfare weights is sustainable development and the weights assigned to preferences of future generations.

17. Dynamic economy is not a euphemism for growth; it refers to changing choices, reframing issues, changing perceptions of markets and resources, and changing values.

18. One of the goals of community level decision-making is sustainable development. The classic definition of sustainable development is "that which ensures the needs of the present are met, without compromising the ability of future generations to meet their own needs" (World Commission 1987).

13
The Practice of Community Economic Development

Now that we have dealt with the key components of decision-making and have provided the institutional context for community economics, we proceed to how development work takes place. In this chapter, we review and discuss the practice of community economic development. In essence, our discussion is akin to viewing the Shaffer Star (see Fig. 1.1) in its entirety. The integration of each of the five elements provides a more complete picture of the community practitioner's task. In addition, motivating individuals and groups within a community to learn and strategize is complex in its own right. The practice of community economic development involves both application of theory and people-oriented process. Although this book has given much attention to the theoretical context, particularly as applied to small open economies, we need to strongly emphasize the importance of people-oriented process.

The complex art of community economic development practice attempts to draw together a diverse range of activities initiated at the local level. These activities involve the mobilization and development of local resources to tackle local economic and social problems. The practitioner essentially uses internal and external resources to bring about local change. But resources alone will not solve the key problems at hand. The astute practitioner realizes quickly that truly effective development practice occurs from the bottom up, initiated by local residents and interest groups to address locally identified needs and desires.

This chapter has three parts. First, we discuss aspects of the people-oriented process of community economic development. Key aspects covered in the process are how to tell if a community is ready for economic development, how to get a community organized, and the different roles for the practitioner involved in the activities associated with econom-

ic development. Specific process elements include strategic planning and visioning. We then examine how important economic data about community economic conditions are to practitioners. This has both comparative and descriptive importance in addressing the issues of community economic development. The final topic is a review of two community economic development programs used by the U.S.D.A. Cooperative Extension Service: *Community Economic Analysis* and *Take Charge*.

THE PROCESS OF COMMUNITY ECONOMIC DEVELOPMENT

Issues of community economic development process revolve around whether the community is ready for economic development, how to get organized for economic development, and ways to work in the community. Convenient approaches to moving groups forward include strategic planning and visioning.

Readiness

A good first step to the practice of development is to determine whether or not the community is ready to move forward with some development initiative. It is often said that community economic analysis attempts to pave the way for a community to realize its full economic development potential. Developing strategies and approaches to achieve this potential are, in large part, motivational tasks. In essence, the practitioner is trying to engage both willing and unwilling conspirators in the process.

Chapter 12 discussed the variety of strategies in which a community could engage. The balance of the earlier sections of the book was essentially a discussion of the economics behind those strategies. What we want to do now is to determine who the willing and unwilling conspirators are apt to be and whether or not they are ready to engage in initiatives that foster community economic development.

The term *unwilling conspirators* has two fundamental components. First, most people are unwilling to engage in a long-term strategy and are more interested in short-term projects. Second, people need to become engaged and encouraged in building a strategy. Chapter 12 implicitly and explicitly talked about citizen participation in decision-making. There are a variety of ways to encourage people to become engaged in the civic process of decision-making and implementation of those decisions. One way of encouraging people's engagement and action is through consistency of contact and promise of feedback. In other words, it is important to ensure that the same person is in touch with the community group and is consistent in keeping them informed about what is happening in the community. This helps the practitioner build relationships and trust. Another civic engagement strategy is to use groups in the community that are already in place and that meet regularly. An example is eliciting assistance through existing communication vehicles such as business associations, churches, craft groups, interagency organizations, parent groups, and Community Action Program (CAP) sites. Yet another strategy for engaging people is through the use of nontraditional meeting places—holding meetings in places such as coffee shops, local restaurants, and private residences that are familiar to local people and provide comfortable gathering places where interested individuals will take time to talk.

Since a large part of the practice of community economic development focuses on people, when planning group meetings it is often a good idea to remember the types of things people enjoy. To encourage attendance and/or involvement in meetings, it can help to have good food. Make it enjoyable and relaxing. Have things that people can do instead of relying solely on talking. Participatory activities and action-based learning are often more effective means of communicating ideas. For inclusiveness, it also helps to remember that local residents have outside commitments and family issues. Often, it helps to provide child care and/or transportation for key groups of participants. This can help to ensure that everyone is welcome, regardless of life status.

Beyond getting people to engage is the question of whether or not the community is ready for economic development. *Community readiness* is the relative level of acceptance of a program, action, or other form of decision-making activity (Collins and

Powers 1976; Donnermeyer et al. 1997). Three types of readiness relate to different decision-making elements—individuals, groups, and communities—that represent three different levels of decision-making and readiness. *Individual readiness* is an individual's dissatisfaction based on perceptions of discrepancy about what is desired and what is reality. *Group readiness* is similar to individual readiness in the dissatisfaction between what exists now and what is desired in the future. However, group decision-making and group leadership can modify the way problems are perceived and identified, alternative solutions are examined, and action is taken. For a group to make a decision requires sharing some consensus on values, norms, and vision. This implies a role for the practitioner in forming that consensus. Community readiness differs from group readiness in that place and locality are dominant sources of membership to the group.

As you contemplate working in a community, there are key indicators of whether the community is really ready or interested in conducting an analysis of community economic development (Neuendorf 1987; Walzer et al. 1995). Neuendorf (1987) uses the following diagram (Fig. 13.1) as a procedure to determine whether or not a community is ready for community economic analysis. Interest is really the capacity of a community to promote economic development. What is the basis of the interest in the community? Has it been a job loss? Are not enough new jobs being created? Are incomes low or not growing fast enough? Is there too much fluctuation in employment and income? Is the community too dependent on one or two businesses? Were you approached by an individual or group? Are they aware of the time or financial commitment? What's the history of completing efforts?

Some suggest the overriding problem in most community-based efforts is a lack of managerial or organizational capacity to pursue development. "Ready" communities have some form of organization that is devoted to promoting the achievement of the community's economic development goals. The organization is vital, active, representative, and effective. The organizations and their leadership have the capacity to identify, collect, and direct the resources available to pursuing the community's economic development goals. You might ask these questions concerning organization: What formal organizations are dedicated to economic development? Are these organizations active? Do these organizations have clear mission statements, goals,

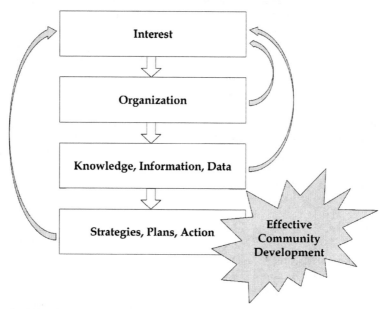

Figure 13.1.   The building blocks of community readiness.

and organizational structures required to be effective? Are these organizations representative of the community? Can these organizations influence the allocation of resources in the community? If there is more than one organization, do they coordinate their activities? What's the history of engaging several organizations in a common agenda? If not you, who can the community turn to for support on organization development and group processes? Who or what organizations are willing to accept responsibility to keep the effort moving forward? What's the pattern of using local resources?

To effectively promote economic development in a community there must be a sound base of knowledge and information. In ready communities, key people have a basic understanding of community economic development principles, their local economy, and the resources available to them. Also, they have on hand current and accurate information about their community. If they don't, they are willing to remedy such a situation.

In terms of strategies/policies/plans, how a community decides to pursue its economic development should be well thought out and documented. A ready community has adopted or is implementing community development strategies/policies/plans appropriate for their goals and situation

(Chapter 12). The strategy/policy/plan is communicated to the whole community. Some questions to ask: Have all the alternative strategies for community economic development been identified and accessed? Have community involvement and input been encouraged during the formulation of strategies/policies/plans? Has the local government adopted policies regarding its role in promoting economic development?

If you can't answer these to your satisfaction, then either you must spend more time on getting the community ready or you move on to a different topic or community. It is extremely important for the practitioner to worry about whether the community really is ready for the time, energy, and commitment required to start and complete a community economic analysis and development initiative. Thus the first task of the practitioner is to determine readiness.

Different Roles and Approaches

The community economic developer can serve three basic functions in communities (Cary 1972). The first function is as an *encourager*, where the practitioner encourages the community to examine its economic conditions from an analytical perspective. The second is as an *educator*, where the practitioner performs the role of helping the community look at itself realistically and examining what can be done. The third is as an *organizer*, where again the practitioner helps the community organize around the issue of economic development and attempts to implement a decision that has been made.

Our assumptions about the community and the practitioner include the following:

- Communities are decision-making and implementing units.
- Communities are a lot like multi-enterprise firms in that they have multiple interests and objectives that are often competing.
- Communities need to think in terms of linked problems rather than a plethora of symptoms.
- Community strategies are composed of projects, but projects don't necessarily add to a strategy.
- Dynamic communities blend hard insights with intuition; specifically, there is place for both right- and left-brain people.
- The practitioner needs to understand and appreciate the history (economic as well as social) and culture of the community.

As the practitioner works in a community, there are several approaches that can be taken. This section is a distillation of five basic approaches: (1) self-help, (2) technical assistance, (3) conflict, (4) asset-based, and (5) self-development. It is naive to think that a given development practitioner would follow any of these in their pure form as described here. Rather, they are more typically blended together in accordance with the specific situation at hand.

The self-help approach is based on the premise that the people of the community can, should, and will solve their own problems (Christenson 1989). The practitioner is a facilitator of the process. The self-help approach requires the practitioner to act differently depending on whether or not the community is well defined. If the community is not well defined, or lacks organization, the practitioner serves as a facilitator or organizer and as a proxy leader. On the other hand, the practitioner injects the right kind of information to key participants if the community is well defined or organized but may not be forward thinking.

The self-help approach has some advantages. It often builds a stronger sense of community, and it often evolves into a holistic approach. It builds a self-sustaining ability to deal with new problems, and it allows for community-specific solutions. Likewise, there are some disadvantages. It works best in smaller communities but change is often

slow. Special interests may cloud issues and cause the true community to take a longer time to appear. Since the practitioner is concerned about the community learning to do it itself, accomplishing specific tasks may be secondary. Finally, decisions may be based on impression rather than on fact.

The *technical assistance approach* is based on the premise that the community is well defined, it has identified a problem or goal, and it is moving toward a plan of action (Christenson 1989). The practitioner supports task-oriented actions. For the practitioner, approaches to technical assistance vary with whether one is doing policy development or implementation. In policy development, the practitioner uses the *scientific method* to identify strengths and weaknesses of the community. These analyses are then used to help formulate policy. In policy implementation, the approach is based on the premise that the community has identified policies to achieve defined goals or objectives. The practitioner helps in the technical implementation of the policy. Advantages of the technical assistance approach include the notions that change can be rapid, that it works in any size community, that it is task driven (easier to "sink your teeth into"), and that decisions are based on fact. Some of the disadvantages include the notions that it gives the illusion of finality of the process, that the process may be lost to task accomplishment, that it often loses the *holistic* view, and that it presumes the practitioner has, or can obtain, the necessary technical skills.

Conflict is the third approach (Christenson 1989). It is based on the premise that the community is fragmented and gridlocked. The practitioner works to break the gridlock. Here the practitioner works either as an advocate or a mediator. As an advocate, the practitioner works with a segment (perhaps the silent majority) of the community assumed to be suppressed by the leadership of the community or other more vocal groups. The role of the practitioner is to act as an advocate for the oppressed group. As a mediator, the practitioner acts as a facilitator to open lines of communication between and within subgroups, then works toward compromise to effect change.[1] Advantages of the conflict approach include change is rapid, communication within the community is opened (silent majority), and future alliances are forged. Some of the disadvantages include the possibility that the practitioner may be viewed as biased and that opponents may become enemies. Also, change is often not sustainable.

Recently, a fourth approach has emerged—*asset-based community development* (ABCD). There are essentially two different ways to do economic analysis according to Kretzmann and McKnight (1993). The first way is the needs/deficiency model, the second is the asset-based model. The needs/deficiency model causes you to focus on what is absent or problematic or what the community needs.[2] Under the needs/deficiency model, the community tends to rely on external resources to solve problems. The needs/deficiency model tends to view the community as a collection of needs or crises.

Asset-based community development is characterized by a community seeing itself as possessing a wide range of resources (assets). Furthermore, reciprocity is a key concept of ABCD, where the development issue is not what the community can do for its residents, but what its residents can do for the community. Community development is a process by which local capacities are identified and mobilized. Mobilization is the connecting of people with capacities to other people, local associations, and local institutions. ABCD sees a community as full of assets: what is present in the community and what the capacities are of its residents and the associational and institutional base of the community. Assets are capacities, gifts, and abilities. The crucial step is identifying and linking assets by creating synergies. It is internally focused in terms of agenda building and relies on the problem-solving capacities of local residents, local associations, and local institutions. ABCD is relationship driven; in other words, it is crucial to build and rebuild relationships among individuals, associations, and institutions.

Assets are owned by three classes in the community:

- *Individuals:* families, youth, elderly, talents, and money
- *Citizen associations:* civic clubs, churches, neighborhood groups, and cultural
- *Local institutions:* businesses, schools, libraries, hospitals, and parks

Each of these can be viewed as a potential way to solve a community problem. Essentially, we need to think of them separately. In other words, an individual who is willing to drive may be used to deliver meals on wheels. A card club could be viewed as a focal point for discussion of community problems and solution building. Churches could be viewed as a mechanism for bringing groups of people together to talk about what the community wants to do.

Finally, the issue of *self-development* needs to be examined. Self-development moves us away from the deterministic model that local economic development is predetermined by location or initial resource endowment or economic/social structure. It reverses the perception that external forces are in total control of change and encourages the community to respond in a proactive manner. Local human capital is crucial for creating opportunity that implies self-reliance. Self-development is counter to the laissez-faire, neoclassic view of the economy. With the latter, a community facing a downturn must accept lower wages and/or out-migration.

The objective of self-development efforts is to gain control of the local economy by the community. This can be accomplished by generating employment for the community and/or inspiring self-help and group-based support. Self-development efforts operate for the benefit of the whole community while promoting the collective management and ownership of enterprises (Blakely 1994; Green, Haines, and Halebsky 2000).

Stöhr (1990) defined local economic development initiative as a local initiative using mainly local resources under local control for predominantly local benefit. In policy terms, self-development is often viewed as a positive means of harnessing endogenous potential and tapping latent skills to promote the interests of the population and exploit development opportunities. The community is seen as an appropriate organization for responding to economic change, providing a basis for mobilizing local populations, and developing social solidarity. New partnerships between public, private, and voluntary sectors provide an effective basis from which to promote development projects and respond positively to the changing external environment.

Regardless of how the practitioner chooses to work in the community, some common themes apply throughout (Dodge 1980). Self-interest is assumed until proven otherwise. This includes not only people in the community, but also local governments and external actors and agency people. It is important to look at the community as a collage of interacting interest groups each seeking the self-interest of its members. The practitioner should not expect people to behave rationally. People may be dealing rationally from their perspectives, goals, values, and understanding of the situation, but not from yours. If practitioners impose their own normative views, they begin to lose their objectivity and are less able to understand why people are doing what they are doing. Also remember that participation is a means, not an end. Most people participate to do something or get something.

Each approach to community economic development should expect conflict and take steps to manage it. The practitioner must learn to negotiate and form coalitions. Negotiation is all about the dynamics and chemistry of community power. The effective practitioner should understand power and be able to harness it for the betterment of the community.[3] These themes tend to apply regardless of approach, and the practitioner who keeps them in mind is more likely to stay sane.

Organizing the Community

To conduct its economic development efforts, the community must create some form of organization, either formal or informal. Community economic development is a time-consuming process that will simply exhaust the energies of one person. Furthermore, its very title, community economic development, suggests a community rather than an individual effort. Research evidence indicates that communities organizing for economic development are not guaranteed success, but organized communities seem to be more successful (Smith, Deaton, and Kelch 1980; Williams, Sofranko, and Root 1977). Appreciate that some form of community organization must sustain the effort over time, and that existing organizations are likely to have a vested interest in the topic but may have a limited perspective.

Economic development organizations serve two roles: (1) to create a community institution to mobilize local resources, including public support, to initiate change, and (2) to provide continuing support for the effort. A committee structure permits individuals to specialize in their knowledge about the community. This specialization allows quick response with current and accurate information to specific and unusual questions rather than general or delayed responses.

Strategic Planning and Visioning

As you start a community economic analysis, it is important that you remember what you are doing: You are examining a community's economy to find both the strategic problems and solutions. This says that you are now in a strategic planning mode. Let us briefly examine elements of strategic planning (Luke et al. 1988; Walzer et al. 1995).

An initial task involved with strategic planning is a diagnosis of social, political, and economic trends.

You do this to gain awareness of how major external forces influence the local community. This involves information gathering and interpretation as well as exploration of alternatives. Assessment of strengths and limitations is always relative to time, technology, and what you seek to accomplish. This phase also involves examining internal conditions and trends. Once the trends assessment is done, the group needs to tease out key strategic issues or problems. The easy thing to do is to focus on symptoms rather than problems (Chapter 12).

The teasing out of problems largely depends on where the community wishes to go. The community needs to develop a vision of potential futures (Ayres 1996; Green, Haines, and Halebsky 2000). This involves a fundamental change from the short-term perspective to strategic initiatives that capture dynamics of change. Remember *visioning* is not projecting trends, but conceptualizing what might be possible. It requires intuitive leaps that are appreciative of current constraints, but it chooses not to be overwhelmed by them. Eventually, visioning requires buy-in by others.

The next step in visioning is the creation of goals and objectives. The practitioner needs to be aware of the relationship among goals and articulate intermediate mileposts to enable midcourse checks and corrections. Several approaches can be used to attain goals. Changing external conditions can uncover opportunities. The presence of alternatives then forces the practitioner and the community to make decisions among emerging opportunities. *Realistic* implementation of programs is important. Action should emphasize local strengths and be realistic with respect to what the community can deliver. This involves a review of different ways to achieve the goals selected. There is a need to persevere and to stay the course; it takes time to implement the strategy and achieve results. The key is to not get discouraged. Finally, the most important element is continual evaluation and correction. By this we mean learn from others and prior efforts, check and celebrate progress, and appreciate that changing external events may open/close options.

The Changing Logic of Community Economic Development

Keep in mind how the theory and practice of economic development has changed and how this change affects our thinking about community strategies. Figure 13.2 is one representation about how the transition in paradigms from "old thinking" to "new thinking" has taken place. The old thinking paradigm

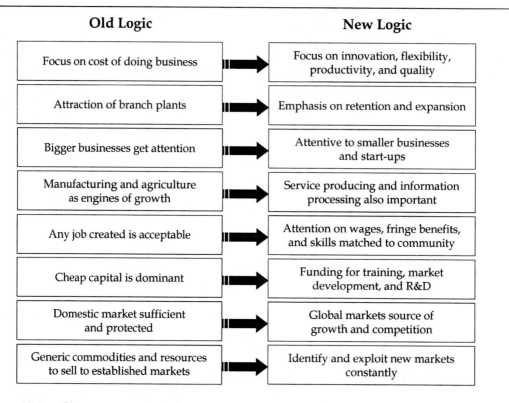

Figure 13.2.   Old versus new logic in community economic development.

tended to be linked to the backward notion of trying to recreate a community in its former self. The new paradigm challenges the community to rethink itself and what it could be in the future. The elements in Figure 13.2 clearly shift the focus from past events to future intuitive leaps.

As global economies continue to exert influence on local economic characteristics, it is important to keep flexibility, adaptability, and innovation in the front and center of people's minds. This globalization extends beyond economic considerations and includes a globalization of cultures, institutions, and ways of thinking about problems. Continuing to do the same old thing just does not cut it anymore. We live in an increasingly competitive realm. Those who succeed will tend to be the smartest, most flexible, and most willing to think and act outside the old paradigm. Communities, as well as firms, need to continually strive for adaptation to changing conditions and competition.

Community Decisions

The ultimate result for the practitioner is that the community makes a decision and in all probability takes action. In Chapter 12 we discussed decision-making in detail, but here we want to review community opinion about the decision. It's generally conceded that if the community can reach consensus on decisions, that is where they want to be. Consensus, however, only means general agreement or accord, not unanimity. There are essentially six levels of consensus.

1. Everyone is enthusiastically in favor of the decision and actions to be taken.
2. Everyone is okay with the decision.
3. Everyone can live with the decision.
4. Not everyone agrees with the decision, but no one blocks it.
5. Those in disagreement with the decision actively seek to block the decision.
6. There is no sense of unity.

As you think about where the community is relative to the decision, you need to strive for number one, but that may not be possible. If the community is at one of the latter levels of consensus, it may be best to revisit the decision and see if you can resolve the objections.

Four things the practitioner must always consider in reaching consensus. (a) Seek and inquire about different perspectives. (b) Find the areas of agreement and disagreement. On the areas of disagreement, work at trying to reach some type of understanding. (c) Seek the best thinking group members have to offer. (d) Try to reach decisions that can be supported by all group members. The practitioner needs to emphasize creating win-win solutions to the problems facing the community. The practitioner also needs to value the diversity of opinion and perspectives with an eye to inclusiveness. This implies that the practitioner values everyone's contribution and takes everyone's concern seriously. Finally, the practitioner is building relationships based on shared values and increased understanding of each other. In making community-level decisions, it is important to be inclusive to the variety of visions, try to get some consensus on vision and how to attain it, and then implement it.

ECONOMIC DATA

The issue of analytics basically revolves around data[4] and measuring economic activity in the community. The idea here is that the community economic development practitioner must worry about the structure of the local and regional economy. In this chapter, we limit ourselves to a very elementary discussion of the economy and defer more detailed discussions of tools and methods to Chapters 14, 15, and 16.

Economic Accounts

Economists focus on certain measures of the local economy. Examples of these measures include population, employment, income, and property values. Before we can develop models that organize such data, we must discuss the data used in community economic analysis. Any comprehensive community economic analysis starts with a compilation of background data on the community. Social accounts organize community data for the specific purpose of facilitating decision-making (Bendavid 1974; Czamanski 1972; Hochwald 1961; Perloff 1961, 1962). These accounts provide a means of recording and organizing community economic activity and represent the aggregation of individual households and businesses. The rudiments of an economic accounting system must precede any community economic analysis, but it is not necessary to prepare a complete set of accounts.[5] An example of a rudimentary set of economic accounts is provided in Table 13.1.

The theoretical concepts used to analyze community economic development questions determine the major elements and structure of economic accounting systems. (We remind the reader that most of this book is a discussion of the theoretical concepts that are measured empirically by using economic accounts [data] and relevant models [or tools].)

It is crucial to keep in mind what insights into the community's economy can be gleaned from the data. Also remember that there are two approaches to collecting and analyzing the data making up the economic accounting system. The first centers on the community practitioner having a question or problem that requires additional insights. The second is to approach the analysis with an open mind and allow the data to tell a story about the community. In either case, the economic accounting system is the first step in confirming or refuting widely held perceptions about the community and the local economy.

Comparing Communities

Demographic and economic statistics are especially useful if they are presented in comparison with other places. To see how your community differs from other places, it is useful to provide two comparison sets of data: comparable communities and a larger reference region, such as the state or nation. Without comparisons, looking at an economic accounting system of the community is akin to using a rubber yardstick. In essence, the community practitioner has no idea of whether conditions in the community are good, bad, or indifferent.

Comparing your community with others allows baselines to be established. These baselines will help determine whether you have low, median, or high values in each data category. For instance, after examining demographics for the community, it may appear that there are a high proportion of white-collar workers. This observation cannot be verified, however, until you know what constitutes an average number of white-collar workers. Comparing your demographics to those of other areas will allow you to make these statements.

Comparable communities can include five or six communities of similar size in the same region or

Table 13.1. Summary of the general social accounts

Account 1. Population and Human Resources

Population
    Number of people
        Age
        Gender
    Sources of change in population
        Natural
        Migration
    Household characteristics
    Place of residence
Labor force/employment
    Total
    Skills
    Training
    Sector of employment
Unemployment
    Skills
    Duration
Job vacancies
Labor participation
Workweek
Labor attitudes
Labor commuting

Account 2. Income, Capital, and Investment

Income
    Level
    Distribution
    Change
Physical resource stocks
    Private plant and equipment
Natural resources
    Public
    Undeveloped land
Investment
    Private
    Public
Financial wealth

Account 3. Attitudes

Private sector aspirations
Private sector evaluation of performance
Public sector aspirations
Attitudes toward financing public programs
Attitudes of community leaders

(continues)

Table 13.1. *(continued)*

Account 4. Public Revenues and Expenditures

Sources of funds
Current expenditures
Capital expenditures
Tax base

Account 5. Flow of Funds

Industry indicators
Output
Distribution of output
    Factor purchases
    Purchases from outside the region
    Industrial organization
Areawide indicators
    Net output produced
    Income produced
    Productivity

state. The communities chosen should reflect similar distances from metropolitan statistical areas (MSA) of the region. Comparable communities also should have a similar economic structure. In other words, you do not necessarily want to compare a recreational-based community with a manufacturing community. At the same time, however, you want to ensure some diversity in the comparable communities. For example, a small farming community may appear to be average when compared to other small farming communities. Common problems, however, facing all small farming communities may be masked, leading to the illusion that everything is fine. Many times when selecting comparison places, the community of interest has identified other places that they would like to emulate.

Measuring a Community's Economy

The choice of units to measure economic activity is guided by the question(s) asked and the availability of the data desired relative to the cost of acquiring it (Andrews 1970a; Bendavid 1974; Blumenfeld 1955; Tiebout 1962). One measurement unit is the *number of businesses and/or households* in the community. While this measure has utility for some questions, it does not generate the detail required for in-depth analysis because it equates a business employing five workers with a business employing 500 workers.

Sales are another measure of economic activity. Sales record the total economic transactions occur-

ring in the community. The problem with sales is double counting intermediate product sales among firms within the community and inflating the final figure. The desire to avoid double counting locally purchased inputs leads to the use of value added as a measure of economic activity. *Value added* is defined as the final sales less the cost of materials purchased. Value added includes the income to non-local owners of resources, which leaks out to those owners but is not deducted from the community value added. Value added does not include returns to local resource owners from the nonlocal use of their resource.

A widely used measure is *income*, which includes wages and salaries; proprietor's income or profits; property income, including rents, dividends, and interest payments; and transfer payments. This measure excludes income to nonlocal resource owners but includes the income generated from the nonlocal use of community land, labor, and capital. Income is the best measure of the economic activity within a community, but it is a very difficult concept to measure, and price changes within the community obfuscate individual income changes.

A related measure of economic activity is *employment*, which is frequently used because the ease of measurement enjoys a greater consistency among data series over time and among communities. Despite its almost universal use as a standard measure of economic activity, employment suffers from some very definite limitations. One problem with employment measures is that individuals can hold more than one job at a time. This characterizes another limitation of employment as a measure—how to define a standard unit of employment. Is it a 40- or 35-hour workweek? Part-time jobs and jobs not fully utilizing the worker's skills are counted and are mistakenly presumed to be equal to full-time jobs using workers' skills to their full potential. An important consideration in using employment to measure economic activity is whether business owners and unpaid family workers are included. In most unemployment insurance–derived databases, they are not. The employment of a community is broader than an administratively determined measure created to meet the legal requirements of unemployment insurance laws.

To count just the number of people employed over time ignores the implications of changes in productivity. This presents another limitation of just counting jobs. For example, a community may experience an increase in the productivity of its export sector and an increase in external sales, but there may be no increase in the export employment or it may even decline. The export sector is generating additional local economic activity (i.e., purchase of local non-labor inputs).

Another limitation of employment as a measure is that it treats all jobs as if they paid the same wage. The generation of 10 jobs paying $35,000 per year creates the same community income as 35 jobs paying $10,000 per year. Employment measures suggest that the 35 jobs have a greater impact on the local community. While that may be true, it is not likely this impact would be 350 percent greater than the 10-person change.

Finally, the employment measure of economic activity does not capture the effects of the flow of funds into a community from external sources, such as interest and dividends, social security payments, pensions, and public assistance. Nor does employment data measure the intra-community flow of rents and profits.

One widely used measure of economic development is *property values*. Property here can include fixed land and the improvements placed on the land. Invariably, the use of property value must consider increases in the value of an existing asset or an expansion of improvements. For many practitioners, keeping a pulse on the performance of the community's real estate market is akin to tracking an important indicator of the community's economic health.

The real problem is that increases in values mix speculative with actual market-based assets values.

There is no single comprehensive measure of the community's economic activity. The analyst should use several measures, but the choice will be constrained by cost, data availability, and the question(s) being asked. The practitioner must keep in mind that there is a story to be told about what is happening to the community's economy and the data are just pieces of that story. The challenge is to filter the data in such a way that the story becomes evident.

Definitions of Industrial Sectors

A widely used scheme of classifying businesses is according to the type of product produced. This classification scheme typically starts with a goods versus services dichotomy. Goods are further divided into durable (lasts three or more years) and nondurable. Services are further divided into business services and consumer services. The most obvious version of product classification is the U.S. Bureau of the Census's Standard Industrial Classification (SIC) system and its replacement with the new North American Industry Classification System (NAICS).[6]

NAICS has been uniquely constructed within a single conceptual framework. According to the Census Bureau, "Economic units that have similar production processes are classified in the same industry, and the line drawn between industries demarcate, to the extent practical, differences in production processes."

In practical terms, NAICS makes industrial-type information more significant by offering the organizing tool that can give it better definition. It has added 10 new business sectors to SIC's 10 divisions, bringing the number of total sectors to 20 (Table 13.2). For example, its new information sector covers 34 industries, including 20 that had not been covered by SIC. Among them are software and database publishing, on-line retrieval services, satellite communications, and the motion picture and broadcasting industries.

Under NAICS, all establishments are classified according to their primary activity. This includes concerns such as corporate headquarters, accounting, transportation, and data processing that provide services to other subsidiaries of a company, which under SIC were classified as auxiliary establishments according to the industry of the establishment they served. This became particularly problematic

Table 13.2. A comparison of SIC divisions/NAICS sectors

SIC Divisions

1. Agriculture, Forestry, and Fishing
2. Mining
3. Construction
4. Manufacturing
5. Transportation, Communication, and Public Utilities (TCPU)
6. Wholesale Trade
7. Retail Trade
8. Finance, Insurance, and Real Estate (FIRE)
9. Services
10. Public Administration

NAICS Sectors

1. Agriculture, Forestry, Hunting, and Fishing
2. Mining
3. Utilities
4. Construction
5. Manufacturing
6. Wholesale Trade
7. Retail Trade
8. Transportation and Warehousing
9. Information
10. Finance and Insurance
11. Real Estate and Rental and Leasing
12. Professional, Scientific, and Technical Services
13. Management of Companies and Enterprises
14. Administrative and Support, Waste Management and Remediation Services
15. Educational Services
16. Health and Social Assistance
17. Arts, Entertainment, and Recreation
18. Accommodations and Food Services
19. Other Services, Except Public Administration
20. Public Administration

under SIC as large complex conglomerates began selling services outside their company so the variety of their economic activities could not be fully captured under the old structure.

WHERE TO FIND DATA

The analytics needed by the practitioner involve both data and theoretical issues. It is important to recognize that neither data nor theory is most important. Both are important. The theory tells us what relationships that we need to be exploring. The data, however, may limit our ability to explore the relationships. We often need to find a proxy data element that represents the theoretical item needed in the analysis. The key to analytics is matching available data with theoretical needs.

Data can take on two basic forms. *Primary* data consist of data developed by the practitioner. Examples include survey-based data, personal interviews, and focus groups. *Secondary* data are data collected and reported by standard units based on some outside specified definition. Typically, primary data is unique to the community at a given time, while secondary data allows comparison across communities and over time. In addition, secondary data is often less costly and more readily available than is primary data.

The practitioner can use primary or survey data in a community economic analysis. Communities often jump into implementing a survey before they have a chance to fully consider all their options and truly ask themselves if a survey is the only way of collecting the necessary information. Practitioners who use survey data are faced with several problems.[7] The practitioner must get a representative sample of the population to be able to generalize the responses. The questionnaire must be structured so that the respondents are providing the information needed. Too often a survey is undertaken without thinking through the survey instrument itself. To ensure that the questionnaire is asking the questions appropriately and you are soliciting the information that is really needed, it is generally necessary to pretest the instrument.

So while many rightfully believe primary data is best, it is not without problems. Secondary data represents a clear alternative, and is generally cheaper. The analytics presented in Chapters 14, 15, and 16 are chiefly driven by secondary data. For this reason, we provide the set of commonly used secondary data sources in a separate text box. Of course, the Internet source information is subject to change over time.

COMMUNITY ECONOMIC ANALYSIS

Why do a community economic analysis? The practitioner will likely do a community economic analysis to find out where the community stands relative to resources, markets, decision-making, rules, and space. The stimulus may be a desire to think about future directions for the community or it may be some sort of crisis such as a major business closure.

Why undertake an economic analysis study of the community? The most obvious reason is to acquire a better understanding of the community's economic structure by improved understanding of business sectors such as manufacturing, trade, or construction. A second reason is to understand what is happening in the community's economy, such as understanding the markets the community sells to, and the linkages among various businesses within the community. A third reason is to anticipate potential problems in the community's economy by exposing community constraints or limitations. In addition, new opportunities that might otherwise be overlooked can be uncovered. These reasons are not mutually exclusive.

There are four essential questions to any economic analysis: Where are we now? Where do we want to be? How do we get there? Have we made it?

The first question—*Where are we now?*—examines the community's existing economic activities and resources and determines the potentials for and restrictions on economic development. Economic development objectives are identified, the residents' attitudes are determined, and support is built in the community. This question probably should include an inventory of the economic resources in the community. For smaller communities, a regional perspective may be necessary. This includes human, built, and natural capital currently used or potentially available and the resultant income and employment from the use of resources. It should appraise its strengths, weaknesses, and underutilized resources, focusing on recent conditions, not historical facts with little current relevance.

This also investigates the structure of the local economy by grouping business establishments into various categories to provide a fundamental understanding of the local economy. This component also integrates the various local/nonlocal markets facing local businesses and identifies the relationship between the businesses serving local and nonlocal markets. Asking where we are now analyzes the nonlocal markets facing the community. This provides insight into the effects on the community of changes in internal and external economic forces.

The second question—*Where do we want to be?*—centers on the notion of visioning and identifies potential economic development desired for the community, builds consensus about the type of development desired, and determines how to evaluate the desirability of any given business prospect. This requires the community to think beyond historical trends and in a new paradigm.

The third question—*How are we going to get there?*—translates the analysis into an action program, deciding what specific actions will be undertaken and by whom, and what's going to help/hinder getting there. (See Chapter 12 for strategies that communities can pursue to reach their visions.)

In community economic development, the process boils down to four basic questions. The fourth is *Did we make it?* Too often communities do not take the time to stop and reflect on their accom-

SOURCES OF SECONDARY ECONOMIC DATA

Several excellent sources of secondary or administrative data are available free over the Internet or in libraries. Many of the sites listed provide data from federal sources that address employment, income, wages, and demographics. States are also excellent sources of data, including up-to-date population and employment numbers. We do not list individual state sites but provide links to sites from which you can link to state-generated data.

EconData.Net
http://www.econdata.net/
This site will take you to just about every site on the Web that provides economic, demographic, and other regional data. It provides links to more than 400 links to socioeconomic data sources, arranged by subject and provider. You can access federal, state, and local sources and data provided by private sources. The site also tells you a little about the data source. This might be a good first stop.

Guide to On-Line Sources for Economic Development Data
http://www.hhh.umn.edu/centers/slp/edweb/dataguid.htm
This site includes links to state and federal sources of data and provides ratings and information on the usefulness of the data.

County Business Patterns
http://www.census.gov/epcd/cbp/view/cbpview.html
This is an annual series of data that provides subnational economic data by industry. The series is useful for studying the economic activity of small areas and analyzing economic changes over time, and is a benchmark for statistical series, surveys, and databases between economic censuses.

Bureau of Economic Analysis (BEA)
http://www.bea.doc.gov/
This site provides the following data for states, regions, and counties: gross state product by industry, personal income, annual employment, and area personal income. The site also includes BEARFACTS, nice summary tables for states and counties.

Bureau of the Census
http://www.census.gov
Here's where you can find more demographic data than you will know how to use. You may not find recent data for communities though. Turn to state data sources for recent population data, for example.

Regional Economic Information System (REIS)
http://www.bea.doc.gov/bea/regional/data.htm
This site includes state and county income and employment data from the Regional Economic Information System. The BEA data includes breakdowns by industry and income by source.

Bureau of Labor Statistics
http://www.bls.gov/bls/regnhome.htm
This site contains data on the workforce, employment, and occupation-related information. It can be difficult to use.

Federal Statistics, all agencies
http://www.fedstats.gov/
This is the gateway to statistics from over 100 U.S. federal agencies, but it may be difficult to navigate.

---

plishments and re-evaluate their strategies and visions. Part of the strategic planning process requires benchmarks to be established to judge if the community is on track. In addition, has something changed that precludes the community from reaching its goals?

There are innumerable ways to do community economic development. These two approaches were chosen because they represent clearly different, but similar, ways to do community economic development: the Community Economic Analysis program and the Take Charge program.

The Community Economic Analysis Program

The Community Economic Analysis (CEA) program blends strategic planning, analysis, knowledge transmission, and citizen involvement ideas in the following fashion.[8] With CEA, it is necessary to set any local choices in the context of ongoing changes in the state, national, and international economy. CEA builds a comprehensive base of information about the local economy to enable people to confirm or deny their perceptions. It is crucial for the community to examine comprehensively what they are currently doing regarding economic development. Within the context of newly acquired knowledge, it is important to establish citizen priorities regarding desired changes. The list of citizen priorities must be converted into a set of implementation activities. The four educational phases involved in a standard CEA are (1) strategic trends, (2) examination of the local economy, (3) current activities, and (4) community goals/program implementation.

Strategic community economic development planning demands awareness of the broader environment that directly influences the prospects of even the smallest locality. Thus, the CEA education meetings open with a review of *strategic trends* and their implications for economic development choices in the community. This is *phase 1*. Examples of strategic trends include the relative stability of employment in goods-producing industries and the rapid growth in the services-producing sector during the last half-century and how this alters the types of economic activities which communities may seek (see Fig. 13.2). Another is the relative growth of passive income, such as transfer payments, dividends, interest, and rents, and its resistance to business cycle swings, weather conditions, or commodity price changes. Furthermore, recognition of the growing importance of passive income and who possesses that income may alter a community's economic development strategies. For example, many communities have recognized and pursued retirees as a strategic development direction. The need to be sensitive to the rapid globalization of the economy and the growing importance of small businesses and entrepreneurship are other factors discussed. These trends are reviewed more to challenge people to recognize

untapped possibilities than to give specific direction to local strategies.

Phase 2 of the educational series involves examining local economic conditions and comparing them to nearby communities and state trends to gain insight into both absolute and relative economic conditions in the community.[9] While, the CEA program compiles a large amount of data for the perusal of local citizens, the emphasis is on discerning trends and patterns, unexpected conditions, and identification of development opportunities suggested by the analysis. Heavy emphasis is placed on comparing local perceptions and secondary data to glean insights.[10]

This requires reliable current data. This is partially solved by using data collected for administrative reasons (i.e., unemployment insurance, income taxes). These data are seldom more than 12 months old and, given the caveats about coverage and other administrative nuances, provide current and relatively reliable information. Other data from the Censuses of Population and Business and Bureau of Economic Analysis are used for population, sales, and personal income. Custom-designed computer software available from several sources is used to make the calculations and to summarize the information graphically. A simple spreadsheet can be used to calculate most of the rudimentary tools used in CEA.

The specific CEA tools used include location quotients, population-employment ratios, trade area capture, pull factors, and shift-share (Hustedde, Shaffer, and Pulver 1993). Individually, none of these provides the depth of insight needed, but when several measures are examined, patterns emerge that provide insight into opportunities. Program participants are asked about unexpected results and are encouraged to put results from the analysis into the context of their own experience. Frequently, at this stage some commonly held beliefs are challenged. For example, the analysis of local income sources may demonstrate the growing share of local income held by retirees. This may be in sharp contrast to an existing perception that most local income comes from farming or some other sector. From an economic development strategy perspective, this suggests the need for greater attention by the local retail and construction sectors to the needs of retirees.

Next, the program addresses the tendency of local leaders to limit their economic development strategies to the goods-producing sector—largely expansion of natural resource–based industries (e.g., farming, forestry, and mining) or manufacturing. This tendency is based on the historic belief that to improve the economic well-being of a region, more money must be brought in through the export of goods.[11] This belief fails to recognize that there are many other ways to increase regional income. Options include using existing resources more efficiently, purchasing consumer and industrial inputs locally rather than importing them, finding new uses for local resources, developing new technology and new products, including both goods and services. Another set of options involves changing the institutional environment. Examples of these include encouraging entrepreneurship and influencing the policies of external governments.

In the Wisconsin Community Economic Analysis program these options are shared with the participants as five general economic development strategies:

1. Improve the efficiency of existing firms
2. Improve the ability to capture dollars
3. Attract new basic employers
4. Encourage business formation
5. Increase assistance received from broader governments

In most situations, local economies can be improved through actions taken on all of these strategies. The challenge for local leaders is to develop comprehensive community economic development programs that fit their unique needs and resources.

Phase 3 of the educational effort engages the community group in a systematic review of current local economic development efforts. This is accomplished using the Community Preparedness Index, a survey of development actions that might be taken by local leaders (see Appendix A). Participants are advised that the presence of an activity on the survey does not imply that it should be done; the purpose of the index is to compile an inventory of what *is being* done. This instrument serves as a teaching outline to stimulate discussion of why various activities may or may not be appropriate in that community. Furthermore, it opens participant discussion about current actions and whether they are sufficient. For example, the discussion may disclose that a community promotional brochure has been developed but languishes in a storage cabinet, which causes the group to discuss a brochure distribution plan.

Phase 4, the next aspect of the series of meetings, is the process of identifying the full range of problems local leaders feel need to be considered. The group prioritizes these problems. They then proceed to develop very specific action plans for each of the highest priorities. Planning includes several components, including a discussion of specific goals and objectives, a review of the forces which may assist or hinder the achievement of each goal, what specific tasks will be done, who will perform these tasks, and when these tasks will be done. In this phase, there needs to be clear consideration of prospective cooperating agencies and organizations, both local and nonlocal, whose participation in further planning and action is absolutely essential.

The success of the entire program hinges on the capacity of the local community practitioner and community leaders to engage other federal, state, and local agencies and institutions in the search for solutions. The educational product of the series of meetings is a cadre of local people with strong insight and knowledge about economic development trends and realistic economic development opportunities for their community. In addition, they have the outline of an action plan that they can use to take advantage of those opportunities that they judge appropriate. The development plans are their own and not those of an external consultant/expert.

The Take Charge Program

An alternative approach used by many communities is the *Take Charge* program. This approach was designed by a team of Extension specialists from the North Central Region of the United States (Ayres et al. 1990). Take Charge relies largely on the scientific method approach for analyzing community economic development potentials. This approach uses less data than the CEA approach. In fact, this approach is much more oriented toward process rather than the application of economic analysis in the CEA program. The Take Charge approach generally involves three phases: (1) an assessment of current conditions, (2) a vision about alternative futures, and (3) strategies to achieve the vision. These phases are represented by a set of three meetings, each of which lasts approximately three hours.

Phase 1 of Take Charge involves a meeting to find out where the community is now. To do this, community practitioners using the Take Charge program, in concert with local stakeholder groups, examine current trends and characteristics of the community. The people involved in Take Charge then move to an assessment of opportunities for economic growth through inventories of existing community resources.

Phase 2 involves a meeting that focuses on where the community wants to be in the future. This visioning meeting examines the following 13 strategies for economic development by analyzing community capacity and also develops a vision for the future of the community by focusing on projects that implement the most realistic economic development strategies:

1. Attracting new industry
2. Retaining and expanding businesses and industries
3. Developing or redeveloping commercial and/or retail trade
4. Developing tourism businesses
5. Attracting retirees
6. Pursuing agribusiness opportunities
7. Developing new businesses
8. Cooperating with neighbors
9. Using outside sources, including increased aids and grants received from state and federal government
10. Developing conservation programs
11. Improving the efficiency of existing firms
12. Taking advantage of commuters and bedroom community status
13. Recovering lost resources

Many of these strategies overlap with the five strategies listed under the CEA program.

Phase 3 emphasizes strategies to attain the future community vision. In particular, this meeting focuses on how the community gets where they want to go. The community is engaged in discussion about organizing for economic development, preparing for action through leadership development, and engaging the total community. It sets priorities for short- and long-term actions, develops a plan of action, identifies resources, and mobilizes these resources for action.

This program really relies on three groups of people being involved. In some cases, they are the same people. The groups that need to be involved are a planning committee to initiate the effort, a group of participants to develop and implement the program, and the entire community. The latter is needed to support and sustain the effort. The planning committee really needs to be an active community group willing to provide leadership to the Take Charge process. This planning committee should represent various interests in the community (i.e., churches, civic organizations, health care, retail, manufacturing, etc.). The planning committee's responsibilities

include legitimizing the effort with key influential community members, identifying and involving participants, making arrangements for meetings, informing the community about the program, identifying and involving outside resource people. The planning committee also develops the agenda for the workshops and presides at the workshops, gathers data and prepares materials for the workshops, and secures local sponsors or funding for the program.

The participants who are charged with developing and implementing the program probably need to include three major community stakeholder groups. First is the general citizenry, who can provide community support and new leadership. The second group includes major agencies and organizations in the community from which resources can be obtained. Third, decision-makers in the community who have the authority to allocate resources must be involved. It is important to emphasize the involvement of the entire community. This is because citizens must support the economic development efforts if they are to be successful. Take Charge is a program that focuses on the process of getting a community to start implementing an action program to assist in solving its economic development concerns.

SUMMARY

The practitioner must address two fundamental issues in community economic development. The first involves the full range of options that a community can pursue. Key to justifying this range of options is the alternative metrics and outcomes that measure a community's economy. The second is a process issue of engaging both willing and unwilling conspirators in building a long-term development strategy. The effective practitioner is good at selling good ideas. The effective practitioner also recognizes that incremental victories are important and should be celebrated. Rarely is the change attained in one grand magnificent swoop. Rather, change occurs as a series of small steps toward a long-term goal that the community envisions for itself.

There are several characteristics of successful communities that are worth emphasizing here (Walzer 1996b; Woods 1996). First, the strategy formers recognize that they are ultimately the ones who end up making it happen. Typically, development occurs because of a group of committed and trusted people that takes charge of the effort, are visible in their roles, and serve as local champions for the effort. Second, they focus on what can be accomplished within a few months or years without forgetting their long-term objectives. Third, they don't get to the solution before they know the problem. In other words, many strategies are nothing more than a collection of programs that treats symptoms rather than community problems.

Successful communities focus on strategies that are specific in outcomes and actions. The who, what, where, and how of implementing strategies are explicitly recognized. Furthermore, they tend to break actions into small parts to recognize the achievement of intervening goals and objectives. Implementation strategies make sense in context of community resources and location. Strategies rely on both local resources and external resources in their implementation efforts. Also, both action and reflection are crucial. Doing something and then seeing what happens allows both the practitioner and the community to make better, more realistic decisions by learning in an active and experiential manner. It is also important to remember that most others know very little about the local development efforts. Informing them and seeking their counsel helps to develop networks and cultivate allies.

Finally, there are two issues that at first blush appear contradictory. While the community attempts to discipline itself to stay with a chosen strategy over time, they continually revise and re-revise as they achieve some intervening goals and objectives. Lack of proper implementation and follow-through are perhaps the most destructive parts of strategic planning/visioning (Walzer et al. 1995).

Why don't we complete community projects? Several reasons come to mind. Probably the most frequent reason is that daily living gets in the way. Most people who are involved in development work also have full-time jobs as workers or business people. This is in addition to their family lives. Often these other activities of daily living tend to keep people too thinly spread. Other obvious reasons also include a general lack of resources. These can include a lack of funding, volunteers, leaders, and political influence. In addition, people involved in the process also suffer from burn out and general fears of being challenged or failing. Successful community economic development is an elusive goal with which practitioners are constantly challenged.

Essentially, the good practitioner is always working on creating economic policy literacy. That means you are helping people make wiser decisions.

Wiser decisions include understanding economic change, learning how to analyze economic conditions, examining what can be done (i.e., options), and reviewing cause and effect of policy choices. Linking a scientifically based theoretical approach to development with the art of development practice is the key to the successful community practitioner. Finally, Schuler and Gardner (1990) provided an excellent summation to this discussion of community economic development practice:

- Leaders find a way to make it work. Waiting is not the answer.
- Bring people together, and get organized.
- Expect conflict and manage it, because you'll never get unanimous agreement.
- Don't get hung up on planning.
- There is no panacea or silver bullet.
- Look inward for solutions, but never ignore external offers of assistance.
- Don't rest on your laurels.
- There is more to community economic development than jobs.
- Community development takes time.

STUDY QUESTIONS

1. How might a practitioner determine if a community is ready to conduct an examination of its economic potentials?
2. What are different approaches to working in a community to better understand its economy and develop visions and strategies for improvement?
3. What is unique about the ABCD approach versus the traditional needs/deficiency approach?
4. Why was it argued that visioning required intuitive leaps rather than just projections of past trends?
5. What does community consensus mean to you? How would you approach attaining community consensus?
6. How has the traditional development paradigm shifted in the recent past (last 10 years)?
7. Identify the major broad social account categories that are typically relevant to community economic assessments.
8. What are key differences between the SIC and NAICS sector classifications? Why was it necessary to develop an alternative system?
9. What are the primary economic accounts of a community?

10. What are some of the considerations that need to be of concern when selecting comparison communities?

11. What is the difference between primary and secondary data?

12. Identify the specific strategies used for development in both the CEA and Take Charge programs. How are they the same or different?

APPENDIX A

COMMUNITY ECONOMIC PREPAREDNESS INDEX

Improving the Economic Opportunities of Local Residents

Produced jointly by the Wisconsin Department of Development, Wisconsin State Rural Development Council, and Center for Community Economic Development, University of Wisconsin-Extension.

Instructions

The purpose of the community economic preparedness index is to help citizens analyze and plan action to improve economic opportunities in their community. The index is a list of activities and conditions that can be controlled by the community. It is <u>not</u> implied that all communities will or should be engaged in every item.

To complete the form, fill in the "yes" or "no" blanks for each item, then rank the category as a whole. If you do not know, mark "?". Items marked "no" and categories rated "fair" or "minimal" indicate areas in need of improvement. The index was designed for communities of between 1,000 and 20,000 people.

1. **The community has an economic development plan:**

   Yes   No   ?

   ☐   ☐   ☐   a.   Prepared and/or reviewed by a citizens committee in last three years.

   ☐   ☐   ☐   b.   Formally adopted or revised by the community/local government within the past three years.

   ☐   ☐   ☐   c.   Includes a complete analysis of sources and levels of economic activity in the community.

   ☐   ☐   ☐   d.   Major economic development actors (e.g., local government, development organizations, chamber) coordinate their actions to support the plan.

   Circle one: *Excellent   Good   Fair   Minimal*

---

2. **The community has a comprehensive land use plan and zoning ordinance.**

    Yes   No   ?

    ☐   ☐   ☐   a.   It has been written or formally reviewed by a
                     citizen's committee within the past three years.

    ☐   ☐   ☐   b.   It is actively enforced by appropriate public
                     bodies and officials.

    ☐   ☐   ☐   c.   Provision is made for expansion of commercial and
                     industrial sites.

    ☐   ☐   ☐   d.   Areas where development (e.g., business and
                     housing) would be <u>in</u>appropriate are identified.

    Circle one: *Excellent   Good   Fair   Minimal*

3. **The community has a development organization.**

    Yes   No   ?

    ☐   ☐   ☐   a.   It is in partnership with other local and county
                     organizations.

    ☐   ☐   ☐   b.   Its mission is comprehensive development (i.e.,
                     includes more than economic development, such as
                     housing).

    ☐   ☐   ☐   c.   Regularly revises/updates its promotional
                     brochure.

    ☐   ☐   ☐   d.   Participates regularly in state or national
                     association activities.

    $_____   f.   Annual budget.

    _____   g.   Membership.

    Circle one: *Excellent   Good   Fair   Minimal*

4. **The community has an industrial development corporation.**

    Yes No ?

    ☐   ☐   ☐   a.   There is an organized industrial development
                     prospect contact team.

    ☐   ☐   ☐   b.   There is a one-stop contact for prospective firms.

    ☐   ☐   ☐   c.   An annual update of industrial development
                     information has been filed with the Wisconsin Dept.
                     of Commerce and utility companies.

    ☐   ☐   ☐   d.   The corporation has financed and cooperated with
                     Forward Wisconsin in an industrial prospect search
                     outside of the community within the past three
                     years.

    ☐   ☐   ☐   e.   It has identified the types of businesses desired
                     and fit the needs of the community.

    ☐   ☐   ☐   f.   The community has completed and distributed a
                     "Community Economic Profile" within the past year
                     (e.g., those done by the Wisconsin Dept. of
                     Commerce, public utilities, etc.).

    _____   g.   Budget (amount).

    _____   h.   Membership (number).

    Circle one: *Excellent   Good   Fair   Minimal*

5. **The community is pursuing strategies to make the best use of existing commercial and industrial buildings.**

    Yes   No   ?

    ☐   ☐   ☐   a.   A list of current commercial and industrial vacancies can be provided to walk-in traffic.

    ☐   ☐   ☐   b.   The list includes square footage, photographs, property description, and ownership or agents.

    ☐   ☐   ☐   c.   The list contains only vacant space which is readily usable.

    ☐   ☐   ☐   d.   An annual update of vacant industrial buildings has been filed with the Wisconsin Dept. of Commerce or utility companies.

    Circle one: *Excellent   Good   Fair   Minimal*

6. **The community has an industrial site (with vacancies).**

    Yes   No   ?

    ☐   ☐   ☐   a.   It owns or has an option on a site of 15 acres or more.

    ☐   ☐   ☐   b.   There is adequate water (10" or more) and sewer lines (12" or more) to the property line.

    ☐   ☐   ☐   c.   There are heavy duty streets, not through a residential area, to the boundary of the industrial site.

    ☐   ☐   ☐   d.   A firm site price has been set.

    ☐   ☐   ☐   e.   A copy of site covenants and restrictions is readily available.

    ☐   ☐   ☐   f.   Site compatible industries have been identified.

    Circle one: *Excellent   Good   Fair   Minimal*

7.   **Labor market information for the community has been updated in the last three   years.**

Yes   No   ?

☐   ☐   ☐   a.   Includes an analysis of wages and fringe benefits by occupation.

☐   ☐   ☐   b.   Identifies the existing supply of labor by skill or occupation.

☐   ☐   ☐   c.   Summarizes employers' future demand by number and skills.

☐   ☐   ☐   d.   Local training institutions regularly survey employers and contact the local development organization.

☐   ☐   ☐   e.   Includes information on support services like child care, transportation, etc.

Circle one: *Excellent   Good   Fair   Minimal*

8.   **The community has a promotional brochure.**

Yes   No   ?

☐   ☐   ☐   a.   It describes the recreational opportunities available in the area.

☐   ☐   ☐   b.   It provides a description of services available (e.g., retail, restaurants).

☐   ☐   ☐   c.   It describes public services available (e.g., schools, hospitals).

☐   ☐   ☐   d.   It describes the housing stock and availability.

☐   ☐   ☐   e.   It describes major employers (industry, commerce, government).

☐   ☐   ☐   f.   It has been revised within the past three years.

Circle one: *Excellent   Good   Fair   Minimal*

9. **The local government supports economic development efforts.**

    Yes   No   ?

    ☐   ☐   ☐   a.   Local government has created a revolving loan fund.

    ☐   ☐   ☐   b.   Has created a Tax Incremental Financing (TIF) district.

    ☐   ☐   ☐   c.   Has supported a Business Improvement District (BID).

    ☐   ☐   ☐   d.   Contributes to the local development corporation annual budget.

    ☐   ☐   ☐   e.   Local unit of government works/cooperates with

    ☐   ☐   ☐   County government/organizations
    ☐   ☐   ☐   Neighboring municipalities/organizations
    ☐   ☐   ☐   Local civic organizations
    ☐   ☐   ☐   Local business groups/interests

    Circle one: *Excellent   Good   Fair   Minimal*

10. **Local banks support community economic development.**

    Yes   No   ?

    ☐   ☐   ☐   a.   Local banks have made Small Business Administration and/or WHEDA guaranteed loans within the past three years.

    ☐   ☐   ☐   b.   Bank officials are active in community economic development organizations.

    ☐   ☐   ☐   c.   The local financial institutions' Community Reinvestment Act plan clearly outlines roles they will play in community economic development.

    Circle one: *Excellent   Good   Fair   Minimal*

11.   **The community has a program to encourage existing commercial and industrial businesses.**

Yes   No   ?

☐   ☐   ☐   a.   There is a business retention plan in place not more than three years old.

☐   ☐   ☐   b.   The community has completed a business retention and expansion study in last three years.

☐   ☐   ☐   c.   At least three courses/seminars/workshops in business management were offered last year.

☐   ☐   ☐   d.   The Chamber of Commerce, business development organization, or industrial group makes regular visits to business owners/managers.

☐   ☐   ☐   e.   The development organization actively works with businesses in ownership transition.

☐   ☐   ☐   f.   An annual industrial and commercial recognition event (exhibit, field day) is held.

Circle one: *Excellent   Good   Fair   Minimal*

12.   **The community has a program to encourage new business formation.**

Yes   No   ?

☐   ☐   ☐   a.   One-to-one business management counseling is locally available for potential new business owners.

☐   ☐   ☐   b.   Business management courses are frequently and regularly offered in the community.

☐   ☐   ☐   c.   There is a business incubator in the community.

☐   ☐   ☐   d.   Sources of patient or equity capital are available or identified.

☐   ☐   ☐   e.   There is a school based entrepreneurship training program in the local high school.

Circle one: *Excellent   Good   Fair   Minimal*

13. The community has a chamber of commerce or business organization working on retail sales programs and commercial development.

Yes   No   ?

☐   ☐   ☐   a. Has a paid executive at least on a part-time basis.

☐   ☐   ☐   b. Participates regularly in state and national association activities.

☐   ☐   ☐   c. There is a cooperative advertising program for retail merchants.

☐   ☐   ☐   d. The community has completed a trade area analysis within the past three years, and shared the results with local businesses and the public.

☐   ☐   ☐   e. The findings have been communicated to business prospects outside of the community.

☐   ☐   ☐   f. Has sponsored at least three courses/seminars/workshops in retail management over the last year.

☐   ☐   ☐   g. Provides a regular business forum to deal with local retail issues.

_____   h. Number of members (number).

_____   i. Budget (figure).

Circle one: *Excellent   Good   Fair   Minimal*

14. **The community has an active downtown program.**

Yes   No   ?

☐   ☐   ☐   a.   It has a regular calendar of main street promotion activity (e.g. monthly trade days).

☐   ☐   ☐   b.   Has completed a downtown physical renovation plan within the past 10 years.

☐   ☐   ☐   c.   Merchants are following the plan when renovating.

☐   ☐   ☐   d.   The community has a historic preservation ordinance.

☐   ☐   ☐   e.   Has a uniform billboard and street sign ordinance.

☐   ☐   ☐   f.   Has improved main street lighting, parking, and traffic flow within the past 10 years.

☐   ☐   ☐   g.   There is an adequate number of downtown business area public parking spaces.

☐   ☐   ☐   h.   Actively encourages new retail businesses to the downtown area.

☐   ☐   ☐   i.   Merchants have coordinated and/or extended store hours.

☐   ☐   ☐   j.   Made application to or is involved in the Wisconsin Main Street program.

Circle one: *Excellent   Good   Fair   Minimal*

15. **The community has an active tourism promotion program.**

Yes   No   ?

☐   ☐   ☐   a.   Has a tourist promotion committee and participates regularly in state or national association activities.

☐   ☐   ☐   b.   There is a cooperative advertising program for recreational businesses.

☐   ☐   ☐   c.   The community has completed a tourism assets and marketing analysis within the past three years, and reported the results to local businesses and the public.

☐   ☐   ☐   d.   The community has a tourism promotion brochure listing restaurants; lodging; recreational facilities.

_____   e.   Number of members on tourism promotion committee.

$_____   f.   Budget for tourism promotion committee.

Circle one: *Excellent   Good   Fair   Minimal*

16. **The community has at least one major community event each year** (one which has an impact broader than the community, attracting people from neighboring communities, e.g. pageants, festivals, contests, derbies, fairs).

List events: _____

_____

_____

Circle one: *Excellent   Good   Fair   Minimal*

17. The community has the capacity to grow.

Yes   No   ?

☐ ☐ ☐   a. The community has access and/or control of an environmentally-sound waste disposal site for at least five years.

☐ ☐ ☐   b. Meets all Dept. of Natural Resources sewer discharge requirements or has initiated the facilities planning process.

☐ ☐ ☐   c. Has excess water capacity equivalent to 5% of its current population.

☐ ☐ ☐   d. Has a community recycling plan in place.

☐ ☐ ☐   e. A housing assessment has been done in last three years.

☐ ☐ ☐   f. Has a Capital Improvement Plan (CIP) updated annually.

☐ ☐ ☐   g. Schools can absorb 10% more students without new buildings or overcrowding.

☐ ☐ ☐   h. There is strong public support/opinion for growth.

☐ ☐ ☐   i. There is sufficient labor force to support new growth.

Circle one: *Excellent   Good   Fair   Minimal*

18. **The community has submitted proposals for state and/or federal funding for development programs in past five years.**

    Yes   No   ?

    ☐   ☐   ☐   a.   For affordable housing, or housing assistance, or homeless assistance.

    ☐   ☐   ☐   b.   For two of the following: sewer, water, streets, fire protection, waste management.

    ☐   ☐   ☐   c.   For one of the following: airport, health protection, public parks, community building.

    ☐   ☐   ☐   d.   Has used Community Development Block Grant for housing including rehabilitation.

    ☐   ☐   ☐   e.   Has used Customized Labor Training Funds.

    ☐   ☐   ☐   f.   Has worked to inform people of eligibility for homestead relief, Earned Income Tax Credit.

    ☐   ☐   ☐   g.   Has used funds for special Hospitality, Recreation, Tourism projects.

    Circle one: *Excellent Good   Fair   Minimal*

19. The community presents a positive living environment.

Yes   No   ?

☐   ☐   ☐   a.   There is an organized senior citizen transportation system.

☐   ☐   ☐   b.   There is senior citizen public housing.

☐   ☐   ☐   c.   There are 10 acres or more of public parks per 1,000 people.

☐   ☐   ☐   d.   There are fewer than 1,000 people per physician.

☐   ☐   ☐   e.   Does the community address the needs of persons with disabilities (e.g., curb cuts and ramps, visual or hearing limitations).

☐   ☐   ☐   f.   All educational systems are adequate.

_____   g.   How many youth organizations are there functioning in the community? (number)

☐   ☐   ☐   h.   People are positive in describing the community.

Circle one: *Excellent   Good   Fair   Minimal*

University of Wisconsin-Extension, Cooperative Extension Service, Ayse Somersan, Director, in cooperation with the U.S. Department of Agriculture and Wisconsin counties, publishes this information to further the purpose of the May 8 and June 30, 1914 Acts of Congress; and provides equal opportunities in employment and programming including Title IX requirements.

Produced by the Center for Community Economic Development, University of Wisconsin-Madison-Extension.

APPENDIX B

Twenty Questions about Your Community

1. What firm/sector is the largest employer? Now? Five years ago?

2. How has the number of jobs changed in the last year? Five years?

3. What firm/sector has experienced the greatest turmoil (layoffs) in the last five years?

4. What wage and fringe benefits are associated with most of the new jobs in the county?

5. Which firm/sector pays the highest annual average wages?

6. What proportion of county total personal income is from passive income (dividends, interest, rent, plus transfers)?

7. What proportion of county employment is from farming, manufacturing, retail, services, health, government?

8. What proportion of county income is from farming, manufacturing, retail, services, health, government?

9. If one were to get a first job, where might he or she look in your county? As a teenager? As a middle-aged first-time worker?

10. What share of local employment is contained in nonfarm businesses of less than 20 employees?

11. Name the businesses that started in the last year? Last two years?

12. Name the businesses that closed in the last year? Last two years?

13. What's it like to start a new business?

14. Where do people shop for most of their personal consumption? Has this changed in the last five years? Why?

15. How has population changed in the last five years?

16. What age group has grown the most?

17. What type of job or in what sector do most of the women hold jobs?

18. Is adequate housing (including affordable housing) available in the county?

19. What was the most recent development project undertaken? Was it successful? Why?

20. What do you see as the biggest obstacle for the county and its residents achieving their desired future?

NOTES

1. Daley and Kettner (1981) offered bargaining as an alternative to conflict.

2. Under a strict interpretation of the needs/deficiency model, essentially what one is doing is generating an endless list of problems for the community that needs some type of response by outside agencies or service providers. Much community economic analysis could be viewed this way, except true community economic analysis is looking at the situation in the community to build a community response, not a response by outside agencies.

3. *Power* is the capacity to influence the actions of others.

4. Numerous sources of data can be drawn on (e.g., the Census of Population, Census of Businesses) and some private vendors, such as the IMPLAN Group and Woods and Poole, Inc. The widespread use of the Internet has literally flooded the community practitioner with potential sources of data (e.g., American FactFinder. http://factfinder.census.gov). However, the community practitioner must feel comfortable with the data before using it.

5. The framework outlined in Czamanski's work in Nova Scotia gives an idea of how extensive and complex a social accounting system can be (Czamanski 1972, Chapters 2 and 3). Before a community social accounting system is developed, the conceptual work reported by Isard (1960, Chapter 4) is useful to demonstrate why it is necessary. Both Czamanski and Isard showed that most communities cannot afford the time and cost of such an effort, but the absence of social accounts does not prevent useful analysis of the community's economic situation. Bendavid (1974, Chapters 3 and 4) presented a much less comprehensive system but it is still beyond that required for most communities.

6. NAICS has been approved by the U.S. Office of Management and Budget and was published in the April 9, 1997, issue of the *Federal Register*.

7. It is clearly beyond the scope of this text to discuss survey methods in detail. Interested readers are referred to standard methods research texts.

8. This section draws on Shaffer and Pulver (1995).

9. In smaller communities or ones without data, "20 Questions about Your Community" (Appendix B) is a substitute.

10. An example of insight gained is one community (4,000 people) that was convinced they needed a CEA to improve the effectiveness of their industrial development program and solve their economic development problems. A simple and quick analysis of their employment structure uncovered that the local employment share in manufacturing was more than three times the national average. This stimulated discussion of what other local development strategies might need to be considered.

11. This includes selling higher-valued products and services and developing niche markets as well as increasing the volume of existing export activities.

Section IV
Tools of Community
Economics

In this section, we attempt to distinguish between fundamental concepts of community economic development and the tools used to analyze community economic activity and public policy. This purposive delineation has been done for a variety of reasons. Perhaps most importantly, this separation was driven by our experiential backgrounds as teachers of economic development, as researchers in economic phenomena, and as practitioners in the art of community development. As teachers, we realize the power of individual tools in the toolbox in explaining issues of community economic change. As researchers, we understand the complexities of how these tools work. As practitioners, we have a real-world understanding of how enormously effective these tools are if used properly and how potentially damaging they can be if used or interpreted improperly. Without the fundamental contextual basis in economic concepts, these methods are simply an assortment of tools without a task. Simply stated, if all you have is a hammer, everything now looks like a nail, regardless of the fact that you are working on an automobile engine. By outlining the community economic development problem and describing the intricacies of economic relationships, we now have the opportunity to match the economic tool with its appropriate application.

This section provides a roadmap of tools found in the economic toolbox. These tools provide a scientific framework for objectively assessing economic activity and the modeling of policy using alternative frameworks. We have chosen to separate our tools into two basic categories. First, we outline the tools that are primarily descriptive. Descriptive tools are well positioned to describe the current or past state of community economic affairs, but they tend not to be well suited for prediction. In other words, descriptive tools describe economic situations; they do not infer change that could result from some policy. The descriptive approaches are important for decision-making because they provide an organization to the wealth of data that inundates community decision-makers.

The second broad category of tools relates to a set of tasks that require the analyst to infer something. Inferential models are well suited to predict change given some policy instrument. One could think of this distinction between descriptive and inferential models as being similar to the inductive and deductive reasoning found in Chapter 2.

Inferential tools rely on prespecified cause-and-effect relationships that determine the type and extent of change. Inferential tools require the analyst to specify some causal model. In this book, we have separated causal models into two types that relate to their ability to model economic change. The first set represents the inferential tools whose causal model assumes that prices do not change. The second set represents the more complex inferential tools whose causal model incorporates some price effect. In more exacting terms, our causal models move from fixed-price assumptions to price endogenous models.

14

Descriptive Tools of Community Economic Analysis

Economic decision-making at the community level depends on accurate and timely information about economic characteristics of the community. Specifically, the applied research component of community economic analysis can provide answers to a multitude of relevant questions. What are the current economic conditions of a community? What economic trends has the community experienced in the past? How has the structure of a community's economy changed? How might it change in the future? What are specific opportunities for growth that public policy can target to effectively move a community forward? What types of businesses should a community target for attraction? If a community is successful in attracting a particular type of business, what economic impacts can be expected? These analytical questions drive the development of economic models.

Throughout previous sections of this book, we have discussed mostly theoretical material that builds a foundation, or contextual background, for applied work in community economic development. Until now, applications that allow the practitioner to make empirical estimates and suggest data-driven and quantitatively justified policy instruments (see Chapter 12) have been largely left unspecified. In this chapter we move from the theoretical context into the more applied world of economic modeling.

The descriptive tools relate to the general circular flow of a community economy as outlined in Figure 4.1. Specifically, the tools we develop in this chapter build a descriptive analysis of the market for goods and services and allow us to better understand the nature of firms that operate in a community. Furthermore, our discussion throughout the chapter deals with consumption and related activities. The definition of the relevant economic region is discussed with particular emphasis on the spatial defi-

nition of the community from Figure 1.1. We then turn to tools that help identify the specialization of the community's economy and its division into export (basic) and local (nonbasic) segments.

GENERAL ECONOMIC TRENDS

In addition to adding comparable places to our economic accounts, it is equally important to consider the data over time. Although detailed economic accounts often are presented as a snapshot picture of the community at a given time, it is just as important to compare the community to itself and to others over time. There are two suggested ways of making this comparison: growth indices and shift-share analysis. While there are other approaches, we have found these two approaches (a) present community practitioners with sufficient detail to gain insights into the questions they have asked and/or (b) paint a sufficient picture of the community to allow them to understand what is happening within the community.

Growth Indices

A *growth index* is a cumulative measure of change based on the performance of the community's economy relative to some starting year. Growth indices can be computed for almost any economic variable, such as income, employment, population, retail sales, and even property values. Using the same justification for comparative places above, growth indices should be computed for more than the community of interest. Commonly, growth indices are computed for the state and the nation and they serve as benchmark reference points.

The index is computed for the community of interest and comparative places as

$$Index_{st}^i = \left(\frac{Y_{st}^i}{Y_{s,1990}^i} \right) 100 \qquad (14.1)$$

with subscripts identifying community (s), industry (i), and year (t). The variable Y is the economic variable of interest, such as population or employment. In this example, Y_{1990} is the value of the variable of interest in the base year or beginning of the time period examined. The growth index compares the absolute level of the economic variable under examination to its level at the beginning of the period. For example, if income from farming is $500 in 1990 and $600 in the current year, then the value of the growth index in the current year is (600/500) × 100 = 120. In this example, income from farming for this region increased by 20 percent (120 − 100).

There are three advantages to using this measure of economic performance. (1) Placing all regional data on an index basis allows a direct comparison between regions or, in this case, the community of interest to the state and nation. (2) Change in the value of the growth index from one year to the next can be interpreted as a growth rate. Fast growth, slow growth, stagnant and declining industries can be identified. (3) By examining the growth index over a span of time, one can establish the relative stability of a particular economic variable.

There are, unfortunately, a few disadvantages to using the growth index as constructed. First, the value of the growth index is very sensitive to scaling or, more specifically, initial levels. For example, a small industry account for $10 in income adds an additional $10 for a total of $20 of income. Here the growth index will go from a base of 100 to 200, indicating that this is a rapidly growing industry for the community. Now suppose a larger industry that has $200 income adds $10 more in income for a total of $210. Here the growth index will go from a base of 100 to 105, indicating modest growth. This problem with the growth index hints at the second shortcoming: The index does not speak to the relative importance of a particular industry to a community's economy.

Shift-share Analysis

One of the perplexing problems facing a community practitioner measuring change over time is whether the change is a result of local conditions or some external factors. A tool that breaks down the change into different components is shift-share analysis.[1] While shift-share can be used to measure any type of change, we examine it just for employment change.

Essentially, shift-share breaks employment change into three components. (a) The first component describes the local economy as if it were growing at the same rate as the national economy (*national growth component:* NG). This assumes that local conditions and structure are the same as the national economy. (b) The second component describes the local economy as made up of different sectors from the national economy (*industrial mix component:* IM). What this component describes is the local differential mix of faster-growing/slower-growing sectors compared to the national economy. In other words, more employment is concentrated in faster-/slower-growth components than it is nationally. (c) The third component is referred to as the *competitive share*. The competitive share presumably is the employment change that is due to local conditions and actions. This number represents where community practitioners want to devote their efforts.

The first step is to calculate the national growth component (sometimes also referred to as the national share, NS). It measures the potential change in local employment assuming the local economy is similar to and growing at the same rate as the national economy. Multiplying the base year employment in each sector by the national average employment growth rate, and then summing over all the sectors calculates the national growth component. The results show how many new jobs are created locally due to national economic trends, again assuming the local and national economies are identical. Mathematically, the national share is calculated according to equation (14.2):

$$NS_i = e_i^{t-1}\left(\frac{E^t}{E^{t-1}}\right) \qquad \textbf{(14.2)}$$

where e is community employment, i is the sector under examination, E is national employment, and t is the time period.

The second step in shift-share is to compute the industrial mix component (*IM*). The industrial mix component is determined by multiplying the local employment in each economic sector by the difference in the national growth rate for that sector and the growth rate for the whole economy. A positive industrial mix indicates that the majority of local employment is in sectors that are growing faster than national total employment. A negative industrial mix indicates just the opposite. Using the same notation, the industrial mix is calculated according to equation (14.3):

$$IM_i = e_i^{t-1} \left(\frac{E_i^t}{E_i^{t-1}} - \frac{E^t}{E^{t-1}} \right) \qquad (14.3)$$

The third step is to calculate the *competitive share component (CS)*, which measures the ability of the local economy to capture an increasing (decreasing) share of a particular sector's growth. Sometimes, this is also referred to as the regional share. It is computed by multiplying the local employment in each economic sector by the difference in the growth rate of that sector nationally and locally. After doing this for all sectors, the results are summed to give the community competitive share. A positive competitive share indicates the community gained additional jobs over that due to national growth and its industrial structure. This gain suggests the community is more competitive (efficient) in securing additional employment than the rest of the nation.

It is important to examine the competitive share for both the community and particular sectors. Each yields different information. Again, using the same notation, the competitive (or regional) share is calculated by using equation (14.4):

$$CS_i = e_i^{t-1} \left(\frac{e_i^t}{e_i^{t-1}} - \frac{E_i^t}{E_i^{t-1}} \right) \qquad (14.4)$$

Let us go through a simple example. Assume that the national employment grew at a rate of 5 percent over the last two years and that nationally sector A grew 7 percent and sector B grew 4 percent. In the community, total employment grew 5.6 percent and sectors A and B grew 6 percent and 5 percent, respectively. The initial local employment in these two sectors was 200 and 120. Using equation (14.2), the national growth component (or national share) is calculated as

Sector	Base Employment		National Growth Rate		
A	200	×	5%	=	10
B	120	×	5%	=	6
National Growth Component				=	16

Using equation (14.3), the industrial mix component is calculated as

Sector	Base Employment		Sector Growth Rate		National Average Growth Rate		
A	200	×	(7%	−	5%)	=	4.0
B	120	×	(4%	−	5%)	=	− 1.2
Industrial Mix Component						=	2.8

Finally, using equation (14.4), the competitive share component is calculated as

Sector	Base Employment		Local Sector Growth Rate		National Sector Growth Rate		
A	200	×	(6%	−	7%)	=	− 2.0
B	120	×	(5%	−	4%)	=	1.2
Competitive Share Component						=	− 0.8

In this example, the community gained 18 new jobs over the last two years. Most of that gain (16 jobs) was due to national economic growth. The majority of local employment is in relatively fast-growing sectors because the community has a positive industrial mix (2.8 jobs). But these sectors are not very competitive compared to national standards (a competitive share loss of 0.8 job). Note that sector *B* had a positive competitive share that was offset by the negative competitive share in sector *A*. This suggests the community needs to determine how it can support sector *B*'s continued competitive position and help sector *A* improve its competitive prospects. Likewise, the community must recognize that nationally sector *B* is growing slower than the overall national average. This may be due to business cycles, shifts in demand, or the adoption of employment-saving technology.

The real problem with shift-share analysis is that it is a tautology: The change that occurs is broken into three separate components that equal the change that occurs. Shift-share analysis does not give you any insight to what conditions and actions can cause the local share to take on any particular value. Thus, the tool, while useful in sorting through change, offers no theoretical insight as to why or how the change occurred. Again, this is a descriptive tool that can help the community practitioner draw inductive conclusions about past changes. It cannot or should not be used to make inferences about future changes.

DELINEATING THE RELEVANT ECONOMIC AREA

The community practitioner is concerned with defining the relevant area for spatial analysis because the spatial distribution of labor and the availability of goods and services impact the quality of life of residents.[2] Two concepts of space are reviewed: trade area and labor shed. A *trade area* is a geographically delineated area containing potential customers to purchase goods and services offered for sale by a particular firm or group of firms (Huff 1964). This geographically delineated area corresponds to the central place theory concept of range of a good or service (Chapter 4, Internal Markets). The *range* of a good or service is the maximum distance people will travel to purchase that good or service at a particular location. It is the outer limit of the geographic market for this good or service from a particular location.

The *labor market shed* is the geographic area in which a community draws its labor force. This generally is the maximum distance that workers are willing to commute on a daily basis. As with the range of a good or service, the commuting zone varies with personal characteristics of the labor force, types of jobs, and transportation networks. For example, for high-paying jobs, people will be willing to commute greater distances than for lower-paying jobs. The problem for the community practitioner is that the relevant economic area fluctuates by the question being asked. In the end, there is no single relevant economic area that captures all dimensions of the local economy.

Over the past few years, the introduction of geographic information systems (GIS) has revolutionized the way in which the community practitioner can visualize the relevant economic area. A formal definition of GIS offered by the U.S. Geological Survey is "a computer system capable of assembling, storing, manipulating, and displaying geographically referenced information (i.e. data identified according to their locations)." In other words, GIS allows data from a variety of formats and sources to be depicted, manipulated, and analyzed spatially. The output of a GIS is often a map, but it also can be statistics, charts, tables or, reports allowing users to view the data in different formats and uncover relationships about the data. GIS has become a powerful tool for the community practitioner.

Defining Trade Areas by the Analysis of Customer Origins

Any community market analysis must start with a delineation of the geographic area from which the community draws its retail/service customers.[3] This delineation can be accomplished through several techniques (Goldstucker et al. 1978; Turner and Cole 1980; Wagner 1974).

Postal Codes

One approach uses post office routes. The presumption is that a communications link (mail, newspapers) is the major determinant of retail/service trade information and the range of a good or service. Mail routes or newspaper circulation areas, however, may not always correspond to the trade area for most of the goods and services offered by a community. A variant of this approach is the use of zip codes. Because of the convenience of secondary data collected by zip codes, it has become a very popular way to report the data. But keep in mind that the construction of the postal zip codes has more to do with efficient mail delivery than proxying a particular economic area. A variant on zip codes is the telephone area code, but again this does not conform to any relevant economic region.

Customer zip codes can be collected in a number of ways. Some businesses, such as video stores, drug stores, and grocery stores, already collect this information for their daily operation. Often, zip code collection can be built into point-of-sale machines. A customer's zip code can be input to a cash register and then downloaded into a spreadsheet format. If these methods are unavailable, a trade area definition sheet, such as the one in Table 14.1, can be used to simplify recording zip codes by hand. As customers visit a store, their zip code can be tallied on the sheet. While it is more labor intensive, the trade area definition sheet has an advantage, as additional information can be recorded along with the zip code (e.g., day, time of day, gender, age, and amount of sale).

A spreadsheet of zip codes allows customers to be summarized by the number and percentage of people originating in each zip code. The results can be reported in a table (e.g., Table 14.2). If you are creating both convenience and comparison shopping trade areas, separate tables should be used.[4] Most often, the trade area includes those zip codes that aggregate into about 70 percent to 80 percent of the customers (Beyard and O'Mara 1999). Regardless of the method used, it is necessary to associate the zip code trade area with geographical boundaries. There are a variety of Internet sources useful in examining the zip code geography of a trade area.[5]

Table 14.1. Sample trade area definition sheet for customer data collection

*Date*_____ *Community*_____

*Business*_____

Trade Area Definition Data Collection Sheet

ZIP	Amount of Sale	Sex		Age						Day		Time Day		
		M	F	1	2	3	4	5	6	1	2	1	2	3
				<20	20-29	30-39	40-49	50-59	60+	Wkday	Wkend	Morn	12-5	Eve
		M	F	1	2	3	4	5	6	1	2	1	2	3
		M	F	1	2	3	4	5	6	1	2	1	2	3
		M	F	1	2	3	4	5	6	1	2	1	2	3
		M	F	1	2	3	4	5	6	1	2	1	2	3
		M	F	1	2	3	4	5	6	1	2	1	2	3
		M	F	1	2	3	4	5	6	1	2	1	2	3
		M	F	1	2	3	4	5	6	1	2	1	2	3
		M	F	1	2	3	4	5	6	1	2	1	2	3
		M	F	1	2	3	4	5	6	1	2	1	2	3
		M	F	1	2	3	4	5	6	1	2	1	2	3
		M	F	1	2	3	4	5	6	1	2	1	2	3
		M	F	1	2	3	4	5	6	1	2	1	2	3
		M	F	1	2	3	4	5	6	1	2	1	2	3
		M	F	1	2	3	4	5	6	1	2	1	2	3
		M	F	1	2	3	4	5	6	1	2	1	2	3

Table 14.2. Sample zip code customer summary

Zip Code	Customers (number)	Customers (%)
50123	661	55.1
50124	122	10.2
50125	86	7.2
50126	80	6.7
Total	949	79.2

GIS is a useful tool for displaying zip code market data. Mapping the data allows the viewer to examine relationships among customers, such as areas of high and low percentages of customers, directional nature of customers, and contiguous versus fractured origin patterns. Figure 14.1 shows an example of a trade area definition performed in Tomah, Wisconsin.

Notice how the customer origins are biased toward the east due to the proximity to a major highway and the competitive presence of the similar-sized community, Sparta, to the west. These are the types of relationships that would not be readily apparent in a table summarizing customer zip codes. With GIS technologies, the notion that a picture is worth a thousand table elements comes to life.

Address Tracking

Another method of delineating a trade area is to find the geographic location of existing customers by asking customers where they live or by tracing the addresses on checks and charge accounts. Ideally, customer origins will be defined by a street address. While zip codes can be used, knowing the street address allows a more accurate trade area definition.

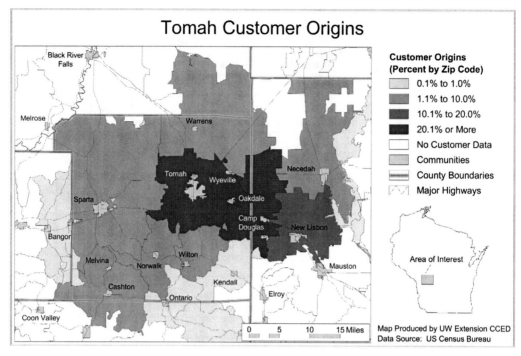

Figure 14.1. Trade area based on customer zip codes.

Using the street address and a process known as *geocoding*, a GIS can map the origin of each customer.[6] After mapping each address, another GIS technique can be used to define rings based on the percentage of overall customers. These rings, or *customer penetration polygons*, can be drawn according to different customer percentages.

As an example, a GIS could draw a customer penetration polygon based on a predetermined customer percentage (e.g., 75 percent). This polygon could then be used as the trade area boundary for a business. Using actual data, Figure 14.2 shows a trade area definition example based on customer addresses. The map shows customer origins for a store in downtown Rapid City, South Dakota, with sample customer penetration polygons of 70 percent and 90 percent.

Defining Trade Areas Based on Travel Time

Trade areas based on travel distance assume that customers will not travel longer than a predetermined distance to visit any given market. Using travel distance to define trade areas is more useful for some product and store categories than for oth-

ers. More specifically, convenience-based stores are particularly affected by travel distance, as consumers are not as willing to travel longer distances for these types of goods.

The traditional method for examining travel distance is by using GIS to construct simple rings around a business district based on distances (e.g., 3-, 5-, 10-mile radii). While the use of rings is an easy method to use, rings fail to recognize travel barriers such as natural features (e.g., mountains, rivers) and cultural elements (e.g., road networks). Subsequently, rings do not create an accurate depiction of a trade area.

In overcoming the deficiencies of rings, GIS is used to examine travel or drive times around a business district. A road network with associated travel time allows the GIS to create polygons based on total travel time from a store. Figure 14.3 shows 5-, 10-, and 15-minute drive times constructed around downtown Missoula, Montana. Notice how the terrain and the subsequent road network influence drive times. If 5-, 10-, and 15-mile rings were used, the trade area would be greatly exaggerated in some directions.

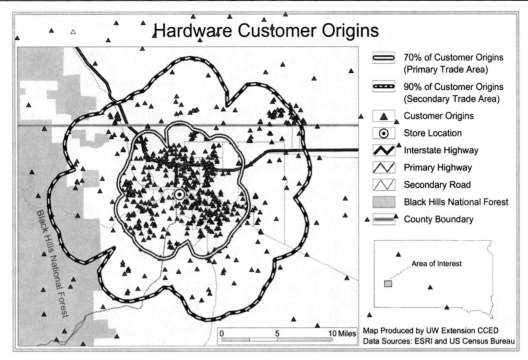

Figure 14.2. Trade area based on customer street addresses.

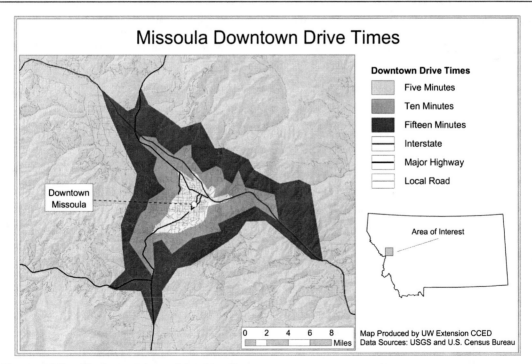

Figure 14.3. Trade area based on drive times.

Defining Trade Areas Based on Travel Distance: Gravity Models

While GIS is a powerful mapping tool, nothing inherent to GIS tells the community practitioner how the individual data points are related economically. The *gravity model* provides a means to predict spatial decisions by relating the rates at which consumers will make use of an activity or commodity provided at a given location by considering the effect of the distance separating the user from the activity. It incorporates the influence of mass and distance on interaction, thus the name gravity (Carrothers 1956; Colwell 1982; Isard 1975). The logic is that the larger a place is, the greater the attraction, and the farther away it is, the less the attraction. The specific measures of mass and distance vary with the analytical questions raised. Some measures of mass are population, employment, income, total sales, and retail space. Examples of distance measures are miles, time, and travel cost. These are actual road miles, rather than as the crow flies.

The gravity model, drawn directly from the study of physics, postulates that the interaction between two population centers varies directly with some function of the population size of each center and varies inversely with some function of the distance between the centers (Deller et al. 1991; Shepard and Thomas 1980). Equation (14.5) is a generalized version of the gravity model:

$$I_{ij} = K \left[\frac{(A_i^a \times A_j^b)}{D_{ij}^c} \right] \qquad \textbf{(14.5)}$$

In equation (14.5), I is the expected interaction between places i and j or the influence of place j on place i. A_i and A_j represent the size (mass) of the places i and j. D_{ij} represents the distance between places i and j. K is a constant and reflects any barriers to trade or interaction. Exponents a, b, and c are estimated parameters for the gravity model and vary with the type of economic activity being considered. For example, c is a measure of the disutility or cost of distance to perform the economic activity. In other words, c is a measure of distance decay. The general value of c is 2.0, but it will vary depending on the specific activity (Bucklin 1971; Goldstucker et al. 1978). The exponents a and b are generally assumed equal to 1.0, but they need not be.

Reilly's law of retail gravitation, which is derived from the gravity model, is used to identify market boundaries (Batty 1978; Reilly 1931). Reilly's law states that customers living at any point between two markets (i and j) will be attracted to the market in accordance with the relative drawing power of the two markets (A_i/A_j), and inversely with the square of the relative distance of the point from the two markets (D_j/D_i)[2] (Krueckeburg and Silvers 1974). This presumes people will shop only in the town with the greatest attraction and customers will not cross market boundaries.

To identify a market boundary requires only setting the formula equal to 1.0 as follows:

$$\left(\frac{A_i}{A_j} \right) \left(\frac{D_j}{D_i} \right)^2 = 1 \qquad \textbf{(14.6)}$$

At the point where the formula is equal to 1.0, the customer is indifferent about which market to shop. In other words, if the market boundary is impenetrable, the customer must be sitting on the market boundary. The boundary of a community's trade area is the point in space where the potential customers flow equally in opposite directions; it is the trade area breaking point.

Population is a common measure of the market's attractiveness and ability to influence its surrounding countryside or tributary area. Alternative measures of attractiveness are advertising expenditures, sales volume, and retail floor space (Wagner 1974). Distance impedes the flow of customers to the market and acts in a negative fashion on retail/service sales. All this says is that a market will draw a higher proportion of the residents closer to it than residents farther away. Some measures of distance are miles, time, and costs.

Reilly's law has been adapted to estimate the breaking point or maximum distance from point i that customers will travel to shop at place i rather than go to place j.[7]

$$D_i = \frac{D_{ij}}{1 + \sqrt{\dfrac{A_j}{A_i}}} \qquad \textbf{(14.7)}$$

Given its ability to easily calculate distances, GIS is ideal for applying Reilly's law. Figure 14.4 shows how GIS can be coupled with Reilly's law to delineate market areas/tributary areas for the Waupaca, Wisconsin, area. The map shows the location of Waupaca (in the center), along with the surrounding communities. The populations of each of these towns and cities are provided beside the community name. By knowing each of the populations and the distribution of communities, a simple trade area (shown by the dark line) can be drawn using the concepts of Reilly's law (equation 14.7).

The results of the GIS calculations are intuitive. For instance, residents of Amherst or Nelsonville live closer to Stevens Point than to Waupaca. Furthermore, the Stevens Point area is larger in population and is in the same county. Therefore, people living in these communities will more likely travel to Stevens Point than to Waupaca. In contrast, residents of Ogdensburg are closer to Waupaca and will most likely shop accordingly. Making observations such as these allows the entire trade area to be estimated. While theoretical, Reilly's law provides a general sense of the community's trade area as a whole. The method requires little effort and resources.

The ease of application of Reilly's law requires some simplifying assumptions that the analyst must acknowledge (Batty 1978; Wagner 1974). Reilly's law assumes that the population in the two places is homogeneous except for size. Thus, there are no demand differences caused by cultural, economic, or social factors, such as income or education. Reilly's law should be used to compare communities of similar size. For example, it should not be used to delineate the trade area boundary between two communities, one with 180,000 people and another with 5,000 people. The analyst should compare boundaries only between places at the same level of the central place hierarchy (see Figs. 4.2, 4.4, and 4.5) because such places offer a similar mix of goods and services. More explicitly, to use Reilly's law to calculate market boundaries requires assuming each of the communities involved offers the same package of goods and services, and this package does not change dramatically over time.

Reilly's law is not very useful in delineating the trade area of urban shopping centers or neighborhood shopping centers in metropolitan areas. Reilly's law is most useful when applied to trading centers surrounded by rural areas where the number of alternative shopping areas is relatively limited (Huff 1964). Using population to measure both origin and destination in Reilly's law assumes that the number of people has a symmetrical influence on

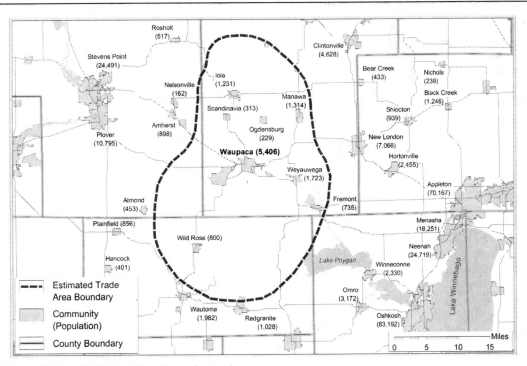

Figure 14.4. Reilly's delineation of a trade area.

both. Yet population is not a symmetrical measure and has no correlation with shoppers' perceptions of a shopping area (Batty 1978). The use of distance to measure spatial relationships does not capture consumers' perceptions of distance. The model does not permit adjustments by businesses in an attempt to respond to or alter consumer shopping patterns (Blomnestein, Nijkamp, and Veenendaal 1980). Reilly's law tends to overestimate the potential customers in an area because it assumes the consumers inside the trade area boundary shop only in the community. This generates a discontinuous trade area because all shoppers within a trade area are assumed to shop only at the nearest shopping area. However, market boundaries are not impenetrable, and trade areas are not discontinuous. Rather, most consumers at i generally shop at i, but some occasionally will shop at j, or vice versa.

Responding to this last criticism, Huff (1964) offered an alternative specification of the deterministic gravity model by recognizing that customers consider and shop at numerous markets (Holden and Deller 1993). Huff addressed the deterministic nature of the gravity model by describing market areas in terms of the consumers' probability to gravitate toward a particular market. By calculating a probability value of customers in the community, surrounding a given market, the market area may be described by probability contours. A probability value for a particular customer at location i describes the likelihood of customers at i going to a particular market j based on the customers attraction to all markets in the surrounding region. In essence, equation (14.5) replaces the distance between two places, i and j in the denominator, with the sum of all the distances between a customer at i and all potential markets in the region. Figure 14.2 could be interpreted as a GIS representation of Huff's formulation. Rather than the mapping contours found in Figure 14.3, the market is surrounded by a set of customer penetration rings. Each ring is interpreted as a probability of the customer at i shopping in j.

Reilly's law and Huff's alternative should be used for the highest-order good or service a market has to offer. Thus, when one is analyzing comparison-shopping goods (e.g., furniture, automobiles, health care), one must compare with a similar mix of goods and services. Gravity models have a considerable amount of appeal in their logic and relative simplicity and have some economic justification. This is particularly true today with the radical advances of GIS technologies. But these types of models cannot

be used in isolation. It is necessary to incorporate the perceptions of merchants about other community linkages, such as school district boundaries and physical features, to identify trade area boundaries.[8]

Defining Labor Market Areas

In defining labor market areas, community practitioners have a number of tools at their disposal. Many of the tools described above for trade areas can be easily modified to define labor market areas. For example, rather than mapping where their customers are located (Figs. 14.1 and 14.2) a business can map where its employees reside. The travel time and gravity models can also be used to assess the area people might be willing to travel for employment opportunities. The most widely used method centers on the use of the Census of Populations commuting data. These data detail the number of people commuting out of a given municipality to surrounding municipalities. By assuming some type of commuting-level threshold, municipalities can be grouped into labor markets. For example, if the commuting flow between municipality i and j is greater than, say, 20 percent of total labor within each municipality, then municipalities i and j can be said to be in the same labor market area. In practice, generally one is interested in looking at a central place and the commuting flow into that central place.

SPECIALIZATION: BIFURCATING THE LOCAL ECONOMY

The structure of the community's economy can be described by simply calculating the percentage of economic activity in various sectors. This provides a crude measure of specialization. Using a simple export base theoretical framework (Chapter 4), specialization can be thought of as dividing the local economy into that which is oriented to the rest of the world (export or basic) and that which is orientated to local markets (nonexport or nonbasic). This is important because of the linkage to the rest of the world in export base multiplier analysis (see Chapters 4 and 15). There are four ways of bifurcating the local economy: direct measure, assumption, location quotient, and minimum requirements.

Direct Methods

Direct methods include an actual survey in the community to determine the flow of economic activity (Tiebout 1962). An example is measuring the physical flow of goods and services to markets outside the community. Problems with this approach include selecting the sample with awareness of firm types and the time of study to avoid seasonal variation, affording the cost of doing the survey, and converting physical units into uniform dollar units or employment.

A second direct approach measures the flow of funds into the community, rather than the physical flow of products from the community. Businesses are asked to divide their sales and purchases according to geographic location along the simple local (nonbasic, nonexport)–nonlocal (basic, export) spectrum. Tracing checks and credit card slips from customers at businesses in the community is one way to collect this information. Individual households are asked to report total income, where it is earned, and where spatially it is spent. Again, the problem is cost, time, representiveness of the sample, and accuracy of responses.

Indirect Methods

Indirect measures of the export base provide inexpensive and relatively accurate estimates of the basic/nonbasic sectors. The three major indirect measures are the assumption approach (sometimes called the assignment approach), the location quotient approach, and the minimum requirements approach.[9]

Assignment Approach

With the *assumption* or *assignment* approach, the analyst simply assumes certain economic sectors produce for the export market (Tiebout 1962). This approach is largely dependent on the community practitioner's personal knowledge of the community. Typically, agriculture, forestry, mining, and manufacturing are assumed to be export sectors, while services and trade are assumed to support local (nonbasic) markets. The assumption approach is definitely the simplest approach available, but it possesses three major sources of error. (1) In many communities the trade sector performs an export function by serving nonlocal markets, including tourists and regional shopping centers. (2) The assumption approach underestimates the export sector by excluding indirect exports.[10] (3) The assumption approach forces the analyst to allocate a specific economic sector to either the export or nonexport sector. There is no provision to simultaneously produce for local and nonlocal markets. Yet, some manufacturing is locally oriented, such as bakeries and some printing and publishing, and some retail and

service businesses are export oriented, including insurance companies and shopping centers. The assumption approach probably is adequate for small, simple economies with minimal cross-hauling.[11]

Location Quotient Approach

Another indirect measure of exports or measure of specialization is known as the *location quotient* approach. The idea is that a community highly specialized in a given sector is exporting that good or service. A location quotient (LQ) for sector i in community s is computed in equation (14.8a):

$$LQ_s^i = \frac{\text{Percent of local economic activity in sector } i}{\text{Percent of national economic activity in sector } i} \quad \textbf{(14.8a)}$$

or mathematically, if we assume that economic activity can be represented by e, with i denoting the sector in community s or for the nation (n), this is equivalent to equation (14.8b).

$$LQ_s^i = \frac{\left(\dfrac{e_s^i}{e_s^t}\right)}{\left(\dfrac{e_n^i}{e_n^t}\right)} \quad \textbf{(14.8b)}$$

Location quotients indicate self-sufficiency and have been widely used as a means to operationalize export base theory as described in Chapter 4 (Andrews 1970b; Isserman 1980b; Mayer and Pleeter 1975). The logic of self-sufficiency becomes more obvious with equation (14.9):

$$X_s^i = \left(\frac{E_s^i}{E_n^i} - \frac{E_s}{E_n}\right) E_n^i \quad \textbf{(14.9)}$$

where E is a measure of economic activity such as employment or income, i is the sector or product, s is the community, and n is some larger reference economy, often the nation. The proportion of national economic activity in sector i located in the community (E_{is}/E_{in}) measures the community's production of product i, assuming equal labor productivity. The proportion of national economic activity in the community (E_s/E_n) proxies local consumption, assuming equal consumption per worker. The difference between local production and consumption estimates production for export (i.e., production > consumption).

The key assumptions to operationalize the location quotient approach are that the local production technology is identical to national production technology (i.e., equal labor productivity) and that local tastes and preferences are identical to national tastes and preferences (i.e., equal consumption per worker). Assuming the national economy is self-sufficient, then the comparison between the community and the national benchmark gives an indication of specialization or self-sufficiency. Equation (14.9) converts export production into export employment (X_{is}):

$$\text{Percent of export economic activity} = \left(1 - \frac{1}{LQ_s^i}\right)100 \quad \textbf{(14.10)}$$

$$\text{Level of export economic activity} = \left(1 - \frac{1}{LQ_s^i}\right)e_s^i \quad \textbf{(14.11)}$$

Three important location quotient values derive from the self-sufficiency interpretation of location quotients. An LQ of 1.0 means the community has the same proportion of its economic activity in sector i as does the nation (the denominator in equation 14.8). The community just meets local consumption requirements through local production of the specified good or service. If the LQ is less than 1.0, the community is not producing enough to meet local needs. This becomes a key indicator for an import substitution strategy. If the LQ is greater than 1.0, the community has a larger proportion of its economy in sector i than does the nation. This excess proportion is for export purposes. If the LQ is 3.0, the community has two-thirds of its sector i activity devoted to export production (equation 14.10).

As with any economic indicator, there are some caveats about the use of location quotients (Greytak 1969; Isard 1960; Isserman 1980b; Leigh 1970; Mayer and Pleeter 1975; Richardson 1985; Tiebout 1962). The location quotient expresses the degree of specialization and dependency of a community economy on a given sector. This degree of specialization can be false because differences between the community and the nation arise from fundamental assumptions rather than reality. Location quotient analysis assumes identical local and national demand and supply functions. Specifically, there are no permitted variations in tastes and preferences, no different marginal propensities to consume locally, no different income levels, no different economies of size and employment efficiency, and no different production practices and technologies.

An LQ greater than 1.0 implies that the community's economy is specialized, but just because the community's economy is specialized does not necessarily mean it is exporting. Differences in local demand (e.g., income, tastes and preferences) may lead to relatively more production and employment than the national average just to meet unique "excessive" local demand. Thus, the specialization implied by an LQ greater than 1.0 may just meet local needs rather than produce for export. The location quotient implicitly assumes no differences in productivity. If local productivity exceeds the national average, relatively fewer people are required to meet local needs, and the community could be exporting more than the location quotient implies. If local productivity is less than the national average, an LQ greater than 1.0 implies that the community is exporting, even though it may be just meeting local needs. If local and national productivity differs greatly, the interpretation of the location quotient becomes suspect.

Location quotient values vary because of data aggregation (Isserman 1977). For example, NAICS 336 of the North American Industry Classification System includes transportation equipment manufacturing, while NAICS 3368 includes only the manufacture of pleasure boats. A community LQ of 1.0 or less at the three-digit NAICS code level suggests no exports, but disaggregating the data to the 4-digit NAICS code level may yield an LQ substantially greater than 1.0, indicating the community has a major export sector. This highlights a need for every community practitioner to examine dimensions of the local economy not reported in official statistics. Any analyst visiting the small community in this example and noting a pleasure boat manufacturer should anticipate those boats are exported, even though the location quotients derived from the reported data did not necessarily indicate it. If an on-site visit is impractical, the data must be disaggregated to avoid this problem. While the science of the LQ may be clear, the interpretation of the LQ may require some art.

Another problem with location quotients, as with other methods of indirect export base determination, is cross-hauling. *Cross-hauling* occurs when a community simultaneously produces goods and services for export while the same goods and services are imported for local consumption (Blumenfeld 1955). An accurate estimate of a community's export sector requires an estimate of gross export activity. In other words, the volume of exports estimated presumes local consumption is satisfied by local production first: No local needs are met by importing the same goods that are exported. Equation (14.9) provides an estimate of gross exports as total local production less local consumption. If local consumption were completely supplied by local production, then the difference estimated in equation (14.9) would be gross exports. But since some local consumption comes from the import of goods and services that also are exported by the community, more local consumption is deducted from local production than is justified. Rather than estimating gross exports, the equation estimates net exports (gross exports less imports of that product).

Self-sufficiency of the reference economy is also an assumption that can cause pause in interpretation of the location quotient. When the community is compared to the national economy, the assumption of self-sufficiency is reasonable, but when the reference economy is a state, one must be careful in interpretation. Self-sufficiency essentially implies that the reference economy is closed in that there are no imports or exports. These caveats must be emphasized since location quotients are subject to misuse because they are inexpensive to calculate, relatively easy to use, rely on readily available data, and capture indirect exports.

Minimum Requirements Approach

The *minimum requirements* approach is another indirect method to determine a community's export base (Isserman 1980b; Moore 1975; Moore and Jacobson 1984; Pratt 1968; Richardson 1985; Ullman 1968; Ullman and Dacey 1960). The minimum requirements approach assumes the community exports nothing until local consumption needs are met. This requires estimating production for local consumption needs. The key assumption that is challenged centers on what is the appropriate refer-ence economy. As argued early in this chapter, the notion of comparative places in community economic analysis is important. The procedure starts with the selection of a group of similar-sized communities. Next, the distribution of economic activity by economic sector is calculated for each community. Then the communities are ranked from highest to lowest on the proportion of their total economic activity in a given sector. The community with the lowest percentage of its total economic activity in a given sector determines the percentage of activity required to serve local market needs. Other communities with a higher proportion of their economic activity in that sector have the difference devoted to export production. This process is repeated for all economic sectors until the total economic base is determined.

The level of economic activity as determined using minimum requirement is a rather straightforward variant of the location quotient export employment, with the difference being the substitution of a peer group of communities for the national reference economy. This is mathematically represented in Equation (14.12).

$$X_s^i = \left(\frac{E_s^i}{E_s} - \frac{E_{min\ peer}^i}{E_{min\ peer}} \right) E_s^i \qquad (14.12)$$

Five hypothetical communities, A through E, are provided in Table 14.3. The proportion of total economic activity measured by employment in industry *i* varies from 7 percent to 35 percent. All of the communities with more than 7 percent of their total employment in industry *i* (i.e., the minimum percentage) are assumed to have the excess in basic employment.

Table 14.3. Hypothetical minimum requirements analysis, five communities (A-E)

Community	Percent Employment Industry *I*		Minimum percent		Percent Basic Employment		Total Community Employment Industry *i*		Basic Employment
A	35	–	7	=	28	×	200	=	56.0
B	23	–	7	=	16	×	80	=	12.8
C	17	–	7	=	10	×	150	=	15.0
D	14	–	7	=	7	×	80	=	5.6
E	7	–	7	=	0	×	55	=	0.0

The minimum requirements approach entails four critical assumptions that are similar to those of the location quotient discussed above: (1) Consumer tastes, incomes, and demands are identical among all communities; (2) production and supply functions are identical among all communities; (3) local consumption demand is met with local production; and (4) every community satisfies its own local consumption needs with the minimum share of its labor force devoted to a particular activity (Isserman 1980b; Pfister 1980; Pratt 1968). The first three assumptions are the same as for location quotients; only the last assumption is an additional restriction, but it is comparable to assuming that national employment distribution represents self-sufficiency.

The minimum requirement and location quotient approaches assume no cross-hauling. While the bias created can be a concern with the location quotient approach, it is exacerbated with the minimum requirements. The minimum requirements approach leads to the implicit situation where every community meets its consumption needs (i.e., minimum requirements) and all communities (except those with just the minimum) export some portion of their output. A paradox occurs because the no cross-hauling assumption means every community exports and no community imports (Pratt 1968; Ullman 1968).

Another problem inherent with the minimum requirements approach is how the analyst selects the cutoff or minimum point (Pratt 1968). While it is typically assumed the smallest percentage is the minimum point, specifically the proportion of local employment needed for local consumption needs, there is no theoretical justification for this choice. An equally legitimate choice is just above the 10th lowest community. There is little reason to believe the minimum percentage city is a *pure nonexporter* (i.e., no cross-hauling) and that other cities do not have larger or smaller proportions of their employment in industry *i* producing for local consumption (Isserman 1980b).

Another problem with the minimum requirements approach is that too much data disaggregation yields numerous NAICS categories with zero or small proportions of total economic activity (Pratt 1968; Ullman 1968). Thus, a very small percentage (e.g., zero) becomes the minimum requirement, implying that a considerable amount of local economic activity is for export. While this may be true in some communities, it need not be universal. This problem appears as the implicit assumption that every community must have some economic activity in every

sector, even though that sector's market may be larger than the community being analyzed. Thus, the minimum requirements approach should not be used for sectors requiring markets larger than the smallest city being analyzed to avoid the zero percent problem.

LOCAL MARKET (NONBASIC) ANALYSIS

In evaluating retail and/or service opportunities,[12] one needs to look at both demand and supply conditions within the trade area.[13] This section provides a procedure to analyze retail/service expansion and promote opportunities in a given trade area. The analysis is intended to be directional in purpose but it should not be used as a substitute for business planning and in-depth feasibility analysis. It is important to remember that while much of the following discussion implies an expansion of the number of retail/service businesses, the analysis is equally adept at examining the existing market for businesses. A comparison of demand and supply by store type can help identify gaps (demand exceeds supply). After considering other more qualitative market factors, including how and where local residents shop, conclusions can be drawn regarding potential business categories worthy of business expansion or recruitment efforts.

The community practitioner must take care when using the tools discussed in this section. Specifically, the tools are aimed at analyzing the local supply of goods and services generally for the residents of the community or trade area. In a crude sense, these tools analyze the nonbasic component of the local economy. As we will see, however, the basic–nonbasic distinction is somewhat artificial. Alternatively, these tools are based on the population makeup of the community, not on the number and mix of businesses. Although local businesses make local purchases, these tools are not designed to capture business-to-business transactions.

Retail and Services Demand Analysis

In retail and service demand analysis, one tries to estimate the amount of sales/demand for a particular type of retail/service business or product line. This typically means that the community practitioner has to estimate the existing trade area (see above).

First, we show methods for estimating this demand. Initially, the assumption is that residents in the local trade area have the same tastes and preferences across the state. This assumption allows the community practitioner to compare the local market to a state average. Then, we look at methods of estimating demand with unique trade area characteristics.

Once the trade area is delineated and described, the question becomes, what proportion of potential sales is actually captured (Goldstucker et al. 1978)? The trade area of a community seldom coincides with the political boundaries of a municipality. Data availability and decision-making, however, are most common at the municipal level. From a community practitioner's perspective, one should be talking about local market analysis at the trade area level. Generally, however, one tends to talk and make decisions at the municipality level. When discussing local retail and service markets, the distinction between trade area and municipal boundaries is particularly relevant.

Sales Retention

Sales retention is an indirect measure of locally available goods and services, assuming people buy locally if possible. While measurement of actual sales is relatively easy, measurement of the sales potential presents some difficulty. This assumes that not only are tastes and preferences identical but also the local trade area is demographically similar to the state. *Local potential sales* can be estimated by statewide average sales per capita adjusted by the ratio of local to state per capita income (Deller et al. 1991; Hustedde, Shaffer, and Pulver 1993; Stone and McConnen 1983):

$$PS_s^i = P_s \left(PCS_{state}^i\right) \frac{PCI_s}{PCI_{state}} \qquad (14.13)$$

where *PS* is potential sales in community *s* for sector *i*, *P* is population, *PCS* is per capita sales, *PCI* is per capita income.

Care must be used in accepting the computed potential sales from equation (14.13). It ignores all of the shopping area and consumer characteristics enumerated in Chapter 4 on internal markets and market area analysis. The potential sales provided from equation (14.13) assume no differences in local consumption patterns except adjusting by relative local income. But this readily calculated estimate represents a realistic initial estimate.

To estimate the sales retention, divide actual sales by sales potential. Actual sales can be gotten from a variety of sources, including census of business, sales tax data, and the merchants themselves. Another approach to sales potential estimates the number

of people buying from local merchants (Hustedde, Shaffer, and Pulver 1993; Stone and McConnen 1983). The *trade area capture* estimates the customer equivalents. Trade area capture used in conjunction with the *pull factor* permits the community to measure the extent to which it attracts nonresidents (e.g., tourists and nonlocal shoppers) and differences in local demand patterns.

Trade area capture estimates the number of customers a community's retailers sell to. Most trade area models consider market area as the function of population and distance. Trade area capture incorporates income and expenditure factors with the underlying assumption that local tastes and preferences are similar to the tastes and preferences of the state. The verbiage here can become somewhat confusing in that the phrase *trade area* discussed above has a definite spatial meaning, but *trade area capture* is aspatial. Thus, the trade area capture estimate suffers from the same caveats enumerated for potential sales estimated:

$$TAC_s^i = \frac{AS_s^i}{PCS_{state}^i \left(\dfrac{PCI_s}{PCI_{state}}\right)} \qquad (14.14)$$

where notation remains the same with the addition of *TAC* as trade area capture and *AS* as actual sales.

The number calculated from equation (14.14) reflects the amount of purchases made for a specific item, such as home furnishings, home renovation, or landscaping. The number reflects current and/or projected people purchased for, not the number of individual people sold to or actual customers in the store (i.e., if one person buys food for a family of four, all four are counted). If trade area capture exceeds the trade area population, then the community is capturing outside trade or local residents have higher spending patterns than the state average. If the trade area capture is less than the trade area population, the community is losing potential trade or local residents have a lower spending pattern than the statewide average. Further analysis is required to determine which cause is more important. Comparison of the trade area capture estimates for specific retail or service categories to the total allows for additional insight about which local trade sectors are attracting customers to the community. It is important to make trade area capture comparisons over time to identify trends.

Trade area capture measures purchases by both residents and nonresidents. The *pull factor* makes explicit the proportion of consumers that a community (the primary market) draws from outside its boundaries (the secondary market, including residents in neighboring areas or tourists). The pull factor is the ratio of trade area capture to municipal population. The pull factor measures the community's drawing power. Over time, this ratio removes the influence of changes in municipal population when determining changes in drawing power.

$$PF_s^i = \frac{TAC_s^i}{P_s} \qquad (14.15)$$

A pull factor (PF) greater than 1.0 implies that the local market is drawing or pulling in customers from surrounding areas. A pull factor less than 1.0 implies that the local market is losing customers to competing markets. The pull factor, much like percent sales retention estimate, can also be loosely interpreted like a location quotient. Pull factors significantly greater than 1.0 often indicate an area of specialization for the local market. For example, tourist areas tend to have high pull factors and location quotients for restaurants, hotels, and miscellaneous retail stores.

The use of any tool by itself can often lead to erroneous conclusions. One must use a variety of tools to gain a clearer understanding of the local economy.

Demand Threshold

Demand threshold is the minimum market required to support a particular good or service and still yield a normal profit for the merchant (Berry and Garrison 1958a, 1958b; Deller and Harris 1993; Henderson, Kelly and Taylor 2000; Olsson 1966; Parr and Denike 1970; Shonkwiler and Harris 1996; Wensley and Stabler 1998). The concept of demand threshold is based on the internal economies of the firm and the characteristics of consumer demand (Chapters 3 and 4). Holding tastes and preferences and buying power constant, population drives market demand in a community setting. The cost structure of the firm, however, is completely independent of the community. For the cost of the firm to be covered, the market must be of sufficient size. If market demand, or population, falls below this threshold, the costs of the firm are not covered and the business will close or not open. *Demand threshold* is a numerical estimate of the population required to meet the cost structure of the firm. Because cost structure varies by firm type, the population threshold will also vary by firm type. Graphically, the demand threshold is

where the average revenue curve is just tangent to the firm's average cost curve (Figs. 3.6 and 4.3).

One set of estimates of population thresholds for multiple occurrences of selected retail/service establishments in Wisconsin is provided in Table 14.4. For example, in Wisconsin it requires only 77 persons to support one tavern but 712 persons to support a shoe store. These differences in market population thresholds can be traced back to the underlying cost structure of each type of firm and the frequency of purchases. This analysis clearly shows that firms with higher cost structures require more people to support the business. In communities with a high population density, the effective spatial market (i.e., range) can be small, whereas in rural areas the spatial market can be quite large. Also note that the population required to support an increasing number of businesses is nonlinear. Specifically, two businesses require more than double the population required to support one business. The first reason is that the number of people shopping at the second store will also shop part time at the first store; that is, there is no reason why one firm cannot service more people. A second reason is that the indivisibility of the investment in a business prevents marginal adjustments until some critical market mass is reached and an additional firm appears. One serious shortcoming to simple market threshold analysis is that it assumes equal firm sizes and ignores store cluster agglomerations in store locations (see Chapters 3 and 4).

Lifestyle Data

The methods discussed above essentially assume that the tastes and preferences and social economic characteristics of the population in the community and its trade area are the same across all communities. While this represents a reasonable and simplifying assumption, it does create problems when your community is dramatically different than the average. Does your trade area population consist more of homeowners or renters? Baby Boomers or Gen-Xers? Which ethnic groups are represented in the population? If they are homeowners, how likely are they to purchase home furnishings, renovate their homes, or spend leisure time landscaping their yards? Current and projected demographic and lifestyle data about your trade area can provide you with a starting point for the in-depth analysis of specific businesses.

Adding information on consumer lifestyle takes market analysis a step further. It recognizes that the way people live (lifestyle) influences what they purchase as much as where they live (geography) or

Table 14.4. Population required to support one or more establishments of selected functions, Wisconsin, 1974

	Number of Establishments			
Function	1	2	3	4
Tavern	77	244	478	711
Food store	92	1,104	4,697	29,119
Fuel oil dealer	164	685	1,577	2,850
Gas station	186	459	799	1,135
Feed store	247	4,895	28,106	97,124
Beautician	268	851	1,673	2,702
Insurance agency	293	666	1,077	1,514
Farm implement store	309	3,426	14,004	38,025
Restaurant	316	754	1,253	1,797
Hardware store	372	1,925	5,032	9,949
Auto repair shop	375	1,148	2,209	3,517
Motel	384	2,072	5,557	11,189
Real estate agency	418	1,226	2,301	3,597
Auto dealer	420	1,307	2,937	4,063
Plumber	468	2,717	7,604	15,780
Physician	493	1,352	2,436	3,702
Lawyer	497	1,169	1,927	2,748
Radio-TV sales	521	1,815	3,765	6,316
Drive-in eating place	537	4,851	17,572	43,799
Dentist	568	1,744	3,379	5,402
Supermarket	587	2,968	7,610	14,881
Appliance store	607	3,709	10,691	22,659
Liquor store	613	4,738	15,669	36,509
Barbershop	632	5,297	18,372	44,404
Drugstore	638	4,285	13,053	28,771
Auto parts dealer	642	5,496	19,284	46,991
Laundromat	649	5,665	20,117	49,264
Women's clothing store	678	5,471	18,544	44,133
Department store	691	5,408	18,012	42,295
Dry cleaner	692	4,131	11,746	24,655
Shoe store	712	7,650	30,670	82,146

Source: Foust and Pickett 1974, Table 1.

their age, income, or occupation (demography). Lifestyle data enables us to include people's interests, opinions, and activities and the effect these have on buying behavior in our analysis (Mitchell 1995). These customer profiles (lifestyle data) can help focus product mix, the services offered, and marketing efforts to target specific high-potential customer segments.

Lifestyle data are based on the premise that "birds of a feather flock together" (Mitchell 1995) or self-segregate by "voting with their feet" (Tiebout 1956c). Did you ever notice that the homes and cars in any particular neighborhood are usually similar in size and value? If you could look inside the homes, you'd find many of the same products. Neighbors also tend to participate in similar leisure, social, and cultural activities. Lifestyle data use these tendencies to redefine neighborhoods into smaller similar groups. The lifestyle clusters are based on demographic similarities (income, education, and household type) and the groups' common lifestyle preferences and expenditure patterns (attitudes, product preferences, and buying behaviors). For example, communities that tend to have higher income and education levels will have more coffee shops and few if any auto parts stores.

The quality of a clustering system is directly related to the data that goes into it. High-quality and useful systems allow the community practitioner to predict consumer behavior. In a retail business targeting tourists, for example, it should allow the analyst to identify products and services that might appeal to this market segment. This usefulness depends on how well the data incorporates lifestyle choices, media use, and purchase behavior into the basic demographic mix. This supplemental data comes from various sources, such as automobile registrations, magazine subscription lists, and consumer product-usage surveys. Accordingly, consider exercising caution when using cluster systems, especially in inner-city areas. Some cluster systems tend to underestimate populations, stereotype residents, and ignore local knowledge and data about the market.[14]

Several private data firms offer lifestyle cluster systems. The firms use data from the U.S. Census and other sources to separate neighborhoods throughout the United States into distinct clusters. They utilize state-of-the-art statistical models to combine several primary and secondary data sources to create their own unique cluster profiles. Most systems begin with data from U.S. Census block groups that contain 300 to 600 households. In more rural areas, the data is more typically clustered by zip code.

One particular cluster system, CACI's ACORN, includes a purchase potential index that measures potential demand for specific products or services.[15] It compares the demand for each market segment with demand for all U.S. consumers. The index is tabulated to represent a value of 100 as the average demand. Values above 100 are more likely to purchase those products or participate in the respective activity. Conversely, values below 100 are less likely to purchase the given product. Private marketing data firms that offer lifestyle segmentation systems can be found through the "Supplier Listing Sourcebook" available through the American Demographics web site.[16] Also, the Consumer Expenditure Survey (CES) provides a starting point for examining demographic purchasing patterns. The CES is published annually by the Bureau of Labor Statistics

and describes consumer purchases by age, income, race, and education.

Retail and Services Supply Analysis

The preceding discussion of the demand side of local markets presumes the assortment of goods and services offered by the community remains fixed and only quantity varies. Community market analysis provides guidance about what different goods and services might be offered. This depends on costs to the firm, access to alternative sources, and consumer demand. (See Chapters 3 and 4 on internal markets and market area analysis for further discussion.) Temporarily ignoring cost differences among merchants, the question becomes, what good or service currently purchased elsewhere (i.e., imported) by residents can be provided locally (i.e., import substitution)? Retail and service supply refers to the amount of sales actually captured or square feet of space serving the trade area. It can be compared with retail and service demand to help identify potential gaps in the market where demand exceeds supply.

To begin the supply analysis, a database of existing businesses needs to be constructed for each of the business categories under investigation. The database for each business category should include all of the retail and service businesses within the primary trade area. The analyst also may want to note other major competitors outside of the trade area even though they will not be included in the demand and supply square foot comparison. In addition, other types of businesses that compete for business in this category should also be included in the database. With the advent of big-box stores such as Wal-Mart, the notion of competing businesses becomes less clear. The database should include a list of the names and addresses of all the current businesses in the primary trade area. For downtown businesses, a complete list could be obtained from the building and business inventory. For trade area businesses that are located outside of the downtown area, a list can be generated from chamber of commerce membership (and nonmember) lists, Internet yellow-page listings, private data firms that sell business lists, and in smaller communities, personal knowledge of local businesses.[17]

The most obvious method for identifying import substitution potential is to compare the goods and services offered by other communities on the same level of the central place hierarchy and to identify the differences (see Chapter 4 and Comparing Communities in Chapter 13). To do this, review the discussion of thresholds and multiple functional units and the mix of goods and services present at a given level of the central place hierarchy (Figs. 4.4 and 4.5). Any import substitution analysis must examine nearby and accessible alternative sources of supply, because there may not be sufficient market to support multiple firms.

Location Quotients

In addition to measuring specialization, location quotients can be directly used in assessing local retail and service markets. Specifically, LQs less than 1.0 also identify import substitution potential (Isserman 1980b; Murray and Harris 1978). LQs with a value less than 1.0 signify that the community has less economic activity than the national average in that particular trade sector. As above, one interpretation of LQs is that a value of 1.0 measures self-sufficiency. If the community LQ is less than 1.0 and it is at least 1.0 for similar communities, there is a potential local market for that trade function. Alternatively, an LQ significantly greater than 1.0 suggests a strength that the community can capitalize on.

Population:Employment Ratio

Murray and Harris (1978) supplement location quotient analysis with what they call the population:employment (P:E) ratio. This ratio is population divided by employment in a particular retail or service sector. Municipal population is used rather than trade area population.[18] In this simple market measure, population is a proxy for demand and employment is a proxy for supply. The ratio then is a crude approximation of where demand and supply are equated. Unfortunately, since there is no critical value (such as an LQ centered on 1.0), P:E ratios require intercommunity comparisons to determine whether a community's employment in a certain activity is unusually high or low. A relatively high P:E ratio indicates that demand is greater than supply, implying a potential for expanding employment. The P:E ratio avoids the computational bias of location quotients arising from the requirement that both the numerator and denominator must sum to 100 percent. This creates a subtle bias, especially in smaller communities, because a dominant sector distorts all the sector shares. Furthermore, the P:E ratio incorporates the total population, not just those who are employed. Counting just employed people produces distorted results if the community has a high proportion of younger and/or older residents.

A simple example illustrates how to identify import substitution potential. Compare five cities of similar population size, each a free-standing community that is not closely linked to a major urban center or to each other, each a county seat, and none with a sizeable state or federal employer. The information to determine the import substitution potential for furniture retailing (NAICS 442) is summarized in Table 14.5.

LQs of less than 1.0 indicate some possibility for import substitution, while LQs greater than 1.0 point to local market strengths. Dividing the municipal population by employment in furniture retailing in each city produces the P:E ratios. The average ratio for the five cities is 314. The P:E ratio for cities A, B, and D is less than the average P:E ratio, while the opposite is true for cities C and E. An above average P:E ratio indicates a potential to increase retail furniture sales and employment in cities C and E. Only city E, therefore, has an abnormally high P:E ratio. The LQ and P:E provide two sources of somewhat different insight to import substitution potential. An LQ less than 1.0 coupled with a high P:E ratio indicates good import substitution potential (e.g., city E). An LQ less than 1.0 and a low P:E ratio is a mixed signal requiring further analysis (e.g., city A). An LQ greater than 1.0 and a low P:E ratio indicate minimal import substitution possibilities (e.g., cities B, C, and D).

Table 14.5. Estimating import substitution potential for furniture retailing (NAICS 442)

	City				
	A	B	C	D	E
Population	6,000	6,092	7,920	9,071	6,039
Employment in NAICS 442	24	28	24	47	11
LQ for NAICS 442	0.98	3.01	1.75	2.70	0.70
P:E ratio for NAICS 442	250	249	330	193	549

One of the critiques of population:emplyment ratios and location quotients is the subjective nature of their critical values. At what point is the computed P:E or LQ sufficiently different from the average to warrant attention? For example, is 0.8 critically different than 1.0 in making a decision? The simple average for a P:E is not rigorous enough a standard to make a judgment. Some community practitioners have turned to statistical analysis to attempt to build confidence intervals to obtain some insights into what constitutes a critical value of either a P:E or LQ.[19]

Location quotients and population:employment ratios are only one step in a local market analysis. The community must also examine local conditions: What are unique local demands? What are the productivity factors affecting the number of workers in this particular service or trade sector? What are the alternative sources of supply in the area? More specifically, an understanding of the lifestyle of the residents of the community is important.

Other Market Considerations and Gap Analysis

Examining quantitative aspects of demand and supply is only part of the analysis. There are strong agglomeration economies in local markets (Chapters 3 and 4). Similar types of firms, such as new car dealerships and shoe stores, tend to cluster together. Customers also tend to locate in similar clusters, and the characteristics of customers vary across clusters. Customers also engage in multipurpose shopping. Two-wage-earner families may turn Saturday into a family outing that includes shopping for multiple goods, dinning out, and going to a movie, and they may be willing to travel great distances for this activity. There are a number of qualitative considerations that require local knowledge and insight about the market. The following considerations also can add to the analysis of each category:

- *Retail mix in comparable communities.* How many businesses in the category are located in the business districts of comparison communities?
- *Quality of existing competitors.* Are existing stores in this category providing the merchandise and service local shoppers demand?
- *Competition from outside the trade area.* Do surrounding communities with regional shopping centers and big-box stores siphon business in this category out of the trade area?
- *Consumer behavior in this retail category.* Are purchases driven by convenience or comparison shopping?

- *Demand from nonresidents.* Is there significant market potential from nonresident customer segments such as tourists and commuters?
- *Demographic and lifestyle information.* Does lifestyle segmentation data indicate that local residents are more likely to purchase goods within this store category?
- *Survey and focus group findings.* What has been learned from local research about consumer behavior and perceptions of the subject business district?
- *Competition from other types of stores in the primary trade area.* Do local discount department stores or supermarkets already fill the niche of more specialized store types?
- *Demand from other businesses.* Are business-to-business sales an important consideration?

Much of the frontier research in community economics has centered on expanding on the simple tools described in this chapter by attempting to incorporate these other market considerations into a more rigorous analysis.

While a community is justifiably concerned about the aggregate package of goods and services offered, often a single good or service is the key that initially attracts customers. All the merchants in the community benefit from that key good or service. In shopping centers, that key merchant is referred to as the shopping center anchor. From a community perspective, the anchor might be a hospital, major employer, college, or other institution. When a community examines its retail and service sectors, there are three types of businesses to consider (Kivell and Shaw 1980). (1) The *generative business* produces sales by itself or attracts customers to the community, such as the shopping center anchor. (2) A *shared business* secures spillover sales from the generative power of nearby businesses; an example is a small specialty shop located near a large general merchandise store. (3) The *suscipient business* is where sales are a coincidental occurrence to other activities; it does not generate sales itself nor from association with nearby shops. Examples are small ice cream shops and tearooms in shopping malls or a hotdog vendor at a ballpark. For the community to realize its retail and service potential requires a balance among the different categories of retail and service businesses.

The quantitative and spatial comparison of retail and service space demand and supply by business type must be analyzed in combination with an understanding of many other market considerations. If there appears to be a significant amount of unmet demand, there may be opportunity for an existing business to expand or a new business to be developed. Business development opportunities may also exist in areas where supply is greater than demand, especially in those communities that are successful in drawing customers from outside their trade area. For example, a community with large LQs and pull factors in hotels and restaurants hints of a strong tourist economy and may point to additional business opportunities.

SUMMARY

We opened the chapter with these questions: What are the current economic conditions of a community? What economic trends has the community experienced in the past? How has the structure of a community's economy changed? How might it change in the future? What are specific opportunities for growth that public policy can target to effectively move a community forward? What types of businesses should a community target for attraction? *If a community is successful in attracting a particular type of business, what economic impacts can be expected?*

The descriptive approaches described in this chapter provide some answers to all but the last question. These tools enable the community practitioner to describe the current situation in the community. (The last question leads us into the next chapter, in which we describe tools that discuss key structural relationships and model change.)

Growth indexes and shift-share analysis provided insight into trends and general growth analysis. These two tools are clearly descriptive in their orientation and should not be used to make predictions about the future. Measuring a community's relationship with external markets was done by using location quotients and minimum requirements. Export base is a fairly simple method of dividing a community's economy into the export portion and the non-export portion. The export portion relates to external markets and brings money into the local economy. The nonexport portion—the local retail/service portion of the economy—is the subject of the market analysis portion of this chapter.

The relevant community was defined as being either the trade area or the labor shed. In particular, the trade area was identified via several approaches, including zip codes, travel time, and gravity models. Once the relevant trade area is identified, one can go through a demand and supply analysis by specific retail/service sector to identify opportunities for expansion. Community market analysis provides answers to three questions about capturing existing markets. First, communities have to know the size, shape, and composition of their market area in order to function efficiently, especially in terms of merchandising and advertising. Second, each community should have some idea of the kinds of goods and services similar communities support to determine potential gaps in the local trade service sector. Third, the community needs to understand what proportion of its total business comes from local residents and what proportion from the surrounding market areas. This will assist in identifying key markets and maybe redirect local marketing strategies. Communities compete for customers in a multitude of ways, including prices, access to stores, variety of goods or services, store hours, clerks' attitudes, credit policies, delivery policies, and personal knowledge of the customer.

Great care must be taken when using these tools in isolation. The use of any tool by itself can often lead to erroneous conclusions. One must use a variety of tools to gain a clearer understanding of the local economy. Any single descriptive tool provides insights into only one piece of a larger puzzle. Using a range of descriptive tools helps the community practitioner better understand the entire community economy and all its complexity. When more than one tool is suggesting that the same story is being told about the community's economy, greater confidence is instilled in that story.

STUDY QUESTIONS

1. Why is specialization such an important concept? How might you measure it?
2. What is trade area capture? What information does it provide? How is it used with pull factor?
3. Can the tools of location quotients, population:employment ratios, and demand thresholds be used for import substitution analysis? If so, how?
4. What are various ways to define a trade area and what are the advantages/disadvantages of each?
5. Describe Reilly's law of retail gravitation. Under what conditions is it a useful tool for analyzing a community's retail market?

6. Under what conditions might an import substitution strategy be a viable alternative development strategy?
7. When discussing trade area demand, why is lifestyle data so important?
8. How can you assess change in a community's economic structure?
9. What key local elements are not included in a shift-share approach to estimating change?
10. How has the introduction of GIS technologies aided the community practitioner?
11. Describe how more than one tool can be used to better understand the local economy.

NOTES

1. See Hustedde, Shaffer, Pulver (1993) for more detail. Another good review of shift-share analysis can be found in Stevens and Moore (1980).
2. The market analysis section draws on work done by Bill Ryan and Matt Kures. The interested reader is referred to their work at the University of Wisconsin-Extension, Center for Community Economic Development (http://www.uwex.edu/ces/cced/dma/).
3. In the area of market analysis, *retail/services* and *retail* are used interchangeably. Note that the services we are referring to are consumer services (e.g., theater, hair styling), not business services (e.g., advertising, engineering consultants).
4. The purchaser buys *comparison goods* (furniture, cars, TVs) and services only after comparing price, quality, and type among stores and places. The purchaser buys *convenience goods* and services (groceries, gasoline) with a minimal amount of effort and usually at the most convenient and accessible store. Convenience goods or services typically have a small unit value, and purchases are frequent and are made soon after the idea of the purchase enters the buyer's mind. *Intermediate goods* possess characteristics of both shopping and convenience goods: The purchaser will spend some time shopping, although the time is minimal and typically the purchase is made close to home. Examples of intermediate goods are drugs, hardware items, banking, and dry cleaning services.

5. Two of these web sites are http://www.usps.com/ (United States Postal Service City and Zip Code Associations) and http://tiger.census.give/ (Census Bureau Mapping Service, Census Units and Zip Codes).
6. Geocoding is the process of assigning locational references to attribute information, in this case the street address of the customer. Using this address and a GIS-based street database with address ranges, the customer's address can be dynamically positioned on a map.
7. This formulation is reached by recognizing that the distance between places i and j (or D_{ij}) is fixed, thus $D_j = D_{ij} - D_i$. This can be substituted into equation (14.6) and then solved for D_i.
8. A wide body of literature exists on trade area and consumer choice modeling (including multitrip models). One overview of the modeling process can be found in O'Kelly (1999).
9. The econometric approach is also available but is not discussed here. The interested reader is referred to an article by Mathur and Rosen (1974).
10. An example of indirect exports would be a vegetable processing plant in a small community whose direct export is canned vegetables. A can manufacturing plant in the community sells cans to the vegetable cannery. It is an independent operation, but has only one customer, the cannery. Since it is serving the export sector, it could be considered a nonbasic activity. The can manufacturer, however, is so closely tied to the export business that it is difficult to classify. The production of cans can be called indirect exports. As an industrial organization becomes more complex and specialized, there is an increasing tendency to subcontract parts of the production process to other divisions of the same firm to capture economies of scale. The subcontracting of production can confuse the distinction between direct and indirect exports.
11. Cross-hauling occurs when a community simultaneously produces goods and services for export while the same goods and services are imported for local consumption (Blumenfeld 1955).

12. Whereas this is written specifically for retail and service sectors, an analogous presentation could be made for more traditional agricultural and manufacturing sectors.

13. This section draws on work done by Bill Ryan and Matt Kures, the authors of portions of this section. http://www.uwex.edu/ces/cced/dma/

14. A discussion of the problems and limitations associated with cluster systems is presented in Pawasarat and Quinn (2001).

15. CACI's ACORN web site: http://www.esribis.com/products/data/db.html #acorn

16. American Demographics web site: http://www.inside.com

17. InfoUSA (American Business Information): http://www.infousa.com

18. Municipal, not trade area, population is used because of the difficulty in consistently estimating total population and employment over numerous trade areas. Using municipal population assumes the same population density around communities.

19. Some would argue that the average location quotient and population:employment ratios for comparable communities are the standard. This would mean that you would compare the community's ratios to comparable places rather than a statewide or national average. The counter argument is that 1.0 is the standard for self-sufficiency in location quotients.

15

Inferential Tools of Community Economic Analysis

Fixed-Price Models

Modeling the effect of change within a community's economic structure requires the analyst to establish assumptions and key relationships that direct how change occurs. If a policy is developed to attract a certain type of firm to a community, how will this new business affect that specific sector's structure? If demand or supply in this sector is affected, how will other sectors that sell to or buy from the affected sector react? How will changes in the supply of this good or service affect production? Namely, how will increased production affect increases in the inputs needed to produce this increased production? How will increased employment opportunities affect the income of households? Once generated, how will this income be distributed among those who own factor resources? These are questions that we now begin to address from the perspective of community economic analysis tools. These models build from the concepts and theories discussed in Chapters 1 through 4.

Recall from our discussion of decision-making (Chapter 12), the Tinbergen policy model requires the policy analyst to specify structural relationships that identify cause and effect. The descriptive approaches to analyzing community economies generally lacked this set of causal relationships that infer change (Chapter 14). This chapter begins to establish some basic causal, or structural, relationships that allow us to build community economic models. These models, constructed by using these structural relationships, thus allow us to infer change in an economy as a result of some specified shock.

Recall from the discussion of the general circular flow of regional economies (Chapter 4) that households and firms are inextricably linked in two basic markets: the market for goods and services and the market for factors of production. Models built to estimate change must somehow simultaneously account for how firms produce goods and services and how households supply land, labor, and capital resources. Likewise, models need to account for how households consume goods and services and how firms demand production inputs of land, labor, and capital. Furthermore, the specific aspects associated with how households and firms interact needs to be specified as part of the set of structural relationships upon which the model is built. The market operations include both price and quantity effects. Models constructed to estimate economic change need either to incorporate these market effects or to make strong assumptions about how markets react to changes in demand and/or supply.

Finally, let's recall the discussion about economic growth, neoclassical production relationships, and income distribution in Chapter 2. As the cliché goes, the devil is in the details. The manner in which firms and households produce and consume is driven by our conceptions of how people and groups of people behave. Our economic understanding of individual and firm behaviors, market effects, and structural linkages is not a precise science and has evolved dramatically during the past 50 years (as presented in the development of growth theories in Chapter 2). As you can see, building a model that accounts for structural economic change is a tricky endeavor, indeed.

Each of the modeling approaches presented in this chapter and the next relies on different aspects of economic growth theory and focuses attention on different economic characteristics. For organization, we have chosen to outline the discussion based on a key set of market-driven relationships. This chapter begins by making some rather strict assumptions about supply and demand components. Namely, we present input-output analysis and social accounting matrices. These two modeling approaches generally restrict supply-demand equilibrium to fixed-price effects. As we move forward into more contemporary modeling approaches in Chapter 16, these strict supply-demand equilibrium conditions are relaxed.

This chapter is organized into three additional sections. First, we present a detailed discussion of input-output (IO) analysis that emphasizes the calculation of income, employment, and output multipliers, and how input-output information provides a better understanding of a community and its economic activities. A key shortcoming of input-output analysis is its overly aggregate approach to modeling change. This shortcoming leads into a section on social accounting matrix (SAM) analysis. SAMs place increased focus on households and the distribution of income. We conclude with a summary that focuses on limitations of fixed-price models and leads us into the topics addressed in Chapter 16.

INPUT-OUTPUT ANALYSIS

Input-output analysis encompasses a broad category of models that estimate economic change built on the basic premise that production in a region is comprised of inextricably linked firms that interact with one another. Also known as *inter-industry analysis,* this modeling approach represents an accounting structure that characterizes economic activity in a given time period and uses strict assumptions about production and supply-demand equilibriums to predict reaction of a community economy to stimulation resulting from shock. This shock is most often the result of some change in consumption, or demand. Other types of shocks that can be assessed using IO analysis include changes in government policies (e.g., spending, redistribution, tax policy), market-oriented demand changes, and changes in production by a given sector. IO provides an important tool to address questions of "economic impact" resulting from some prespecified exogenous shock. Depending on the manner in which IO is set up, it can model economic change based on both demand and supply shocks. Throughout the past 50 or so years, it has been widely used to "plan" for economic effects in both free-market economic structures as well as those structures that are more centrally planned or socialist based. IO has been applied at all spatial scales, ranging from local community impacts to broader regional and national-level geographies.[1]

Viewing IO analysis as a tool to generate employment, income, or output multipliers is far too limiting and fails to recognize other insights the technique offers the community economic analyst. Input-output provides a framework to collect, categorize, and analyze data on the inter-industry structure and interdependencies of the community's economy.

While IO can account for economic activity and existing interrelationships, it is not strictly comparable to traditional economic accounting systems. Conventional accounting of economic activity measures the total output of the economy as the value of all final products or all final sales. If a community produces $100 million of output, that number includes the cost of all the inputs and intermediate products used in the production process. Input-output, on the other hand, measures the value of each transaction (sale) that occurs (Miernyk 1965). The result exceeds the value of output derived through conventional accounting methods by the value of all intermediate products included in the accounting total but counted separately in the IO approach. Inclusion of the flow of intermediate goods and services among economic sectors is the heart of the IO approach.

In this section, we review the basic assumptions of the input-output model and how the model structure relates to a community's economy. A simplified example is used to calculate some multipliers.

The Input-Output Descriptive Accounts

The basic IO accounting framework begins by assuming an economy with n industry sectors. For elaborating relationships, we can specify X_i as total output of sector i in value (monetary) terms, Y_i as total final demand for sector i's product, and z_{ij} as the inter-industry demands from sector i (intermediate input) to sector j (intermediate demand). Thus, demand for each industry can be written as

$$X_i = z_{i1} + z_{i2} + z_{i3} + \ldots + z_{in} + Y_i \quad \textbf{(15.1)}$$

In the IO structure, production takes place under very strict linear conditions. Specifically, IO analysis assumes that inter-industry flows from i to j are wholly dependent on the output of j. This leads to a set of production relationships that are generally referred to as *technical coefficients*. Specified mathematically, these technical coefficients are defined as

$$a_{ij} = \frac{z_{ij}}{X_j} \quad \textbf{(15.2)}$$

where a_{ij} is the technical coefficient that translates value in dollar units into a proportion. A technical coefficient implies that production takes place under rather strict conditions of constant input to output ratios. Once derived, use of a technical coefficient also implies fixed prices. This technical coefficient reflects a specific production function that is unique from the more classical production functions described in Chapter 2 (see Fig. 15.1).

Provided with this manner of specifying output, inputs, and production, the IO approach progresses into a valuable accounting structure that captures important economic relationships. Operationally, we can take these relationships and build a numerical representation of how a region's economic activities relate to one another. These are captured in a set of tables that progress from description to proportional views of community economic activity.

Crucial Assumptions of IO

Using IO analysis requires some crucial assumptions (Bendavid 1974; Miller and Blair 1985; Miernyk 1965). The ability to recognize and understand these assumptions affects the usefulness of the results. A critical assumption of IO is that the econ-

omy represented by a balanced IO table is initially in an equilibrium state. This means that the amount of output produced in a given sector is just equal to the amount of inputs purchased by that sector. Again, we need to remember that IO accounts are comprehensive in the sense that all purchases by both sellers and buyers are captured. When we refer to IO tables as being "balanced," this assumption of equilibrium is stated explicitly within the construction of the table. When IO is used as a predictive tool (more on this later in the chapter), the assumption of equilibrium further states that the economy automatically returns to an analogously defined equilibrium.

Another obvious assumption is that input-output assumes the industry expansion path is linear and has constant returns to scale. This assumption implies that doubling inputs leads to doubling output. Constant return in production also means that there exists no external economies or diseconomies associated with any changes in production. Increases in output neither create a shortage of skilled labor (i.e., external diseconomy) nor generate an

Production functions

- **Classical production functions**
    - Cobb-Douglas forms
        - Negative slopes represent the fact that if we decrease the input of one, we must increase the input of the other for equal output.
        - The fact that isoquants "bulge" toward origin (convex) represents the law of diminishing marginal returns.

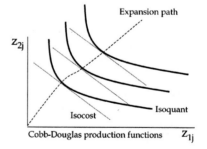

- **Input-output production functions**
    - Leontieff forms
        - Increasing the input of one is useless unless we also increase the input of the other.
        - It doesn't matter what the price ratio is, we need inputs in fixed proportions.
        - Production of $X_j = \min(z_{ij}/a_{ij})$.
        - The expansion path, in the Leontieff case, is a straight line.

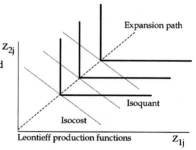

Figure 15.1. Specification of the type of production characterized within input-output analysis.

improved distribution or financial system (i.e., external economy) that further alters output.

A related assumption in IO is that changes in relative factor prices will not affect the proportion of factors used. The only way more of or less of a given factor or intermediate product will be used is through a change in final demand.

In IO, the way each sector produces output is simplified into an aggregate representation of production. For example, a community with several similar firms combined into the same economic sector requires assuming they all use a similar production process. For example, a furniture manufacturer and iron foundry are separate manufacturing sectors, but an IO model with only one manufacturing sector means that the furniture manufacturer is considered the same as the iron foundry. Clearly this is not the case, but when individuals use an IO table from another study, they make that implicit error if the sectors are aggregated differently.

While some of the basic assumptions of IO tend to be quite restrictive, the major difficulty is failure by the analyst to recognize their significance to the results generated.

Transactions Table

The basic components of the IO model are the transaction table, direct requirements table, and the total requirements table. The transactions table, the foundation for the other two tables, displays the flow of goods and services among suppliers and purchasers in a specific locale for a specific time, typically a year (Bendavid 1974; Miernyk 1965; Miller and Blair 1985). The data in the transactions table represent the total sales (output) of every unit in the local economy for that study period.

The transactions table gives a complete accounting of the flows of goods and services within the economy. Before discussing some transformations increasing the analytical usefulness of the transactions table, we examine the main components of the transactions table and extract their analytical value.

The simplified transactions table in Figure 15.2 has four quadrants—A, B, C, and D. Quadrant A is the inter-industry or processing sector. This quadrant shows the sales by local industries (suppliers or selling sectors) to other local industries (users or buying sectors). This portion of the transactions table displays the local economy and the interrelationships within itself. The processing sector is the prime concern here and represents all the sales and purchases undergoing further use by local economic units within the accounting period being studied.

Quadrant B is the final demand or final sales or final users quadrant. This quadrant includes all the output destined for economic sectors that will not process it further in this time period or in this locale. It contains both local and nonlocal users, but most will be nonlocal users. Some local users are households, sales for local investment (gross and net), and sales to local government. Investment by local processing sectors is considered a final use because the product experiences no further processing nor is it resold in this time period. Sales to local households are considered a final sale if households consume the output and do not process it further. If households alter their consumption patterns, they will be included in the processing sector. Local government is part of the final demand sector, assuming it does not process local inputs into outputs or its level of output is invariant to local economic activity.

Some nonlocal users in the final demand sector are nonlocal households, nonlocal businesses, and central government. The sales to nonlocal businesses can be for further processing, investment, or final use. It makes no difference because the further use occurs outside the local economy represented by the processing sector (Quadrant A). The final segment of most final demand sectors is the net addition to the inventory of local businesses. It represents no further processing of the good or service in the current accounting period (i.e., a final use).

Quadrant C represents final payments or primary inputs, the result of sales by primary input suppliers to local processing sectors. Composition of the primary inputs sector depends on the analytical questions asked. The two major variations are whether local households and local governments are included as primary inputs or as part of the processing sector.[2] Final payments to households include wages, salaries, proprietor income, property income, and other sources of personal income. Other final payments can be to local and nonlocal suppliers, including gross business savings,[3] payments for imported goods and services, and taxes to local and nonlocal government.[4] For small communities, the purchase of imported inputs is a major portion of the primary inputs sector. These imports are the difference between total inputs needed and what is supplied locally.[5]

Quadrant D represents the final payment to the final demand sector. This section of the transactions

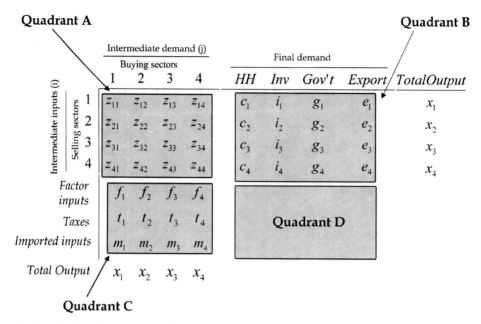

Figure 15.2. A simplified transactions table. Note: Inv = invest; HH = households.

table records total sales by suppliers to purchasers (i.e., volume of total transactions). At this point, Quadrant D is empty. As we move into the discussion of social accounting matrices, this quadrant becomes more important.

When the primary inputs and final demand sectors are added to the processing sector, an equivalency of total inputs and outputs occurs. In other words, the economy must generate enough output to use all the inputs purchased during the period under study, or the economy purchases only enough inputs to produce the outputs generated during the time studied.

The preceding discussion of input-output is one way of examining the elements of the transactions table. Another way is to divide the transactions table into the two basic economic actors it represents— suppliers and demanders.

The *rows* of quadrants A and C represent suppliers. The rows in Quadrant A are the intermediate supply sectors. These rows represent local producers who purchase and transform inputs into products sold to other economic sectors (intersecting columns of the row) for further processing or final use. These rows represent the economic sectors in the community using local and nonlocal inputs to

produce goods and services for further production or final use.

Quadrant C represents primary suppliers who produce inputs not requiring the purchase of intermediate goods and services from the local processing sector in this time period to generate those inputs. Payment for these inputs is a final payment and generates no further economic transactions in the local economy. In some studies, local households are considered a primary input supplier (i.e., labor), and payments for labor do not generate changes in consumption by community households.

The *columns* of quadrants A and B are the portion of the transactions table representing the purchasers. Quadrant A represents both the suppliers and purchasers because the interpretation of the data changes with how the table is read. Reading across a row in Quadrant A gives an indication of the distribution of sales (i.e., a supplier). Reading down a column gives an indication of what inputs the economic sector requires for its production.[6] The columns of Quadrant A report the purchases of intermediate goods and services for further processing.

Quadrant B represents the final users who purchase the outputs from the processing sector in their final form and for their final use. The "rest of the world" (alternatively identified as exports in Fig. 15.2) determines level of final demand for the local economy. At this simple level, input-output is analytically similar to our perspective of export base theory as presented in Chapter 4. This analogy strengthens if the shock that drives economic impact is representative of exportable goods and services. The IO approach, however, is not as simplistic as the export base approach, since the export (final demand) and nonexport sectors (processing sector) are analyzed in much greater detail.

Delineating the supplier-purchaser elements of the transactions table highlights the interrelationships in the table. Intermediate suppliers and purchasers (i.e., processing sector) are the same economic sectors (e.g., retail trade, agriculture, manufacturing, transportation). The sales and purchases among these economic sectors are interrelated because the purchase of inputs depends on the sector's sales. Primary suppliers and final purchasers need not be the same economic parties, although they can be, because of nonlocal purchasers and suppliers. Households are the major economic sector included in both the primary input and final demand sector. Households supply the primary input of labor and also purchase goods and services for final consumption. Even though the households may be identical, they or any other identical economic sector included in both the primary inputs and final demand sectors are treated as if their demand and supply decisions are independent.

Additional information can be gleaned from the transactions table by triangularization (Bendavid 1974; Bromley 1972). Triangularization simply rearranges the rows in the processing sector on the basis of the number of linkages with (sales to) other local economic sectors. This explicitly demonstrates the level of interdependence among the various elements of the processing sector and gives insight to the role any economic sector plays with the rest of the local economy as measured by the degree of economic interconnectedness.

A variation of triangularization is to rank each sector, including the final demand (primary supply) sectors, according to the portion of sales (inputs) that go to (come from) the processing sectors and final demand (primary suppliers) sectors. Ranking processing sectors according to the proportion of sales to either further processing or final demand indicates how sensitive a given sector is to local versus nonlocal demand forces. The ranking of each

sector's purchases from other local businesses or as imports gives an idea of the dependence on local versus nonlocal suppliers. Both insights help the community economic analyst determine the relative importance of external and local forces to a particular sector.

Direct Requirements Table

The first modification of the transactions table for IO analysis yields the direct inputs required for the output of any processing sector (Bendavid 1974; Miernyk 1965; Miller and Blair 1985). The ratio of direct inputs to output, called the technical coefficient, measures the technical relationship between inputs and outputs in an economic sector. Every business knows these numbers because that is how it adjusts its purchases of materials and labor as sales change. Dividing each number in a processing sector column by the total output produced by that sector computes the technical coefficients in the processing sector (Quadrant A in Fig. 15.2). This calculation is not performed for the economic sectors in the final demand sector because sales to final demand do not represent sales to other local firms for further local processing in this accounting period.

An Example

Tables 15.1 and 15.2 give a simple numerical example of the transactions table and direct requirements table.

Table 15.1, a simplified IO transactions table, indicates this economy's total output is $7,020 during the accounting period under study. Reading across the Agriculture *row* of Table 15.1 indicates agriculture sells its $741 of output in the following manner: $202 to other local agricultural producers, $182 to manufacturers, $10 to trade businesses, $12 to local service businesses, $100 to households, and $235 to nonlocal users. Thus, over half ($406) of the agricultural production goes to other local businesses for further local processing and uses, and households consume $100. Reading down the Agriculture *column* of Table 15.1 indicates the economic sectors providing inputs for agricultural output. Over half ($374) of the total inputs required to produce the agricultural output comes from households ($200) and imported inputs ($174); of the balance, $32 comes from manufacturing firms, $47 comes from trade firms, and $86 comes from service firms.

Table 15.2 displays the direct requirements or technical coefficients. The technical coefficients for

Table 15.1.   Simplified input-output transactions table (in dollars)

	Agriculture	Manufacturing	Trade Businesses	Local Service Businesses	Households	Nonlocal Users	*Total Output*
Agriculture	202	182	10	12	100	235	*741*
Manufacturing	32	68	2	26	39	300	*467*
Trade	47	35	991	334	1,200	172	*2,779*
Service	86	59	565	561	1,500	262	*3,033*
Households	200	40	205	1,250	1,698	100	
Imported inputs	174	83	1,006	850	333	1,053	
Total inputs	*741*	*467*	*2,779*	*3,033*			*7,020*

Table 15.2.   Direct requirements table

	Agriculture	Manufacturing	Trade	Service
Agriculture	0.27	0.39	0.00	0.02
Manufacturing	0.05	0.15	0.00	0.01
Trade	0.06	0.07	0.36	0.15
Service	0.12	0.13	0.20	0.17
Households	0.27	0.08	0.36	0.24
Imported inputs	0.23	0.18	0.36	0.24
Total inputs	*1.00*	*1.00*	*1.00*	*1.00*

BOX 1

Sector	Computation		Technical Coefficient
Agriculture	202/741	=	.27
Manufacturing	32/741	=	.05
Trade	47/741	=	.06
Service	86/741	=	.12
Households	200/741	=	.27
Imported Inputs	174/741	=	.23

the Agriculture sector (column) are computed from Table 15.1 as shown in Box 1.

Table 15.2 shows that every dollar of agricultural production requires $0.27 of inputs from other agricultural firms and $0.06 from trade firms. Table 15.2 displays how the local economy responds to a change in final sales of a particular sector. A $1.00 increase in agricultural sales for final demand causes a direct increase in agricultural production of $1.27 or $1.00 + $0.27 (see direct requirements table). This increase means additional local inputs must be purchased to permit an increase in production as calculated from the direct requirements table shown in Box 2.

The $0.292 represents the amount of additional output generated in other sectors of the local economy to produce the $1.27 change in agricultural output. However, this figure does not include the effects on other sectors because manufacturing increased its output by $0.064 and so on. Table 15.3 shows the total effect (direct and indirect).

The Predictive Input-Output Model

Until this point, our description of the IO approach has been limited to the valuable descriptive elements that help us understand how a local economy is structured. We have yet to describe the predictive model to infer change given some prespecified shock. Historically, development of this predictive form did not occur until the 1930s when Wassily Leontief, in his work on national accounts, discovered a very interesting set of mathematical relation-

BOX 2

Sector	Direct Output Change in Initiating Sector		Direct Requirements		Direct Output Change in Other Sectors
Manufacturing	$1.27	×	0.05	=	$0.064
Trade	$1.27	×	0.06	=	$0.076
Service	$1.27	×	0.12	=	$0.152
Total					$0.292

ships that moved IO accounts forward to be a useful predictive modeling tool. For this, Leontief was awarded the Nobel prize in the early 1970s. In large part, Leontief's most important contribution occurred as he worked through the following descriptive identities and was able to reduce and transform a purely descriptive phenomenon into a predictive tool.

Leontief's predictive form of IO analysis relates change in total output to the macroeconomic components of demand (intermediate and final). Basically, he stated that if we can estimate changes in final demand, we can predict how an economy will react as measured in change in output. To understand his logic, we return to our original specification of output found in equation (15.1) and reorganize the technical coefficient found in equation (15.2) into the following:

$$z_{ij} = a_{ij}x_j \qquad (15.3)$$

Substituting equation (15.3) into equation (15.1), we arrive at the following statement for sectoral output in a simple two-sector model:

$$\begin{aligned} x_1 &= a_{11}x_1 + a_{12}x_2 + y_1 \quad (\text{sector 1}) \\ x_2 &= a_{21}x_1 + a_{22}x_2 + y_2 \quad (\text{sector 2}) \end{aligned} \quad (15.4)$$

This set of equations can be rearranged to be represented as

$$\begin{aligned} x_1 - a_{11}x_1 - a_{12}x_2 &= y_1 \quad (\text{sector 1}) \\ x_2 - a_{21}x_1 - a_{22}x_2 &= y_2 \quad (\text{sector 2}) \end{aligned} \quad (15.5)$$

Further reorganization can represent these relationships alternatively as

$$\begin{aligned} (1 - a_{11})x_1 - a_{12}x_2 &= y_1 \quad (\text{sector 1}) \\ -a_{21}x_1 + (1 - a_{22})x_2 &= y_2 \quad (\text{sector 2}) \end{aligned} \quad (15.6)$$

When we extend this simple model into more than a couple of sectors, our scalar representation gets rather complex and computationally challenging. Our discussion is greatly simplified if we move to matrix notation. Recognize that the equations for the two sectors found in equation (15.6) can be rewritten in matrix form as

$$\begin{bmatrix} 1 - a_{11} & -a_{12} \\ -a_{21} & 1 - a_{22} \end{bmatrix} \begin{bmatrix} x_1 \\ x_2 \end{bmatrix} = \begin{bmatrix} y_1 \\ y_2 \end{bmatrix} \quad (15.7)$$

and that the first array can be rewritten as

$$\begin{bmatrix} 1 & 0 \\ 0 & 1 \end{bmatrix} - \begin{bmatrix} a_{11} & a_{12} \\ a_{21} & a_{22} \end{bmatrix} \quad (15.8)$$

that, when written in matrix notation, can be represented as $(I - A)$.

In matrix notation, Leontief's contribution walked through the following rearrangements (equations 15.9–15.11) to set up the predictive IO form found in equations (15.12) and (15.13). Each equation is presented in matrix form with the analogous scalar form identified from above.

$$X = AX + Y \,(\text{analogous to 15.4}) \quad (15.9)$$

$$X - AX = Y \,(\text{analogous to 15.5}) \quad (15.10)$$

$$(I - A)X = Y \,(\text{analogous to 15.7}) \quad (15.11)$$

Dividing $(I - A)$ by both sides results in Leontief's prime contribution. In matrix form, the quotient operation is represented as an inverse. Giving deference to its originator, this particular inverse is often referred to as the *Leontief inverse*.

$$X = (I - A)^{-1}Y \quad (15.12)$$

Now we are able to predict change in a community's economic output (X) by specifying changes in demand (Y). Formally, the IO predictive form is given as

$$\Delta X = (I - A)^{-1}\Delta Y \quad (15.13)$$

This set of relationships can be restated in words, which is sometimes helpful for the student learning this for the first time. The story goes as follows. A community's economic output for each of many sectors is produced with a unique set of inputs. In other words, output (X_j) is produced using intermediate purchased inputs and primary factors. These are fixed and reflective of the year in which the tableau was constructed. The amount of input purchased by a sector is determined solely by its level of output, which is reflected within its technical coefficient. An example might be that the number of pigs required by a bratwurst manufacturer and purchased from local hog farmers is determined by how many bratwursts the firm produces.

Embedded within this set of relationships are important economic assumptions. Namely, we must assume that no external economies of scale exist, the in-region and out-of-region distribution of purchases and sales is fixed, as well as that there are *no* constraints on resources (what is needed in production is readily available without constraint) and that local resources are efficiently (and fully) employed.

The Causal Relationship

The mathematical specification created by Leontief represents an important causal relationship that can be described in words: Final demand change in sector *i* affects the output of sector *i* but also creates demand for other sectors. These other sectors change their output through increasing inputs, which creates further change in a diminishing manner. This continues in an increasingly diminishing fashion until the effect is fully exhausted. Also, note that if we set up the tableaus correctly, output change is accomplished through additional use of labor resources, which creates additional income, and government revenue, which also proceeds until the effect is exhausted.

There is another important inference that is a mathematical tautology but numerically represents what we just described. This inverse inference rep-

resents the "magic" of the Leontief inverse. It mathematically reflects the above stated set of causal relationships and identifies a total effect of change as represented by a power series approximation:

$$[I - A]^{-1} = I + A + A^2 + A^3 + \dots + A^n \quad \text{(15.14)}$$

To restate, this power series approximation reflects the round-by-round impacts associated with a pre-specified (or exogenous) shock to the system in a diminishing fashion. Indeed, this is the causal effect embedded within IO analysis.

Total Requirements Table

To return to our example, a final transformation computes the total requirements table. The total requirements table sums the direct and indirect input requirements per unit of output and is operationalized by using the Leontief inverse (Table 15.3). Total requirements recognize that if output for final demand increases, not only must purchases of direct inputs increase, but firms supplying those direct inputs also must increase their purchases of inputs. Thus, the total requirements table displays the additional production (i.e., input supply) increases responding to the initial stimulus.

The calculation of total requirements requires matrix inversion as specified in equations (15.3–15.14). (An excellent discussion for review can be found in Bendavid 1974, Miernyk 1965, and Miller and Blair 1985.)

The total requirement table represents all of the repeated increases in output and input purchases. Readers are encouraged to derive this result by using the equations presented above. This can be accomplished easily with most spreadsheet software packages.

Input-Output Multipliers

The most obvious result from an IO analysis is the estimate of multipliers. The three types of multipliers

Table 15.3. Total requirements table

	Agriculture	Manufacturing	Trade	Service
Agriculture	1.44	0.68	0.04	0.12
Manufacturing	0.12	1.26	0.05	0.16
Trade	0.21	0.29	1.67	0.34
Service	0.28	0.37	0.42	1.33
Output multipliers	2.05	2.60	2.18	1.95

generally estimated are *output, income,* and *employment* (Miller and Blair 1985; Richardson 1972).

The basic multiplier estimated with IO analysis is the output multiplier since the data in a transactions table is gross sales (output) data. The *output multiplier* indicates the total output change in all processing sectors in an economy resulting from a $1.00 change in the final demand for the output of a specific sector. The output multiplier is computed by summing the total (direct and indirect) coefficients in the processing sector only. The last row of Table 15.3 reports the output multipliers.

The *income multiplier* is the total change in local income resulting from a $1.00 change in output for a specific sector (Miernyk 1965). The basic assumption is that a change in output for a processing sector results in a local income change that is some fraction of the output change. The *direct* income change is the direct (technical) coefficient of the household row intersecting with the column representing the processing sector experiencing the initial change in output. This direct coefficient is the wages, salaries, proprietors income, and so on per dollar of output (Table 15.2). The *indirect* income change is the conversion of indirect output changes into household income. In other words, as output of intermediate producers changes to meet the change in final demand, some proportion of these indirect output increases are income for workers and business owners.

The direct income effect is the row of household coefficients in the direct requirements table (see Table 15.2). For example, if agricultural production increases by $1.00, then direct household income would increase by $0.27. The direct change in agri-cultural household income as a result of the $1.00 change in final demand for agricultural products is $0.39. The total requirements table (Table 15.3) indicates that each $1.00 increase in final demand increases total (direct and indirect) agricultural output by $1.44. This $1.44 increase in agricultural production generates $0.27 of household income for each dollar change in output as shown in Box 3.

But, the increase in agricultural production (output) also affects the output of other economic sectors and the households (workers) in those sectors. The total requirements table (Table 15.3) indicates that a $1.00 increase in agricultural output causes a $0.12 increase in output of the manufacturing sector, a $0.21 increase in the output of the trade sector, and a $0.28 increase in the output of the service sector. These supply the total output stimulated ($1.44) in the agriculture sector. The household row of the direct requirements table (Table 15.2) indicates how much income is created in each sector of the local economy as a result of changes in their output. This row indicates that households in the respective sectors receive the following amounts of income from each $1.00 increase in output for that sector: manufacturing $0.08, trade $0.07, and services $0.41. Using this information, the indirect income change from the $1.44 change in agricultural output creates $0.14 of income in other sectors as shown in Box 4.

The total income change from the $1.00 change in the final demand for agricultural output is $0.53. The total income change is the sum of the direct and indirect income changes ($0.39) in the initiating sector (agriculture) and the income changes from the indirect output effects ($0.14). The income multiplier in this case is not .53 but 1.963 and is computed as follows:

BOX 3

Total Change in Output for Sector		Direct Income Coefficient for Sector		Income change for Households in Sector
$1.44	×	0.27	=	$0.39

BOX 4

Sector	Total Output Change		Direct Income Coefficient		Indirect Income Change
Manufacturing	$0.12	×	0.08	=	$0.010
Trade	$0.21	×	0.07	=	$0.015
Services	$0.28	×	0.41	=	$0.115
Total					$0.14

$$\frac{\text{Total income change}}{\substack{\text{Direct income change} \\ \text{in initiating sector}}} = \text{Income multiplier}$$

For this particular example, it is calculated as:

$$\frac{\$0.53}{\$0.27} = 1.963$$

The *employment multiplier* measures the total change in employment due to a $1.00 change in final demand in a specific economic sector. The measure used is typically total number of jobs or annual full-time equivalents (FTE). The basic assumption is a linear relationship between annual FTEs of employment and dollars of output generated. An increase in output leads to a linear and proportional increase in labor required to produce the additional output. The household row in the direct requirements table becomes FTEs of labor supplied rather than income received. The procedure for estimating the employment multiplier parallels the income multiplier.

Graphically, we can illustrate the round-by-round relationships modeled using IO analysis. This is found in Figure 15.3. The direct effect of change is shown in the far left-hand side of the figure. For simplification, if this direct effect is a $1.00 change in the level of exports, the indirect effects will spill over into other sectors and create an additional $0.66 of activity. In this example, the simple output multiplier is 1.66. A variety of multipliers can be calculated using IO analysis. The type of multiplier calculated is a function of the manner in which the direct requirements table is closed for purposes of inversion. In other words, the amount of transactions included in the inversion will determine the type of multiplier calculated.

Type I versus Type II Multipliers

The issue of whether households and/or local government are included or excluded from the processing sector depends on the questions analyzed and the assumptions made. A major difference among IO studies is which sectors are included in the processing sector and final demand/payments sector. The most frequent difference is whether households and local government are in the processing sector or in the final demand/payments sector. If households are included in the processing sector, the model is a closed IO model. This has an important role in the nature of the multipliers produced by the model. In models that are closed through only the inter-indus-

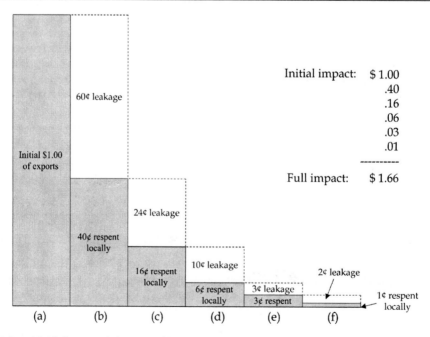

Figure 15.3. Multipliers and the round-by-round impacts estimated by using input-output analysis. (Adapted from Lewis et al. 1979.)

try transactions, the multipliers that are produced are referred to as *Type I multipliers*. These multipliers capture the direct and indirect affects of any given shock. If the model is *closed through labor* or households, then the multipliers that are produced are referred to as *Type II multipliers*. Type II multipliers capture direct, indirect, and induced impacts. Closing the IO model permits induced changes in the consumption/spending patterns.

One can think of the differences between Type I and II multipliers as what is included in the production function of the industry. If labor is not included in the production function, we have Type I multipliers. If labor is included in the production function (i.e., the model is closed through labor), we have Type II multipliers. The difference can be described as labor spending its wages in the local economy. The numerical difference between the Type I and II multipliers is the induced effect, or the impact of labor spending wages. Note that Type II multipliers will always be bigger than Type I multipliers.

The community practitioner can use these multipliers to gain valuable insight into the makeup of industries in the local economy. For example, in labor-intensive industries, the Type II multiplier will generally be much larger than the Type I multiplier. Similarly, industries that pay higher wages will tend to have larger induced effects. By decomposing the multiplier into its separate components (direct, indirect, and induced), the practitioner can gain a better understanding of the local economy.

Common Sense and Input-Output Multipliers

Some obvious elements of IO analysis lead to the need for reflection on the use of IO multipliers. First, it is imperative to understand that *IO tableaus are constructed for and unique to regions and communities.* Therefore, it is not feasible to use one region's tableau for another region because economic characteristics will most likely be dramatically different. Input-output multipliers are region specific and therefore cannot be standardized. To reinforce this statement, note from Figure 15.3 that the amount of leakage depends on the size of the region. Typically, larger regions (both economic and geographic) tend to have smaller leakages because more economic activity is captured within the region. Thus, multipliers calculated for large regions tend to be larger than those for smaller regions. In any event, there is no such thing as a "standardized" IO multiplier.

One multiplier cannot be interchanged with another. This includes both multiplier type *and* the industry for which the multiplier is calculated. An employment multiplier is very different than an output multiplier. Furthermore, an employment multiplier for the pulp and paper sector could easily be quite different when compared to an employment multiplier for the retail sector. Once again, *IO multipliers evade standardization.* In addition to being region specific, IO multipliers are industry and type specific.

Multipliers include direct effects. The initial shock used to derive impact is already counted in the multiplier. Practioners interpreting the multipliers should be wary of double counting direct effects because overestimates will occur when basic and nonbasic sector multipliers are combined. Interpretations should focus on the use of single multipliers. Multipliers are *not* additive in the sense that you cannot count the direct impact more than once.

Input-output multipliers do NOT represent the turnover of dollars (number of times a dollar changes hands) in a local economy, rather they are the net result of inter-industry purchases and increases in income from outside. A common misuse of multipliers is confusing a multiplier with turnover. Turnover is the number of times a dollar changes hands in a community before it escapes. This is not the same as the income multiplier. The difference in the value of a multiplier and turnover is evident in Figure 15.3. The $1.00 of initial exports is respent or turned over five times in the community (i.e., turnover is 5), but the multiplier is 1.66. Note that only $0.03 are involved in the last exchange, thus the last turnover does not involve the full dollar initiating the change.

There will be a dramatic difference between the use of employment multipliers based on an FTE and those based on total numbers of jobs. This is particularly true for sectors that use large numbers of seasonal and/or part-time employees. The analyst should be wary of employment multipliers that are based on total numbers of jobs for certain types of retail and service sectors. A good example of the need for careful interpretation is found in the industry sectors that comprise tourism since a large proportion of these jobs are seasonal and/or part time.

The time to accomplish economic impact is rarely addressed in IO analysis. Typically, IO multipliers are to be considered static. The assumption in the use of static multipliers is that if demand changes *now*, the community impacts will eventually occur sometime in the future.

Input-output analysis rests on some rather strong assumptions that tend to be violated if the shock introduced to the system is extremely large relative to the size of the community economy and the sector being shocked. Small amounts of change, rather than large changes, tend to reflect more realistic economic impacts as modeled using IO.

Input-Output Analysis in Practice

Input-output offers several advantages in community economic analysis. An IO model disaggregates changes into individual economic sectors of the local economy. Both the effect of a change in a specific sector on all other sectors, and the effect of change in the output of all sectors on a specific sector can be examined. The approach distinguishes direct changes, indirect changes, and if the model is closed appropriately, induced changes.[7]

The most obvious disadvantage of IO analysis is the immense amount of data necessary to detail the interrelationships within an economy. This data requires considerable time to generate and is very expensive to assemble. The time and costs involved in assembling data for community and small area IO models has led to a subfield within IO studies shortcutting the data collection requirements (Davis 1976; Miernyk 1976; Miller and Blair 1985; Morrison and Smith 1974; Otto and Johnson 1993; Richardson 1972; Richardson 1985; Round 1983; Schaffer and Chu 1969).

Constructing Input-Output Tables

As hinted above, IO tables contain a vast array of data that evade easy compilation. Historically, input-output tables were developed based on surveys. These survey-based tables were, and continue to be, superior because of their sensitivity to community anomalies in production and data on economic flows (or transactions). They are often preferable due to regional specificity and accuracy. The downside of survey-based IO tables is that they are, in general, prohibitively expensive to operationalize. Many of the early efforts were severely criticized because of their myopic focus on construction details to the detriment of their application. It is safe to say that survey-based IO suffers because the effort generally distracts the analyst from the policy analysis driving the effort.

Today, non-survey-based IO tables are common. These IO tables are built using secondary data and they typically apply national IO coefficients to subnational regions. Secondary data on community activity is used to "benchmark" (control total) subregional economic activity. Interregional flows are adjusted by using a variety of methods. Examples include the use of regional purchase coefficients and supply-demand pooling (see Braschler and Devino 1993).

Accuracy in Input-Output

A topic related to survey- versus non-survey-based IO tables has to do with accuracy. According to Jensen (1980), there are two types of accuracy with respect to IO analysis. The first, *partitive accuracy,* is the accuracy of individual IO table components. Partitive accuracy is concerned with statistical significance and accuracy of cell contents and addresses the question of whether each cell accurately represents truth.

The second, and perhaps more important, type of accuracy in IO analysis is termed *holistic accuracy.* Holistic accuracy is concerned about the accuracy of model structure and its use and focuses on the relationship between the model results and reality. This type of accuracy addresses the question of whether the overall model makes sense. In other words, does the predicted change affect community activity in some reasonable fashion? Holistic accuracy addresses a key need in policy-making. Namely, is the model developed to predict change capable of capturing the unique attributes of the region or community being examined? Does the model capture unique community production characteristics? Does the model capture unique export/import values? These values are critical for smaller geographical units because they are more dependent on outside influences for economic change. Large regions tend to be less dependent because more of the economic activity is captured internally. Another key aspect of holistic accuracy deals with how different regions interact. Does the model developed for policy analysis realistically capture multiregional interaction?

Regional Input-Output Modeling

During the past 40 years, several attempts have been made to operationalize IO modeling. Early models were typically cumbersome and generally not user-friendly. Technical extensions and refinements to non-survey-based modeling, coupled with adaptations that improved user-friendliness, have produced two rather strong competitors that can provide practitioners with a fairly detailed IO analysis at the community level.[8] These two competitors provide very different approaches to community economic modeling and have various strengths and weaknesses.

IMPLAN (IMpact Analysis for PLANning) is a non-survey-based IO modeling software system that couples with county-level databases to provide analysts with an ability to generate useful economic impact reports at flexible spatial scales. It was originally developed in the 1970s by Greg Alward and others within the U.S.D.A. Forest Service to address important issues pertinent to National Forest planning (Alward et al. 1989). Over time, it has had several technical refinements in both the software and databases to improve its holistic and partitive accuracy. Also, important adaptations from the original mainframe version to microcomputer versions and improvements in its user-friendliness have made this an operational tool for community economic analysis. A private consulting group known as Minnesota IMPLAN Group maintains the current version. The IMPLAN software and databases provide a relatively easy to understand and adjust IO modeling system. The databases that are used with the MicroIMPLAN software are constructed from top to bottom using standardized secondary data sources. They are internally consistent and use regional purchase coefficients for trade adjustments. Applications typically introduce shocks as changes to final demand. A rather innovative extension has adapted the general IO structure into a social accounting framework. Like standard IO analysis, it is static and reflects the key limitations of IO as outlined above. Its relatively inexpensive packages combined with its user-friendliness provide a realistic and useable tool for the community economic development practitioner.

REMI (Regional Economic Models, Inc.) is a county-level non-survey-based conjoined input-output/econometric modeling software system originally developed and maintained by George Treyz and others (conjoined models are discussed in Chapter 16). It combines the beneficial detail of inter-industry analysis with an ability to forecast important components of economic change to capture a more dynamic set of characteristics in its impact assessments. This latter component is an important limitation to standard analyses that rely only on IO to estimate change. This said, REMI is relatively less easy to understand or to adjust and represents a more "black box" system for estimating economic impacts. Technically, REMI uses input-output for first round indirect results, then passes these results on to an econometric forecasting model to estimate change. It relies on readily adjustable "policy" variables and is more theoretically consistent in that it explicitly accounts for important economic structure attributes, including supply-demand linkages, cost linkages, and wage-determination linkages. Compared to MicroIMPLAN, REMI tends to be regarded as a more sophisticated and theoretically consistent modeling approach. Along with this benefit, however, REMI also tends to be a considerably more expensive option, which could limit its use by community economic development practitioners.

Applications of IO analysis are often dependent on the quality of questions asked. Prior to undertaking an IO analysis, the analyst should first consider some important contextual issues. The first issue builds on the notion that with IO analysis, our geographic definition of *community* matters. The analyst needs to think carefully about a reasonable definition of the region that encompasses the community. The policy being analyzed will need to be specific to this region, which in itself has unique economic characteristics. Does the community analysis coincide with the location of policy shock? What forward and backward linkages will result from the policy? How will these linkages be analyzed? Is the location of policy impact related to some adjacent functional economic area? What regional homogeneity is important in delineating a modeling area? Does the regional delineation capture the overarching uniqueness of the community being modeled? These questions should be carefully assessed before proceeding with IO analysis.

Another issue that needs careful consideration has to do with the relevant policy question. How well does IO modeling relate to the specific public policy objective being analyzed? Specifically, how does the public policy instrument relate to the economic shock introduced into the IO analysis? How realistic are the assumptions with respect to specific policy instruments that effect change? What unintended consequences might develop that may or may not be modeled using an IO approach? The analyst must be careful to identify *reasonable* final demand changes. Relevant questions should critique the appropriate level of sectoral definition. Is the driver of impact (the shock) of an appropriate magnitude and are the results reflective of appropriate units of analysis?

Finally, before embarking with an IO impact analysis, the analyst must consider how the results will be used. Is there an appropriate level of specificity with respect to effects? Are the results sensitive to changing parameters? Are there an accounting for and an acceptance of key assumptions, limitations, and/or caveats?

Extending the Traditional IO Framework

To reiterate, standard IO analysis tends to be static and demand driven. Extensions of standard demand-driven IO analysis may be dictated by alternative conceptions of what drives a community economy and/or general shortcomings of demand-driven IO analysis in answering the policy question at hand. This latter motivation to extending demand-driven IO models is substantive. Shortcomings of demand-driven models include a general failure to capture unique supply constraints. A good example of this is the general failure of demand-driven models to allow useful analysis of factor markets to model real-world situations reflective of labor strikes, land production policies (set-asides or production incentives), and capital shocks resulting from unique windfall profits. These supply constraints are better modeled by using extensions to traditional IO analysis developed throughout the years. An example is supply-determined (mixed exogenous/endogenous) IO analysis, which is beyond the scope of this chapter. (The reader interested in this specific extension is referred to the classic text in IO written by Miller and Blair 1985, Chapter 9.)

Extensions to traditionally defined IO may also be called for because of an inability to accept the underlying assumption of staticness. For instance, an analyst may wish to focus on tracking change over time or may need more temporal specificity as to how change occurs through time. In addition to the previously described extension built into REMI (conjoined IO econometric analysis), an extension to standard IO that more clearly specifies change through time is captured by a method known as dynamic IO analysis. (This topic is also beyond the scope of this chapter, but Miller and Blair 1985, Chapter 9, is a good source).

Finally, another important extension to traditionally defined IO is motivated by the overly aggregate nature of IO results. Increasingly, development analysis is interested in more than aggregate economic growth and change. Astute decision-makers often have interests that focus on the distribution of income resulting from some policy. To address this need, IO analysis has been extended into the broader social accounting framework of social accounting matrix analysis.

SOCIAL ACCOUNTING MATRIX ANALYSIS

Social accounting matrix (SAM) analysis represents a category of modeling approaches that extend IO analysis to more fully address development issues that focus on income distribution. General dissatisfaction with IO analysis and its rather myopic focus on aggregate job growth, aggregate income growth, and incomplete accounting for sources of income led to a need for this specific adaptation.[9] SAM extensions were initially developed during the late 1960s and early 1970s as a result of general dissat-

isfaction with the manner in which income flows were treated. A good overview of SAM development and analytical background can be found in Pyatt and Round (1985), Hewings and Madden (1995), and Isard et al. (1998).

Building on our earlier discussion of economic growth theory in Chapter 2, *income distribution* has been and continues to be a leading problem in community economic development. It is no longer satisfactory to develop policies based on estimates of aggregate change without identifying and estimating impacts on key stakeholder groups. A significant component of contemporary public policy debate frames the question from more of a "who wins" and "who loses" perspective. This statement refers back to our discussion in Chapter 1 that distinguished economic development from economic growth. Many argue that growth may be a necessary but insufficient condition for development. Also, many continue to argue that there is more to community economic analysis than estimates of aggregate jobs and income. SAM analysis provides a logical extension to IO that focuses on distribution and development issues and responds to these contemporary arguments associated with public policy impact.

Input-output tables provide one data framework but lack the comprehensive accounting of income flows. Base data on these income flows are necessary to address labor components, production structures, and government interaction necessary to conduct policy analysis. This more comprehensive accounting structure for community economies is provided through social accounting matrix analysis. SAMs shift attention from industries (firms) to households (individuals). Relevant social accounting questions deal with how households generate income and, once generated, how this income is distributed. In particular, SAMs recognize the simple reality that different household income groups act differently and are affected in different ways. In addition, SAMs tend to be more theoretically consistent with general circular flow of community economies. SAMs act to disaggregate "institutions" in the economic accounting structure.

The SAM Accounting Structure

SAMs generally include four basic components:

1. Production (*n* sectors)
2. Consumption (*k* household groups) of households that is supported by provision of factor inputs (*f* factors of production)
3. Accumulation in institutions (*m* types of institutions)[10]

4. Trade, which is specified (or exogenous) in the models

Thus, IO accounts are extended by explicitly specifying household income, as generated through the use of factor inputs. An example of the accounting structure employed through a social accounting matrix is shown in equation (15.15). For ease, we only show that portion of the SAM that is typically considered endogenous (for purposes of analysis, it represents the closed portion of the SAM tableau).

$$A^{*SAM} = \begin{matrix} & (n) & (f) & (m) & (k) \\ (n) & A^{(n \times n)} & 0 & 0 & C^{(n \times k)} \\ (f) & F^{(f \times n)} & 0 & 0 & 0 \\ (m) & 0 & Y^{(m \times f)} & 0 & 0 \\ (k) & 0 & 0 & H^{(k \times m)} & 0 \end{matrix} \quad \textbf{(15.15)}$$

where

> A = technical coefficients on intermediate transactions
> F = factor income (value added) coefficients
> Y = factor income distribution (institutional) coefficients
> C = household expenditure coefficients
> H = household income coefficients
> A^{*SAM} = the SAM coefficient matrix:

$$(n + f + m + k) \times (n + f + m + k)$$

As an accounting structure, the SAM extends IO by more completely specifying the tableau found in Figure 15.2. Until now, we have not dealt with Quadrant D nor have we specified how households earn income. Furthermore, we have not specified how that income is distributed among groups of people in a community. SAMs provide a more complete accounting for household income.

As predictive models, SAMs can be closed in a variety of ways. Most typically, though, SAMs are endogenized through households. In this way, SAMs track structural linkages that involve households and specify distribution of nominal income. In essence, SAMs track the use of factor inputs owned by households. These factor inputs, when employed, generate income for households that support the demands for goods and services, thus linking households and firms more completely.

Constructing a Regional SAM

SAMs can be constructed in a variety of ways. The manner in which a SAM is specified is typically driven by the problem being addressed. A thorough assessment of the various types of SAM structures is beyond the scope of this chapter. For this discussion, a generic SAM structure will be discussed.

Data Elements for Constructing a SAM

An illustrative SAM framework was provided in equation (15.15). From an IO perspective, the rows and columns that correspond to industry and commodity are the focus. Whereas IO generally focuses on the firm and how it relates to other firms, SAMs extend the data set to more fully capture income distribution resulting from returns to primary factors of production (land, labor, and capital). In this way, the circular flow of goods and services to households from firms and the corresponding factor market flows to firms from households are captured.

In the SAM, row totals and column totals are equal, thus representing a regional economy in equilibrium. For example, total industry output equals the outlay used in its production. Institutional income (to households, for example) equals the outlay required for the use of institutionally owned land, labor, and capital in the factor markets. In general, total income equals total cost of inputs. SAM accounts are constructed to balance outputs with inputs.

Data Sources for SAM Building

Although the specific data requirements for constructing a regional SAM vary depending on the type of problems being addressed, some generalizations can be made. In addition to standard IO data (industry production, inter-industry transactions, final demands, factors of production and imports/exports), typical SAMs require additional data on total factor payments, total household income (by income category), total government expenditures and receipts (including intergovernmental transactions), institutional income distribution, and transfer payments. The latter element is particularly interesting as traditional IO is ill equipped to handle transfer payments. SAMs more clearly identify income to households and businesses that originate from government sources. Examples include Social Security, unemployment insurance, crop and/or land set-aside payments to farmers, and subsidies paid to firms. SAMs are typically built as static snapshots of a region, thus data elements will need to be generally consistent in temporal and geographic specificity.

Interpreting SAM Results

When closed through households, a SAM now captures a wider array of economic transactions. In addition to its ability to estimate the direct and indirect effect of changes among industries, a SAM is now capable of explicitly capturing the linkage between the generation of income into households and the reaction of households as their income changes. Economic impact linkages (from some policy shock) are now modeled with respect to household effects. The generation of income more completely captures returns to primary factor inputs (land, labor, capital), transfer payments by household income group, and unearned income. This ability of households to combine earnings of different types allows the analyst to then directly estimate how income is distributed. In constructing a regional SAM, the distribution of income will tacitly account for alternative ownership of primary factor inputs by household income groups. In doing so, specific distributional patterns of transfer payments and unearned income can be captured in the household income stream.

Strictly speaking, a SAM that is closed (endogenized) through the households account and then used to calculate a SAM inverse generates an expanded set of "total" multipliers. Again, round-by-round impacts are analogous to a power series approximation, with SAM multipliers acting to explain an increasingly diminishing (and now more comprehensive) stream of impacts. Decomposing SAM multipliers can be complex. (For a complete discussion of the complexities of SAM multiplier decomposition, see Pyatt and Round 1985, Part III, Chapters 8 and 9, and Holland and Wyeth 1993).

Like IO, SAM multipliers are still sensitive to regional attributes. Community size and unique regional conditions will still dictate what elements of economic activity found within a region and what elements must come from the outside. Examples include payments to exogenous variables (leakages) that reflect how a region is linked to its adjacent regions. They are also still limited to the previously described IO caveats that include fixed prices, strict production assumptions, and staticness. They now, however, more clearly incorporate the manner in which accounts pass income from industries to institutions (households) and back to industries. Also, SAMs provide a greatly improved accounting structure that is more comprehensive than IO.

The Usefulness of SAM Analysis in Community Economics

Like IO accounts, SAMs provide an accounting structure of regional market-based productive activities and utilize similar double-counting bookkeeping entries. Unlike input-output, however, social accounts focus on the household as the relevant unit of analysis and provide a comprehensive, and additional, set of accounts that track how household income is generated and distributed. Where IO tables are focused on industries and their respective relationships with regional output, SAMs extend this into a more complete range of market mechanisms associated with generating household income. The relevant focus thus shifts from how regional output is produced to also address how regional income is generated and distributed. This comprehensive element is particularly important in assessments that focus on both production processes and the economics of household factor supply, commodity demand, and government interaction.

Social accounting matrices have been employed in a wide array of situations arising in policy development to address key issues of economic structure and impact assessment. A good overview of SAM applications in policy analysis by Erik Thorbecke is in Isard et al. (1998). Basically, SAMs are useful in assessments that require a more comprehensive accounting of circular flows of an economy.

Particularly useful for addressing issues of income distribution, SAMs have been widely employed in assessing development effectiveness in attaining equity-based outcomes of policy.[11] Applications, however, are not limited to assessing redistributive income policies. This is particularly true in the United States as national- and state-level policies that support the redistribution of income to the poor are largely out of favor. Increasingly, welfare reform legislation has emphasized the role of private markets to provide for individual welfare. SAMs have been employed to assess the relative impacts of alternative market-based changes on the distribution of income within regions. Thus SAMs will continue to be relevant tools to address a wide array of policy situations and development issues.

The major strength of regional SAMs is accounting comprehensiveness. Although still widely used, it is important to note that SAMs, like IO analysis, have some rather serious theoretical shortcomings when used to model economic change. Chief among these is their inability to develop models that capture price changes. These modeling caveats have, in part, driven the movement toward developing more flexible modeling systems. As such, their use in computable general equilibrium models is the focus of Chapter 16.

SUMMARY AND KEY LIMITATIONS OF FIXED-PRICE MODELS

Input-output analysis offers the community economic analyst a powerful tool to examine the community's economy. In its simplest descriptive form, it concisely presents an enormous amount of quantitative information on the inter-industry relationships in the local economy. It highlights the strategic importance of various sectors to the local economy.

The most obvious use of IO analysis calculates the output, income, and employment multipliers to evaluate the total effects of various development options. While the direct coefficient of the households row gives some idea about the direct income (employment) effect of development options, the income (employment) multiplier measures direct and indirect (total) changes.

The total requirements table gives the analyst insight into several community economic development questions. The total requirements table permits the analyst to (1) simulate the effects of a change in total final demand on a specific processing sector, (2) simulate the effect of a change in a specific portion of final demand on all processing sectors, and (3) simulate the effect of changing technology (technical coefficients) for a specific processing sector on the rest of the local economy. The first allows examination of the effects on a particular sector as total community output changes. The second permits analysis of changes in output of a particular local sector on all other businesses in the community. The third allows tracing the impacts from one sector of the local economy adopting new technology or from introducing a new economic sector. These permit anticipating the supporting businesses and manpower needs from a development event.

Key limitations with IO involve how tables are constructed, production is assumed to occur, and impacts are modeled. A key limitation of IO that led to development of social accounting matrices is the overly aggregate nature of income accounting. SAMs provide a more complete picture of income and refocus our attention on the activities and characteristics of the household. SAM models more completely capture the flows of community economic activity in generating income and allow the analyst to focus on how income is distributed among households and/or institutions.

At this point, our organization in separating the three toolbox chapters in this section makes very tacit delineation between community economic description and prediction (or inference). The first set of tools found in Chapter 14 was wholly descriptive and had very limited predictive abilities. Our discussion of IO and SAMs in this chapter allowed us to present models that combined community economic description with a predictive component.

Please notice, however, that this predictive component is severely constrained in its ability to model the microeconomic foundations of supply and demand outlined in Chapters 1 and 2. As an illustration, consider the two alternative regional supply and demand situations found in Figure 15.4. Our predictive ability

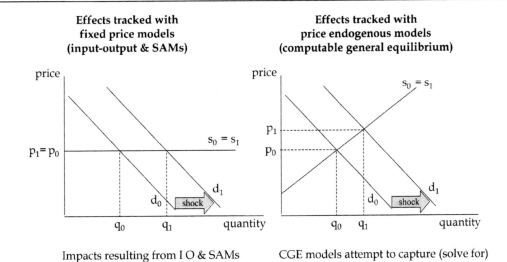

Effects tracked with fixed price models (input-output & SAMs)

Impacts resulting from I O & SAMs underestimate price and overestimate quantity

Effects tracked with price endogenous models (computable general equilibrium)

CGE models attempt to capture (solve for) re-equilibrating mechanisms of supply and demand

Figure 15.4. Effects tracked by different economic models.

with both IO and SAM analyses rests on the severe assumptions about supply. The standard use of IO and SAMs are to model demand shifts (movement from d_0 to d_1) assuming that supply is flat, or nonresponsive ($s_0 = s_1$). Results of IO and SAMs are thus biased toward overpredicting quantity change (difference between q_0 and q_1) and completely ignoring (or underestimating) price change ($p_0 = p_1$).

Economic theory would lead us to consider modeling community economic change in a more flexible manner. This more flexible approach is illustrated in the right-hand side of Figure 15.4. Given some policy shock, the real world would react to demand change more realistically with some price effect. In IO and SAM analyses, our assumption of resource availability is severely constraining. If demands were to increase (d_0 to d_1), the factors used in producing that which is demanded would create scarcity in that good, service, or production input. Namely, the supply of the good, service, or production input (s) would logically increase with quantity produced (s is an increasing function). This price effect would relate to economic scarcity in both consumption and production. A more realistic model would capture a dampened quantity change (compared to models assuming flat supply functions) shown as q_0 to q_1 in the right-hand side of the figure, while simultaneously predicting a price effect (p_0 to p_1). This ability to model economic re-equilibration between decreasing demands and increasing supplies provides the focus for the community economic modeling tools discussed in Chapter 16.

STUDY QUESTIONS

1. What does an IO multiplier that captures direct effects represent?
2. In IO analysis, what is the difference between direct, indirect, and induced effects?
3. What is the difference between a transactions table, a direct requirements table, and a total requirements table?
4. In a standard IO table, what is contained in Quadrants A, B, and C?
5. In a standard IO table, what is contained in Quadrant D?
6. Derive the mathematical form that represents the predictive element of IO analysis.
7. What is the importance of a power series approximation?
8. What are the general assumptions of IO analysis?
9. What elements must an analyst be careful to specify in IO analysis
10. Why was IO analysis extended into SAM frameworks?
11. What aspects are the same and what aspects differ between IO and SAM tables?
12. IO and SAMs are fixed-price models. What does this mean?
13. Do IO and SAM multipliers represent economic turnover? Why or why not?

NOTES

1. Input-output analysis has a fairly interesting history of development as an economic modeling tool. The early work of two distinguished Frenchmen developed input-output as a descriptive mechanism in France during the 1700s and 1800s. François Quesnay is recognized as the father of IO accounting for his work in developing and analyzing the first general expenditure pattern for the French countryside. His accounting structure was first published as *Tableau Economique* in 1758. The other notable French economist is Leon Walras, who is perhaps best known for his work on general economic equilibrium, production coefficients, and factor inputs used in producing output thus extending the accounting structure to a set of macroeconomic relationships. The usefulness of IO accounting increased as its predictive form was developed by Russian-born Wassily Leontief, who was awarded the 1973 Nobel Prize in Economics for "the development of the IO method and for its application to important economic problems." Early supply-side IO models were key tools used for central economic planning in the USSR during the 1950s, 1960s, and 1970s. Demand-driven IO models continue to be used to model free-market economic structures across the globe.

2. If included as primary inputs, they will be included in the final demand sector.

3. Gross business savings include capital consumption allowances that are also sometimes referred to as depreciation.

4. Taxes to local government will be included in the primary inputs sector only if sales to local government are included in the final demand sector.

5. An implicit assumption is that local production will be used as intermediate inputs for local production before any selling to nonlocal markets occurs (i.e., final demand).

6. The distinction between Quadrant A inputs and Quadrant C inputs can be local or nonlocal suppliers.

7. *Closed* refers to including the household sector in the processing sector, whereas *open* is where the household sector is part of the final demand and primary inputs sectors.

8. These comments and the following descriptions are based on our experience in assisting communities with economic impact analysis. Our views are not intended to infer endorsement of specific products nor are they provided as a comprehensive overview of all products available. The interested reader is referred to an excellent overview of microcomputer-based IO modeling in Otto and Johnson (1993) and Brucker, Hastings, and Latham (1987) and an evaluation of the modeling software found in Crihfield and Campbell (1991).

9. Like IO analysis, SAM analysis has a rather interesting history of development. The aggregate nature of IO analysis and its inability to distinguish issues of income distribution led an Englishman named Sir Richard Stone to develop the key extensions. Stone's early life was spent in India with his father. He did his studies at King's College under Kahn and Keynes and was particularly interested in distributional implications of economies in transition. Stone was awarded the 1984 Nobel Prize in Economics "for having made fundamental contributions to the development of systems of national accounts and hence greatly improved the basis for empirical economic analysis." In addition to Stone, key figures in the development of SAM analysis included Benjamin King and Graham Pyatt of the World Bank, Jeffrey Round of the University of Warwick, and Erik Thorbecke from Cornell University.

10. Institutions can take on a variety of meanings depending on the analyst's empirical interests. For example, groupings of individual ownership categories for land, labor, or capital endowments can often be represented as institutions. Given the overarching importance of income distribution, households are often a separate analytical unit.

11. A good compilation of the SAM application literature can be found in an annotated bibliography by Jepson et al. (1997). In addition to SAM applications, this bibliography puts the array of regional economic models in context with respect to modeling market and nonmarket intricacies.

16
Inferential Tools of Community Economic Analysis

Price Endogenous Models

To an economist, prices are the driving force for markets. Firms and consumers adjust production and consumption patterns in response to prevailing market prices (Chapter 1). Under strong assumptions of utility and profit maximization, perfect competition, and no externalities, market forces will result in a Pareto optimal allocation of resources with prices playing the central role.

Unfortunately, most of the analytical tools used by economists have serious shortcomings. The tools presented in Chapter 14 were purely descriptive, lacking causal elements that are characteristic of economic theories. They also examined only a narrow part of the economy, ignoring the interlinkages among and between sectors. Some would call these tools *partial* in nature. The tools presented in Chapter 15 have consistent causation embedded within their modeling framework, but like those in Chapter 14, assumed that prices are fixed. But unlike the descriptive tools of Chapter 14, input-output (IO) analysis and social accounting matrix (SAM) analysis are *general* in the sense that they model all interlinkages among and between sectors.

The distinction between *partial* and *general* theories is important: Partial theories only look at small subcomponents of the overall economy, but general theories attempt to capture all elements of the economy at once. For example, central place theory is a partial theory in that it tries to explain and predict market location on an economic plane. Other parts of the economy are ignored. Input-output and SAMs, on the other hand, model the linkages between product and factor markets, hence they are more consistent with general equilibrium theory.[1] In essence, the notion of ceteris paribus, meaning "all else held constant," separates partial and general theories. We now start to develop general equilibrium models that allow prices to be determined by the model.

In one sense, if markets are perfectly competitive and communities have little influence over prices, or are price takers, assuming that prices are exogenous and fixed does not seem unreasonable. But as we have seen repeatedly, community prices are not fixed, and community policies can and do influence prices. For example, as we saw in our discussions of firm location (Chapter 3) and land markets (Chapter 5), land prices can be strongly influenced by internal factors. In addition, a commonly stated goal of community economic development efforts is to raise prevailing wages. The goal of this chapter is to provide an introduction to the range of analytical tools available to the community practitioner where prices are allowed to adjust and indeed to play an integral role in the functioning of the economy.

Most models of regional economic structure and policy analysis tend to be demand driven where supply is perfectly elastic, or flat (Chapter 15). In the case of export base and its derivative input-output and social accounting matrices, the perfectly elastic supply curve translates into a fixed-price model; prices are not allowed to adjust to changes in the economy. The implications of fixed-price modeling on policy analysis are discussed in detail at the close of Chapter 15. In addition, many of the modeling tools widely used by economists are partial equilibrium in that one market is analyzed at a time. For example, the tools of trade area analysis, including market threshold, pull factors, and potential sales, are partial in the sense that the retail and service sectors are viewed in isolation. The interactions between retail and service markets with the labor, capital, or land markets are assumed away.

But the community practitioner is not limited to the fixed-price models of export base, input-output, or social accounting matrices in a partial equilibrium world. Two extensions that are gaining significant use in community economic analysis are computable general equilibrium (CGE) and hybrid models that conjoin IO or SAM with econometric models (EC). Sometimes these are referred to in the literature as IO/SAM-EC models. The idea here is to not only allow prices to adjust and to play a more central role in the empirical representation of the economy but also to more explicitly recognize that all markets are interrelated. But computable general equilibrium and hybrid conjoined models are currently state of the art, and their complexity can be daunting. Our discussion here is intended to be an introduction to the ideas, approaches, and limitations of these approaches. It is beyond the scope of this book to provide a detailed discussion of CGE and hybrid conjoined modeling issues.

Another modeling issue the practitioner needs to be aware of is the increasing use of spatial econometrics. The idea here is that our theory (e.g., central place theory) infers that communities are linked spatially on the economic landscape. Traditional statistical methods used in econometrics have tended to ignore these spatial relations. *Spatial econometrics* is an explicit attempt to improve the performance of our econometrics by introducing space into the statistical analysis. We close our discussion of hybrid conjoined models with an elementary introduction to the ideas of spatial econometrics.

COMPUTABLE GENERAL EQUILIBRIUM MODELS

The computable general equilibrium (CGE) approach, like fixed-price models such as IO and SAM, take into account interactions throughout the economy in a consistent manner; they are *general* in nature. The major step forward is that prices are not only allowed to fluctuate but are endogenous to (or determined by) the model. As suggested by the circular flow model of the economy (Fig. 4.1), if one part of the economy changes then there will be effects on other parts of the economy; these are automatically taken into account when the economist uses a CGE.[2]

Computable general equilibrium models are a natural extension of IO and SAM models that have been widely used for decades in policy analysis. Indeed, the introduction of user-friendly IO modeling systems such as IMPLAN has made the use, and to some extent abuse, of IO widespread and increasingly common. CGE models extend these older models to take into account substitution possibilities in terms of labor, capital-intensive technology

choices, and the circular flow of income across households and firms. Although it is possible to make ad hoc extensions of IOs and SAMs to incorporate some of these concerns, one is effectively trying to indirectly construct a CGE. Thus, a CGE model provides a straightforward generalization of earlier models used for studying local economies (Kraybill, Johnson, and Orden 1992; Morgan, Mutti, and Partridge 1989; Schreiner et al. 1996).

Another reason for using CGE models or indeed any multisectoral models, including IOs and SAMs, addressing the economic effects of policy or some exogenous change is that they provide comparative results for a variety of sectors. Many times community economic development policy is sector specific and any economic analysis of the policy is limited to that sector. No industry or sector operates in isolation from the whole of the economy, and informed decision-makers, as well as the community practitioner, need to more fully understand the nature and extent of economic linkages. The notion of industrial clusters as an economic development strategy builds on strengthening economic linkages (Chapters 3 and 4). Because these linkages are vast and inherently complex, some type of multisectoral model should be used.

Keep in mind, however, that a CGE is a tool to help the community practitioner and policy-makers think through how the local economy functions, how separate sectors are linked, and how policy can impact the economy. This returns us to our *partial* versus *general* equilibrium discussed previously. One must be careful to use IOs, SAMs, and CGEs as tools, not an end in themselves or *black boxes* providing the answers. Indeed, many economists who worry about the construction of these types of models find that the insights on the local economy they gain by building these models prove to be more powerful in the end than the model itself. There is as much art as science in the construction and application of these models, and the intuition and insights—the art—is just as important as the science.

How a CGE Model Works

A CGE model, along with IOs and SAMs, work by using data to describe the economy in a benchmark time period, usually the most current year for which data are available, and by varying one or more elements in the model to shock the economy. Once the shock or policy change has worked its way through the model, the new and benchmark values are compared and contrasted for the economy as a whole and as individual sectors.

The CGE really is composed of three parts:

1. *Endogenous variables* predicted by the model
2. *Exogenous variables,* including policy instruments that are determined outside the model
3. *Structural parameters* of the model that determine the magnitude and direction of linkages, thus specifying causation

The *benchmark* of a CGE is an actual solution of the model that replicates the observed economic data for some base time period. Replicate means that the model reproduces the observed economy. The task of calibrating the parameters of the CGE model involves modifying the parameters of the model so that it successfully replicates the observed economic data. For the benchmark time period, the modeler has data on all endogenous and exogenous variables. Input-output tables and SAMs provide the accounting framework in which the benchmark data are collected and organized.

Early work with CGE models focused on methods for calibration and solving the model itself, something that is not an easy computational task.[3] But the advent of more powerful personal computers and the expansion of modeling systems like IMPLAN have caused a rapid expansion of CGE model applications. This expansion of CGE models follows the logic of technological adaptation discussed at length in Chapter 8. Initially a small handful of economists was focusing on the basic research of CGE construction, calibration, and solution methods. As computational and data barriers have been reduced, the use of CGE models has expanded rapidly. Today, much of the work on CGE model construction and use focuses on specific policy applications.

What a CGE Model Looks Like

A CGE model is basically a large set of supply and demand functions that cover every market, both commodities in the output market and factors of production in the input market. In essence, CGEs are a rigorous empirical representation of the circular flow model (Fig. 4.1). The demand side of the commodity markets is composed of private households, governments, and firms. Some of these economic agents are local or within the community and some are nonlocal or outside the community. Most regional or local CGEs are called *open models* because they allow imports and exports.[4] Conversely, *closed models* do not allow for imports or exports and are generally more suited to large national economies where imports and exports are a relatively small part of the overall economy.

Private households are generally assumed to maximize utility subject to a budget constraint, hence, they buy commodities following the general rule outlined in equation (1.1) and Figure 1.2 in Chapter 1. In a simple world of two goods (X_1 and X_2) and perfect competition in which all consumers face prices P_1 and P_2, all consumers will maximize their individual welfare (utility) by equating their individual marginal rates of substitution to the ratio of prices:

$$MRS_{X1,X2} = P_{X1}/P_{X2} \qquad (16.1)$$

again subject to an income budget constraint. Households receive this income as they sell the factors of production to firms.

As households sell their factors of production to firms, it enables firms to produce goods and services. In addition to buying the primary factors of production—land, labor and capital—from households, firms also buy intermediate inputs from each other. In an IO or SAM model, inter-industry transactions in the transaction and direct requirement tables capture these intermediate firm-to-firm transactions. Much like consumers, firms will purchase goods and services following the allocation rule stated in equation (16.1):

$$MRTS_{L,K} = w/r \qquad (16.2)$$

where w is the price of labor, or wages, and r is the price of capital, or interest rates. This condition was represented visually in Figure 1.3 in Chapter 1.

There are important substitution opportunities incorporated in CGE models. If there is a change in relative prices, such as the cost of labor relative to capital, then theory predicts that firms will adjust their production input mix to become more or less labor intensive, again following the rule outline in equation (16.2). Similarly, consumers will alter their purchasing decisions based on changing relative commodity prices. Because most of the market analysis tools used by community practitioners assume prices are fixed, these tools or models may provide biased policy analysis results (see Fig. 15.3).

Consider a small community that has a major employer opening or significantly expanding operations. Clearly, a large influx of workers into the local labor market will put downward pressure on local wages that will ripple throughout the local economy.

How might this affect the local housing market? On the one hand, downward pressure on wages might harm the local housing market. But more people living in the community will place upward pressure on the housing market. How will these two countervailing forces balance out? The only way to capture these impacts is through the use of CGEs; an IO- or SAM-type model will ignore these important wage effects.

The extent of these responses to changing prices, or substitution possibilities, is captured by the specified elasticities that are drawn from the existing literature. The *elasticity* is a numerical estimate of price sensitivity. Consider the demand curve presented in Figure 16.1. The price elasticity of demand captures the sensitivity of quantity demanded to changes in price. The steeper the demand curve, the smaller the elasticity, and large changes in price result in small changes in quantity demand. A perfectly elastic demand curve is flat or horizontal to the quantity axis; a small change in price results in an infinite drop in quantity demand. For a CGE to be operationalized, likely elasticities must be obtained for a range of economic linkages.

Elasticities must be assembled for primary factor substitution, such as labor for capital, import demand, import sources, local demand by commodity, and the transformation of local supply into local and export products. Since, in practice, virtually all the elasticity values are drawn from the relevant economic literature, they may be subject to error. These errors can be from statistical errors in the estimation of the elasticities, mismatches between what is available in the literature and the modeling effort at hand, and the appropriateness of national or regional estimates for the community of interest. Indeed, one of the widest criticisms of CGE models is the lack of reliable elasticity estimates (Bilgic et al. 2002). In many cases, no estimates exist and the modeler must assume a reasonable value.

Once a set of elasticities is collected, the modeler must calibrate the model as described above. In that discussion the modeler must alter the parameters, or more correctly elasticities, to ensure that the benchmark data are replicated by the model. At this stage, the model can also be tested for sensitivity of model results with respect to a plausible range of elasticity estimates. While the strength of the CGE is in its theoretical appeal, its weakness comes in what can be called the art of model calibration.

A Simple Example

For purposes of illustration, the following example outlines the basic conditions under which CGE estimates economic effects. It is a classic economic example that portrays an economy that consumes two goods and produces these goods with a single input. Note that this is an extremely simple example

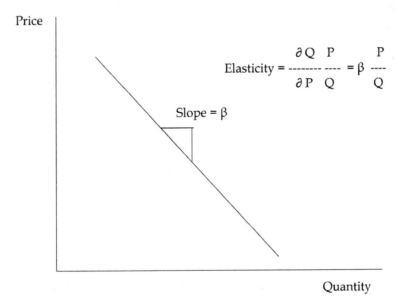

$$\text{Elasticity} = \frac{\partial Q}{\partial P} \cdot \frac{P}{Q} = \beta \frac{P}{Q}$$

Slope $= \beta$

Figure 16.1. Price elasticity.

of a closed economy CGE and is used here for illustration only.[5]

Consider an economy with two firms and two consumers. Firm 1 is owned by consumer 1 and it produces guns (*g*) from oil (*x*) using the following production technology:

$$g = 2x \qquad (16.3)$$

Firm 2 is owned by consumer 2. It produces butter (*b*) from oil via a production function specified as

$$b = 3x \qquad (16.4)$$

Furthermore, let us assume for this problem that each consumer initially owns 10 units of oil. Consumer 1's preferences are given by the Cobb-Douglas utility specified as

$$u^1(g, b) = g^{.4}b^{.6} \qquad (16.5)$$

Consumer 2's preferences are given by

$$u^2(g, b) = 10 + .5 \ln g + .5 \ln b \qquad (16.6)$$

Market-clearing Prices

Finding the market-clearing prices for oil, guns, and butter is a rather straightforward task. We start by assuming competitive conditions between profit-maximizing firms that drive economic profits to zero. Specifically, firms face the simple profits relationship equal to revenues minus costs. Revenues are equal to the price of the good times the number of units sold, while costs are calculated as the per unit costs of inputs times the number of units produced. This is mathematically represented as

$$\Pi = p_g g - p_x x = 0 \qquad (16.7)$$

Substituting in the production functions into each firm's profit equation yields the following relationships:

$$\Pi = p_g(2x) - p_x x = 0 \qquad (16.8)$$

To simplify this solution, we must identify a *numeraire* good on which results will be based.[6] Let us assume that $P_x = 1$. Thus, we can solve equation (16.8) for p_g:

$$p_g(2x) = x \qquad (16.9)$$
$$P_g = \tfrac{1}{2}$$

Thus, the market-clearing price for guns is 1/2 (based on the price of oil as the numeraire good equal to 1). In a similar fashion, the astute student can confirm that the market-clearing price for butter is equal to 1/3.

Determination of Equilibrium Consumption

The model can be further analyzed to estimate the quantity of guns and butter consumed by each consumer. To begin this calculation, we will assume utility-maximizing behavior of both consumers. Specifically, consumers maximize utility subject to a budget constraint. The budget constraint specifies that the amount consumed cannot exceed the income level of the consumer. Since each consumer is endowed with 10 units of oil, the income is 10 times the price of oil. Furthermore, their consumption of guns and butter can be represented as the amount of each consumed times each good's respective price. Finally, at its limit, all income will be consumed in the purchase of these goods (income minus expenses equals zero). Thus, this budget constraint for consumer 1 can be written

$$10p_x - p_g g - p_b b = 0 \qquad (16.10)$$

Substituting in our previously calculated prices, the equation simplifies to

$$10(1) - 1/2\, g - 1/3\, b = 0, \text{ or } 10 - 1/2\, g - 1/3\, b = 0 \qquad (16.11)$$

To maximize utility, we will use the Lagrangian approach. Remembering consumer 1's utility, we can state the Lagrangian as

$$L^1 = g^{.4}b^{.6} + \lambda(10 - 1/2\, g - 1/3\, b) \qquad (16.12)$$

This is differentiated with respect to each term as follows:

$$\delta L_1/\delta g = .4b^{.6}/g^{.6} - 1/2\,\lambda = 0$$
$$\delta L_1/\delta b = .6b^{.4}/g^{.4} - 1/3\,\lambda = 0 \qquad (16.13)$$
$$\delta L_1/\delta \lambda = 10 - 1/2\, g - 1/3\, b = 0$$

Solving the first two conditions of equation (16.13) simultaneously, straightforward manipulation yields

$b = 9/4 \, g$, which, when substituted into the third condition, yields consumer 1's consumption of guns to be 8 with her consumption of butter to be 18. Following similar procedures, consumer 2 must then consume 10 units of guns and 15 units of butter. Thus, in total, there will be 18 units of guns consumed and 33 units of butter.

Production Inputs

Finally, we can calculate an equilibrium amount of input used to produce guns and butter. To determine the amount of oil used in their production, we can use the previously calculated information on market-clearing prices. To make 18 guns, we will need 9 barrels of oil. To make 33 units of butter, we need 11 barrels of oil. Total oil needs involve 20 barrels, which is exactly equal to the combined initial endowment.

Empiricizing a CGE for a specific region is a difficult task, indeed. This task is fraught with a dearth of usable relationships unique to the region under study. As we saw in Chapter 12 on the consumption side, aggregate region-specific utility functions do not exist. Given this lack of region-specific economic information, the typical empirical CGE either (1) uses production and consumption relationship estimates derived by other studies for other regional specifications or (2) specifies these relationships in an ad hoc fashion. The community economic analyst should remember that the CGE modeler has enormous opportunity to specify the model to generate almost any result. A widely held joke among many CGE modelers is that given sufficient time, they can provide any policy result desired.

Operational CGE Models

Because of the relatively wide use of CGE models now economists have a wealth of prior information to draw on. Seldom is a new model completely built from scratch. Rather, a modeler draws on existing models and often alters an existing model to address a specific policy question. The ORANI modeling system developed in Australia is an example of such a system (Dixon et al. 1982). The ORANI model of the Australian economy became operational in 1977 and has been used for many policy-oriented analyses. The model is national in scope and includes disaggregation of results to the regional (state) level. The model grew out of the Johansen (1960) class of multisectoral models. ORANI allows for multiprod-uct industries and multi-industry products. In standard applications, the ORANI model has 115 commodities and 113 industries and contains 113 parameters just for the elasticity of substitution between domestic and foreign sources of supply.

There are positive sides to such a large modeling system as well as significant drawbacks. Because of the well-established performance of the ORANI system, modelers have at hand a ready-built system that could be easily added to as interest in new sectors or policy issues arise.[7] On the other hand, the system can become so complex and unwieldy that any one modeler cannot possibly understand all that is happening in the model itself. In all honesty, there are numerous examples of analysts modifying small parts of the model that have unintended and unknown consequences on other parts of the model, which are only uncovered in later unrelated analyses.

The trade-off that economists face when moving from a simpler modeling approach like IO/SAMs to CGEs is sectoral detail and model specificity. As we have seen with IOs and SAMs, we can develop models with extreme sectoral detail. In the widely used IMPLAN modeling system, there are 528 industrial sectors, 8 household sectors, and a handful of public (government) sectors. Because of this detail, we can focus very closely on specific sectors of the economy. Most CGEs, on the other hand, have only a small number of sectors, such as agriculture, durable and nondurable manufacturing, trade, services, and maybe government. CGE models that tend to have more than just the basic sectors can become computationally complex in very short order. Remember that the CGE must be able to replicate the benchmark data as well as provide reasonable simulations.

Because most CGE models work with only a handful of sectors, to keep the model tractable they tend to be subject to aggregation bias.[8] One of the central purposes of building regional models is to explore how different sectors of the economy respond to shocks or policy changes. Clearly different sectors will respond differently, sometimes significantly differently. *Aggregation bias* is the treatment of dissimilar sectors as being the same. Take for example a typical regional CGE model that may have three sectors: agriculture, manufacturing, and services. In this case, the model explicitly assumes that all agricultural sectors will respond in the same manner. A small dairy farm processes the same technology, production processes, and market behaviors as a large cotton farm. Similarly, a hospital behaves the same way as a general merchandise store or a paper mill functions the same as a furniture manufacturing firm.

This is the trade-off that economists face when deciding between using an IO or SAM versus a CGE model. The power of the CGE modeling approach is that it more accurately reflects our theoretical frameworks, but the weakness is aggregation bias. In addition, the economist is faced with the selection of appropriate parameters, or elasticities, with which to calibrate the model. For example, the price elasticity of French wine is different than a gallon of milk. When using aggregated sectors, what are the appropriate parameters to choose?

Computable general equilibrium models can be very powerful when seeking insights to very specific sectoral or policy questions. If the question is how the dairy industry might react to changes in federal pricing policies, CGE models can provide detailed insights. The health-care sector is likely not to be impacted by changing federal dairy pricing policies, so ignoring or aggregating health care into a general services sector may not present a serious problem. While IOs and SAMs may be best suited for looking at general questions across a range of sectors, CGE models may be better suited to being asked focused questions. At the community level, CGE models may provide valuable insight into specific questions, such as What may happen to land prices if restrictive land use policies are put into place?

HYBRID CONJOINED MODELS

One area of regional modeling that has drawn significant attention from economists is statistical modeling,[9] built on the early work of Klein and Goldberger (1955) and the Wharton model (Preston 1975) of the U.S. economy. Our Keynesian representation of the economy outlined in Chapter 4 has served as the foundation of a whole genre of models that use statistical approaches at the national, regional, and community levels (Bennett and Hordijk 1986; Bolton 1985; West 1995). Take, for example, equation (4.3), which relates income to total consumption. Here we define consumption (C) as a simple linear function of income (Y), or $C = a + bY$. Given data on income and consumption levels, it is a straightforward problem to estimate the Keynesian consumption function where "a" would be some minimal level of consumption required to live and "b" is the marginal propensity to consume. Statistical studies have suggested that the marginal propensity to consume is close to .8, meaning that for every dollar of additional income $0.80 is spent on consumption.

Expanding this type of basic Keynesian model to include equations across a range of consumption categories along with investments by type and different levels of government, we can create a detailed statistical model of a given economy. These models have proven to be very powerful in terms of helping economists better understand how an economy functions and can be affected by policies, such as government spending and taxation, and what the economy may look like in the future. Keynesian-type models, particularly at the national or macro level, have been widely used for economic forecasting. Because these models are often estimated using time-series data (annual, quarterly, and sometimes monthly data), economists have used these models to gain surprisingly reliable insights into what the future of the economy looks like.

This is often accomplished by introducing lags into the structure of the models. Reconsider our simple consumption function: $C = a + bY$. This formulation assumes that consumption today depends on income today. Suppose instead that there is a lag in our spending; our level of income today determines our level of spending or consumption tomorrow. We can restate our consumption function as $C_{t+1} = a + bY_t$ where C_{t+1} is consumption tomorrow and Y_t is income today. What I earn today, I spend tomorrow. Given data on income today, an economist can make a fairly accurate forecast of what consumption will be tomorrow. In turn, if consumption tomorrow determines production tomorrow, it also determines income tomorrow. This type of logic allows for long-term forecasts.

Economic forecasts have become extremely important in the functioning of a modern economy. Households, businesses, and governments make future plans based in part on what the future of the economy holds. If households and businesses think that the economy will be strong tomorrow, they will make certain decisions today. In an economy where households and businesses buy or invest on credit or debt, this type of a lag structure makes intuitive sense. If the automobile industry thinks that demand, or consumption, for cars will be strong tomorrow, they will invest in plants for production today. Similarly, if a household feels confident about what the future holds in terms of employment and income potential, they will make purchasing decisions today. If I feel confident in the security of my job, I may purchase that new car today with plans of using future income to pay for it.

Conversely, if there is risk and uncertainty about the future of the economy, households and businesses may hold off on investment and consumption decisions today. If I am not sure about the security of my job tomorrow, I will not purchase that car today. If the automobile industry thinks that the economy may be slowing down and the demand for cars will be lower in the future, they will delay investments in new plants and may indeed cut back on production today.

Governments also depend on economic forecasts of the future because expenditure decisions are based on expected future revenue streams. Federal, state or provincial, and local governments lay out budgets for the fiscal year based on revenue forecasts. Budget officers (forecasters) put tremendous amounts of time and effort into these forecasts. An economist can be very pleased with a forecast that is within 5 percent of what actually happened, but for a state government with a $1 billion annual budget, a 5 percent error means a deficit or surplus of $50 million.

Because of the importance of generating reasonable economic forecasts, significant time and energy have gone into the construction of forecasting models at the national and regional level. The earliest attempts were at the national level and were direct attempts to build on Keynesian models by using national time-series data. Later, regional economists attempted to replicate the national Keynesian models by using annual data for states or large cities (Klein 1969). As statistical methods improved and data became more widely available, economists moved up the learning curve developing very detailed and reliable models.

Community development practitioners are now starting to tap into the wealth of information that has been developed at the national and state levels. One fundamental problem facing economists who were interested in building statistical models of the community is data availability. Economists who have built national- and state-level Keynesian-type models have access to a wealth of time-series data, whether it is annual, quarterly, or even monthly. Data at the community level, unfortunately, is seldom available over any reasonable length of time in sufficient detail to directly estimate even the simplest of Keynesian models. At best, economists and community practitioners have census data that is available only every 5 to 10 years.[10] To compound

problems, disclosure rules to ensure confidentiality mean that complete data is available for only the largest places.

With the widespread use of IOs and SAMs at the local level, economists are exploring ways to leverage the power of IOs and SAMs with the power of statistical models. Because both approaches, IO/SAM and econometric, have attractive features, it was inevitable that economists would try to select the most promising attributes of each in developing a hybrid model. These models that use both IO/SAM and econometrics are often referred to as being *integrated* or *conjoined*. In his 1977 presidential address to the American Economics Association, Nobel Laureate Lawrence Klein suggested that models that integrate IO/SAMs and econometric methods might become a new method or approach to guide our thinking about the performance of the economy.

While this approach was not new at the national level—Preston's Wharton Model debuted several years earlier—little work had been done at the subnational or regional level. Several economists took Klein's challenge to heart and built a number of such models, including but not limited to Kort and Cartwright (1981), and Coomes, Olson, and Glennon (1991). Perhaps the most widely used conjoined model is REMI (Regional Economic Models, Inc.), originally developed by George Treyz and Ben Stevens (Stevens, Treyz, and Kindahl 1981; Stevens et al. 1983; Treyz 1993).

In an integrated or conjoined model, the IO/SAM component is used to determine industry supply and primary factor demands. The econometric component determines final demands, factor prices (hence, price endogenous), primary factor supplies, and financial market variables. The aim is to retain the sectoral detail of the IO/SAM and close the model with a system of endogenous econometric relationships (Dewhurst and West 1991). The closure mechanism links primary factors and final demand. Returning to our discussion of model closure from Chapter 15, closure here does not involve bringing households into the direct requirements table, rather the behavior of households is determined econometrically. In other words, the IO/SAM details interindustry linkages, and all other components of the economy are modeled econometrically.

Hybrid conjoined models offer several important benefits from an applied perspective.

1. The IO/SAM component allows for more sectoral detail, or dissaggregation, than the

traditional econometric or CGE approaches. This is important because aggregation bias can greatly distort policy analysis results.

2. The econometric equations provide a better mechanism for introducing policy simulations by fully specifying the underlying components of final demand.

3. The full employment assumption of IO/SAM and CGE can be relaxed—the econometric component allows for unemployment or underutilization of local resources.

4. The econometric components can accommodate spatial issues that are of particular importance to communities, such as migration and commuting.

5. The econometric components explicitly allow prices to fluctuate.

How a Conjoined Model Is Structured

The simplest way a conjoined IO/SAM-econometric model can be structured is found in Figure 16.2 (Shields and Deller 1998; Shields, Deller, and Stallmann 2000). The representative model presented in

Figure 16.2 belongs to a family of community-level models widely known as CPAN-type models (Community Policy Analysis Network) (Scott and Johnson 1998). For now, let us keep with a simple model; alternative approaches of structuring a conjoined model are highlighted below.

The standard IO/SAM model is in the upper left-hand corner of the figure (the boxes labeled "final demand" and "input-output"). This drives the rest of the model. Those portions below these two boxes are estimated econometrically. Such economic variables as wages, unemployment, commuting patterns, housing, property values, local government revenues and expenditures, and retail sales are potential endogenous variables predicted by the econometric component of the model. This is but a partial listing of economic variables examined. The analyst can reduce or expand the list depending on the level of detail desired.

A shock to the economy is presented to the model as a change in final demand, identical to shocking an IO or SAM. The model outline in Figure 16.2 takes the information flowing from the IO to estimate

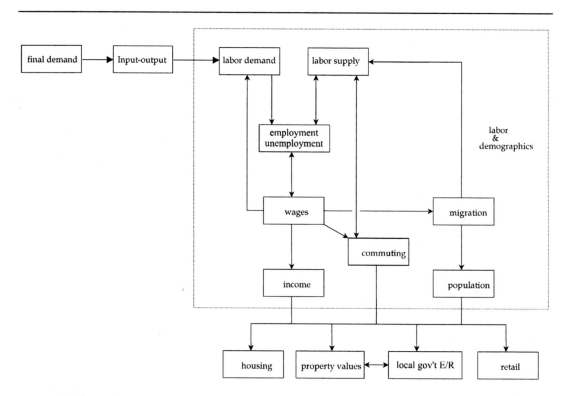

Figure 16.2.   Example of a hybrid conjoined IO/SAM-econometric model.

changes in the demand for labor. If it is a positive shock, such as a local firm expanding its operations, the IO will predict how much output will be increased by sector. The change in industry output derived from the IO is then used to predict econometrically changes in the demand for labor.

The supply of labor in the local community can come from four potential sources: (1) changes in the unemployment rate, (2) changes in commuting patterns, (3) in-migration, and (4) changes in labor force participation rates. Because the econometric specification of the model allows for downward-sloping demand curves as well as upward-sloping supply curves, we can expect to see changes in the local wage rates. The change in wages is an important predictor in how the supply of labor will respond to the initial shock.

Once we have a predicted change in employment, wages (income), and population, the analyst can use that information to look at changes in the local housing and retail markets, property values, as well as changes in the demand for local government services and the flow of revenues to the local government. These *modules* can be simplistically composed of a single equation, such as total retail sales, or composed of many equations, such as retail sales by sector. Again, the level of detail is based on the questions asked and the level of insight desired by the analyst and/or decision-maker.

One of the challenges facing the analyst when building a hybrid conjoined model is deciding on how to formally link the IO/SAM and the econometrics (West and Jackson 1998). In the example we just walked through, the linkage is straightforward: Type I industry output multipliers provide a computed change in industry output. This change can be introduced into the econometric labor market component. But this is only one way to integrate the different modeling approaches. Beaumont (1990) suggests a spectrum of integration methods with pure econometric models at one end and IO/SAMs at the other. The integrated models occupy the middle portion. At some point, a model is too close to an end point and cannot be considered an integrated model. The positioning of the integrated models on this continuum depends largely on integration strategies.

In summarizing integration strategies, Rey (1997) introduced two important aspects of integrated models. The *integration regime* refers to the nature and extensiveness of the interaction between the IO/SAM and the econometrics. The nature of the interaction considers whether the models are *recursive*, reflecting a sequential ordering of the direction of causality between the two components, or *simultaneous*, containing feedback between the two components. In practice, there are nearly as many methods to integrate the components as there are models themselves; the choice of technique usually relies on data availability and the focus of the modeling effort. Unfortunately, the consequences of the choice are not trivial: The specific structure of the model can have important effects on error propagation throughout the entire model.

Recognizing that hybrid conjoined models vary in their choice of integration regime and structure, Rey (1997) suggestd a taxonomy identifying three fundamentally different classes if integration: embedded, linked, and coupled. In *embedded* models, the econometric component is predominant; the IO/SAM is encompassed in the regression equations. Typically, embedded models are the least complex and offer the fewest interactions between the two components. In *linked* models, there is little overlap between the econometric and IO/SAM components, the models retain a great degree of independence and interact in a recursive manner, with outputs from one component serving as inputs to the other. Our simple model presented in Figure 16.2 is a recursive-linked model. *Coupled* models are the most complex and are highlighted by a relatively large degree of feedback between the IO/SAM and econometric components, but the IO/SAM and econometric modules do maintain some independence. In the end, the idea of a conjoined model is to build on the strengths of two modeling approaches.

INTRODUCTION TO SPATIAL STATISTICS

When building the econometric components of a hybrid conjoined model, great care must be taken in the statistical estimation of the equations that make up the components. The most widely used statistical method is that of *ordinary least squares,* or OLS, often referred to as *classical regression analysis.* There are some fundamental underlying assumptions that OLS is based on that may cause problems for the community analyst. Let us track through these classical assumptions and their implication for hybrid models.

Suppose that we wish to estimate the demand equation for labor:

$$L_i = \beta_0 + \beta_1 O_i + \beta_2 R_i + \epsilon_i \qquad (16.14)$$

where L_i is labor demand, O_i is industry output, and R_i is the wage rate for the *i*th community, and β_0, β_1, and β_2 are a constant and slope parameters, respectively, that are to be estimated. The characteristics of the error term (ϵ_i) are of particular importance. With traditional classical regression analysis the error term is assumed to be independently, identically disturbed (*iid*) following a normal distribution, or

$$\epsilon_i \sim N(0, \sigma^2 I) \qquad (16.15)$$

with expected value of the error equal to zero [E(ϵ) = 0], constant variance [E($\epsilon'\epsilon$) = σ^2] and the distribution of the error terms are independent [E($\epsilon_i\epsilon_j$) = 0 for all $i \neq j$]. The simplest way to think of the last assumption is that the error terms are not correlated. In this notation $\sigma^2 I$ is the variance of the error term times an identity matrix (*I*). The ones along the diagonal of the identity matrix capture the constant variance assumptions and the zeros (0s) off the diagonal of the identity matrix capture the independence assumption. In other words, the error term is "well behaved."

As long as the condition outlined in equation (16.15) holds concerning the structure of the error term, OLS will provide us with estimates that are

- *Unbiased:* The expected value is equal to the true parameter.
- *Efficient:* The variance is less than that of other estimators of that parameter for a given sample.
- *Consistent:* The probability limit approaches the true value of the estimator as the sample size gets large.

If the first two properties are met, we can say that the OLS estimator is *BLUE*:

- *Best* (it has minimum variance and is efficient)
- *Linear*
- *Unbiased*
- *Estimator*

Given data on labor, industry output, and wages across a sample of communities we can apply OLS to equation (16.14). If we are given a change in industry output from the IO/SAM, we can plug the new level into the estimated equation (16.14) and derive a new level of labor demand.

While there are several potential problems with classical regression analysis, the one we want to focus on is the *error independence* [E($\epsilon_i\epsilon_j$) = 0 for all $i \neq j$] assumption. For the community analyst, this assumption explicitly states that two communities

sitting side by side are independent of each other. Clearly, all of our spatial theories ranging from classical firm location theory to central place theory say this cannot be the case. Community economic theory tells us that the assumption of independence is violated. This problem is widely referred to as *spatial autocorrelation*. Using OLS in the presence of spatial autocorrelation results in biased and inconsistent parameter estimates and the *BLUE* characteristics are not present. In addition, doubt is cast on our ability to make inferences about the statistical validity of the estimates. This has direct ramifications on hypothesis testing.

The violation of the error independence assumption centers on the makeup of $\sigma^2 I$ in equation (16.15). Specifically, the off-diagonals of the I matrix are no longer zero. The field of *spatial econometrics* has arisen to address this specific problem. (A complete discussion of spatial econometrics is beyond the scope of this book; the interested reader can look at the classic work of Anselin (1988) and Cliff and Ord (1973, 1981).) While spatial econometrics can become rather daunting, there are some simple approaches that shed light on the overall logic of the general approach. Let us walk through this simple logic.

Consider an economic plane that is composed of five communities (A-E) in Figure 16.3. Let us assume that communities that are physically adjacent to one another influence each other, but communities that are not adjacent are sufficiently far enough away from each other that they are independent. In this simple example, community A is adjacent to communities B and C but is nonadjacent to communities D and E. From a statistical perspective, the error terms of communities A, B, and C would be related or correlated, but the error terms of A, D, and E would be independent. A similar logic applies to all the communities in this example.

A widely used test to see if spatial autocorrelation is indeed a problem is the Moran's I:

$$I = \epsilon'W\epsilon/\epsilon'\epsilon \qquad (16.16)$$

where ϵ is the vector of OLS error terms and W is a spatial weight matrix where w_{ij} is an element of the spatial weight matrix equal to 1 if spatial units *i* and *j* share a border and 0 otherwise.[11] For our simple graphical example with five communities, the corresponding spatial weight matrix is in the lower left-hand corner of Figure 16.3. Unfortunately, the interpretation of the Moran's I is not always straight-

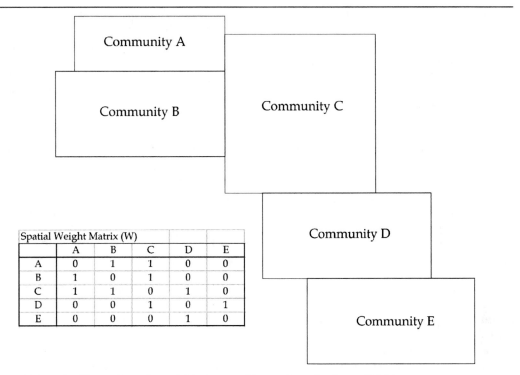

Figure 16.3. An illustration of spatial linkages. (Drawn from LeSage 1997.)

forward. But the logic of the Moran's I is pretty straightforward. The idea is to compare the error of the classical OLS with the spatially weighted errors. If the difference is significant, spatial autocorrelation is said to exist.

If the analyst wishes to correct for spatial autocorrelation, there are three standard approaches: the spatial error model (SEM), the spatial lag model (SLM), and the spatial autoregressive model (SARM), which is simply the combination of the first two. Consider our simple demand equation (16.14) expressed as a SEM:

$$L_i = \beta_0 + \beta_1 O_i + \beta_2 R_i + \epsilon_i \qquad \textbf{(16.17a)}$$

$$\epsilon_i = \rho W \epsilon_i + u_i \qquad \textbf{(16.17b)}$$

Here, the demand equation is the same as before, but the error structure takes on a slightly more complex form. Recall that the problem is that the OLS error terms are correlated; equation (16.17b) makes that correlation explicit, where ρ is the correlation coefficient, the spatially weighted OLS error ($W\epsilon_i$) is the independent variable, and u is a second error term that

is assumed to be well behaved. The spatial weight matrix here is the same as with the Moran's I.

The second alternative is the SLM model:

$$L_i = \beta_0 + \rho W L_i + \beta_1 O_i + \beta_2 R_i + \epsilon_i \qquad \textbf{(16.18)}$$

In this case, we have captured the spatial dependency not through the error structure but through the dependent variable. Indeed, the error independency assumption can also be thought of as independence of the dependent variable across observations. In essence, they are two sides to the same coin. Again, OLS is applied to equation (16.18), but the inclusion of $\rho W L_i$ on the right-hand side explicitly captures the spatial dependency.

The final approach, the SARM, combines the SEM (equation 16.17) and SLM (equation 16.18) approaches:

$$L_i = \beta_0 + \rho W L_i + \beta_1 O_i + \beta_2 R_i + \epsilon_i \qquad \textbf{(16.19a)}$$

$$\epsilon_i = \rho W \epsilon_i + u_i \qquad \textbf{(16.19b)}$$

where all the terms are the same as before. Unfortunately, theory does not really tell us that one spatial model is better than another and the final selection often comes down to an empirical question.

In the end, the community development practitioner needs to be aware that when regression analysis is used and the unit of observation is the community, the error independence assumption may be violated, resulting in less than desirable estimates. Because we know that space, or spatial proximity, is the culprit, we can adjust our statistical methods by introducing what we have called the spatial weight matrix. As we have seen, there is more than one way to introduce this correction and the "correct" approach is often an empirical question. Over the past few years, spatial econometrics has increasingly moved from the classroom (theory) to the field (practice).

SUMMARY

In this chapter we introduced price endogenous general equilibrium models. Two approaches were discussed, computable general equilibrium models (CGEs) and hybrid conjoined models. The idea has been to move the analyst one step forward by allowing prices to fluctuate and indeed be determined by the model. Partial and general equilibrium theories and models, along with several approaches to provide the analyst the flexibility to more realistically model the community economy, were discussed. From a theoretical perspective, CGEs and hybrid conjoined models present more "correct" ways of modeling the community economy.

These models are not without their problems. Model complexity can become a concern in the real world. Issues that need to be acknowledged include aggregation bias, errors in model calibration, limitations in data, inappropriately used statistical methods, and general misunderstanding of how the economy functions. All of the models we have discussed in Chapters 15 and 16 presume that the economic measures for any given point in time represent an economic equilibrium. Kaldor (1979) suggested that economies are always in some form of disequilibrium or transition, and therefore the models start from a misleading place. This represents a larger problem that we have between theory and practice. Often, the data that we use do not fully represent the conceptual basis that we are trying to analyze. In essence, the economy is constantly moving toward an equilibrium and that equilibrium itself is constantly moving.

Another impediment to the expanding use of price endogenous models to community-level questions is the general lack of community-level price data. To circumvent this problem the *small region* assumption is usually imposed: The smaller the region is relative to the rest of the nation, the more open the local economy will be. The essence of this assumption is that the community activity is assumed to be too small to affect prices, and producers and consumers in the community can be considered price takers. In the long run, perfect factor mobility, an assumption of neoclassical growth theory, ensures that the supply of inputs, including labor, is perfectly elastic. Thus, local price will converge to the national average. The practical implication for those interested in studying and understanding the behavior of community economies is a return to the horizontal local supply curves, which returns us to the fixed-price models of IO and SAM. In many situations, fixed-price models can be perfectly reasonable and theoretically sound representations of the local economic condition.

Finally, from the standpoint of econometric modeling, we have introduced the reader to the simple notion that "place in space" matters. Analytically, using ordinary least squares regression on spatial data opens the analyst up to concerns of spatial autocorrelated errors and less-than-best linear, unbiased estimates of the underlying relationships. One solution to this modeling dilemma that was outlined employed a spatial weight matrix to explicitly account for where data elements fall within and across space. This is an important correction for econometric analysis of spatial data.

From the perspective of the community practitioner, CGE, hybrid conjoined models, and spatial econometrics may appear to be beyond their reach. But not long ago, IO and SAMs were seldom brought down from the ivory tower. With programs such as IMPLAN and other user-friendly software, community practitioners are now conducting such analyses. It is our contention that in the not too distant future, CGEs and hybrid conjoined models will be as widely used as are IO and SAMs. It is important to recognize that the movement from theory to practice includes movement from science to art. When a judgment call is required of the modeler is when art enters the science. The practitioner must always be careful with this subtle but important distinction.

STUDY QUESTIONS

1. Why are economists so concerned about prices when modeling the economy?
2. Discuss some of the trade-offs economists face between using IOs, SAMs, and CGEs.

3. What determines prices in a CGE model?

4. What fundamental problem does an economist interested in building a community-level model face that an economist building a national-level model does not face?

5. What types of problems might practitioners run into if they use classical regression analysis with cross-community data?

6. What are some general components of a hybrid conjoined model?

7. How does an analyst ensure that a model provides reasonable estimates?

8. Discuss how "art" enters into community model building.

9. How might IO/SAM models be conjoined or integrated with econometric models?

10. Discuss the difference between partial and general equilibrium theoretical models.

11. Differentiate between descriptive economic tools, fixed-price general equilibrium models, and price endogenous general equilibrium models.

NOTES

1. Much of classical microeconomics addresses partial equilibrium where the analysis focuses on comparative statics. In comparative statics, price is a key metric that is measured in a partial equilibrium sense.

2. The overview of CGE modeling presented here is only intended to introduce the topic. Detailed discussions are provided by Dervis, Melo, and Robinson (1982); Friesz, Miller, and Tobin (1988); Shoven and Whalley (1984).

3. It is helpful to think about IO/SAMs and CGEs as being computed rather than estimated as we will see with the hybrid conjoined models later in the chapter. While construction of a CGE draws on the econometric literature for parameter estimates, the CGE itself is not estimated.

4. The use of the terms *open* and *closed models* in this chapter differ from our discussion of IO and SAMS in Chapter 15.

5. An interesting historical side note is worth mentioning. Although this example comes from Varian (1992), it originally was based on a historic speech by Adolf Hitler where he promised the German people that building a large army (guns) would not distract the economy from producing consumer goods (butter). In essence, his argument was that the German economy was inside the production possibilities frontier and that there was significant waste in the economy. This was also an argument used by the Johnson administration to explain the economic effects of large-scale military spending during the Vietnam War while simultaneously waging a costly War on Poverty and other social programs at home.

6. A *numeraire* is a base reference point and can be thought of as an index number. By using the price of oil as an index, we can set it to any value we wish, generally 1. This removes one unknown from the problem.

7. The model has grown to contain literally millions of equations. The current joke is that there are three Australians for each equation in the model.

8. Clearly, the ORANI system could be an exception.

9. This section draws on the dissertation work of Martin Shields (1998).

10. In the United States the detailed housing and population census data is collected every 10 years and business census data every 5 years.

11. The spatial weight matrix that we have offered for discussion here, a simple adjacency-nonadjacency rule, is but one of many ways to construct the weight matrix. More sophisticated approaches use gravity models or distance decay functions (Chapters 4 and 14) to specify the off-diagonal weights. But again, theory does not shed much light on what is the best way to construct the weight matrix.

17
Looking to the Future

There is a bright future in community economics. This future lies largely in building the practice of community economics on the foundation of theory. Theory and practice—each has something to offer to the other. Community economics is a field that has many practitioners, most of whom "practice with little awareness or knowledge of the theory on which they conduct their day-to-day activities. These practitioners often want to dispense with theory and get down to earth" (Hustedde and Ganowicz 2002). They want studies to shed light on issues such as how to attract a manufacturing plant, how to get more people shopping downtown, or how to get working capital in business, or a range of other issues that need immediate attention. Hence, there is more interest in empirical research or how-to initiatives than in theory itself. Theoreticians often accuse practitioners of worrying about symptoms of the problem and never the root problem itself, while practitioners accuse theoreticians of being isolated in the ivory tower. Community economic development, as a field of study, is practice based and practice driven, but that practice must have its basis in theory.

The purpose of this book is to bridge theory and practice by placing the practice of community economics within a theoretical construct. In particular, the construct involves the disciplines of economics, planning, sociology, political science, geography, community development and regional science—the whole complex of disciplines that are involved in the relationships between community and the economy. Community economics, by definition, is an interstitial discipline; it is a bridging discipline. In that sense, it is similar to biochemistry rather than biology or chemistry. Its variables are derived from two or more social science disciplines. The problems faced by the practitioner are very seldom constrained by a particular discipline.

There are several different personal perspectives one can bring to the practice of community economic development. The *economic growth perspective* views community economic development as creating more jobs, sales, property value, and income. It is the perspective of economists, development planners, and most business people. The *natural resource management perspective* of community economic development is about managing the land and natural resources surrounding the community in a sustainable way. Professional resource managers, land use planners, and environmentalists hold this view. The *perspective of human services* is about getting human services effectively provided in areas of underserved populations. This view is held by those involved in workforce training, education, health-care provision, and support programs. It is the perspective of educators, social workers, and health-care professionals. The *perspective of infrastructure financing* is about making loans or grants available to a community so a tangible infrastructure project can be built. Local grant writers, engineers, and administrators who use these programs hold this view. It is the prospective of the physical planner, the public works person, and the accountant. A person holding the *local public administration prospective* views community economic development as balancing state and federal mandates on local government with the needs of local constituents. Many county commissioners, financial planners, and mayors hold this view. The *community activist* views community economic development as empowering the disenfranchised, advocating for social change, and reducing poverty. This view is held by many leaders of nonprofit organizations, advocacy planners, and by political activists. The challenge to the community economic development practitioner is to balance these many perspectives.

In this book, the concept of community is a very important consideration. *Community* is an organization of people in a physical setting with geographic, political, and social boundaries and discernable communications linkages. When thinking about the theoretical ideas and perspectives outlined in this book, it is convenient to think about a community as a stand-alone place, such as a city, village, or town, but remember that the community does not coincide with the political boundaries of a municipality. Many of the concepts also apply to neighborhoods or a collection of smaller places within larger urban places. The important things are a commonality, such as trade area or labor shed, and the ability to implement decisions.

REVISITING OUR DEFINITIONS OF COMMUNITY ECONOMIC DEVELOPMENT

Community development is about local leadership, citizen participation, collective decision-making, and community organization. *Community economic development* is about how economic forces and theory explain community change and how policies and actions can affect that change. It is about how economic structure influences the choices that we can make. It is about increasing community wealth in both monetary and nonmonetary forms. It is about creating economic opportunities for residents. It is about how dynamics and the resultant disequilibrium or how changing circumstances create tensions within the community that require choices. This means consideration of the interaction of the economic, political, social, and institutional components of a community. Implementing decisions and strategies means the people are intervening in the economy with the idea that they can achieve some type of desired outcome.

Community economic development is sustained progressive change to attain individual and group interests through expanding, intensifying, and adjusting the use of resources; identifying new or expanding markets; altering rules of economic activity to facilitate adjustment to changing conditions or altering the distribution of rewards; and improving insights into the choices available.

Community economic development *decision-making capacity* is the ability of a community to initiate and sustain activities that promote local economic and social welfare. The overall purpose of community economic development policy is to reduce and/or abolish the barriers in product and factor markets that prevent the positive culmination of economic development processes.

A strong distinction is drawn between growth and development. *Growth* is simply more of the same. It generally represents quantitative differences such as real estate transactions or more employment and income, to name a few. *Development* is about dynamics, creative destruction, transformation, structural change, increasing opportunity sets; it is long term and permanent. Development in most cases is more than just economics. Development tends to imply more understanding, more insight, more learning, more nuances and structural change. Development implies a change in the capacity to act and innovate and expands the set of choices for both individuals and the community as a whole.

Community economic development is dynamic; it is concerned with movement and change, with overcoming obstacles and capturing opportunities. It creates and closes opportunities depending on the community's preparedness, willingness, and capacity to respond. It recognizes a need to use and maintain the resource base of a community. The community is thinking long term, not just short-term exploitation of its resources. Community economic development tries to ensure that the needs of the present are met without compromising the ability of future generations to meet their own needs. The community is sensitive to the sustainability of its actions. Important to the concept, as we view it, is the sense that the community controls its own destiny or is self-reliant. Clearly, this sense of control means the community does not view itself as a victim. This means, as we have emphasized, that communities have choices and can make those choices and implement their decisions. Sometimes communities make wrong decisions, but they learn from past mistakes.

Community economic development recognizes development is not a fixed state of harmony, rather it is a process of change in which use of resources, directions of investments, orientations of technological development, and institutional change are made consistent with present and future needs. One could think of the community's economy as never really being in equilibrium but always moving toward equilibrium. Community economic development not only influences the location of the equilibrium but also the path toward that equilibrium.

Communities aggressively manage and control their destiny based on a realistic and well thought through vision. Such a community-based management and control approach requires that a process be instituted within the community that effectively uses knowledge and knowledge systems (theory) to direct change and determine appropriate courses of action. The process must be comprehensive and address social, economic, physical, and environmental concerns in an integrated fashion while maintaining central concern for present and future welfare of individuals and the community. An important consideration in community economic development is that it implies that all groups are given an opportunity to participate in defining problems, setting goals, and analyzing results. This means that the practitioner may need to aggressively ensure that all members of the community have the opportunity for input. It is important to remember that the analysis and group process will be for naught without action.

While development implies that the community is resilient to adverse shocks, it recognizes that development is an inherently uneven process in terms of its sectoral distribution, its timing, and its geographic distribution. Disruption is a necessity in a dynamic economy and is part of the dynamic process associated with development. Schumpeter referred to this dynamic process as "creative destruction," where old resources become less valuable, new resources become more valuable. Gaps in knowledge and spillovers of externalities are generated among markets as some institutions or industries lead and some lag in the development process. In reality, the economy does not adjust instantaneously and painlessly to the disruptions caused by development. There are inherent structural barriers, inconsistencies in the market, which make this process less than smooth. One goal of development policy is to facilitate this adjustment. The goal of development policy should not be to avoid disruption and adjustments but to facilitate the adjustment process.

This leads us to our goal of community economic development. We think that community economic development is about producing wealth, opening economic opportunities, and facilitating adjustment. It is important to include both monetized wealth (jobs and income) and nonmonetized wealth (opportunity, adjustments to changing conditions, resilience to external changes, or even maintaining a nice view of the lake). The interpretation of nonmonetized wealth as increasing choices or opportunities means that the range of choices and opportunities for individuals has expanded. Facilitating change or adjustment means that development is a dynamic process in which new resources, new markets, and new rules are appearing both inside

and outside the community. The change or adjustment means that the community is able to absorb those new factors, thereby increasing its economic resilience.

Economic development exists in an interconnected web of community stakeholders, business managers, nonprofit agencies, government departments, and multinational corporations, among others. Managing contemporary local economic development requires the ability to facilitate interaction, to mobilize stakeholders, and to reconcile different goals and values among key development actors. This role too often leads some practitioners to emphasize group processes at the expense of analytics. While this distinction is often expressed in terms of process versus content, the truly effective community development economist must be able to bring both process and analytical skills to the table.

Economic development takes teamwork. You need both collaborators and unwilling conspirators. While willing collaborators may be easy to find, many people are unwilling to get involved. To engage unwilling conspirators in building a long-term economic development strategy takes skill on the practitioner's part. It means that often people are not interested in building long-term strategies, they are more interested in worrying about short-term projects.

WHERE WE HAVE BEEN

We hope this book increases economic literacy about the choices, decisions, and problems faced by communities and reduces the chance of drawing inappropriate conclusions. We have tried to explicitly state the variety of assumptions that are critical in making the various theories and models work. Too often people tend to forget the very explicit assumptions that are made when thinking through a problem resulting in inappropriately applied models.

As we have seen throughout the book, the purpose of a market economy is to allocate scarce resources of society across private and public goods and services in a manner that maximizes the well-being of citizens within that economy. Within a market economy, the questions of maximizing are answered by the interplay of buyers and sellers through the forces of market demand and supply. This all occurs within a set of explicit and implicit rules governing how the market functions. A *Pareto optimum* is said to occur when resources can find no better uses or places. The power of competitive markets, or Adam Smith's *invisible hand,* is often touted without fully reflecting on some basic and powerful assumptions.

Following are the basic assumptions of the perfectly competitive model that are violated and represent why there is a subdiscipline of community economics.

1. Community economics does not necessarily accept the distribution of income among people and places as acceptable and immutable. One must keep in mind that a Pareto optimal equilibrium says nothing to the concepts of equity or fairness.
2. In community economics, information is not free or available everywhere, nor is it fully comprehended.
3. Community economics recognizes resources are not totally mobile in a spatial sense both across communities and across uses. In the real world, labor, capital, and land cannot easily shift from one use to another at the whims of markets forces.
4. Externalities are present in community economics.

In a typical economic model, men and women are considered independent, rational, and all knowing. *Independent* means that each individual makes his or her decisions regardless of the actions of someone else.[1] *Rational* means that the individual orders all possibilities from best to worst and picks the best by using some objective criteria. As we have seen, many of our economic theories presume utility and profit maximization. But again, in the real world, people and firms may be pursuing numerous objectives simultaneously. *All knowing* means that the individual is aware of all of the options and can comprehend all of them. In reality, most men and women do not approach decision-making in such an ideal fashion. In particular, community economics is concerned with decisions in which individuals are not independent and often are not all knowing in their decisions.

We discussed at length the practice of community economic development. Many maintain that the practice tends to be more art than science. Included in this art are ways in which to practice community economic development: determining whether a community is really ready to engage in long-term economic development and managing conflicting goals or criteria held within the community. What we have advanced in this book is an economic science that underlies much of the art. By having a theoretical construct in mind, the artful practitioner knows which relationships are important.

Community economic development *decision-making capacity* is the ability of a community to initiate and sustain activities that promote local economic and social welfare. Decision-making capacity means that the community thinks holistically or systemically. The community thinks in terms of problems rather than just symptoms. The community is able to create and implement appropriate responses. Community economic development encourages communities to think in the long run and to understand how they should function. It encourages communities to think strategically. The community is able to reflect and learn from past decisions. The community has a broad spectrum of people involved in the decision, and this involvement to a large degree is determined by community culture.

Society and culture define values that establish how a community views and defines legitimate problems and legitimate solutions to those problems, and determines what is desirable and what is undesirable. Culture defines what behaviors are acceptable and is often referred to as social norms. Institutions/rules place into law those social norms. Culture is the largely unwritten rules about how people deal with each other, deal with conflict, define fairness, and identify acceptable behavior. Institution/rules are formal and govern how we bring markets and resources together through decision-making across space. Rules are the social, political, and legal tenets that must be accounted for in the use of resources in exchange and in the distribution of rewards. Rules and institutions define who receives the income generated from the use of resources, thus facilitating or hampering economic development.

The creation of an institutional framework supportive of community economic development, however, is not automatic and may be the critical element in a community's economic development efforts. Institutional arrangements evolve in the face of new scarcities, new knowledge, new technology, and new tastes and preferences. Any community economic development effort must be conscious of institutions and culture. While the economic forces that are at play are similar between developing and developed economies, the fundamental differences are the unique characteristics of institutions and society.[2]

Communities function in economic space, complementing, and at times competing with, neighboring communities. The narrow perception of location theory is that it explains decisions to locate or relocate a business. But location theory goes further and helps explain how spatially separated economic units interact among themselves and their input and output markets. Since every economic transaction, including consumption and where to live and work, has a spatial dimension, each transaction represents a location decision. Thus, *location decision* refers to any economic transaction with a spatial dimension, not just the traditional relocation decision.

The contribution of specific resources to community economic development depends on where that use occurs in space. Likewise, the shifts in demand for products affect specific geographic locations of production, including communities. A *community* represents the operating environment for economic units interacting in space: businesses and households buying and selling output, labor, raw materials, and capital. *Location theory* provides insight into how location decisions are made and why economic activities occur where they do. By gaining insight about how location decisions are made and what forces influence location decisions, communities can consciously try to influence those decisions.

We argue that in the short run, less than satisfactory economic development is not the result of inadequate productive capacity because there is unemployment, unused capital, and out-migration from the community. Rather, the lack of development results from inadequate demand. What needs to be done is increase the demand for goods and services produced by the community, or change the types of goods and services produced in the community to those where demand exists.

Traditional community economic development theory focuses on the export sector bringing income into the community. Traditionally, economic development policy has focused on identifying export activity and promoting it. The type of export sector sought should have good internal linkages with the rest of the community so the external economic forces are transmitted effectively to the local economy. The nonexport sector serves markets within the community and generally supports the export sector, although it can lead development. A focus on just the export sector is a far too narrow view for a comprehensive approach.

Central place theory is appropriate when considering nonexported trade and services. The significance of central place theory to community economic analysis is its recognition that the community is part of a system. No community, especially a smaller community, can provide all the goods and services necessary and desired. Residents in smaller communities and their surrounding tributary areas must relate to larger communities for many goods and services.

In the long run, community economic development is more likely to focus on factor markets, including land, labor, and capital. This focus should include factor productivity and allocation across sectors for which market demand is growing. Included in this are latent factors of production, such as amenities, quality of life, public sector goods and services, and technology.

Regardless of the range of factors of production used in the functioning of a modern economy, society places great emphasis on *labor or jobs*. The possession of a job in a modern economy provides the income that determines, to a large extent, the capacity to pursue a particular lifestyle. Because jobs are central to society and personal perception of worth, preparing people for work, placing and keeping them in jobs, and providing opportunities for advancement are critical.

The exchange of labor services between worker and employer occurs in the context of demand for and supply of labor, within an overarching institutional framework that affects the interchange. In theory, a labor market is an institution where labor services are bought and sold and therefore allocated to various occupations, industries, and geographic areas to yield the greatest output. Market forces establish the price of labor and allocate labor to its most productive use. A job can be judged in many ways, including skills required, wages, fringe benefits, opportunity for advancement, working conditions, and commuting distance. Employers do not desire workers because of their intrinsic value. Rather, employers' demand for labor derives from labor's productivity and the price of the product labor produces. The three theories examined are human capital, center-periphery and, segmented labor markets.

Two subtle shifts in labor market analysis in recent years are a focus on inclusion and a focus on job quality. The first shift explicitly recognizes that some individuals, because of race, gender, age, or other factors, have been systematically excluded from full participation in the labor market and its associated rewards. The second shift recognizes that a job is more than mere employment and that some jobs possess characteristics that workers appear to value more than others. Yet not everyone desires jobs having the characteristics so highly valued by many workers. For example, some people do not want a job that requires them to solve problems or to work full time. Thus it is important to maintain a mix of jobs in designing any community employment program. One must keep in mind that a broad goal of economic development is the creation of economic opportunities, and this is perhaps clearest in the range of jobs offered in the community.

A key element in community economic development activities is the availability and use of financial resources within private capital markets. The basic theoretical model of financial markets contends that capital is allocated among uses, regardless of place or type of use, in such a fashion that the return on capital in all uses and places is equal. Under ideal conditions, if there are demand/supply inequities, financial markets, at any given point in time or space, create a set of interacting forces to bring the market into balance. The equilibrating force is interest rates or return on capital. This type of movement between uses is easily seen in the spatial neoclassical growth model.

Capital by itself is not sufficient for community economic development, although it is a necessary component. It is important to realize that markets, management, and labor are equally if not more important to community economic development. A *supply gap* could be a situation where a community's capital market cannot attract sufficient funds to support local investment. A *demand gap* exists when an area lacks the management and entrepreneurs capable of transforming existing financial capital into physical capital required for production, such as buildings and equipment. Possibly as important as any concern about the shortage of capital is the shortage of talent to mobilize and utilize capital in a fashion both publicly and privately rewarding while protecting the parties against loss.

Land, a primary factor of production, has served as a critical production input for both household subsistence and income generation while acting as an important tradable asset. Economic growth has led to an array of land use changes within and around cities, towns, and villages. The increased demands for land and its distance to the urban core point out the notion that distance, not soil productivity, plays a primary role in determining land use within rapidly growing regions. The essence of growth management is to balance the need for economic growth (conversion of open space to built-up urban uses) with other less market-oriented objectives of society (maintaining open space). Many of the open space attributes demanded by society are of a nonmarket nature. The values of these non-market goods and services are difficult to quantify. Our ability to balance economic growth with other societal-determined wants and needs provides the challenge associated with managing growth.

The role of the public as a key stakeholder in providing, maintaining, and improving resource endowments helps characterize the economic condition of communities. *Amenities* such as open green spaces serve as important latent inputs to production in communities. As such, they present a complex mixture of market-based and nonmarket goods and services to the challenge of community economic development.

The economic values we place on natural amenity resources and publicly provided infrastructure is, in large part, governed by ownership and/or rivalness. *Ownership* of a resource allows the owner to exclude use to those who are not willing to pay. This excludability often presents problems with valuing natural amenity resources and publicly provided infrastructure. What is the economic value of a stretch of uncongested road? Another aspect that affects an amenity resource has to do with how competitive the environment is with respect to using the resource. We sometimes refer to this as the level of *rivalness* that a good exhibits. A *rival good* is one in which the addition of another user acts to diminish its value for the original users. A nonrival good is one that all can enjoy without diminishing its value to other users. Nonmarket goods and services can be arrayed with respect to their levels of exclusivity, rivalry, or congestion.

Governments at the community level play a vital role in all economic activity. Local government policies related to taxation, spending, and borrowing impact all actors in the economy in their various roles of consumers, workers, employers and producers. *Public goods* include the construction and maintenance of streets, roads, public utilities (sewer/water, energy, and others), and public buildings. Publicly provided services include fire and police protection, sanitation services, and education. Decisions about where to make investments in infrastructure can have significant impacts on the location decisions of people and firms. Governments also monitor, regulate, and stimulate economic actors. Many communities, for example, have aggressive land use ordinances regulating signage, building designs, and development impact fees. For the community, local governments are important partners in the economic development and growth process.

In the simplest sense, technology is how land, capital, and labor are brought together to produce an output. Technology and innovation in a broader sense capture not only new production processes but also ideas and the way of conducting business. Technology and innovation are critical in a dynamic economy. As we discussed, much of the current thinking on modern growth theory focuses on technology and why growing economies should invest in technology. In this dynamic economy, you invariably create winners and losers when new technology is applied. When a community or industry experiences technological change, the inputs required in the production process are likely to change. When these processes change, other areas may gain comparative advantage in the delivery of those inputs. Another way to look at technology is in the products being produced. Here again, as the products being produced change, the communities and industries that produce them probably gain some type of competitive advantage (i.e., profits and wealth).

There are three forms of technology: idea or process, production, and output. The latter two are sometimes referred to as things. Things are rival in that "things," such as a new robot technology, can be patented. Process technology is how we do things or deal with each other. Broadly speaking, ideas are nonrival and easily copied and communicated, but implementation processes can be made rival. For example, the idea of business computer networks is nonrival, but the specific approach in implementing the idea can be rival.

Innovation is the successful introduction of ideas perceived as new into a community or as the first commercial or genuine application of some new development outside of experimentation. The innovation process is not linear; it is complex, interactive, iterative and characterized by uncertainty, trial, and error. Adopting a new technology depends on the firm's technical capacity and organizational ability. For the individual firm or community, the adoption process is based on access to information, the estimation of profitability or usability, and the evaluation of adjustment costs.

One of the major critiques of community economic analysis is that it is a random collection of analytical tools ranging from simple descriptive tools, such as location quotients and pull factors, to more complex tools, such as input-output, computable general equilibrium models, and econometrics. While the level of sophistication of the tools can be impressive, the community economist must keep in mind that these are empirical representations of vastly complex economic, political, and social systems. Indeed, most of the analytical tools outlined ignore the political and social elements of the community system. As with any theory of how a system works, the practitioner must keep in mind that our analytical tools are intended to provide insights. As you may suspect, an economist without models and tools is an economist searching for a way to make a contribution.

A COMPREHENSIVE APPROACH TO COMMUNITY ECONOMIC DEVELOPMENT

In Figure 17.1, we display what we think is a comprehensive model of community economic development. The Shaffer Star outlines the fundamental elements required for an inclusive view of community economic development: space, decision-making, resources, rules, society, and markets. While all elements are equally important, in any given community setting practitioners may find themselves focusing on one or two dimensions, but to be effective, they must keep all elements in mind.

When you look at Figure 17.1, you can see that *space* is central to the concept of community economic development. Space influences markets, resource use, and even decision-making. Space plays a critical role in community economic development because one is concerned about overcoming distance in terms of reaching markets, moving resources, or bringing together the actors critical in reaching community decisions. Our discussion of spatial markets also drives home the point that communities do not function in isolation but are part of a larger spatial economy. In Figure 17.1, space enters the prongs not only as distance but also as the ability to meet and discuss issues and ideas face to face.

Resources are important because they represent the inputs to community outcomes. The welfare of the community (outcome) is directly related to the extent and characteristics of community resources and how individuals and groups use these resources. The important aspect of resources is that as you look at a community's economy you naturally think of the labor, capital, and land that enter into the production function. Yet, frequently it is the public sector and the nonmeasurable latent resources that are

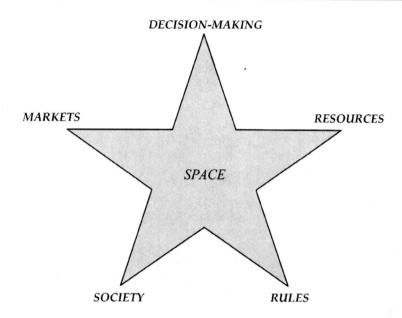

Figure 17.1. The Shaffer Star of community economic development.

the driving forces in local development. Resources are used to produce community output, such as jobs, income, health care, manufactured goods, and agricultural products, which in turn leads to community outcomes, or overall community welfare or well-being.

Outcomes are distinct from outputs and are really the generation of community wealth. *Community wealth* is defined as both monetary and nonmonetary. The monetary dimensions of wealth are such things as income, jobs, sales, and property values among others. The nonmonetary dimensions of wealth are those aspects of a community that fundamentally are immeasurable. They could include air quality, personal relationships, the vista of a beautiful sunset, or simply greater economic opportunities.

The *markets* prong incorporates both external and internal markets. The external markets represent the general avenue in which national and regional economic impulses are transferred into the community. The internal markets, however, represent the linkage between the external market and local markets and are the way the external impulses play out in the community. Export base and central place are the two major theories that we use to explain communi-

ty goods and services markets. When examining the internal markets of a community, a process referred to as market area analysis represents a way that was suggested for developing appropriate strategies. Keep in mind that although the export base is the carrier of external economic activity into the community, the linkages with the non-export sector are what allow you to transfer external effects into local change. These linkages can be either local or nonlocal. Another powerful way of thinking about markets is to visualize the circular flow of economic activity. This includes the market for factors and the market for goods and services. There are three general actors in this market: firms, households, and government.

The two prongs represented by *society* and *rules* are important but seldom recognized. The society, or culture, prong is frequently not recognized because it is so pervasive and tends to be background in local communities. The culture of the community can vary among communities. There are communities that have a culture where they feel like they are victims to external forces, while other communities take the external forces and amplify or deflect them to their advantage. The latter communities are what we refer to as being more entrepreneurial in their outlook. The idea of rules includes the enforcement of contracts or the idea of property rights. Institutions in our model generally represent some type of legal mandate. Institutions can also mean an organization such as a municipal, state, or federal government, chambers, churches, unions, or housing co-ops. These organizations provide a mechanism for establishing and enforcing social norms and formal rules.

The reason the notion of society in our model is so difficult to grasp is that to a large extent it represents the local culture and how things have typically been done. Most people living in a specific community fail to recognize the pervasiveness of forces to maintain the status quo. A community that is seeking to change its culture must challenge the status quo. This means that the community, or individuals within the community, must ask perceptive questions about why things have always been done the way they have been. It also means that the askers of the questions must be willing to offer alternatives.

To a large extent, the society/culture prong determines the scope of the transactions organized by markets, which is what this book is largely about. The divergent interests and pursuits of actors in the market do not automatically mesh to form a harmonious whole; specific mechanisms are needed to keep competition as contained conflict from escalating into all-out conflict. The starting point of community economic analysis is that the economy is a subsystem of the societal system.

This brings us to *decision-making,* the final prong of the Shaffer Star. We think that decision-making may be the most important prong among equals. Decision-making is really an implicit idea in most other models of community economic development. It can be, although we think it should be much more, very process oriented. Here the practitioner worries about who's involved in decisions, group processes, and how decisions are made. We think decision-making should be based on information, not perceptions. The analytics that we present in this book provide a foundation for economics and planning information. The analytics that are implied are that the community can sort its way through a variety of symptoms and tackle the root problem, and we tend to think that most community problems are subject to economic theory and analysis at their roots.[3]

The decision-making prong is where the practitioner can really enter into changing the conditions within a community. Decision-making is critical because it means that the community is willing to engage in addressing problems rather than symptoms and the community is willing to act on the decisions that it has reached. These two elements also are critical because communities often do not put both of them together. Some communities may spend their efforts on addressing symptoms. Other communities may discuss problems and reach tremendous decisions only to fail to follow through.

It is important to recognize the interplay between decision-making and society/culture. If society/culture is interpreted as the way things are done, then they say a lot about who is involved and how decisions are made. What we are asking is, Do only a few people become involved in making community-wide decisions, or is there open and frank discussion about what the problems may be and potential solutions to them? The involvement of a wide spectrum of the community in problem definition and solution options opens those choices to several perspectives and opens the definition of solutions to a wider resource base.

No theory of human (community) behavior can be developed without a core assumption about the intellectual capabilities of the person or the collective community. We suggest that people are poor processors of information—just the opposite of the neoclassical economists' assumption that information flows and absorption is instantaneous, all without any costs. Indeed, strong evidence shows that people start with little knowledge and can comprehend limited amounts. Clearly, we are not saying people are stupid. We are saying that they are not the all-knowing and all-comprehending rational economic person. In our model, people are poor processors of information and are defective decision-makers. According to our economic theories, many decisions appear to be irrational. Decisions are often based on values and emotions rather than on hard analysis. Choices are made without processing information, drawing inferences, or deliberating. Most choices, including the selection of goals and also of means, are made to a significant extent on the bases of values, emotions, or social norms. Values, emotions and social norms often limit the range of those options that are considered; in these cases, knowledge plays a limited role. But we argue that knowledge/analytics can and should play a bigger role. Indeed, one of our basic strategies is for the community to act smarter.

Implicit in decision-making is that the community needs to establish its values and set priorities. Emotions and social norms do play an important role here. Values, to a large extent, say what you see as problems. Each community faces a plethora of problems, and community priorities go a long way in determining which ones will be addressed first, second, or not at all.

One of the essential things to keep in mind in reviewing this text is that our basic assumption is that the community can use its own volition to reach decisions regarding its economic future. A quick review of the discussion in the first section of the book brings home both regional and national theories as to how the economy is structured and changes over time and space. It is through understanding the ramification of these theories and the insights they provide that communities can make their choices.

STRATEGIES FOR COMMUNITY ECONOMIC DEVELOPMENT

We began our discussion of community economic development with a review of a very abstract stylized model of the economy, specifically the neoclassical model of markets and economic growth. We saw that under a set of very strong assumptions the forces of Adam Smith's *invisible hand* lead to an efficient allocation of resources. The policy or strategy implication for communities is that the market in essence knows what it is doing and that collective intervention will lead to suboptimal market outcomes. Indeed, barriers to the efficient operation of the economy, like restrictive land use policies, should be avoided.

We also saw that our perfectly competitive view of the world does not match up with reality. Firms can assert monopoly powers in the short term, externalities in production and consumption lead to market imperfections, and economic agents lack full information or the skills to fully process the information they have. Some have challenged the fairness of the Pareto allocation that follows from perfect competition. In other words, there is a very real role for policy in affecting the outcome of the economy.

Beginning with the Mississippi Balancing Agriculture with Industry (BAWI) policies of the 1930s, virtually every state and community has sought to attract export or basic employers. Clearly, BAWI-type policies moved away from the neoclassical view of policy formation in that government assumed a proactive role in economic development and growth. Today, the BAWI policy is widely referred to as "smoke-stack chasing" and/or the

"new war between the states."

Pulver (1979) offered the first set of comprehensive policies or strategies that can be acted on by communities:

- Attract new basic employers (BAWI-type policies).
- Improve efficiency of existing firms.
- Improve ability to capture dollars.
- Encourage new business formation.
- Increase aids/transfers received.

While these five strategies laid a foundation for much of the current thinking on community economic development policy, they have been critiqued for being ad hoc and not rooted on any theory of the local economy. Indeed, Pulver offered them as an alternative to the narrow thinking of firm recruitment but clearly appreciated the economic theory that supported these strategies. Perhaps Pulver's greatest contribution was that he used layperson language to describe what we speak of more technically in our seven strategies.

We offer an alternative way of thinking about community economic development policies or strategies. Rather than a laundry list of ideas, we turned to the economic theories discussed throughout the book. For example, the state-of-the-art thinking on economic growth theory emphasizes the role of technology and innovation in not only growth but also as part of profit-maximizing behavior on the part of the firm. Given the enforceability of patents and copyrights, firms have an incentive to invest in the research and development of new ideas, products, and processes. From the community's perspective, the natural strategy that follows *is to work* with new and existing business in using existing resources differently.

Generally, this means that you apply new technology. You find new ways to combine existing capital and labor to produce greater output per worker. It could also mean that you use existing capital and labor to produce a new good or service that previously had not been produced locally. It could also mean that you will have local jobs for workers who previously commuted elsewhere for work. Or it could mean that workers receive training to be able to do different tasks than before.

Our theoretical discussion of external and internal markets also points the way to a range of economic development policies. Indeed, the BAWI-type policies that dominated economic development thinking for so long were based on export-base theory. In our terminology, we are suggesting that the community should be **increasing the flow of dollars into the community.** This means that the community essentially brings income to the community by attracting new basic employers. Our discussion of firm location theory provides insights into which types of businesses a particular community may have a comparative advantage with. Other ways of bringing dollars into the community include existing basic employers increasing their sales outside the community, the community increasing its visitors, or the community increasing its intergovernmental aids.

These policies focused solely on the inflow of dollars into the community with no attention to internal markets. People and businesses must have some place to spend the money. They spend it either here or elsewhere. Too often communities think all they have to do is attract a basic employer (i.e., the buffalo hunt or smoke-stack chase) and their problems are solved. Firm location theory, specifically demand maximization, and central place theory open the door to numerous policy options, including but not limited to **increasing the recirculation of dollars in the community**. This approach means that the community is plugging leakages out of the local community's economy. In other words, the community is actively seeking ways to get people and businesses to spend more locally. It could be doing customer surveys, altering store hours, encouraging new or different store types, physically renovating downtowns or even talking to businesses about buying inputs locally rather than from nonlocal sources. The end result is that dollars turn over at least one more time locally. Firm location and central place theory points us in the correct direction.

Our discussion of growth theory and the circular flow model of the economy points to the importance of local factor resources. We have seen that technology as a factor is extremely important to the growth and development process, but equally important is the amount of land, labor, and capital. **Increasing the amount of resources available** simply means that the community increases the amount of land, labor, and capital available for producing output. This could be local financial institutions making more loans available locally, an outside business making a local investment, or forming a credit union. It could be people moving or commuting into the community or working more hours. The important thing to remember is the spatial component. The resources are increased here rather than there.

We also discussed at length the importance of noneconomic factors in community economic development. Returning to the Shaffer Star, the idea of decision-making and rules come to the forefront. **Acting smarter** translates into how the community makes decisions and sets up and implements strategies. Does it involve a broad spectrum of interests or just a select few? Does the community really get at the problem(s) or just treat symptoms? Does the community integrate sound analysis with community perspectives and desires?

Changing the rules means that the community seeks a change in rules that would benefit the community or seeks a change in interpretation of rules. For example, a land use plan might encourage further development on some land by ensuring that incompatible uses do not occur next to one another. On the other hand, changes into some land use regulation can impose major costs on some firms. Or maybe the community gets the state to re-interpret eligibility rules on some type of manpower training fund, thus making some community residents eligible. Remember that rules are societal constraints that govern how we either use resources or exploit markets.

We hope readers come away from our discussion with an appreciation of the fact that economic theories are by necessity abstract and simplifying. Indeed, the whole reason to develop theories in the first place is to help think through complex issues. As such, our theories always leave some element of the real world out. Even with this weakness, economic theory lends the practitioner one last policy option: **getting lucky**. This may seem like an unusual item, but think about it for a second. A small rural community could be located within the commuting shed of a growing metro area or 50 years ago could have been the birthplace of a budding entrepreneur. While we like to think more than luck is involved, and it is, it also explains a lot of current economic activity.

A comprehensive strategy that any community can pursue entails all of these seven elements. Clearly, most communities will pursue vigorously only one or two elements at once, but over time one should see the community considering all seven. Many communities use these strategies as building blocks, beginning with a simple foundation and moving upward over time with more sophisticated approaches. At each stage, the prior building blocks are reinforced. Remember that an economic development undertaking is part of a larger effort by the community to achieve some desired outcome or change.

Inherent in all of our discussions about community economic development are four basic tenets.

1. A community is a logical economic unit that can exert some control over its economic future.
2. Intervention in the form of conscious group decisions and actions will affect local economic welfare more than the sum of individual actions.
3. The action or policy must be designed and undertaken within the larger perspective of broader community goals.
4. The required resources will be available or can be found to implement the policy. Here resources are more than monetary and include all factors of production, especially social capital.

Our view of the economic system in a community hinges on the actions of people. In most communities, the success or failure of any specific economic development activity centers on the level of and commitment of volunteer citizens.

There are several characteristics associated with successful communities. Strategy formers recognize they are the ones who will make it happen, not someone else. They focus on what can be accomplished within a few months without forgetting their long-term objectives. They do not get to the solution before they know the problem. In other words, many "strategies" are nothing more than a collection of programs that treat symptoms rather than community problems. The strategy is specific in outcomes and actions. This means that who, what, where, and how are explicitly recognized. They tend to break actions into smaller parts to recognize the achievement of intervening goals and objectives. They remember that most others know very little about the effort, and they seek to find and cultivate allies. They also discipline themselves to stay with their strategy over time while revising, revising, and revising the strategy as they achieve some intervening goals and objectives.

One almost philosophical question that all community development practitioners struggle with is, What defines a "successful community"? Throughout our discussions we have implicitly presumed that the notion of a successful community is well understood. Unfortunately, in the real world, success in one community could be viewed as failure in another. In some communities, the social capital or sense of community is non-existent. The only thing that ties the residents of the place together is a common mailing address. Here, a collection of community meetings without the meetings deteriorating into shouting matches could be deemed a huge success. In other communities, there is a desire to measure success in much more concrete terms, such as number of jobs created or new businesses started. Both of these examples exhibit what we think are measures of success. We would label a *successful community as any community that has achieved movement toward some common goal.* In the end, the community itself must define success.

WHAT IT'S ALL ABOUT

We agree with Hart and Murray (2000, pp. 4-6), who argue that community economic development incorporates several important characteristics:

It is bottoms up. While the sensitivity to local needs and opportunities is important, it is also appropriate that development priorities should be determined in an environment that reflects the interest of local governments, the business sector, community groups, and voluntary organizations.

It is integrative. Local development concerns itself with making connections vertically and horizontally between stakeholders and across programs. Integration seeks to enhance the capacity for seamless policy-making and smooth management while recognizing that innovative organizational approaches must have regard to variations in authority and responsibilities, relationships with government, strategic preferences, and bureaucratic culture.

It is strategically driven. Local development, in order to be effective, rises above an association with a series of ad hoc initiatives in any locality. A clear direction based on local understanding of local issues and supported by a confident but realistic vision of the future is vital. Local development is about long-term targeted action to create change, both in places and with people.

It is collaborative. Local development requires the involvement of multiple stakeholders working together rather than on an individual basis. It is an inclusive activity that embraces the volunteerism within the community and voluntary sectors, elected representatives, public officials, and private sector participants.

It is interactive. Local development should not be perceived as solely a technical activity better left to others who appear more qualified. The hegemony of public officials and a dependency on consultants for expertise can be tempered by a local development approach that recognizes the knowledge-based input of local people and businesses into agenda setting and implementation. It requires an ongoing investment in local capacity through a combination of community development processes and initiative management skills.

It is multidimensional. Local development embraces a wide range of concerns. It does deal with job creation, business growth, and connecting people to jobs within the locality. But it also extends across a wide range of social action; it reaches out to the most marginalized in local society, but requires the participation of those who may, in relative terms, be asset rich.

It is reflective. Local development is always willing to learn from experience regarding what works well under different circumstances and what could work better.

It is asset based. The community has a wealth of assets at its disposal. These include formal assets such as public and private resources as well as informal assets such as personal networks, personal skills, and interests. Local development embraces and builds on all assets of the community.

We would add to Hart and Murray's thinking by including these two characteristics:

It is based on theory. Before any undertaking proceeds, there must be a framework to think about the problems at hand. In economic development, there must be an awareness of how a community's economy functions and changes through time and how the community can affect that change.

It is based on analytics. Communities are able to quantify, confirm, or refute perceptions of their community. Local knowledge of the community is of utmost importance, but analytics opens new doors to ways of thinking about problems and potential strategies. Too often there is too much emphasis on group processes (it must be considered) versus appropriate consideration of analytics. The misconception is that a room full of flipcharts is all you need.

A COMPREHENSIVE THEORY

We have pointed out that economists prefer deductive over inductive theories when thinking about the economy. Our policies of laissez-faire and the power of Adam Smith's *invisible hand* come from a deductive theory of how competitive markets function. In many of our theoretical discussions of the economy, we have advanced deductive ways of thinking about the problem at hand. But as we moved into our discussion of community economic development practice, we slowly started talking in terms of inductive approaches. We did this because of the multidisciplinary nature of community economic development. From the Shaffer Star, the rules, society, and decision-making nodes are typically assumed away in deductive economic theory. Our challenge in creating a comprehensive theory of community economic development is in melding deductive and inductive approaches into our theoretical understanding. Clearly, we are still searching for a general theory of community economic development. While such a theory should contain a reference to group decision-making, let us focus on some of the more traditional economic forces it must explain.

The theory must recognize that risk and uncertainty are present within the environment in which communities operate. Communities are faced with a plethora of problems but are limited in the resources, including time and interest, to solve them. The risk is devoting resources (time and interests) to problems of little consequence or that are intractable. The theory needs to recognize that information is neither free (time and costs involved in acquiring and incorporating) nor equally available everywhere. Another aspect of this is that information from different sources is perceived differently and not all of it is comprehended. The theory needs to recognize that the structure of the local economy goes a long way in explaining the ability of the community to develop. While it must recognize the dynamics of entrepreneurship and innovation in changing structure, where you start economically goes a long way in explaining where you will end up.

The theory needs to explain how a community responds to and amplifies/deflects external forces. It appears that in the short run, demand is the fundamental stimulus to community economic development. This recognizes that individual communities are influenced by outside forces over which they have little control. These outside forces can affect the local markets for both factors of production as well as output markets. Another cause for short-term community economic development is reinterpretation of rules. These demand factors have long-term implications, but they really require little time to influence the local economy.

In the long run, it is crucial that supply forces be considered. What forces either promote or hinder the movement of capital or labor among uses and places? What explains the diffusion and adoption of technology? How about the use of land? What about explaining changing tastes and preferences for amenities? Or how do you go about altering public sector? How do you explain the process of a community gaining knowledge or capacity? Even though it is slow, how do you go about explaining changing culture?

The theory must be dynamic. The dynamics mean that you cannot only explain how a community moves from A to B through time but also predict it. This is an important aspect because it explains how technology is adopted, how investments are made, and how decisions are reached. No theory does this presently.

The theory must discuss what policy levers can be pulled and how they affect the community's economic development potential. The theory must distinguish between short-term and long-term policies and their effects. It clearly needs to differentiate policies by whether they are community or extra-community in control.

In the end, the challenge for a comprehensive community economic development theory is that it must not only explain the patterns we observe but also predict them. As we have seen, we have excelled, to some extent, in our ability to describe and explain observable phenomena. But our ability to predict with any degree of reasonableness falls far short of the mark.

Since the first edition of *Community Economics* was written in 1989, two contrasting phenomena have occurred. First is how much more we understand about the functioning and change in small open economies. Second is how much we still need to understand. Community economics is still a work in progress.

NOTES

1. The recent surge of interest in game theory within microeconomics speaks directly to the ramifications of this assumption.

2. Some call community development the building of local institutions and capacity, while economic development presumes those institutions and capacity are in place. The real world is somewhere in the large gray zone in between.

3. One philosophy offered to community economic development practitioners is to aid communities in making more-informed decisions. This involves bringing information as well as process skills to the table.

References

Ahn, Sanghoon. 1999. *Technology Upgrading with Learning Cost: A Solution for Two Productivity Puzzles*. Organization for Economic Cooperation and Development, Economics Department. August.

Albright, S. Christian, and Wayne L. Winston. 2001. *Practical Management Science*. Second edition. Belmont, CA: Duxbury Press.

Allen, Kevin, Chris Hull, and Douglas Yuill. 1979. Options in Regional Incentive Policy. In *Balanced National Growth*. Kevin Allen (ed.), pp. 1–34. Lexington, MA: Lexington Books.

Althauser, Robert. 1990. Internal Labor Markets. *Annual Review of Sociology* 15: 143–161.

Alward, G.S., L.E. Siverts, D. Olson, J. Wagner, D. Senf, and S. Lindall. 1989. *MicroIMPLAN Software Manual*. Fort Collins, CO: USDA Forest Service.

Alyea, Paul E. 1967. Property Tax Inducements to Attract Industry. In *Property Taxation, USA*. Richard W. Lindholm (ed.), pp. 139–158. Madison, WI: University of Wisconsin Press.

Ammons, David N. 1996. *Municipal Benchmarks: Assessing Local Performance and Establishing Community Standards*. Thousand Oaks, CA.: Sage Publications.

Andersson, Äke. 1985. Creativity and Regional Development. *Papers of Regional Science Association* 56: 5–20.

Andrews, Richard B. 1970a. The Problem of Base Measurement. Reprinted in *The Techniques of Urban Economic Analysis*. Ralph W. Pfouts (ed.), pp.50–59. West Trenton, NJ: Chandler-Davis.

Andrews, Richard B. 1970b. Mechanics of the Urban Economic Base: Special Problems of Base Identification. Reprinted in *The Techniques of Urban Economic Analysis*. Ralph W. Pfouts (ed.), pp. 40–49. West Trenton, NJ: Chandler-Davis.

Andrews, Richard B. 1970c. Mechanics of the Urban Economic Base: The Problem of Terminology. Reprinted in *The Techniques of Urban Economic Analysis*. Ralph W. Pfouts (ed.), pp. 60–70. West Trenton, NJ: Chandler-Davis.

Anselin, L. 1988. *Spatial Econometrics: Methods and Models*. Dordrecht: Kluwer Academic Publishers.

Anselin, L. 2003. Spatial Externalities, Spatial Multipliers, and Spatial Econometrics. *International Regional Science Review* 26 #2: 153–166.

Arrow, K.J. 1951. *Social Choice and Individual Values*. New York: Wiley.

Arrow, K.J. 1962. The Economic Implications of Learning by Doing. *Review of Economic Studies* 29: 155–173.

Arrow, Kenneth. 1973. Higher Education as a Filter. *Journal of Public Economics* A #2: 193–216.

ASPO (American Society of Planning Officials). 1976. *Subdividing Rural America: Impacts of Recreational Lot and Second Home Development*. Washington, DC: Council on Environmental Quality and the Office of Policy Development and Research, Department of Housing and Urban Development.

Audretsch, David B. 2003. Innovation and Spatial Externalities. *International Regional Science Review* 26 #2: 167–174.

Averitt, Robert T. 1968. *The Dual Economy: The Dynamics of American Industry Structure*. New York: W.W. Norton.

Ayres, Janet. 1996. Essential Elements of Strategic Visioning. In *Community Strategic Visioning Programs*. Norman Walzer (ed.), pp. 21–36. Westport, CT: Greenwood Publishing.

Ayres, Janet, Robert Cole, Claire Hein, Stuart Huntington, Wayne Kobberdahl, Wanda Leonard, and Dale Zetocha. 1990. *Take Charge: Economic Development in Small Communities*. Ames, Iowa: North Central Regional Center for Rural Development.

Bacon, Robert W. 1984. *Consumer Spatial Behavior: A Model of Purchasing Decisions Over Time and Space*. Oxford: Clarendon Press.

Bahl, R., M. Johnson, and M. Wasylenko. 1980. State and Local Government Expenditure Determinants: The Traditional View and a New Approach. In *Public Employment and State and Local Government Finance*. R. Bahl, J. Burkhead, and B. Jump, Jr. (eds). Cambridge, MA: Balinger Publishing.

Bailey, S.J., and S. Connolly. 1998. The flypaper effect: Identifying areas for further research. *Public Choice* 95: 335–361.

Barkley, D.L., M.S. Henry, and S. Bao. 1994. Metropolitan Growth: Boon or Bane to Nearby Rural Areas? *Choices* 4th Quarter: 14–18.

Barkley, D.L., M.S. Henry, and S. Bao. 1996. Regional Economic Areas. *Land Economics* 72 #3: 336–357.

Barkley, David L., Mark S. Henry, and S. Bao. 1998. The Role of Local School Quality and Rural Employment and Population Growth. *The Review of Regional Studies* 28 (Summer): 81–102.

Barkley, David L., Deborah M. Markley, David Freshwater, Julia Sass Rubin, and Ron Shaffer. 2001a. *Establishing Nontraditional Venture Capital Institutions: Lessons Learned*. P2001-11A, Part 1 of 4 of the Final Report of RUPRI Rural Equity Capital Initiative's Study of Nontraditional Venture Capital Institutions. Rural Policy Research Institute, University of Missouri: Columbia.

Barkley, David L., Deborah M. Markley, David Freshwater, Julia Sass Rubin, and Ron Shaffer. 2001b. *Establishing Nontraditional Institutions: The Decision Making Process*. P2001-11C, Part 3 of 4 of the Final Report of RUPRI Rural Equity Capital Initiative's Study of Nontraditional Venture Capital Institutions. Rural Policy Research Institute, University of Missouri: Columbia.

Barlowe, R. 1986. *Land Resource Economics: The Economics of Real Estate*. Englewood Cliffs, NJ: Prentice-Hall.

Barr, J., and O. Davis. 1966. An elementary political and economic theory of local governments. *Southern Economic Journal* 33: 149–165.

Barro, R., and X Sala-i-Martin. 1992. Public Finance in Models of Economic Growth. *Review of Economic Studies* 59 #4: 645–661.

Barro, R., and X Sala-i-Martin. 1995. *Economic Growth*. New York: McGraw-Hill.

Barron, W.F., R.D. Perlack, and J.J. Boland. 1998. *Fundamentals of Economics for Environmental Managers*. Westport, CT: Quorum Books.

Bartik, T. 1989. Small Business Start-Ups in the United States: Estimates of the Effects of Characteristics of States. *Southern Economic Journal* 4: 1004–1018.

Bartik, Timothy J. 1990. The Market Failure Approach to Regional Economic Development Policy. *Economic Development Quarterly* 4 #4 (November): 361–370.

Bartik, T. 1991. *Who Benefits with State and Local Economic Development Policies?* Kalamazoo, MI: W.E. Upjohn Institute for Employment Research.

Bartik, T.J. 1992. The Effects of State and Local Taxes on Economic Development: A Review of Recent Literature. *Economic Development Quarterly* 6 #1 (February): 103–111.

Bartik, Timothy J. 1994. The Effects of Metropolitan Job Growth on the Size Distribution of Family Income. *Journal of Regional Science* 34 #4 (November): 483–501.

Barzel, Y. 1997. *Economic Analysis of Property Rights*. Cambridge: Cambridge University Press.

Batty, M. 1978. Reilly's Challenge: New Laws of Retail Gravitation Which Define Systems of Central Places. *Environment and Planning A* 10(February): 185–219.

Baumol, William J. 1965. *The Stock Market and Economic Efficiency*. New York: Fordham University Press.

Beale, Calvin L., and Kenneth M. Johnson. 1998. The Identification of Recreational Counties in Nonmetropolitan Areas of the USA. *Population Research and Policy Review* 17: 37-53.

Bearse, Peter. 1979. Influence Capital Flows for Urban Economic Development: Incentives or Institution Building? *Journal of Regional Science* 19 (February): 79–91.

Beaton, W.P. 1983. The Demand for Municipal Goods: A Review of the Literature from the Point of View of Expenditure Determination. In *Municipal Expenditures, Reviews, and Services: Economic Models and Their Use by Planners*. W.P. Beaton (ed.). New Brunswick, NJ: Center for Urban Policy Research, Rutgers University.

Beaulieu, L.J., and D. Mulkey. 1995. Human Capital in Rural America: A Review of Theoretical Perspectives. In *Investing in People: The Human Capital Needs of Rural America*. L.J. Beaulieu and D. Mulkey (eds.), pp. 3–21. Boulder, CO: Westview Press.

Beaumont, P. 1990. Supply and Demand Interaction in Integrated Econometric and Input-Output Models. *International Regional Science Review* 13 #1-2: 167–181.

Beauregard, R.A. 1993. Constituting Economic Development: A Theoretical Perspective. In *Theories of Local Economic Development: Perspectives from Across the Disciplines*. R.D. Bingham and R. Mier (eds.), pp. 267–283. San Francisco: Sage.

Beck, Roger, and Frank Goode. 1981. The Availability of Labor in Rural Communities." In *New Approaches to Economic Development Research in Rural*

Areas. Roger Beck (ed.), pp. 39–61. Ithaca, NY: Northeast Center for Rural Development.

Becker, Gary S. 1962. Investment in Human Capital: A Theoretical Analysis. *The Journal of Political Economy* 70 #5-2: 9–49.

Becker, Gary S. 1965. A Theory of the Allocation of Time. *Economics Journal* 75 (September): 493–517.

Becker, Gary S. 1985. *Human Capital.* New York: National Bureau of Economic Research.

Becsi, Z. 1996. Do State and Local Taxes Affect Relative State Growth? *Economic Review* (April/March): 18–36.

Behrman, Jack N., and Dennis A. Rondinelli. 1992. The Cultural Imperatives of Globalization: Urban Economic Growth in the 21st Century. *Economic Development Quarterly* 6 #2 (May): 115–126.

Belous, Richard. 1989. *The Contingent Economy: The Growth of a Temporary, Part-time and Sub-contracted Workforce.* Washington, DC: National Planning Association.

Bendavid, Avrom. 1974. *Regional Economic Analysis for Practitioners.* Revised edition. New York: Praeger.

Bennett, D. Gordon. 1996. Implications of Retirement Development in High-Amenity Non-metropolitan Coastal Areas. *Journal of Applied Gerontology* 15 #3: 345–360.

Bennett, R.J., and L. Hordijk. 1986. Regional Econometric and Dynamic Models. In *Handbook of Regional and Urban Economics.* Volume 1: *Regional Economics.* Amsterdam: North-Holland.

Benson, Earl D., J.L. Hansen, A.L. Schwartz, Jr., and G.T. Smersh. 1998. Pricing Residential Amenities: The Value of a View. *Journal of Real Estate Finance and Economics* 16: 55–73.

Bergstrom, G., D. Rubinfeld, and P. Shapiro. 1982. Micro-based Estimates of Demand Functions for Local School Expenditures. *Econometrica* 50: 1183–1205.

Bergstrom, T., and R. Goodman. 1973. Private Demand for Public Goods. *American Economic Review* 63: 286–296.

Bernat, G. Andrew, Jr. 2001. Convergence in State Per Capita Personal Income 1950-99. *Survey of Current Business.* USDC June 2001: 36–48.

Bernstein, Paul, and Richard Hall (eds.). 1997. *American Work Values: Their Origin and Development.* SUNY Series in the Sociology of Work. Albany, NY: State University of New York.

Berry, Brian J.L., and William L. Garrison. 1958a. Recent Developments in Central Place Theory. *Proceedings of the Regional Science Association* 4: 107–121.

Berry, Brian J.L., and William L. Garrison. 1958b. A Note on Central Place Theory and the Range of a Good. *Economic Geography* 34 (October): 304–311.

Bewley, T.F., 1981. A Critique of Tiebout's Theory of Local Public Expenditures. *Econometrica* 49: 713–740.

Beyard, Michael D., and W. Paul O'Mara. 1999. *Shopping Center Development Handbook.* Third edition. Washington, DC: Urban Land Institute.

Beyers, William B., and Peter B. Nelson. 2000. Contemporary Development Forces in the Nonmetropolitan West: New Insights from Rapidly Growing Communities. *Journal of Rural Studies* 16: 459–474.

Bilgic, A., S. King, A. Lusby, and D.F. Schreiner. 2002. Estimates of U.S. Commodity Trade Elasticities of Substitution. *Journal of Regional Analysis and Policy* 32 #2: 31–50.

Black, D. 1958. *The Theory of Committees and Elections.* Cambridge: Cambridge University Press.

Blair, John P. 1973. A Review of the Filtering Down Theory. *Urban Affairs Quarterly* 8 #3: 303–316.

Blair, John P., and Carole R. Endres. 1994. Hidden economic development assets. *Economic Development Quarterly* 8 #3 (August): 286–291.

Blakely, E.J. 1983. Community Attitudes Toward Change. Paper at American Collegiate Schools of Planning. San Francisco (October).

Blakely, E.J. 1994. *Planning Local Economic Development: Theory and Practice.* Thousand Oaks, CA: Sage.

Blank, Rebecca M., and Ron Haskins (eds.). 2001. *The New World of Welfare.* Washington, DC: Brookings Institute.

Blaug, M. 1985. Where Are We Now in the Economics of Education? *Economics of Education Review* 4: 17–28.

Blomnestein, H., P. Nijkamp, and W. Van Veenendaal. 1980. Shopping Perceptions and Preferences: A Multidimensional Attractiveness Analysis of Consumer and Entrepreneurial Attitudes. *Economic Geography* 56 (April): 155–174.

Bluestone, Barry, W.M. Murphy, and M. Stevenson. 1973. *Low Wages and the Working Poor.* Ann Arbor: University of Michigan, Institute for Labor and Industrial Relations.

Blumenberg, Evelyn. 2002. On the Way to Work: Welfare Participants and Barriers to Employment. *Economic Development Quarterly* 16 #4: 314–325.

Blumenfeld, Hans. 1955. The Economic Base of the Metropolis. Reprinted in *The Techniques of Urban*

Economic Analysis. Ralph W. Pfouts (ed.), pp. 227–277. West Trenton, NJ: Chandler-Davis.

Bolton, R. 1985. Regional Econometric Models. *Journal of Regional Science* 25 #4: 495–520.

Bolton, Roger. 1992. Place Prosperity vs. People Prosperity Revisited: An Old Issue with a New Angle. *Urban Studies* 29 #2: 185–203.

Booth, Douglas E. 1999. Spatial Patterns in the Economic Development of the Mountain West. *Growth and Change* 30: 384–405.

Boothroyd, P., and H.C. Davis. 1993. Community Economic Development: Three Approaches. *J. Planning Education And Research* 20: 230–240.

Borcherding, T.E., And R.T. Deacon. 1972. The Demand for the Services of Non-federal Governments. *American Economic Review* 62: 891–901.

Borchert, John R. 1963. The Urbanization of the Upper Midwest: 1930-1960. Upper Midwest Study. Urban Report No.2. Minneapolis, MN: Department of Geography, University of Minnesota.

Borts, George. 1971. Growth and Capital Movements Among U.S. Regions in the Postwar Period. In *Essays in Regional Economics.* John F. Kain and John R. Meyer (eds.), pp. 189–217. Cambridge, MA: Harvard University Press.

Bowles, Samuel, and Herbert Gintis. 1976. *Schooling in Capitalist America.* New York: Basic Books.

Boyle, Kevin J., and Richard C. Bishop. 1988. Welfare Measurements Using Contingent Valuation: A Comparison of Techniques. *American Journal of Agricultural Economics* 70 #1: 20–28.

Bradshaw, T.J., and E.J. Blakely. 1999. What Are "Third-Wave" State Economic Development Efforts? From Incentives to Industrial Policy. *Economic Development Quarterly* 13 #3 (August): 229–244.

Bramley, Glen, Murray Steward, and Jack Underwood. 1979. Local Economic Initiatives: A Review. *Town Planning Review* 5 (April): 137–147.

Braschler, Curtis H., and Gary T. Devino. 1993. Nonsurvey Approach to I/O Modeling. In *Microcomputer-based Input-Output Modeling.* Daniel M. Otto and Thomas G. Johnson (eds.). Boulder, CO: Westview Press.

Brealey, R., and S. Myers. 1988. *Principles of Corporate Finance.* Third edition. New York: McGraw-Hill.

Breschi, Stefano, and Francesco Lissoni. 2001. Localized Knowledge Spillovers vs. Innovative Milieux: Knowledge "Tacitness." *Papers in Regional Science* 80 #3 (July): 255–273.

Bromley, Daniel W. 1972. An Alternative to Input-Output Models: A Methodological Hypothesis. *Land Economics* 48 (May): 125–133.

Bromley, D.W. 1985. Resources and Economic Development: An Institutionalist Perspective. *Journal of Economic Issues* 19 (September): 779–796.

Brown, Amy. 1997. *Work First: How to Implement an Employment-focused Approach to Welfare Reform.* New York: Manpower Demonstration Research Corporation.

Brucker, Sharon M., Stephen Hastings, and William R. Latham. 1987. Regional Input-output Analysis: A Comparison of Five Ready-made Model Systems. *Review of Regional Studies* (Spring).

Brueckner, J.K. 1979. Property Values, Local Public Expenditure and Economic Efficiency. *Journal of Public Economics* 11: 223–245.

Brueckner, J.K., 1982. A Test for Allocative Efficiency in the Local Public Good Sector. *Journal of Public Economics* 11: 311–331.

Brueckner, J.K., 1983. Property Value Maximization and Public Sector Efficiency. *Journal of Urban Economics* 14: 1–15.

Brueckner, J. 2000. Urban Sprawl: Diagnosis and Remedies. *International Regional Science Review* 23: 160–171.

Brueckner, Jan, and L. Saavedra. 2001. Do Local Governments Engage in Strategic Property Tax Competition? *National Tax Journal* 54: 203–229.

Bryson, John M. 1990. *Strategic Planning for Public and Nonprofit Organizations: A Guide to Strengthening and Sustaining Organizational Achievement.* San Francisco: Josey-Bass.

Buchanan, James M., and John E. Moes. 1960. A Regional Countermeasure to the Minimum Wage. *American Economic Review* 50 (June): 434–438.

Buck, T.W., and M.H. Atkins. 1976. Capital Subsidies and Unemployed Labor: A Regional Production Function Approach. *Regional Studies* 10 (June): 215–222.

Bucklin, Louis P. 1971. Retail Gravity Models and Consumer Choice: A Theoretical and Empirical Critique. *Economic Geography* 47 (October): 489–497.

Burchell, R.W., and D. Listokin. 1978. *The Fiscal Impact Handbook: Estimating Local Costs and Benefits of Land Development.* New Brunswick, NJ: Center for Urban Policy Research.

Cain, Glen. 1976. The Challenge of Segmented Labor Market Theories to Orthodox Theory: A Survey. *Journal of Economic Literature* 14 (December): 1215–1257.

Camagni, Robert P. 1985. Spatial Diffusion of Pervasive Process Innovation. *Papers of the Regional Science Association* 58: 83–96.

Carlino, G.A., and E.S. Mills. 1987. The Determinants of County Growth. *Journal of Regional Science* 27 (February): 39–54.

Carroll, R., and M. Wasylenko. 1994. Do State Business Climates Still Matter? Evidence of a Structural Change. *National Tax Journal* 47 #1: 19–37.

Carrothers, Gerald A.P. 1956. An Historical Review of the Gravity Potential Concept of Human Interaction. *Journal of the American Institute of Planners* 22 #2: 94–102.

Cary, Lee J. 1970. *Community Development as a Process.* Columbia, MO: University of Missouri Press: 1–6.

Cary, Lee J. 1972. The Roles of the Professional Community Developer. *Journal of the Community Development Society* 3 #2 (Fall): 36–41.

Castle, Emery N. (ed.). 1995. *The Changing American Countryside: Rural People and Places.* Lawrence, Kansas: The University Press of Kansas.

Cavaye, Jim. 2000a. Social Capital—The Concept, the Context. Unpublished paper, DPI: Brisbane, Queensland, Australia.

Cavaye, Jim. 2000b. *Our Community, Our Future: A Guide to Rural Community Development.* DPI: Brisbane, Queensland, Australia.

Cebula, R.J., and M. Zaharoff. 1975. Interregional Capital Transfers and Interest Rate Differentials: An Empirical Note. *Annals of Regional Science* 8: 87–94.

Chapple, Karen. 2002. I Name It and I Claim It—In the Name of Jesus, This Job Is Mine: Job Search, Networks and Careers for Low Income Women. *Economic Development Quarterly* 16 #4: 294–313.

Chenery, Hollis. 1979. *Structural Change and Development Policy.* Washington, DC: The World Bank.

Chenery, Hollis, Sherman Robinson, and Moshe Syrquin. 1986. *Industrialization and Growth: A Comparative Study.* Washington, DC: The World Bank.

Cherry, Robert, and William Rogers (eds.). 2000. *Prosperity for All: The Economic Boom and African Americans.* New York: The Russell Sage Foundation.

Christaller, Walter. 1933. *Die Zentralen Orte in Suddeneutschland.* Jena Fisher, trans. by C.W. Baskin (1966), *Central Places in Southern Germany.* Englewood Cliffs, NJ: Prentice-Hall.

Christenson, James A. 1989. Themes of Community Development. In *Community Development in Perspective.* James A. Christenson and Jerry W. Robinson, Jr. (eds.). Ames, Iowa: Iowa State University Press.

Christenson, J.A., and J.W. Robinson, Jr. 1993. In Search of Community Development. In *Community Development in America.* J.A. Christenson and J.W.

Robinson, Jr. (eds.), pp. 3–17. Ames, IA: Iowa State University Press.

Clark, Gordon L. 1983. Fluctuations and Rigidities in Local Labor Markets. Part I: Theory and Evidence. *Environment and Planning A* 15 (February): 165–186.

Clark, G.L., and J. Whiteman. 1983. Why Poor People Do Not Move: Job Search Behavior and Disequilibrium Amongst Local Labor Markets. *Environment and Planning A* 15 (January): 85–105.

Clawson, Marion. 1974. Conflicts, Strategies and Possibilities for Consensus in Forest Land Use and Management. In *Forest Policy for the Future, Papers and Discussions from a Forum on Forest Policy for the Future* (May 8-9), pp. 101–191. Washington, DC: Resources for the Future.

Cliff, A., and J. Ord. 1973. *Spatial Autocorrelation.* London: Pion.

Cliff, A., and J. Ord. 1981. *Spatial Processes, Models and Applications.* London: Pion.

Coleman, James, S. 1988. Social Capital in the Creation of Human Capital. *American Journal of Sociology* 94 Supplement: S95–S120.

Collins, John N., and Bryan F. Powers. 1976. Community Developer and Local Problem Solving. *Journal of the Community Development Society* 7 #2 (Fall): 28–40.

Colwell, Peter F. 1982. Central Place Theory and the Simple Economic Foundations of the Gravity Model. *Journal of Regional Science* 22 (November): 541–546.

Coomes, R. D. Olson, and D. Glennon. 1991. The Interindustry Employment Demand Variable: An Extension of the I-SAMIS Technique for Linking Input-Output and Econometric Models. *Environment and Planning A* 23 #5: 1063–1068

Coppock, J.T. (ed.). 1977. *Second Homes: Curse or Blessing?* Oxford, England: Pergamon.

Crihfield, John B., and Harrison S. Campbell, Jr. 1991. Evaluating Alternative Regional Planning Models. *Growth and Change* 22 #2: 1–16.

Cumberland, John H., and Fritz Van Beek. 1967. Regional Economic Development Objectives and Subsidization of Local Industry. *Land Economics* 43 (August): 253–264.

Czamanski, Stan. 1972. *Regional Science Techniques in Practice: The Case of Nova Scotia.* Lexington, MA: DC Health.

Daley, John Michael, and Peter M. Kettner. 1981. Bargaining in Community Development. *Journal of the Community Development Society* 12 #2: 25–38.

Daniels, Belden H. 1979. The Mythology of Capital in Community Economic Development. Policy

Note P79-2. Cambridge, MA: Harvard University, Dept. of City and Regional Planning (June).

Darwent, D.F. 1969. Growth Poles and Growth Centers in Regional Planning: A Review. *Environment and Planning* 1 #1: 5–31.

Davelaar, E.J. 1991. *Regional Economic Analysis of Innovation and Incubation.* Brookfield, VT: Avebury.

Davies, R.L. 1977. Store Location and Store Assessment Research: The Integration of Some New and Traditional Techniques. *Transactions of the Institute of British Geographers* 2NS #2: 141–157.

Davis, H. Craig. 1976. Regional Sectoral Multipliers with Reduced Data Requirements. *International Regional Science Review* 1 (Fall): 18–29.

Davis, Lane E., and Douglass C. North. 1971. *Institutional Change and American Economic Growth.* Cambridge: Cambridge University Press.

Deacon, R., 1978. A Demand Model for the Local Public Sector. *Review of Economics and Statistics* 60: 184–192.

Deavers, Ken L., and Robert A. Hoppe. 1991. The Rural Poor: The Past as Prologue. In *Rural Policies for the 1990's.* C.B. Flora and J.A. Christenson (eds.), pp. 85–101. Boulder, CO: Westview Press.

Deller, S.C. 1990. An Application of a Test for Allocative Efficiency in the Local Public Sector. *Regional Science and Urban Economics* 20: 395–406.

Deller, Steven C. 1995. Economic Impact of Retirement Migration. *Economic Development Quarterly* 9 #1 (February): 25–38.

Deller, Steven C., and John M. Halstead. 1994. Efficiency in the Production of Rural Road Services: The Case of New England Towns. *Land Economics* 70 #2: 247–259.

Deller, S.C., and T.R. Harris. 1993. Estimation of Minimum Market Thresholds for Rural Commercial Sectors Using Stochastic Frontier Estimators. *Regional Science Perspectives* 23: 3–17.

Deller, Steven C., and Carl H. Nelson. 1991. Measuring the Economic Efficiency of Producing Rural Road Services. *American Journal of Agricultural Economics* 72 (February): 194–201.

Deller, Steven C., and Sue Tsung-Hsiu Tsai. 1998. An Examination of the Wage Curve: A Research Note. *Journal of Regional Analysis and Policy* 28 #2: 3–12.

Deller, Steven C., David L. Chicoine, and Norman Walzer. 1988. Economies of Size and Scope of Rural Low-Volume Roads. *Review of Economics and Statistics* 70 (August): 459–465.

Deller, Steven C., Carl H. Nelson, and Norman Walzer. 1992. Measuring Managerial Efficiency in Rural Governments: The Case of Low-Volume Rural

Roads. *Public Productivity and Management Review* 15 (Spring): 355–370.

Deller, Steven C., James C. McConnon, Jr., John Holden, and Kenneth Stone. 1991. The Measurement of a Community's Retail Market. *Journal of the Development Society* 22 #2: 68–83.

Deller, S.C., Tsung-Hsiu Tsai, D.W. Marcouiller, and D.B.K. English. 2001. The Role of Amenities and Quality of Life in Rural Economic Growth. *American Journal of Agricultural Economics* 83 #2: 352–365.

Dervis, K., J. de Melo, and S. Robinson. 1982. *General Equilibrium Models for Policy Development.* Washington, DC: The World Bank.

Dewar, Margaret E., Joan Fitzgerald, and Nancy Green-Leigh. 1994. Introduction: Women's Fortunes and Economic Restructuring. *Economic Development Quarterly* 8 #2 (May): 141–146.

Dewhurst, J., and G. West. 1991. Conjoining Regional and Inter-regional Input-Output Models with Econometric Models. In *New Directions in Regional Analysis: Integrated and Multi-regional Approaches.* J. Van Dijk et al. (eds). Dordrecht, The Netherlands: Kluwer Academic Publishers.

Dissart, Jean Christophe, and Steven C. Deller. 2000. Quality of Life in the Planning Literature. CPL Bibliography 360. *Journal of Planning Literature* 15 #1: 135–161.

Dixon, P., B. Parmenter, J. Sutton, and D. Vincent. 1982. *ORANI: A Multisectoral Model of the Australian Economy.* Amsterdam: North-Holland.

Dodge, Willard K. 1980. Ten Commandments of Community Development or One Middle-aged Graduate's Advice to New Graduates. *Journal of Community Development Society* 11 #1 (Spring): 49–58.

Doeksen, G.A., and J. Peterson. 1987. *Critical Issues in the Delivery of Local Government Services in Rural America.* USDA ERS: Washington, DC.

Doeringer, P.B. 1995. Business Strategy and Cross-Industry Clusters. *Economic Development Quarterly* 9 #3 (August): 225–237.

Doeringer, Peter B., and Michael J. Piore. 1971. *Internal Labor Markets and Manpower Analysis.* Lexington, MA: Heath Books.

Dollery, B.E., and A.C. Worthington. 1996. The Empirical Analysis of Fiscal Illusion. *Journal of Economic Surveys* 10 #3: 261–297.

Donnermeyer, Joseph F., Barbara A. Plested, Ruth W. Edwards, Gene Oetting, and Lawrence Littlethunder. 1997. Community Readiness and Prevention Program. *Journal of the Community Development Society* 28 #1: 64–83.

Doss, C.R., and S.J. Taff. 1996. The Influence of Wetland Type and Wetland Proximity on Residential

Property Values. *Journal of Agricultural and Resource Economics* 21 #1: 120–129.

Dresch, Stephen. 1977. Human Capital and Economic Growth: Retrospect and Prospect. In *U.S. Economic Growth from 1976 to 1986: Prospects, Problems and Patterns.* Vol. 11, *Human Capital.* U.S. Congress Joint Economic Committee, 95th Congress, 1st Session (May): 112–153.

Dresser, Laura, and Joel Rogers. 1997. *Rebuilding Job Access and Career Advancement Systems in the New Economy.* Madison, WI: University of Wisconsin-Madison Center on Wisconsin Strategy (December).

Dresser, Laura, and Joel Rogers. 2000. *The State of Working Wisconsin.* Madison, WI: University of Wisconsin-Madison: Center on Wisconsin Strategy.

Due, John R. 1961. Studies of State–Local Tax Influences on the Location of Industry. *National Tax Journal* 14 (June): 163–173.

duRivage, Virginia L. (ed.). 1992. *New Policies for the Part-time and Contingent Workforce.* Armonk, NY: M.E. Sharpe.

Earnhart, D. 2001. Combining Revealed and Stated Preference Methods to Value Environmental Amenities at Residential Locations. *Land Economics* 77 #1: 12–29.

Edin, K., and L. Lien. 1997. *Making ends Meet: How Single Mothers Survive Welfare and Low-Wage Work.* New York: Russell Sage Foundation.

Ehrenberg, Ronald G., and Robert S. Smith. 1985. *Labor Economics: Theory and Public Policy.* Second edition. Glenview, IL: Scott Foresman & Co.

Eisinger, Peter. 1988. *The Role of the Entrepreneurial State.* Madison: University of Wisconsin Press.

Eisinger, Peter. 1995. State Economic Development in the 1990s: Politics and Policy Learning. *Economic Development Quarterly* 9 #2 (May): 146–158.

Elango, B., V. Fried, R. Hirsrich, and A. Polonchek. 1995. How Venture Firms Differ. *Journal of Business Venturing* 10: 157–179.

English, Donald B. K., David W. Marcouiller, and H. Ken Cordell. 2000. Tourism Dependence in Rural America: Estimates and Effects. *Society and Natural Resources* 13: 185–202.

Enright, Michael J. 2000. The Globalization of Competition and the Localization of Competitive Advantages: Policies Towards Regional Clustering. In *The Globalization of Multinational Enterprise Activity and Economic Development.* Neil Hood and Stephen Young (eds.). New York: St. Martins Press.

Erickson, R.A. 1994. Technology, Industrial Restructuring and Regional Development. *Growth and Change* 25 #3 (Summer): 353–380.

Esparza, Adrian, and John Carruthers. 2000. Land Use Planning and Exurbanization in the Rural Mountain West. *Journal of the Planning Education and Research* 20 #1: 23–36.

Fagan, Mark, and Charles F. Longino, Jr. 1993. Migrating Retirees: A Source for Economic Development. *Economic Development Quarterly* 7 #1 (February): 98–106.

Federal Reserve Bank of Boston (FRB-Boston). 1997. The Effects of State and Local Policies on Economic Development. *New England Economic Review* (March/April).

Fesher, Edward J., and Stuart H. Sweeney. 2002. Theory, Methods and a Cross-Metropolitan Comparison of Business Clusters. In *Industrial Location Economics.* Philip McCann (ed.). North Hampton, MA: Edward Elgar.

Fingleton, Bernard. 2003. Externalities, Economic Geography, and Spatial Econometrics: Conceptual and Modeling Developments. *International Regional Science Review* 26 #2: 197–207.

Fisher, R.C. 1997. The Effects of State and Local Public Services on Economic Development. *New England Economic Review* (March/April): 53–67.

Flora, Cornelia. 1997. Enhancing Community Capitals: The Optimization Equation. In *Rural Development News* 21 #1 (March): 1–3.

Flora, Cornelia B., and Jan L. Flora. 1990. Developing Entrepreneurial Rural Communities. *Sociological Practice* 8: 197–207.

Flora, C.B., and J.L. Flora. 1993. Entrepreneurial Social Infrastructure: A Necessary Ingredient. *Annals, American Association of Political and Social Sciences* 529 (September): 48–58.

Flora, C.B., J.L. Flora, J.D. Spears, and L.E. Swanson. 1992a. Community and Culture. In *Rural Communities: Legacy and Change,* pp. 57–78. Boulder, CO: Westview Press.

Flora, C.B., J.L. Flora, J.D. Spears, and L.E. Swanson. 1992b. Social Infrastructure. *Rural Communities: Legacy and Change,* pp. 231–250. Boulder, CO: Westview Press.

Flora, Jan L. 1998. Social Capital and Communities of Place. *Rural Sociology* 63 #4 (December): 481–506.

Foust, Brady J., and Edward Pickett. 1974. Threshold Estimates: A Tool For Small Business Planning in Wisconsin. Unpublished manuscript, University of Wisconsin-Extension.

Foust, Brady J., and Anthony R. de Souza. 1978. *The Economic Landscape: A Theoretical Introduction.* Columbus, OH: Charles E. Merrill Publishing Co.

Fox, W.F. 1980. *Size Economies in Local Government Services: A Review.* Washington, DC: USDA-ERS.

Freilich, Norris. 1963. Toward an Operational Definition of Community. *Rural Sociology* 33 (June): 117–127.

Freshwater, David, David L. Barkley, Deborah M. Markley, Julia Sass Rubin, and Ron Shaffer. 2001. *Nontraditional Venture Capital Institutions: Filling A Financial Market Gap.* P2001-11B, Part 2 of 4 of the Final Report of RUPRI Rural Equity Capital Initiative's Study of Nontraditional Venture Capital Institutions. Rural Policy Research Institute, University of Missouri-Columbia.

Friedmann, John. 1972. A General Theory of Polarized Development. In *Growth Centers in Regional Economic Development.* Niles M. Hansen (ed.), pp. 82–107. New York: The Free Press.

Friesz, T., T. Miller, and R. Tobin. 1988. Competitive Network Facility Location Models: A Survey. *Papers of the Regional Science Association.* 65 #1: 47–57.

Fujita, M., P. Krugman, and A. Venables. 1999. *The Spatial Economy: Cities, Regions, and International Trade.* Cambridge, MA: MIT Press.

Gabszewicz, Jean Jaskold, and Jacques-Francois Thisse. 1986. Spatial Competition and the Location of Firms. In *Location Theory.* Jean Jaskold Gabszewicz, Jacques-Francois Thisse, Masahisa Fujita, and Urs Schweizer (eds.). New York: Harwood Academic Publisher.

Garrod, Guy, and Kenneth G. Willis. 1999. *Economic Valuation of the Environment.* North Hampton, MA: Edward Elgar Publishing.

Gaston, Robert J. 1990. Financing Entrepreneurs: The Anatomy of a Hidden Market. In *Financing Economic Development: An Institutional Response.* Richard D. Bingham, Edward W. Hill, and Sammis B. White (eds.). Thousand Oaks, CA: Sage Publications.

Gaunt, Thomas P. 1998. Communication, Social Networks and Influence in Citizen Participation. *Journal of the Community Development Society* 29 #2: 276–297.

Gillespie, Andrew, and Ronald Richardson. 2000. Teleworking and the City: Myths of Workplace Transcendence and Travel Reduction. In *Cities in the Telecommunications Age: The Fracturing of Geographies.* J.O. Wheeler, Y. Aoyama, and B. Warf (eds.). New York: Routledge.

Gillis, Bill, and Ron Shaffer. 1985. Community Employment Objectives and Choice of Development Strategy. *Journal of Community Development Society* 16 #2: 18–37.

Gillis, Bill, and Ron Shaffer. 1987. Matching New Jobs to Rural Workers. *Rural Development Perspectives* 4 #1 (October): 19–23.

Gillis, William R. 1983. Achieving Employment Objectives in the Nonmetropolitan North Central Region. Unpublished Ph.D. dissertation, Department of Agricultural Economics, University of Wisconsin-Madison.

Giloth, R. 1995. Social Investment in Jobs: Foundation Perspective on Targeted Economic Development during the 1990s. *Economic Development Quarterly* 9 #3 (August): 279–289.

Godbey, Geoffrey, and Malcolm I. Bevins. 1987. The Life Cycle of Second Home Ownership: A Case Study. *Journal of Travel Research* 25 #3: 18–22.

Goetz, Edward G. 1990. Type II Policy and Mandated Benefits in Economic Development. *Urban Affairs Quarterly* 26 #2 (December): 170–190.

Goldstein, Harvey A., and Michael I. Luger. 1993. Theory and Practice in High Tech Economic Development. In *Theories of Local Development: Perspectives from Across the Disciplines.* Richard D. Bingham and Robert Mier (eds.), pp. 147–174. Newbury Park: Sage.

Goldstucker, Jac L., Danny N. Bellenger, Thomas J. Stanley, and Ruth L. Otte. 1978. *New Developments in Retail Trading Area Analysis and Site Selection.* Atlanta, GA: College of Business Administration, Georgia State University.

Gordon, David. 1972. *Theories of Poverty and Underemployment: Orthodox, Radical and Dual Labor Market Perspectives.* Lexington, MA: Lexington Books.

Gordon, David M., Richard Edwards, and Michael Reich. 1982. *Segmented Work, Divided Workers.* Cambridge: Cambridge University Press.

Graham, Stephen, and Simon Marvin. 2000. Urban Planning and the Technological Future of Cities. In *Cities in the Telecommunications Age: The Fracturing of Geographies.* J.O. Wheeler, Y. Aoyama, and B. Warf (eds.). New York: Routledge.

Gray, Ralph. 1964. Industrial Development Subsidies and Efficiency in Resource Allocation. *National Tax Journal* 17 (June): 164–172.

Greason, Michael C. 1989. Here a Parcel, There a Parcel—Fragmented Forests. *The Conservationist* 44 #3: 46–49.

Green, Gary P., and Deborah Klinko Cowell. 1994. Community Reinvestment and Local Economic Development in Rural Areas. *Journal of the Community Development Society* 25 #2: 229–245.

Green, Gary P., Anna Haines, and Stephen Halebsky. 2000. *Building Our Future: A Guide to Community Visioning.* Madison, Wisconsin: Cooperative Extension Service, G3708.

Green, Gary P., Jan L. Flora, Cornelia Flora, and Frederick E. Schmidt. 1990. Local Self-Development

Strategies: National Survey Results. *Journal Community Development Society* 21 #2: 55–73.

Greenhut, Melvin L. 1956. *Plant Location in Theory and Practice.* Chapel Hill, NC: University of North Carolina Press.

Green-Leigh, Nancy. 1995. Income Inequality and Economic Development. *Economic Development Quarterly* 9 #1 (February): 94–103.

Greytak, David. 1969. A Statistical Analysis of Regional Export Estimating Techniques. *Journal of Regional Science* 9 (December): 387–395.

Grossman, G.M., and E. Helpman. 1994. Endogenous Innovation. *Journal of Economic Perspectives* 8 #1: 23–45.

Grzywinski, Ronald A., and Dennis R. Marino. 1981. Public Policy, Private Banks and Economic Development. In *Expanding the Opportunity to Produce.* Robert Friedman and William Schweke (eds.), pp. 243–256. Washington, DC: The Corporation for Enterprise Development.

Haag, Susan White. 2002. Community Reinvestment: A Review of Urban Outcomes and Challenges. *International Regional Science Review* 25 #3 (July): 252–275.

Haidler, Donald. 1992. Place Wars: The New Realities of the 1990's. *Economic Development Quarterly* 6 #2 (May): 127–134.

Halstead, John M., and Deller, Steven C. 1997. "Public Infrastructure in Economic Development and Growth: Evidence from Rural Manufacturers." *Journal of the Community Development Society* 28 #2: 149–169.

Hamberg, Daniel. 1971. *Models of Economic Growth.* New York: Harper and Row.

Hamrick, Karen S. 2001. *Displaced Workers: Differences in Non-metro and Metro Experiences in the Mid-1900's.* Washington, DC: Economic Research Service-USDA. RDRR-92.

Hansen, Derek. 1981a. Expansion—A Program Policy Guideline (Part Two). In *Expanding the Opportunity to Produce.* Robert Friedman and William Schweke (eds.), pp. 217–242. Washington, DC: The Corporation for Enterprise Development.

Hansen, Derek. 1981b. *Banking and Small Business, Studies in Development Policy,* Volume 10. Washington, DC: Council of State Planning Agencies.

Hansen, Niles. 1975. An Evaluation of Growth-Center Theory and Practice. *Environment and Planning A* 7: 821–832.

Hansen, Niles. 1992. Competition, Trust and Reciprocity in the Development of Innovative Regional Milieux. *Papers in Regional Science* 71 #2: 95–105.

Hanson, S., and G. Pratt. 1995. *Gender, Work and Space.* New York: Routledge.

Harmston, Floyd K. 1979. A Study of the Economic Relationships of Retired People and a Small Community. Paper at the annual meeting of the Mid-continent Section of the Regional Science Association.

Harrigan, F.J., J.W. McGilvray, and I.H. McNicoll. 1980. Simulating the Structure of a Regional Economy. *Environment and Planning A* 12 (August): 927–936.

Harris, K.M. 1993. Work and Welfare Among Single Mothers in Poverty. *American Journal of Sociology* 99 #2: 317–352.

Harris, K.M. 1996. Life after Welfare: Women, Work and Repeat Dependency. *American Sociological Review* 6 #3: 407–426.

Harrison, Bennett, and Sandra Kanter. 1978. The Political Economy of States Job Creation Business Incentives. *Journal of the American Institute of Planners* 44 (October): 424–435.

Hart, Mark, and Michael Murray. 2000. *Local Development in Northern Ireland: The Way Forward.* Belfast Northern Ireland Economic Development Office. (December).

Hatry, H.P. 1999. *Performance Measurement: Getting Results.* Washington, DC: The Urban Institute Press.

Haveman, Robert H., and Gregory B. Christiansen. 1978. Public Employment and Wage Subsidies in Western Europe and the U.S.: What We're Doing and What We Know. Disc. Paper 522–578. Institute for Research on Poverty, University of Wisconsin-Madison.

Hayter, R. 1997. *The Dynamics of Industry Location: The Factory, the Firm and the Production System.* New York: Wiley.

Held, J. R. 1996. Clusters as an Economic Development Tool. *Economic Development Quarterly* 10 #3 (August): 240–261.

Helling, Amy, and Patricia Mokhtarian. 2001. Worker Telecommunication and Mobility in Transition: Consequences for Planning. *Journal of Planning Literature* 15 #4: 511–525.

Helms, L.J. 1985. The Effect of State and Local Taxes on Economic Development: A Time Series–Cross Sectional Approach. *The Review of Economics and Statistics,* 574–582.

Henderson, J.W., T.M. Kelly, and B.A. Taylor. 2000. The Estimation of Agglomeration Economies on Estimated Demand Thresholds: An Extension of Wensley and Stabler. *Journal of Regional Science* 40 #4: 719–733.

Henderson, Yolanda. 1989. The Emergence of the Venture Capital Industry. *New England Economic Review* (July/August): 66–79.

Henry M.S., D.L. Barkley, and S. Bao. 1997. The Hinterland's Stake in Metropolitan Growth: Evidence from Selected Southern Regions. *Journal of Regional Science* 37 #3: 479–501.

Hewings, G.J.D., and M. Madden (eds.). 1995. *Social and Demographic Accounting*. Cambridge: Cambridge University Press.

Hill, Edward W. 1998. Principles for Rethinking the Federal Government's Role in Economic Development. *Economic Development Quarterly* 12 #4 (November): 299–312.

Hill, E. W., and N.A. Shelley. 1995. An Overview of Economic Development Finance. In *Financing Economic Development: An Institutional Response*. R.D. Bingham, E.W. Hill, and S.B. White (eds.), pp. 13–28. Newbury Park: Sage Publications.

Hillery, George A. 1955. Definitions of Community Areas of Agreement. *Rural Sociology* 20 (June): 111–123.

Hirsch, S. 1967. *Location of Industry and International Competitiveness*. Oxford: Clarendon Press.

Hirschl, Thomas A., and Gene F. Summers. 1982. Cash Transfers and the Export Base of Small Communities. *Rural Sociology* 47 (Summer): 295–316.

Ho, Yan ki. 1978. Commercial Banking and Regional Growth: The Wisconsin Case. Ph.D. dissertation, Department of Agricultural Economics, University of Wisconsin-Madison.

Hochwald, Werner (ed.). 1961. *Design of Regional Accounts*. Baltimore, MD: Johns Hopkins Press.

Hodge, Ian. 1995. *Environmental Economics: Individual Incentives and Public Choices*. Basingstoke: MacMillan Press.

Hodge, Ian, and Charles Dunn. 1992. *Valuing Rural Amenities*. Paris, France: OECD Publication.

Holden, John P., and Steven C. Deller. 1993. Analysis of Community Retail Market Area Delineation Techniques: An Application of GIS Technologies. *Journal of the Community Development Society* 24 #2: 141–158.

Holland, David, and Peter Wyeth. 1993. SAM Multipliers: Their Decomposition, Interpretation and Relationship to Input-Output Multipliers. Research Bulletin XB1027, Washington State University, Pullman, Washington.

Holupka, C.S., and A.B. Shlay. 1993. Political Economy and Urban Development. In *Theories of Local Economic Development: Perspectives from Across the Disciplines*. R.D. Bingham and R. Mier (eds.), pp. 175–191. San Francisco: Sage.

Honadle, Beth Walter. 1986. Defining and Doing Capacity Building Perspectives and Experiences. In *Perspectives on Management Capacity Building*. B.W. Honadle and A.M. Howitt (eds.), pp. 9–23. Albany, NY: New York State University Press.

Hotchkiss, Julie L., and Bruce Kaufman. 2002. *The Economics of Labor Markets*. Sixth edition. Independence, KY: Thomson South-Western Publishing.

Hoy, F. 1996. Entrepreneurship: A Strategy for Rural Development. In *Rural Development Research: A Foundation for Policy*. T.D. Rowley, D.W. Sears, G.L. Nelson, J.N. Reid, and M.J. Yetley (eds.), pp. 29–46. Westport, CT: Greenwood Press.

Huff, David L. 1964. Defining and Estimating a Trade Area. *Journal of Marketing* 28(July): 34–38.

Hustedde, Ron, and Jacek Ganowicz. 2002. The Basics: What's Essential About Theory for Community Development Practice? *Journal of the Community Development Society* 33 #1: 1–19.

Hustedde, Ron, Ron Shaffer, and Glen Pulver. 1993. *Community Economic Analysis: A How To Manual*. (RRD141) Ames, IA: North Central Regional Center for Rural Development.

Hyett-Palma, Inc. 1989. Business Clustering: How to Leverage Sales. Washington, DC: Hyett-Palma, Inc.

Inman, R. 1979. The Fiscal Performance of Local Governments: An Interpretative Review. In *Current Issues in Urban Economics*. P. Mieszkowski and M. Straszhiem (eds.). Baltimore, MD: John Hopkins University Press.

Institute for Local Self-Reliance (ILSR). 2002. http://www.newrules.org/finance/index.html.

Isard, Walter. 1960. *Methods of Regional Analysis: An Introduction to Regional Science*. Cambridge, MA: The MIT Press.

Isard, Walter. 1975. A Simple Rationale for Gravity Model Type Behavior. *Proceedings of the Regional Science Association* 35: 25–30.

Isard, W., I. J. Azis, M. P. Drennan, R. E. Miller, S. Salzman, and E. Thorbecke. 1998. Methods of Interregional and Regional Analysis. Brookfield, VT: Ashgate Publishing Company.

Isserman, Andrew M. 1977. Location Quotient Approach to Estimating Regional Economic Impacts. *Journal of the American Institute of Planners* 43 (January): 33–41.

Isserman, Andrew M. 1980b. Alternative Economic Base Bifurcation Techniques: Theory, Implementation and Results. In *Economic Impact Analysis: Methodology and Applications*. Saul Pleeter (ed.), pp. 32–53. Boston, MA: Martinus Nijhoff Pub.

Jeep, Edward G. 1993. The Four Pitfalls of Local Economic Development. *Economic Development Quarterly* 7 #3 (August): 237–242.

Jensen, Rodney C. 1980. The Concept of Accuracy in Regional Input-Output Models. *International Regional Science Review* 5 #2: 139–154.

Jepson, Edward J., David W. Marcouiller, and Steven C. Deller. 1997. Incorporating Market and Nonmarket Values into Regional Planning for Rural Development.

CPL Bibliographies 338, 339, and 340. *Journal of Planning Literature* 12 #2: 220–257.

Johannisson, Bengt. 1990. Community Entrepreneurship: Cases and Conceptualization. *Entrepreneurship and Regional Development* 2 #1: 71–88.

Johannisson, B., and A. Nilsson. 1989. Community Entrepreneurs: Networking for Local Development. *Entrepreneurship and Regional Development* 1 #1: 3–19.

Johansen, L. 1960. *A Multi-sectoral Study of Economic Growth.* Amsterdam: North-Holland.

Johnson, Kenneth M., and Calvin L. Beale. 1998. The Rural Rebound. *Wilson Quarterly* 22 #2: 16–27.

Johnson, Kenneth M., and Glenn V. Fuguitt. 2000. Continuity and Change in Rural Migration Patterns, 1950–1995. *Rural Sociology* 65 #1: 27–49.

Johnson, T.G., and J.I. Stallman. 1994. Human Capital Investment in Resource-Dominated Economies. *Society and Natural Resources* 7 #3: 221–233.

Jones, E., J.I. Stallman, and C. Infanger. 2000. Free Markets at a Price. *Choices* 1st Quarter: 36–40.

Kahn, J.R. 1998. *The Economic Approach to Environmental and Natural Resources.* Fort Worth, TX: The Dryden Press.

Kahne, Hilda, and Andrew I. Kohen. 1975. Economic Perspective on the Roles of Women in the American Economy. *Journal of Economic Literature* 13 (December): 1249–1292.

Kain, John F. 1992 The Spatial Mismatch Hypothesis: Three Decades Later *Housing Policy Debate* 3 #2: 371–460.

Kalachek, Edward D. 1973. *Labor Markets and Unemployment.* Belmont, CA: Wadsworth.

Kaldor, N. 1970. The Case for Regional Policies. *Scottish Journal of Political Economy* (November): 337–348.

Kaldor, N. 1979. Equilibrium Theory and Growth Theory. In *Economics and Human Welfare: Essays in Honor of Tibor Scitovsky.* New York: Academic Press.

Kalleberg, Arne L., and Aage B. Sorensen. 1979. The Sociology of Labor Markets. *Annual Review of Sociology* 5: 351–359.

Karlsson, C., and J. Larson. 1990. Product and Price Competition in a Regional Context. *Papers of the Regional Science Association* 69: 83–99.

Keating, M. 1989. Local Government and Economic Development in Western Europe. *Entrepreneurship and Regional Development* 1 #3: 305.

Keating, W.D., and N. Krumholz. 2000. Neighborhood Planning. *Journal of Planning Education and Research* 20 #1: 111–114.

Kerr, Clark. 1982. The Balkinization of Labor Markets. In *Readings in Labor Economics and Labor Relations.* Third edition. Lloyd G. Reynolds, Stanley H. Masters, and Collete Moser (eds.). Englewood Cliffs, NJ: Prentice-Hall.

Kieschnick, Michael. 1981a. The Role of Equity Capital in Urban Economic Development. In *Expanding the Opportunity to Produce.* Robert Friedman and William Schweke (eds.), pp. 373–387. Washington, DC: The Corporation for Enterprise Development.

Kieschnick, Michael. 1981b. *Taxes and Growth: Business Incentives and Economic Development.* Washington, DC: Council of State Planning Agencies.

Kieschnick, Michael, and Belden Daniels. 1979. *Development Finance: A Primer for Policymakers.* Washington, DC: National Rural Center.

Kilkenny, Maureen. 2002. Community Credit. *International Regional Science Review* 25 #3: 247–251.

Kivell, P.T., and G. Shaw. 1980. The Study of Retail Location. In *Retail Geography.* John A. Dawson (ed.), pp. 95–155. New York: John Wiley and Sons.

Klein, L. 1969. The Specification of Regional Econometric Models. *Papers of the Regional Science Association.* 23 #1: 105–115.

Klein, L., and A. Goldberger. 1955. *An Econometric Model of the United States: 1929-52.* Amsterdam: North-Holland.

Knapp, G.J. 1985. The Price Effects of Urban Growth Boundaries in Metropolitan Portland, Oregon. *Land Economics* 61 #1: 26–35.

Kohlhase, J. 1991. The Impacts of Toxic Waste Sites on Housing Values. *Journal of Urban Economics* 30: 1–26.

Kort, J., and J. Cartwright. 1981. Modeling the Multiregional Economy: Integrating Econometric and Input-Output Models. *The Review of Regional Studies* 11 #1: 1–17.

Kraybill, David S., and Bruce A. Weber. 1995. Institutional Change and Economic Development in Rural America. *American Journal of Agricultural Economics* 77 #5 (December): 1265–1270.

Kraybill, D., T. Johnson, and D. Orden. 1992. Macroeconomic Imbalances: A Multiregional General Equilibrium Analysis. *American Journal of Agricultural Economics* 74 #3: 726–736.

Kreps, J.; Martin; R. Perlman; and G.G. Somers. 1980. *Contemporary Labor Economics: Issues, Analysis and Policies.* Belmont, CA: Wadsworth Publishing.

Kretzmann, John P., and John L. McKnight. 1993. *Building Communities from the Inside Out: A Path Toward Finding and Mobilizing a Community's Assets.* Chicago: ACTA Publications.

Kriström, B., and P. Riera. 1996. Is the Income Elasticity of Environmental Improvements Less Than

One? *Environmental and Resource Economics* 7 #1: 37–56.

Krueckeburg, Donald A., and Arthur L. Silvers. 1974. *Urban Planning Analysis: Methods and Models.* New York: John Wiley and Sons.

Krugman, Paul. 1991a. Increasing Return and Economic Geography. *Journal of Political Economy* 99: 483–489.

Krugman, Paul. 1991b. *Geography and Trade.* Cambridge, MA: MIT Press.

Krugman, Paul. 1995. *Development, Geography and Economic Theory.* Cambridge, MA: MIT Press.

Krugman, Paul. 1999. The Role of Geography in Development. *International Regional Science Review* 22: 142–161.

Krumm, Ronald J. 1983. Regional Wage Differentials, Fluctuations in Labor Demand and Migration. *International Regional Science Review* 8 (June): 23–46.

Kuznets, Simon. 1955. Economic Growth and Income Inequality. *American Economic Review* 45 #1: 1–28.

Ladd, Helen F. 1994. Spatially Targeted Economic Development Strategies: Do They Work? *Cityscape: A Journal of Policy Development and Research:* 193–218.

Ladd, Helen F. 1998. *Local Government Tax and Land Use Policies in the United States: Understanding the Links.* North Hampton, MA: Edward Elgar.

Laird, William E., and James R. Rinehart. 1967. Neglected Aspects of Industrial Subsidy. *Land Economics* 43 (February): 25–31.

Laird, William E., and James R. Rinehart. 1979. Economic Theory and Local Industrial Promotion: A Reappraisal of Usual Assumptions. *American Industrial Development Council* 14 (April): 33–49.

Lancaster, K.J. 1966. A new approach to consumer theory. *Journal of Political Economy* 78: 311–329.

Lang, R.E., and S.P. Hornburg. 1997. Planning Portland Style: Pitfalls and Possibilities. *Housing Policy Debate* 8 #1: 1–10.

Lansford, N.H., and L.L. Jones. 1995. Marginal price of lake recreation and aesthetics: An hedonic approach. *Journal of Agricultural and Applied Economics* 27 #1: 212–223.

Lansford, N.H, and L.L. Jones. 1996. Recreational and Aesthetic Value of Water Using Hedonic Price Analysis. *Journal of Agricultural and Resource Economics* 20 #2: 341–355.

Larrivee, John. 2000. Informal Activity and Home Production in Non-metropolitan Wisconsin. Unpublished Ph.D. dissertation, Department of Agricultural and Applied Economics, University of Wisconsin-Madison.

Ledebur, Larry C. 1977. *Issues in the Economic Development of Nonmetropolitan United States.* Economic Research Report, Economic Development Administration, U.S. Department of Commerce (January).

Ledebur, Larry C., and Douglas Woodward. 1990. Adding a Stick to the Carrot: Location Incentive with Clawbacks, Recisions and Recalibrations. *Economic Development Quarterly* 4#3 (August): 221–237.

Leigh, Roger. 1970. The Use of Location Quotients in Urban Economic Base Studies. *Land Economics* 46 (May): 202–205.

Lenzi, Raymond C. 1992. Rural Development Finance Gaps: Bank CDCs as an Alternative. *Journal of the Community Development Society* 23 #2: 22–38.

LeSage, J.P. 1997. Regression Analysis of Spatial Data. *Journal of Regional Analysis and Policy* 27 #2: 83–94

Lesser, J.A., D.E. Dodds, and R.O. Zerbe, Jr. 1997. *Environmental Economics and Policy.* New York: Addison Wesley Publishers.

Lester, Richard T. 1966. *Manpower Planning in a Free Society.* Princeton, NJ: Princeton University Press.

Lever, W.F. 1980. The Operation of Local Labor Markets in Great Britain. *Papers of the Regional Science Association* 44: 37–56.

Lewis, Eugene, Russell Youmans, George Goldman, and Garnet Premer. 1979. Economic Multipliers: Can a Rural Community Use Them? WREP 24 in *Coping with Growth Series.* Corvallis, OR: Western Rural Development Center (October).

Lewis, W. Arthur. 1954. Economic Development with Unlimited Supplies of Labor. *Manchester School of Economic and Social Studies* 22 (May): 139–191.

Lewis, W. Arthur. 1955. *The Theory of Economic Growth.* London: Allen and Unwin.

Liechtenstein, Gregg A., and Thomas S. Lyons. 2001. The Entrepreneurial Development System: Transforming Business Talent and Community Economics. *Economic Development Quarterly* 15 #1 (February): 3–20.

Litvak, Lawrence, and Belden Daniels. 1979. *Innovations in Development Finance.* Washington, DC: Council of State Planning Agencies.

Lloyd, Peter E., and Peter Dicken. 1977. *Location in Space: A Theoretical Approach to Economic Geography.* Second ed. New York: Harper-Row.

Long, Huey B., Robert C. Anderson, and Jon A. Blubaugh (eds.). 1973. *Approaches to Community Development.* Iowa City, IA: National University Extension Association and American College Testing Program.

Lucas, R.E. Jr. 1988. On the mechanics of economic development. *Journal of Monetary Economics* 22 (July): 3–42.

Luke, J.S., C. Ventriss, B.J. Reed, and C.M. Reed. 1988. *Managing Economic Development: A Guide to State and Local Leadership Strategies.* San Francisco: Jossey-Bass Public.

MacIntosh, Jeffrey G. 1997. Venture Capital Exits in Canada and the United States. *Financing Growth in Canada.* Paul Halpern (ed.). Calgary, Alberta: University of Calgary Press.

Mack, Richard S., and Peter V. Schaeffer. 1993. Nonmetropolitan Manufacturing in the United States and Product Cycle Theory: A Review of the Literature. *Journal of Planning Literature* 8 #2 (November): 124–139.

Madden, Janice F. 1977. An Empirical Analysis of the Spatial Elasticity of Labor Supply. *Papers of the Regional Science Association* 39: 157–174.

Malecki, E.J. 1981. Public and Private Sector Interrelationships, Technological Change and Regional Development. *Papers of the Regional Science Association* 47: 121–138.

Malecki, E.J. 1983. Technology and Regional Development: A Survey. *International Regional Science Review* 8: 89–125.

Malecki, E.J. 1989. What About People in High Technology? Some Research and Policy Considerations. *Growth and Change* 20 #1 (Winter): 67–79.

Malecki, E.J. 1994. Entrepreneurship in Regional and Local Development. *International Regional Science Review* 16 #1 and 2: 119–154.

Malecki, E.J., and P. Nijkamp. 1988. Technology and Regional Development: Some Thoughts on Policy. *Environment and Planning C: Government and Policy* 6: 383–399.

Mäler, Karl-Göran. 1998. Environment, Poverty and Economic Growth. In *Annual World Bank Conference on Development Economics 1997.* Washington, DC: The World Bank.

Marcouiller, D.W. 1995. *Tourism Planning.* CPL Bibliography 316. Chicago, IL: American Planning Association.

Marcouiller, D.W. 1998. Environmental Resources as Latent Primary Factors of Production in Tourism: The Case of Forest-based Commercial Recreation. *Tourism Economics* 4 #2: 131–145.

Marcouiller, D.W., and T. Mace. 1999. *Forests and Regional Development: Economic Impacts of Woodland Use for Recreation and Timber in Wisconsin.* Monograph G3694, University of Wisconsin System Board of Regents, Madison, WI.

Marcouiller, D.W., J.G. Clendenning, and R. Kedzior. 2002. Natural Amenity-led Development and Rural Planning. CPL Bibliography 365. *Journal of Planning Literature* 16 #4: 515–542.

Markley, Deborah, and Ron Shaffer. 1993. Rural Banks and Their Communities: A Matter of Survival. *Economic Review* 78 #3: 73–79. Kansas City: Federal Reserve Bank of Kansas City.

Massey, Doreen. 1975. Approaches to Industrial Location Theory: A Possible Spatial Framework. In *Regional Science: New Concepts and Old Problems.* E.L. Cripp (ed.), pp. 84–108. London: Pion.

Mathur, Vijay K. 1999. Human Capital-based Strategy for Regional Economic Development. *Economic Development Quarterly* 13 #3 (August): 203–216.

Mathur, Vijay K., and Harvey S. Rosen. 1974. Regional Employment Multiplier: A New Approach. *Land Economics* 50 (February): 93–96.

Mayer, C., and T.C. Somerville. 2000. Land Use Regulation and New Construction. *Regional Science and Urban Economics* 30 #6: 639–662.

Mayer, Wolfgang, and Saul Pleeter. 1975. A Theoretical Justification for the Use of Location Quotients. *Regional Science and Urban Economics* 5 (August): 343–355.

McCann, Philip. 2002. Classical and Neo-classical Location-production Models. In *Industrial Location Economics.* Philip McCann (ed.). North Hampton, MA: Edward Elgar.

McConnell, Campbell R., Stanley Brue, and David MacPherson. 2003. *Contemporary Labor Economics.* Sixth edition. New York: McGraw Hill Publishing.

McDowell, George R. 1995. Some Communities are Successful, Others Are Not: Toward an Institutional Framework for Understanding the Reasons Why. In *Rural Development Strategies.* David W. Sears and J. Norman Reid (eds.), pp. 269–281. Chicago: Nelson Hall Publishers.

McFadden, D.L., and G.K. Leonard. 1992. Issues in the Contingent Valuation of Environmental Goods: Methodologies for Data Collection and Analysis. In *Contingent Valuation: A Critical Assessment.* Cambridge: Cambridge Economics.

McGranahan, David A. 1999. *Natural Amenities Drive Rural Population Change.* Agricultural Economic Report Number 781. Washington, DC: Economic Research Service, United States Department of Agriculture.

McGuire, T. 1992. Review of "Who Benefits from State and Local Economic Development Policy?" *National Tax Journal* 75 #4: 457–459.

Melkers, Julia, Daniel Bugler, and Barry Bozeman. 1993. Technology Transfer and Economic Development. In *Theories of Local Development: Perspectives from Across the Disciplines.* Richard D. Bingham and Robert Mier (eds.), pp. 232–247. Newbury Park: Sage.

Mier, Robert, and Richard D. Bingham. 1993. Metaphors of Economic Development. In *Theories of Local Economic Development: Perspectives From Across the Disciplines*. R.D. Bingham and R. Mier (eds.), pp. 284–304. San Francisco: Sage Publications.

Mier, Robert, and Robert P. Giloth. 1993. Cooperative Leadership for Community Problem Solving. *Social Justice and Local Development Policy*. Robert Mier (ed.), pp. 165–181. San Francisco: Sage Publications.

Mier, Robert, and Howard M. McGary. 1993. Social Justice and Public Policy. In *Social Justice and Local Development Policy*. Robert Mier (ed.), pp. 20–31. San Francisco: Sage Publications.

Mier, Robert, Thomas Vietoriz, and Bennett Harrison. 1993. Full Employment at Living Wages. In *Social Justice and Local Development Policy*. Robert Mier (ed.), pp. 1–19. San Francisco: Sage Publications.

Miernyk, William H. 1965. *The Elements of Input-Output Analysis.* New York: Random House.

Miernyk, William H. 1976. Comments on Recent Development in Regional Input-Output Analysis. *International Regional Science Review* 1 (Fall): 47–55.

Mikesell, James, and Steve Davidson. 1982. Financing Rural America: A Public Policy and Research Perspective. In *Rural Financial Markets: Research Issues for the 1980's.* Chicago, IL: Federal Reserve Bank of Chicago (December): 159–197.

Miller, Ronald E., and Peter D. Blair. 1985. *Input-Output Analysis: Foundations and Extensions.* Englewood Cliffs, NJ: Prentice-Hall, Inc.

Mincer, Jacob. 1962. On-the-job Training: Costs, Returns and Some Implications. *Journal of Political Economy* 70 (October): 50–79.

Mishan, E. J. 1976. *Cost-benefit Analysis.* New and expanded edition. New York: Praeger Publishing.

Mitchell, Susan. 1995. Birds of a Feather. *American Demographics* (February): 40–48.

Moes, John E. 1961. The Subsidization of Industries by Local Communities in the South. *Southern Economic Journal* 28 (October): 187–193.

Moes, John E. 1962. *Local Subsidies to Industries.* Chapel Hill, NC: University of North Carolina Press.

Mofidi, A., and J. Stone. 1990. Do State and Local Taxes Affect Economic Growth? *The Review of Economics and Statistics* 72 #4: 686–691.

Molotch, H.L. 1976. The City as a Growth Machine: Towards a Political Economy of Place. *American Journal of Sociology* 82 #2: 309–332.

Moore, Craig L. 1975. A New Look at the Minimum Requirements Approach to Regional Economic Analysis. *Economic Geography* 51 (October): 350–356.

Moore, Craig L., and Joanne M. Hill. 1982. Interregional Arbitrage and the Supply of Loanable Funds. *Journal of Regional Science* 22 (November): 499–512.

Moore, Craig, and Marilyn Jacobson. 1984. Minimum Requirements and Regional Economics, 1980. *Economic Geography* 60 (July): 217–224.

Morgan, W.J., J. Mutti, and M. Partridge. 1989. A Regional General Equilibrium Model of the United States: Tax Effects on Factor Movements and Regional Production. *The Review of Economics and Statistics* 71 #4: 139–144.

Moriarty, Barry M. 1980. *Industrial Location and Community Development.* Chapel Hill, NC: University of North Carolina Press.

Morrison, W.I., and P. Smith. 1974. Nonsurvey Input-Output Techniques at the Small Area Level: An Evaluation. *Journal of Regional Science* 14 (April): 1–14.

Morss, Elliott, R. 1966. The Potentials of Competitive Subsidization. *Land Economics* 42 (May): 161–169.

Moses, Leon N. 1958. Location and Theory of Production. *The Quarterly Journal of Economics* 72(May): 259–272.

Mueller, Elizabeth J., and Alex Schwartz. 1998. Leaving Poverty Through Work: A Review of Current Development Strategies. *Economic Development Quarterly* 12 #2 (May): 166–180.

Mulligan, Gordon F. 1984. Agglomeration and Central Place Theory: A Review of the Literature. *International Regional Science Review* 9 #1: 1–42.

Murray, James M., and James L. Harris. 1978. *A Regional Economic Analysis of the Turtle Mountain Indian Reservation: Determining Potential for Commercial Development.* Minneapolis, MN: Federal Reserve Bank of Minneapolis.

Mydral, G. 1957. *Economic Theory and Under-developed Regions.* London: Duckworth.

Nelson, Peter B. 1997. Migration, Sources of Income and Community Change in the Nonmetropolitan Northwest. *Professional Geographer* 49 #4: 418–430.

Neuendorf, Dave. 1987. Community Economic Preparedness. Personal communication.

Newman, R.J., and D. Sullivan. 1988. Econometric Analysis of Business Tax Impacts on Industrial Location: What Do We Know and How Do We Know It? *Journal of Urban Economics* 23: 215–234.

Nickerson, C.J., and L. Lynch. 2001. The Effect of Farmland Preservation Programs on Farmland Prices. *American Journal of Agricultural Economics* 83 #2: 341–351.

Nightingale, Demetra Smith, and Pamela A. Holcomb. 1997. Alternative Strategies for Increasing

Employment. *The Future of Children: Welfare to Work* 7 #1 (Spring): 52–64.

Nord, M., and J. B. Cromartie. 1997. Migration: The Increasing Importance of Rural Natural Amenities. *Choices* 3: 22–23.

North, Douglass C. 1955. Location Theory and Regional Economic Growth. *Journal of Political Economy* 63 (June): 243–258.

North, Douglass C. 1956. A Reply. *Journal of Political Economy* 64 (April): 170–172.

North, Douglass C. 1961. *The Economic Growth of the United States 1790–1860.* Englewood Cliffs, NJ: Prentice-Hall Inc.

North, Douglass C., and Robert Thomas. 1970. An Economic Theory of the Growth of the Western World. *Economic History Review* 23 (2nd series): 1–17.

Oakland, W.H. 1978. Local Taxes and Intraurban Industry Location: A Survey. In *Metropolitan Financing and Growth Management Policies.* Proceedings of a symposium sponsored by the Committee on Taxation, Resources and Economic Development (TRED) at the University of Wisconsin-Madison.

Oates, W.E., 1969. The Effects of Property Taxes and Local Public Spending on Property Values: An Empirical Study of Tax Capitalization and the Tiebout Hypothesis. *Journal of Political Economy* 77: 957–971.

Oates, W.E. 1988. On the Nature and Measurement of Fiscal Illusion: A Survey. In *Taxation and Fiscal Federalism: Essays in Honour of Russell Mathews.* G. Brennnan, B.S. Grewel, and P. Groenwegen (eds.), pp. 65–82. Sydney: Australia University Press.

Oberle, Wayne H., Kevin R. Stowers, and James P. Darby. 1974. A Definition of Development. *Journal Community Development Society* 5 (Spring): 61–71.

OECD (Organisation for Economic Co-operation and Development). 1999. *Cultivating Rural Amenities: An Economic Development Perspective.* Paris: OECD.

Oerlemans, Leon A.G., Marius H.T. Meeus, and Frans W.M. Boekema. 2001. Firm Clustering and Innovation: Determinants and Effects. *Papers in Regional Science* 80 #3 (July): 337–356.

O'Farrell, Patrick N. 1986. Entrepreneurship and Regional Development: Some Conceptual Issues. *Regional Studies* 20 #6 (December): 565–574.

Ofstead, Cynthia M. 1998. Temporary Help Firms and the Job Matching Process. Unpublished Ph.D. dissertation, Department of Sociology, University of Wisconsin-Madison.

O'Kelly, M.E. 1999. Trade Area Models and Choice-based Samples: Methods. *Environment and Planning A* 31 #4: 613–627.

Olsson, Gunnar. 1966. Central Place Systems, Spatial Interaction and Stochastic Processes. *Proceedings of the Regional Science Asssociation* 18: 18–45.

Osterman, Paul. 1984. *Internal Labor Markets.* Cambridge, MA: Massachusetts Institute of Technology Press.

Ostrom, E. 1977. Why Do We Need Multiple Indicators of Public Service Outputs? In *National Conference on Nonmetropolitan Community Services Research.* Washington, DC: Senate Committee on Agriculture, Nutrition and Forestry.

Otto, Daniel M., and Thomas G. Johnson (eds.). 1993. Microcomputer-based Input-Output Modeling. Boulder, CO: Westview Press.

Parr, John B. 1973. Structure and Size in the Urban System of Losch. *Economic Geography* 49 #3: 185–212.

Parr, John B. 1981. Temporal Change in a Central Place System. *Environment and Planning A* 13 (January): 97–118.

Parr, John B. 2002. The Location of Economic Activity: Central Place Theory and the Wider Urban System. In *Industrial Location Economics.* Philip McCann (ed.). North Hampton, MA: Edward Elgar.

Parr, John B., and Kenneth G. Denike. 1970. Theoretical Problems in Central Place Analysis. *Economic Geography* 46 (October): 568–586.

Parsons, Ken H. 1964. Institutional Innovations in Economic Growth. In *Optimizing Institutions for Economic Growth,* pp. 81-106. Raleigh, NC: Agricultural Policy Institute.

Pawasarat, John, and Lois M. Quinn. 2001. Exposing Urban Legends: The Real Purchasing Power of Central City Neighborhoods. Brookings Institution Discussion Paper (June).

Pearce, D.W. 1983. *Cost-benefit Analysis.* Second edition. New York: St. Martin's Press.

Pellenbarg, Pier H., Leo J.G. van Wissen, and Jouke van Dijk. 2002. Firm Migration. In *Industrial Location Economics.* Philip McCann (ed.). North Hampton, MA: Edward Elgar.

Perloff, Harvey S. 1961. Relative Regional Growth: An Approach to Regional Accounts. In *Design of Regional Accounts.* Werner Hochwald (ed.), pp. 38–65. Baltimore, MD: Johns Hopkins University Press.

Perloff, Harvey S. 1962. A National System of Metropolitan Information and Analysis. *American Economic Review* 52 (May): 356–364.

Perroux, François. 1950. Economic Space, Theory and Applications. *Quarterly Journal of Economics* LXIV.

Perroux, François. 1988. The Pole of Development's New Place in a General Theory of Economic Activity. In *Regional Economic Development*. Benjamin Higgins and Donald J. Savoie (eds.). Boston: Unwin Hyman.

Peterman, William. 2000. *Neighborhood Planning and Community-based Development: The Potential and Limits of Grassroots Action*. Thousand Oaks, CA: Sage Publications.

Peters, Alan H. 1993. Clawbacks and the Administration of Economic Development Policy in the Midwest. *Economic Development Quarterly* 7 #4 (November): 328–340.

Pfister, Richard. 1976. Improving Export Base Studies. *Regional Science Perspectives* 6: 104–116.

Pfister, Richard. 1980. The Minimum Requirements Technique of Estimating Exports: A Further Evaluation. In *Economic Impact Analysis: Methodology and Application*. Saul Pleeter (ed.), pp. 59–67. Boston, MA: Martinus Nijhoff Pub.

Phillips, J., and E. Goodstein. 2000. Growth management and housing prices: The case of Portland, Oregon. *Contemporary Economic Policy* 18 #3: 334–344.

Piore, Michael J. 1973. Fragments of a "Sociological" Theory of Wages. *American Economic Review* (May): 377–384.

Plantinga, A.J., and D.J. Miller. 2001. Agricultural Land Values and The Value of Rights to Future Land Development. *Land Economics* 77 #1: 56–67.

Pleeter, Saul. 1980. Methodologies of Economic Impact Analysis: An Overview. In *Economic Impact Analysis: Methodology and Applications*. Saul Pleeter (ed.), pp. 7–31. Boston, MA: Martinus Nijhoff Pub.

Ploch, Louis A. 1978. The Reversal in Migration Patterns: Some Rural Development Consequences. *Rural Sociology* 43 #2: 293–303.

Pommerehne, W. 1978. Institutional Approaches to Public Expenditure: Empirical Evidence from Swiss Municipalities. *Journal of Public Economics* 9: 255–280.

Porter, Michael E. 1990. The Competitive Advantage of Firms in Global Industries. *The Competitive Advantage of Nations*. New York, NY: The Free Press.

Porter, Michael E. 1995. The Competitive Advantage of the Inner City. *Harvard Business Review* (May/June): 55–71.

Porter, M.E. 1996. Competitive Advantage, Agglomeration Economies and Regional Policy. *International Regional Science Review*. 19 #1 and 2: 85-94.

Porter, Michael E. 1997. New Strategies for Inner-city Economic Development. *Economic Development Quarterly* 11 #1 (February).

Potter, Robert B. 1982. *The Urban Retailing System: Location, Cognition and Behaviour*. Aldershot, England: Gower Publishing Company, Ltd.

Pratt, Richard T. 1968. An Appraisal of the Minimum Requirements Technique. *Economic Geography* 44 (April): 117–124.

Pred, Alan R. 1967. *Behavior and Location: Foundations for Geographic and Dynamic Location Theory*. Lund Studies in Geography, Part 1. Uppsala, Sweden: Lund University.

Preston, R. 1975. The Wharton Long-term Model: Input-Output Within the Context of a Macro Forecasting Model. *International Economic Review* 16 #1: 3–19.

Preston, Valerie, and Sara McLafferty. 1999. Spatial Mismatch Research in the 1990's: Progress and Potential. *Papers in Regional Science* 78 #4 (October): 387–402.

Pryde, Paul L. 1981. Human Capacity and Local Development. In *Expanding the Opportunity to Produce*. Robert Freidman and William Schweke (eds.), pp. 521–533. Washington, DC: The Corporation for Enterprise Development.

Pulver, Glen C. 1979. A Theoretical Framework for the Analysis of Community Economic Development Policy Options. In *Nonmetropolitan Industrial Growth and Community Change*. Gene F. Summers and Arne Selvik (eds.). Lexington, MA: Lexington Books.

Pulver, Glen C. 1988. Changing Economic Scene in Rural America. *Journal of State Government* 61 #1 (January/February): 3–8.

Pulver, Glen C., and Ronald J. Hustedde. 1988. Regional Variables Which Influence The Allocation of Venture Capital: The Role of Banks. *The Review of Regional Studies* 18 #2 (Spring): 1–9.

Putnam, Robert D. 1995. Bowling Alone: America's Declining Social Capital. *Journal of Democracy* 6 #1 (January): 65–78.

Pyatt, G., and J. I. Round (eds.). 1985. *Social Accounting Matrices, A Basis for Planning*. Washington, DC: The World Bank.

Ranney, David C., and John J. Betancur. 1992. Labor-force-based Development: A Community-oriented Approach to Targeting Job Training and Industrial Development. *Economic Development Quarterly* 6 #3 (August): 286–296.

Ray, Cynthia T. 1996. Analyzing the Specifics of Retail Markets. Chapter 6 in *Shopping Centers and Other Retail Properties: Investment, Development, Financing and Management*. New York: John Wiley and Sons, Inc.

Rees, John. 1974. Decision-Making, the Growth of the Firm and the Business Environment. In *Spatial Perspectives on Industrial Organizations and Decision-making.* F.E.I. Hamilton, (ed.), pp. 189–211. New York: John Wiley and Sons.

Reese, L. A., and D. Fasenfest. 1996. Local Economic Development over Time. *Economic Development Quarterly* 10 #3 (August): 280–289.

Reese, Laura A., and David Fasenfest. 1997. What Works Best? Values and the Evaluation of Local Economic Development Policy. *Economic Development Quarterly* 11 #3 (August): 195–207.

Reese, Laura A., and David Fasenfest. 1999. Critical Perspectives on Local Development Policy Evaluation. *Economic Development Quarterly* 13 #1 (February): 3–7.

Reich, Michael, David M. Gordon, and Richard C. Edwards. 1973. A Theory of Labor Market Segmentation. *American Economic Review* (May): 359–365.

Reilly, William J. 1931. *The Law of Retail Gravitation.* New York: Knickerbocker Press.

Rey, S. 1997. Integrating Regional Econometric ans Input-Output Models: An Evaluation of Embedding Strategies. *Environment and Planning A* 29 #6: 1057–1072.

Ricardo, David. 1817. On the Principles of Political Economy and Taxation. See web version at: http://www.econlib.org/library/Ricardo/ricP.html.

Richardson, Harry W. 1969a. *Regional Economics: Location Theory, Urban Structure, Regional Change.* New York: Praeger.

Richardson, Harry W. 1969b. *Elements of Regional Economics.* Baltimore, MD: Penguin Books.

Richardson, Harry W. 1972. *Input-Output and Regional Economics.* New York: John Wiley and Sons.

Richardson, Harry W. 1973. *Regional Growth Theory.* New York: John Wiley and Sons.

Richardson, Harry W. 1978b. *Regional and Urban Economics.* New York: Penguin Books.

Richardson, Harry W. 1985. Input-Output and Economic Base Multipliers: Looking Backward and Forward. *Journal of Regional Science* 25 (November): 607–662.

Rinehart, James R., and William E. Laird. 1972. Community Inducements and Zero Sum Game. *Scottish Journal of Political Economy* 77 (February): 73–90.

Robbins, Lionel. 1935. *An Essay of the Nature and Significance of Economic Science.* London: Macmillan and Company.

Roberts, R. Blaine, and Henry Fishkind. 1979. The Role of Monetary Forces in Regional Economic Activity: An Econometric Simulation Analysis. *Journal of Regional Science* 19 (February): 15–30.

Robison, Lindon J., and Peter J. Barry. 1987. *The Competitive Firm's Response to Risk.* New York: Macmillan Publishing Company.

Robison, Lindon J., and S.D. Hanson. 1995. Social Capital and Economic Cooperation. *Journal of Agricultural and Applied Economics* 27 #1: 43–57.

Rogers, Joel, and Laura Dresser. 1996. *Dane County Community Career Ladders Feasibility Study.* University of Wisconsin-Madison: Center on Wisconsin Strategy (May).

Rogers, Joel, and Laura Dresser. 2000. *Pulling Apart: The Strong Wisconsin Economy Masks Growing Inequality.* University of Wisconsin-Madison: Center on Wisconsin Strategy (January).

Romans, J. Thomas. 1965. *Capital Exports and Growth Among U.S. Regions.* Middleton, CT: Wesleyan University Press.

Romer, P. 1984. Dynamic Competitive Equilibria with Externalities: Increasing Returns and Unbounded Growth. Unpublished Ph.D. dissertation, University of Chicago.

Romer, Paul. 1986. Increasing Returns and Long-run Growth. *Journal of Political Economy* 94: 1002–1038.

Romer, Paul. 1987. Growth Based on Increasing Returns Due to Specialization. *American Economic Review* 77: 56–62.

Rosenfeld, Stuart A. 1997. Bringing Business Clusters into the Mainstream of Economic Development. *European Planning Studies* 5 #1: 3–23.

Rostow, W.W. 1962. *The Process of Economic Growth.* Second edition. Oxford, England: Oxford University Press and Clarendon Press.

Rostow, W.W. (ed.). 1965. *The Economics of Take-off into Sustained Growth.* New York: St. Martins Press.

Rostow, W.W. 1991. *The Stages of Economic Growth: A Non-communist Manifesto.* Third edition. Cambridge: Cambridge University Press.

Round, Jeffrey I. 1983. Nonsurvey Techniques: A Critical Review of the Theory and the Evidence. *International Regional Science Review* 8 (December): 189–212.

Rowley, Thomas D., and David Freshwater. 2001. The Rural Dilemma. *Terrain: A Journal of the Built and Natural Environments* #10 (Fall/Winter).

Rubinfeld, D.L. 1987. The Economics of the Local Public Sector. In *Handbook of Public Economics*, Vol. II. A.J. Aurbach and M. Feldstein, (eds.). Elsevier Science Publishers B.V., North-Holland.

Rudzitis, Gundars. 1999. Amenities Increasingly Draw People to the Rural West. *Rural Development Perspectives* 14 #2: 9–13.

Ryan, Bill, and Gerry Campbell. 1996. Retail Development Strategies and Your Community: Strategies That Work Elsewhere May Not Work Here. *Let's Talk Business, No. 2.* Center for Community Economic Development, University of Wisconsin-Extension. http://www.uwex.edu/ces/cced/publicat/letstalk.html

Ryan, Bill, and Dave Muench. 1997. Business Clustering to Build Retail Sales. *Let's Talk Business, No. 11.* Center for Community Economic Development, University of Wisconsin-Extension. http://www.uwex.edu/ces/cced/publicat/letstalk.html

Ryan, Vernon D. 1994. Community Development and the Ever Elusive "Collectivity." *Journal of the Community Development Society* 25 #1: 5–19.

Sahlman, W. A. 1990. Structure of Venture-Capital Organizations. *Journal of Financial Economics:* 473–521.

Salamon, Sonya. 1989. What Makes Rural Communities Tick? *Rural Development Perspectives* (June): 19–24.

Samuelson, P.A. 1954. The Pure Theory of Public Expenditures. *Review of Economics and Statistics* 36: 387–389.

Sanders, Irwin T. 1966. *The Community: An Introduction to a Social System.* New York: The Ronald Press.

Sawhill, Isabel. 1998. From Welfare to Work: A new Anti-poverty Agenda. In *Getting Ahead: Economic and Social Mobility in America.* Isabel Sawhill and Daniel P. McMurrer (eds.), pp. 27-30. Washington, DC: Urban Institute Press.

Schaffer, William A., and Kong Chu. 1969. Nonsurvey Techniques for Constructing Regional Interindustry Models. *Papers of the Regional Science Association* 23: 83–104.

Schmid, A. Allan. 1972. Analytical Institutional Economics: Challenging Problems in the Economics of Resources for a New Environment. *American Journal of Agricultural Economics* 54 (August): 893–901.

Schmid, A. Allan, and Lindon J. Robison. 1995. Applications of Social Capital Theory. *Journal of Agricultural and Applied Economics* 27 #1 (July): 59–66.

Schmookler, J. 1966. *Innovation and Economic Growth.* Cambridge, MA: Harvard University Press.

Schreiner, D., H.S. Lee, Y.K. Koh, and R. Budiyanti. 1996. Rural Development: Toward an Integrative Policy Framework. *Journal of Regional Analysis and Policy* 26 #2: 53–72.

Schuler, Galen, and Richard Gardner. 1990. Pacific Northwest Strategy Community Development Guide. USFS-Pacific Northwest Region (October).

Schultz, Theodore W. 1961. Investment in Human Capital. *American Economic Review* 51 (March): 1–17.

Schultz, T.W. 1968. Institutions and the Rising Eco-

nomic Value of Man. *American Journal of Agricultural Economics* 50 (December): 1113–1127.

Schumpeter, J.A. 1942. *Capitalism, Socialism and Democracy.* New York: Harper and Row.

Schumpeter, Joseph A. 1961. *The Theory of Economic Development.* Cambridge, MA: Harvard University Press.

Schweke, W. 1990. *The Third Wave in Economic Development.* Washington, DC: Corporation for Enterprise Development.

Scott, A.J. 2000. Economic Geography: The Great Half-Century. *Cambridge Journal of Economics* 24 #4: 483–504.

Scott, J.K, and T.G. Johnson. 1998. The Community Policy Analysis Network: A National Infrastructure for Community Policy Decision Support. *Journal of Regional Analysis and Policy.* 28 #2: 49–63.

Scott, Loren C., Lewis H. Smith, and Brian Rungeling. 1977. Labor Force Participation in Southern Rural Labor Markets. *American Journal of Agricultural Economics* 59 (May): 266–214.

Shaeffer, Peter. 1985. Human Capital Accumulation and Job Mobility. *Journal of Regional Science* 25 (February): 103–114.

Shaffer, Ron. 1990. Building Economically Viable Communities: A Role for Community Developers. *Journal of the Community Development Society* 21 #2: 74–87.

Shaffer, Ron. 2002. Community Economic Development and Smart Growth. *Community Economics Newsletter* (January) #303.

Shaffer, Ron, and Glen Pulver. 1990. Rural Nonfarm Business Access to Debt and Equity Capital. In *Financial Market Intervention as a Rural Development Strategy.* Patrick Sullivan (ed.), pp. 39–58. ARED-ERS-USDA, ERS-Staff Report No. AGES 9070 (December).

Shaffer, Ron, and Glen C. Pulver. 1995. Building Local Economic Development Strategies. In *Rural Development Strategies.* David Sears and J. Norman Reid (eds.), pp. 9–28. Chicago: Nelson-Hall Publishers.

Shaffer, Ron, and Bill Ryan. 1997. Attracting Retailers to Your Community. *Let's Talk Business, No. 5.* Center for Community Economic Development, University of Wisconsin-Extension. http://www.uwex.edu/ces/cced/publicat/letstalk.html

Shaffer, Ron, and Luther Tweeten. 1974. Measuring the Net Economic Changes from Rural Industrial Development in Oklahoma. *Journal of Land Economics* 50 (August): 261–270.

Shapero, Albert. 1981. The Role of Entrepreneurship in Economic Development at the Less-Than-National Level. In *Expanding the Opportunity to Produce.* Robert Friedman and William Schweke (eds.), pp. 25–35. Washington, DC: The Corporation for Enterprise Development.

Sharp, E.B., and M.G. Bath. 1993. Citizenship and Economic Development. In *Theories of Local Economic Development: Perspectives from Across the Disciplines.* R.D. Bingham and R. Mier (eds.), pp. 213–231. San Francisco: Sage.

Shepard, Eric S. 1980. The Ideology of Spatial Choice. *Papers of the Regional Science Association* 45: 197–213.

Shepard, I.D., and C.J. Thomas. 1980. Urban Consumer Behaviour. In *Retail Geography.* John A. Dawson (ed.), pp. 18–94. New York: John Wiley and Sons.

Shields, M. 1998. *An Integrated Economic Impact and Simulation Model for Wisconsin Counties.* Unpublished Ph.D. dissertation, Department of Agricultural and Applied Economics, University of Wisconsin-Madison.

Shields, M., and S. Deller. 1998. Commuting's Effect on Local Retail Market Performance. *Review of Regional Studies* 28 #2: 71–90.

Shields, M., S. Deller, and J. Stallmann. 2000. Comparing the Impacts of Retiree versus Working-age Families in a Small Rural Region: An Application of the Wisconsin Economic Impact Modeling System. *Agricultural and Resource Economics Review* 30 #1: 20–31.

Shonkwiler, J.S., and T.R. Harris. 1996. Rural Retail Business Thresholds and Interdependencies. *Journal of Regional Science* 36: 617–630.

Shoven, J.B., and J. Whalley. 1984. Applied General Equilibrium Models of Taxation and International Trade: An Introduction and Survey. *Journal of Economic Literature* 22 (September): 1007–1051.

Shumway, J. Matthew, and James A. Davis. 1996. Nonmetropolitan Population Change in the Mountain West: 1970-1995. *Rural Sociology* 61 #3: 513–529.

Simon, Herbert. 1982. *Models of Bounded Rationality.* Cambridge, MA: The MIT Press.

Skidmore, M., and M. Peddle. 1998. Do Development Impact Fees Reduce the Rate of Residential Development? *Growth and Change* 29 #4: 383–400.

Smith, Adam. First published 1759. *The Theory of Moral Sentiments.* The reprints of Economic Classics. New York: Augustus M. Kelley Publishers, 1966.

Smith, David. 1971. *Industrial Location.* New York: John Wiley and Sons.

Smith, D.M. 1975. Neoclassical Growth Models and Regional Growth in the US. *Journal of Regional Science* 15 #2: 165–181.

Smith, Eldon B., Brady Deaton, and David Kelch. 1980. Cost Effective Programs of Rural Community Development. *Journal of the Community Development Society* 11 (Spring): 113–124.

Smith, Stephen L.J. 1987. Defining Tourism: A Supply-side View. *Annals of Tourism Research* 14 #1: 179–190.

Smith, Stephen L.J. 1994. The Tourism Product. *Annals of Tourism Research* 21 #3: 582–595.

Solow,R. 1956. A Contribution to the Theory of Economic Growth. *Quarterly Journal of Economics* 71 #1: 65–94.

Sorensen, Aage, and Arne Kalleberg. 1974. Jobs, Training and Attitudes of Workers. Disc. Paper 204–274. University of Wisconsin-Madison, Institute for Research on Poverty.

Spalatro, Fiorenza, and Bill Provencher. 2001. An Analysis of Minimum Frontage Zoning to Preserve Lakefront Amenities. *Land Economics* 77 #4: 469–81.

Spence, Michael. 1974. *Market Signaling: Informational Transfer in Hiring and Screening Processes.* Cambridge, MA: Harvard University Press.

Steinbrink, Stephen R. 1992. Keynote address to Building Healthy Communities Through Small Business Financing. In *Building Healthy Communities Through Bank Small Business Financing.* Washington, DC: Administrator of National Banks, Office of the Comptroller of the Currency (December).

Steiner, Michael. 2002. Clusters and Networks: Institutional Settings and Strategic Perspectives. In *Industrial Location Economics* Philip McCann (ed.). North Hampton, MA: Edward Elgar.

Sternberg, E. 1987. Practitioners Classification of Economic Development Policy Instruments. *Economic Development Quarterly* 1 #2 (May): 149–161.

Stevens, B. G. Treyz, and J. Kindahl. 1981. Conjoining an Input-Output Model and a Policy Analysis Model: A Case Study of Regional Effects of Expanding a Port Facility. *Environment and Planning A* 13 #4: 1029–1038.

Stevens, Benjamin H., and Craig L. Moore. 1980. A Critical Review of the Literature on Shift-share as a Forecasting Technique. *Journal of Regional Science* 20 #4: 419–437.

Stevens, B. G. Treyz, E. Ehrlich, and J. Bower. 1983. A New Technique for the Construction of Non-survey Regional Input-Output Models. *International Regional Science Review* 8 #3: 271–294.

Stier, J.C., K.K. Kim, and D. W. Marcouiller. 1999. Growing Stock, Forest Productivity and Land Ownership. *Canadian Journal of Forest Research* 29 #6: 1736–1742.

Stiglitz, J.E., 1977. The Theory of Local Public Goods. In *Economics of Public Service.* M. Feldstein and R. Inman (eds.). New York: Macmillan Press.

Stöhr, W.B. 1986. Regional Innovation Complexes. *Papers Regional Science Assn.* 59: 29–44.

Stöhr, Walter B. 1989. Local Development Strategies to Meet Local Crisis. *Entrepreneurship and Regional Development* 1 #3: 293–300.

Stöhr, W.B. (ed). 1990. *Global Challenge and Local Response.* London: Mansell.

Stöhr, Walter, and Franz Todtling. 1977. Spatial Equity: Some Antitheses to Current Regional Development Strategy. *Papers of the Regional Science Association* 38: 33–54.

Stone, Kenneth E., and James C. McConnon. 1983. Analyzing Retail Sales Potential for Counties and Towns. Paper at American Agricultural Economics Association Meetings, Ames, IA: Iowa State University.

Straszheim, Mahlon R. 1971. An Introduction and Overview of Regional Money Capital Markets. In *Essays in Regional Economics.* John F. Kain and John T. Meyer (eds.), pp. 218–242. Cambridge, MA: Harvard University Press.

Summers, Gene F. 1986. Rural Community Development. *Annual Review of Sociology* 12: 347–371.

Summers, G.F., S.D. Evans, F. Clements, E.M. Beck, Jr., and J. Minkoff. 1976. *Industrial Invasion of Nonmetropolitan America: A Quarter Century of Experience.* New York: Praeger.

Swan, T.W. 1956. Economic Growth and Capital Accumulation. *Economic Record* 32 #3: 334–361.

Swanson, L. 1996. Social Infrastructure and Rural Economic Development. In *Rural Development Research: A Foundation for Policy.* T.D. Rowley, D.W. Sears, G.L. Nelson, J.N. Reid, and M.J. Yetley (eds.), pp. 103–121. Westport, CT: Greenwood Press.

Taft, Stephen J., Glen C. Pulver, and Sydney D. Staniforth. 1984. *Are Small Community Banks Prepared to Make Complex Loans?* R3263. College of Agriculture and Life Sciences, University of Wisconsin-Madison (April).

Taha, Hamdy. 2002. *Operations Research: An Introduction.* Seventh edition. New York: Prentice Hall.

Thomas, Morgan D. 1972. Growth Pole Theory: An Examination of Some of Its Basic concepts. In *Growth Centers in Regional Economic Development.* Niles Hansen (ed.), pp. 50–81. New York: The Free Press.

Thomas, Morgan D. 1985. Regional Economic Development and the Role of Innovation and Technological Change. In *Regional Economic Impacts of Technological Change.* A.T. Thwaites and R.P. Oakey (eds.), pp. 13–35. London: Frances Pinter.

Thompson, James H. 1962. Local Subsidies for Industry: Comment. *Southern Economic Journal* 29 (October): 114–119.

Thompson, Wilbur R. 1965. *A Preface to Urban Economics.* Baltimore, MD: Johns Hopkins Press.

Thompson, W.R. 1987. Policy-based Analysis for Local Economic Development. *Economic Development Quarterly* 1 #3: 203–213.

Thorsnes, P., and G.P.W. Simons. 1999. Letting the Market Preserve Land: The Case for a Market-driven Transfer of Development Rights Program. *Contemporary Economic Policy* 17 #2: 256–266.

Thünen, Johann Heinrich von (1783-1850). 1966. *Isolated State* (an English edition of *Der isolierte Staat,* translated by Carla M. Wartenberg). New York: Pergamon Press.

Thurow, Lester C. 1975. *Generating Inequality: Mechanisms of Distribution in the U.S. Economy.* New York: Basic Books, Inc.

Thwaites, A.T., and R.P. Oakey. 1985. Editorial Introduction. In *Regional Economic Impacts of Technological Change.* A.T. Thwaites and R.P. Oakey (eds.), pp. 1–12. London: Frances Pinter.

Tickamyer, Ann R., and Cynthia M. Duncan. 1991. Work and Poverty in Rural America. In *Rural Policies for the 1990s.* Cornelia B. Flora and James A. Christenson (eds.), pp. 102–113. Boulder, CO: Westview Press.

Tiebout, Charles M. 1956a. Exports and Regional Economic Growth. *Journal of Political Economy* 64 (April): 160–169.

Tiebout, Charles M. 1956b. The Urban Economic Base Reconsidered. *Land Economics* 31 (February): 95–100.

Tiebout, Charles M. 1956c. A Pure Theory of Local Expenditures. *Journal of Political Economy* 64: 416–424.

Tiebout, Charles. 1962. *The Community Economic Base Study.* Suppl. Paper No. 16. New York: Committee for Economic Development.

Tietenberg, Tom. 2002. *Environmental and Natural Resource Economics.* Boston, MA: Addison-Wesley.

Tinbergen, Jan. 1967. *Development Planning.* New York, McGraw-Hill.

Todaro. Michael P. 2000. *Economic Development.* New York: Addison-Wesley.

Tornquist, Gunnar. 1977. The Geography of Economic Activities: Some Viewpoints on Theory and Application. *Economic Geography* 53(April): 153–162.

Townroe, Peter M. 1974. Post Move Stability and the Location Decision. In *Spatial Perspectives on Industrial Organization and Decision-making.* F.E.I. Hamilton (ed.), pp. 287–307. New York: John Wiley and Sons.

Townroe, Peter M. 1979. The Design of Local Economic Development Prices. *Town Planning Review* 50 (April): 148–163.

Treyz, G. 1993. *Regional Economic Modeling.* Boston, MA: Kluwer Academic Publishers.

Turner, Robyne. 1999. Entrepreneurial Neighborhood Initiatives: Political Capital in Community Development. *Economic Development Quarterly* 13 #1 (February): 15–22.

Turner, R., and H.S.D. Cole. 1980. The Estimation and Reliability of Urban Shopping Models. *Urban Studies* 17 (June): 140–150.

Ullman, Edward L. 1968. Minimum Requirements after a Decade: A Critique and An Appraisal. *Economic Geography* 44 (October): 364–369.

Ullman, Edward L., and Michael F. Dacey. 1960. The Minimum Requirements Approach to the Urban Economic Base. *Papers of the Regional Science Association* 6: 175–194.

Van Kooten, G. Cornelius. 1993. *Land Resource Economics and Sustainable Development: Economic Policies for the Common Good.* Vancouver, BC: University of British Columbia Press.

Varian, H. 1992. *Microeconomic Analysis.* New York: Norton.

Vaughan, Roger J. 1985. *The Wealth of States: The Political Economy of State Development.* Washington, DC: Council of State Planning Agencies.

Vaughan, Roger J., and Peter Bearse. 1981. Federal Economic Development Programs: A Framework for Design and Evaluation. In *Expanding the Opportunity to Produce.* Robert Friedman and William Schweke (eds.), pp. 307–329. Washington, DC: The Corporation for Enterprise Development.

Vernon, R. 1966. International Investment and International Trade in the Product Cycle. *Quarterly Journal of Economics* 80: 190–207.

Waddell, Steven J. 1995. Emerging Social-Economic Institutions in the Venture Capital Industry: An Appraisal. *American Journal of Economics and Sociology* 54 #3 (July): 323–354.

Wagner, John E., and Steven C. Deller. 1998. Measuring the Effects of Economic Diversity on Growth and Stability. *Land Economics* 74 #4: 541–556.

Wagner, William B. 1974. An Empirical Test of Reilly's Law of Retail Gravitation. *Growth and Change* 6 (July): 30–35.

Walzer, Norman (ed). 1996a. *Community Strategic Visioning Programs.* Westport, CT: Greenwood Publishing.

Walzer, Norman. 1996b. Common Elements of Successful Programs. In *Community Strategic Visioning Programs.* Norman Walzer (ed), pp. 183–198. Westport, CT: Greenwood Publishing.

Walzer, Norman, Steve C. Deller, Hal Fossum, Gary Green, John Gruidl, Scott Johnson, Steve Kline, David Patton, Alice Schumaker, and Mike Woods. 1995. *Community Visioning/Strategic Planning Programs: State of the Art.* Ames: North Central Regional Center for Rural Development, Iowa State University.

Wasylenko, M.J. 1980. Evidence of Fiscal Differentials and Intrametropolitan Firm Relocation. *Land Economics* 31 #8: 1251–1278.

Wasylenko, M.J. 1981. The Location of Firms: The Role of Taxes and Fiscal Incentives. *Urban Affairs Annual Review* 20: 339–349.

Weber, Rachel, and Nik Theodore. 2002. Introduction: Focus Section on Low Wage Labor Markets. *Economic Development Quarterly* 16 #4: 291–293.

Wensley, M.R.D., and J.C. Stabler. 1998. Demand Threshold Estimation for Business Activities in Rural Saskatchewan. *Journal of Regional Science* 38: 155–177.

West, G.R. 1995. Comparison of Input-Output, Input-Output-Econometric and Computable General Equilibrium Impact Models at the Regional Level. *Economic Systems Research* 7 #2: 209–227.

West, G.R., and R.W. Jackson. 1998. Input-Output + Econometric and Econometric + Input-Output: Model Differences or Different Models? *Journal of Regional Analysis and Policy* 28 #1: 33–48.

White, M.J. 1998. Comment. In *Local Government Tax and Land Use Policies in the United States: Understanding the Links.* H.F. Ladd (ed.) North Hampton, MA: Edward Elgar.

White, Sammis B., and Lori A. Geddes. 2002. The Impact of Employer Characteristics and Workforce Commitment on Earnings of Former Welfare Recipients. *Economic Development Quarterly* 16 #4: 326–341.

Wiewel, W., M. Tietz, and R. Giloth. 1993. The Economic Development of Neighborhoods and Localities. In *Theories of Local Economic Development: Perspectives from Across the Disciplines.* R.D. Bingham and R. Mier (eds.), pp. 80–99. San Francisco: Sage.

Wilkinson, Ken. 1992. The Process of Emergence of Multicommunity Collaboration. In *Multicommunity Collaboration: An Evolving Rural Revitalization Strategy, Conference Proceedings.* Ames, IA: North Central Regional Center For Rural Development, RRD 61.

Williams, James, Andrew Sofranko, and Brenda Root. 1977. Industrial Development in Small Towns: Will Social Action Have Any Impact? *Journal of the Community Development Society* 8(Spring): 19–29.

Wilson, J. 1999. Theories of Tax Competition. *National Tax Journal* 52: 269–304.

Wolkoff, Michael J. 1992. Is Economic Development Decision Making Rational? *Urban Affairs Quarterly* 27#3 (March): 340–355.

Wolman, Harold, and David Spitzley. 1996. The Politics of Local Economic Development. *Economic Development Quarterly* 10 #2 (May): 115–150.

Woods, Michael D. 1996. Preconditions for Successful Program Implementation. In *Community Strate-* *gic Visioning Programs. Norman* Walzer (ed), pp. 75–92. Westport, CT: Greenwood Publishing.

World Commission on Environment and Development. 1987. *Our Common Future.* Oxford: Oxford University Press (generally referred to as Brundtland Commission).

Zider, Bob. 1998. How Venture Capital Works. *Harvard Business Review* 76 #6 (November/December): 131–39.

Index

CPSIA information can be obtained at www.ICGtesting.com
Printed in the USA
BVOW06*0440090215

386617BV00013B/110/P